THE PERIPHERAL ARTERIAL CHEMORECEPTORS

THE PERIPHERAL ARTERIAL CHEMORECEPTORS

Proceedings of an International Workshop
edited by

M. J. PURVES
Department of Physiology, University of Bristol

CAMBRIDGE UNIVERSITY PRESS

Published by the Syndics of the Cambridge University Press
Bentley House, 200 Euston Road, London NW1 2DB
American Branch: 32 East 57th Street, New York, N.Y. 10022

Library of Congress Catalogue Card Number: 74–16996

ISBN: 0 521 20522 0

First published 1975

Printed in Great Britain
at the Alden Press, Oxford

CONTENTS

THE EFFERENT PATHWAY

VASCULAR AND METABOLIC ASPECTS OF CHEMORECEPTORS

CHEMOREFLEXES AND THE CIRCULATION

LIST OF PARTICIPANTS

ABBOUD, F. M.
Cardiovascular Division, Department of Medicine, University of Iowa College of Medicine, Iowa City, Iowa, U.S.A.

ANGELL JAMES, JENNIFER
Department of Physiology, Medical College of St Bartholomew's Hospital, Charterhouse Square, London, E.C.1.

ACKER, H.
Max-Planck Institut für Systemphysiologie, 20 Rheinlanddamm, Dortmund 46, West Germany.

BANISTER, JEAN
University Laboratory of Physiology, Parks Road, Oxford, England.

BINGMANN, D.
Physiologischen Institut, Universität Münster, Westring 6, Münster, Westfalen, West Germany.

BISCOE, T. J.
Department of Physiology, The Medical School, University Walk, Bristol, England.

BULLER, A. J.
Department of Physiology, The Medical School, University Walk, Bristol, England.

BYSTRZYCKA, EWA
Department of Physiology, The Medical School, University Walk, Bristol, England.

CASPERS, H.
Physiologischen Institut, Universität Münster, Westring 6, Münster, Westfalen, West Germany.

DALY, M. DE BURGH
Department of Physiology, Medical College of St. Bartholomew's Hospital, Charterhouse Square, London, E.C.1.

EYZAGUIRRE, C.
Department of Physiology, College of Medicine, University of Utah Medical Center, Salt Lake City, Utah, U.S.A.

FILLENZ, MARIANNE
University Laboratory of Physiology, Parks Road, Oxford, England.

GOODMAN, N. W.
University Laboratory of Physiology, Parks Road, Oxford, England.

HEISTAD, D. D.
Cardiovascular Division, Department of Medicine, University of Iowa College of Medicine, Iowa City, Iowa, U.S.A.

HESS, A.
Department of Anatomy, College of Medicine and Dentistry of New Jersey, Rutgers Medical School, University Heights, Piscataway, N.J., U.S.A.

HOWE, A.
Department of Physiology, Chelsea College of Science and Technology, Manresa Road, London.

JONES, J. V.
Department of Cardiology, The Royal Infirmary, Glasgow, C.4, Scotland.

KEEN, P.
Department of Pharmacology, The Medical School, University Walk, Bristol, England.

KOBAYASHI, S.
Department of Human Morphology, University Park, Nottingham, England.

LANDON, D. N.
M.R.C. Research Group in Applied Neurology, Institute of Neurology, 8/11 Queen Square, London, W.C.1.

McDONALD, D. M.
Cardiovascular Research Institute, U.C. Medical Center, San Francisco, California, U.S.A.

MILLS, E.
Department of Physiology & Pharmacology, Duke University Medical Center, Durham, North Carolina, U.S.A.

MITCHELL, R. A.
Cardiovascular Research Institute, U.C. Medical Center, San Francisco, California, U.S.A.

NAIL, B. S.
Department of Physiology, The Medical School, University Walk, Bristol, England.

O'REGAN, R. G.
Department of Physiology, University College, Upper Merrion Street, Dublin 6, Ireland.

PALLOT, D.
Department of Physiology, The Medical School, University Walk, Bristol, England.

PURVES, M. J.
Department of Physiology, The Medical School, University Walk, Bristol, England.

SAMPSON, S. R.
Department of Physiology, University of California Medical School, U.C. Medical Center, San Francisco, California, U.S.A.

SEIDL, ELLEN
Max-Planck Institut für Systemphysiologie, 201 Rheinlanddamm, Dortmund 46, West Germany.

SHEPHERD, J. T.
Department of Physiology, The Mayo Foundation, Rochester, Minnesota, U.S.A.

SPECKMANN, E.-J.
Physiologischen Institut, Universität Münster, Westring 6, Münster, Westfalen, West Germany.

TORRANCE, R. W.
University Laboratory of Physiology, Parks Road, Oxford, England.

VERNA, A.
Laboratoire de Cytology, Université de Bordeaux II, Avenue des Facultés, 33-Talence, France.

WILLSHAW, P.
Sherrington School of Physiology, St Thomas's Hospital Medical School, London, S.E.1.

WOODS, R. I.
Physiological Laboratory, Downing Street, Cambridge.

PREFACE

The peripheral arterial chemoreceptors continue to engage the interest of a wide variety of workers and this simply reflects the wide range of problems which will have to be solved before their structure, mode of excitation and reflex role is properly understood. The problems range from the elucidation of the cellular and intracellular structure of the receptor complex on the one hand to the role of the receptors in the respiratory and circulatory response to hypoxia on the other. And it is evident that the rate at which progress has been made in answering these questions varies considerably. Perhaps the question which has received most detailed consideration has been the chemoreflex control of respiration though it should be noted that the dynamic aspects of this question and the crucial questions of how chemoreceptor fibres project centrally and affect respiratory neurones have scarcely been tackled. As is shown in the present volume, there has been substantial progress in determining the chemoreflex control of various vascular beds and in consequence, the protective and homeostatic role of the circulation and the central strategic function of the chemoreceptors in this is now fairly well appreciated.

By contrast, we know very much less about which structure is the receptor, how it is excited, which transmitters if any are involved and, indeed, what the effective stimulus is which excites the receptor. This is due in part to the lack to date of really good quantitative studies on the morphology of the cellular components of the receptor complex, the pattern of their innervation and their vesicle content, which is essential before sensible physiological experiments can be carried out to determine the synaptic sequence, and how or where the potential is generated. In part the slow progress has been due to the relative inaccessibility of the receptor to intra- or extra-cellular recording and in this volume there are a number of hopeful indications of new techniques being developed which are likely to resolve this particular question. In part, our uncertainty about these fundamental properties of the receptor is due to the recent publication of experimental results from various workers which are almost diametrically opposed and this has led to a situation where it has been virtually impossible to give a coherent account – even at an elementary level – of what the receptors are and how they work.

This confusion in the literature was the main reason for convening the small meeting held in Bristol in July 1973. As originally intended, it was to be an opportunity for the workers most closely involved with various aspects of chemoreceptor physiology and morphology to meet and, by reviewing the literature critically either to attempt some form of reconciliation of the evidence or to suggest future experiments which might do just this. As often happens, this intention was overtaken by events and the scope of the meeting was very considerably broadened by the inclusion of the results of a number of important studies which had matured earlier in the year. These experimental results not only helped to resolve much of the earlier contentious evidence but also indicated that the physiology of the chemoreceptors was even more interesting than had been thought previously. Of particular value was the paper given by Professor Carlos Eyzaguirre which was the result of much patient examination by him of Professor Ferdinand De Castro's original slides which confirmed how comprehensive and meticulous his studies had been with the techniques then available and how perceptive were his projections.

It was largely because of the possibility of including this historical and recent material that it was decided to publish the proceedings of the meeting since it was felt that those who are interested in the chemoreceptors could have an up-to-date, even if temporary, statement between one set of covers. The decision to publish the discussion was taken because it was felt that it showed, more eloquently than is possible in a formal review, how statements made with some confidence in presentations were amplified, modified or even retracted in the face of informed criticism. These decisions will have been amply justified if the contents of this book lead to further imaginative and decisive experiments.

M. J. Purves
Bristol, 1973

Acknowledgements

The initiative for this meeting and much of the planning came from Dr Frank Abboud of the University of Iowa and Dr John Shepherd of the Mayo Foundation and the meeting was made possible only by generous grants from the American Heart Association, The British Heart Foundation, Hoescht Pharmaceuticals, Hoffman LaRoche Inc, Imperial Chemical

Industries, The Physiological Society, The University of Bristol and its Publication Committee, The Wates Foundation, William Warner Ltd and the Wellcome Trust.

I am grateful to the publishers and authors as indicated in the text for permission to reproduce figures or plates and to my secretary, Mrs Myra Cook, for assistance in the transcription of tapes and for the preparation of the manuscript.

THE ULTRASTRUCTURE OF THE TYPE I CELL AND ITS INNERVATION

An examination of De Castro's original slides*

C. EYZAGUIRRE† AND A. GALLEGO‡

† Department of Physiology, University of Utah College of Medicine, Salt Lake City, Utah, USA and

‡ Departamento de Fisiología y Bioquímica, Facultad de Medicina, Universidad Complutense, Madrid, España

The purpose of this presentation is to re-examine De Castro's original work since there is a renewed interest in the morphology and physiology of the carotid body chemoreceptors. Furthermore, his original work (which established morphologically the chemoreceptor nature of the carotid body in 1928) has not received due recognition amongst investigators working on these problems in recent years. It is our belief that this has been due, in part, to the fact that he published all his papers either in Spanish or French, languages which are not familiar to many investigators interested in this field. The authors of this communication, being rather conversant with the Spanish language and moderately proficient in French, have been able to review De Castro's original publications. More important, we have been able to examine Professor De Castro's slides utilizing modern optical techniques (including Nomarski optics) and have confirmed the veracity of his drawings which were made at a time when photography was not fully employed by histologists. They had good reasons for this; their sections were relatively thick and a drawing gives a more faithful representation of the distribution of nerve fibres and their relations to other structures. They did not have to take hundreds of photographs to make a final composition with results similar to those obtained in a drawing.

* We are very grateful to Mr Fernando de Castro, Jr for his kindness in allowing us to examine his father's slides which, fortunately, are in his possession, thus avoiding the too-often practised pilferage and destruction of famous works.

One remarkable thing about De Castro's preparations (both silver and methylene blue stained) is that they remain in excellent condition and the details described in his papers can be fully recognized. This is particularly impressive in the case of the sections stained with methylene blue which are in excellent condition even though more than 40 years have passed since they were made. This may be due, at least in part, to the fact that De Castro used a relatively impure chemical which seemed to work better than the newer and more purified methylene blues. Alas, purity does not always help!

De Castro (1928) concluded that 'the glomus represents an organ with multiple sensory receptors containing receptor cells very similar to those found in other sense organs'. He tentatively inferred that 'the "glomus caroticum" represents a sense organ (the only one known at present) that detects qualitative variations in the blood, a function which by reflex action will have an influence on the functional activity of other organs'. These ideas led Heymans and collaborators to study respiratory reflexes originating in the carotid body (see Heymans & Neil, 1958).

The material we examined consisted of histological sections of carotid bodies of cat, dog and man. We paid special attention to the cat carotid body and this paper refers exclusively to it. The preparations were stained either by the methylene blue or reduced silver methods. Normal preparations of young or adult cats were examined, in addition to material obtained nine days after severance of the glossopharyngeal nerve either beyond its emergence from the jugular foramen or intracranially at its emergence from the medulla.

NERVE ENDINGS ON GLOMUS CELLS

The glomus cells are grouped in glomerules surrounded by a periglomerular nerve plexus from which fine nerve fibres are distributed inside the glomerules. Branches of these fibres give origin to the terminals which are found on glomus Type I cells. The structure of these terminals was examined by De Castro (1926) using primarily methylene blue vital staining. He wrote:

> it is clear that fine, unmyelinated fibres leave the periglomerular plexus, to become intra-glomerular fibres. These fibres have varicosities which are often located at the point of division of the fibres. The latter divide into many very thin collaterals which, frequently, are hard to see.... One may easily confuse the varicosities with the true nerve endings.... The small fibres terminate as annulae or as small reticulated masses upon the cells.

Later on, De Castro (1940) expands further on the description of the nerve endings (Plate 1). He found that at birth the nerve endings were not fully developed:

> some cells seem to be privileged to be innervated before others. . . . On them, the final terminal expansion forms a cup or thimble enveloping the glomus cells in a manner similar to the terminals seen in the ganglion of Scarpa innervating the epithelial cells of the crista ampullaris of the semicircular canals and maculae of the saccule and the utricule. We made similar observations in the carotid body of the young rat and referred to them as 'fibres à terminaisons en calice . . . from these endings, delicate fibres emerge to terminate on neighbouring cells'.

In normal, adult cats one can see nerve fibres originating in the carotid nerve that divide inside the carotid body into two or more terminal

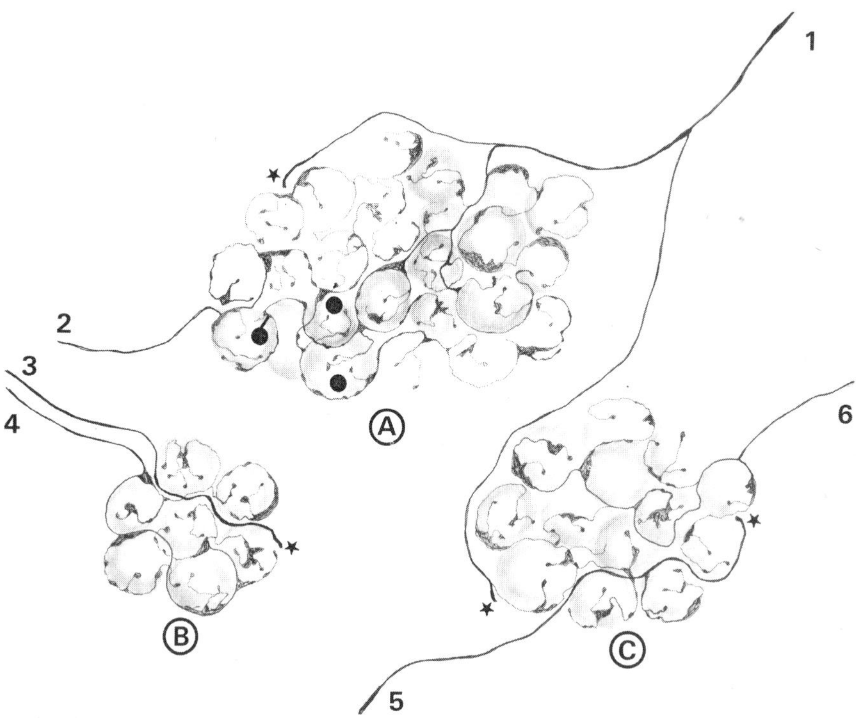

Fig. 1. Schematic representation of innervation of glomerules in the carotid body. Fibre 1 divides to give off branches to different glomerules (A and C). Fibre 1 in (★) leaves the optic field to join the periglomerular plexus, probably branching again. On glomerule A, fibres 1 and 2 converge and end occasionally on the same glomus cells (●). Fibre 3 crosses glomerule B, apparently without branching. Fibres 1, 5 and 6 converge on glomerule C. One cannot exclude the possibility that some of the fibres in this drawing originate from a single carotid nerve fibre.

branches going to different glomerules (Fig. 1; 1). These fibres were called 'pluriglomo-receptor fibres' by De Castro. It is common to see that terminal fibres dive into the glomerules after leaving the periglomerular plexus, to form one or more fascicles of axons at one pole of the glomerule (Plate 2*f*, 3, 4*c* and *d*).

The general innervation of glomerules is represented in Fig. 1, which shows that some glomus cells receive terminals from different fibres, a fact that was described by De Castro (1940). Conversely, one carotid nerve fibre (that undergoes extensive division both extra- and intra-glomerularly) may innervate numerous glomus cells even if they are located in different glomerules (also Plate 5).

It is impossible to determine in well-stained and fine histological sections the distribution of the 'sensory unit' in the carotid body. We were able to see in De Castro's preparations that from 10 to 15 cells were innervated by one nerve fibre. It is likely, however, that the real number may be much higher. But, preparations like De Castro's are unlikely to be useful in carefully measuring the sensory unit. For that purpose, different techniques such as Golgi staining of thick sections should be used.

Examination of the nerve endings on glomus Type I cells appears to confirm fully De Castro's original description: a single terminal nerve fibre shows enlargements in the form of annuli, masses, disks or reticulated calices and bouton-like endings.

We only found nerve endings on the glomus Type I cells. To study them we drew them directly from the preparations using camera lucida and photomicrographs taken on different focal planes. Drawings were made also from the photographs. Plate 5 is a composition from photographs of preparations stained with methylene blue. One can see terminal calices (Plate 5*a*,*b*), photographed at two focal planes, and three enlargements (like rosary beads) ending on the same cell on which the calix or meniscus ends also. In other instances, fine collaterals, which take off from a meniscus or calix, end as boutons on a neighbouring cell (Fig. 6*c*,*d*). Menisci (calices) are sometimes small and very delicate, presenting a reticulated aspect (Plate 5*e*,*f*), despite the age of the preparations and the tendency of methylene blue, especially after fixation, to show dark and solid-looking endings. The frequency of this combination of structures, such as calices and boutons of different size, can be appreciated when two different planes of two neighbouring cells show the terminal nests that envelop them.

We utilized Nomarski optics to see the relationship between the nerve endings and Type I cells (Plate 5*d*). The nerve endings are apposed to the surface of the cells and the menisci (calices) appear as described by De Castro on several occasions.

Plate 6 shows that a periglomerular fibre gives a branch (clear in *c* and *d*) that penetrates a glomerule while innervating some peripherally located Type I cells. The endings envelop the glomus cells, the same fibre having an enlarged bifurcation (visible in *b*), a calix (plate)-type ending (*c*) and fine fibres, some of which emerge from the large plate to give boutons of different sizes and larger 'plate-like' endings (*d*).

Nerve endings apposed to different cells can often be followed in more than one focal plane. Plates 7 and 8 show drawings made from preparations and photomicrographs. Of the latter we are presenting two and three focal planes respectively of the eight planes photographed. The terminal fibres show enlargements, calices or disks, and terminal boutons as shown also by De Castro (see Plate 1). It is interesting to notice that one cell has plates and terminal boutons originating from the same fibre. Also, some cells receive endings originating from more than one fibre, as described also by De Castro. However, in these preparations one cannot exclude the possibility that fibres which appear to be independent are really branches of a common parent fibre which has divided outside the glomerule.

De Castro's preparations show clearly that nerve endings terminate only on Type I cells and this is in agreement with more recent electron microscope studies (Biscoe & Stehbens, 1966; Hess, 1968; Hess & Zapata, 1972). However, studies with the light microscope may have missed other, and less frequent, nerve endings terminating elsewhere in the glomera. Also, it is of interest that a single nerve fibre may terminate in different forms on the glomus cells giving rise to boutons, calices or menisci (Plate 9). These various forms of endings do not necessarily come from different nerve fibres as may be assumed from electron micrographs showing bouton-like terminals having abundant clear-core vesicles and calices (enveloping cells) with fewer vesicles as shown by K. Nishi & L. J. Stensaas (unpublished; see also, Eyzaguirre, Nishi & Fidone, 1972).

Studies with the electron microscope show clearly that nerve endings are apposed to the surface of glomus cells. De Castro, until 1940, insisted that nerve endings did not penetrate into the cytoplasm. It is curious then that in 1951 he shows, schematically, nerve endings penetrating into the cytoplasm. Perhaps this was due to pressure from other histologists or to his concept of bipolarity of the glomus cells. In fact, in 1951 he shows terminals in the form of a meniscus inside the cytoplasm. Our study of his own preparations shows that his earlier interpretations were correct and well detailed in his papers of 1926, 1928 and 1940. However, in this last paper he appears to waver:

One question that worries considerably some histological schools is whether nerve endings are superficial or penetrate into the cell. We think that they are on the surface of the cell. In fact, this is the position of the terminal meniscus, mass or annulus on the larger glomus cells, at some distance from the nucleus when this is not located at the periphery of the cell. In young or newborn animals the ending is almost in contact with the nucleus, the ending forming a cup-like structure which almost tends to impale the nucleus as can be seen in our paper of 1926 when a young rat was used. We do not have any trouble in accepting that the ending is in the cytoplasm in this particular case; however, this does not mean that the neuronal theory has received a serious setback, as assumed by Boeke (1932); the behaviour of the nerve cell is the same. The nerve endings of striated muscle fibres are hypolemmal (penetrate the cell). It is quite a different thing to pretend that there is cytoplasmatic fusion of the nervous element and the innervated cell, which is contrary to the thought of the sagacious neurologist from Utrecht; when the carotid nerve is sectioned, the fibres degenerate rapidly because their continuity with their trophic centre has been discontinued and this should not happen if their cytoplasms had been continuous. From that moment there is absence of that 'dynamic influx that irradiates from the nerve cell body which prevents its wilting' and that stimulates the nutrition of the fibre, as was admirably conceived by Cajal.

AFFERENT NATURE OF NERVE ENDINGS ON GLOMUS TYPE I CELLS

De Castro (1926) described the nerve endings on the glomus cells. In 1928 he tackled the problem of studying the afferent or efferent nature of these endings and these experiments are again referred to in his paper of 1940:

> In view of so many contradictory opinions with regard to the function of the glomus caroticum . . . because of the results obtained with organ extracts or after its removal, we decided to resolve the one question that had to be answered to be sure whether or not the glomus is or is not a gland. This consisted of demonstrating whether the fibres innervating it were centrifugal or motor. The analysis was done in 1928. It consisted of sectioning the glossopharyngeal roots at their entry into the medulla. Naturally, the fibres of the glossopharyngeal and carotid nerves should degenerate after several days if they were centrifugal since their trophic center is in the medulla; centripetal or sensory fibres should remain intact because their trophic center is below the section. In fact, . . . the carotid nerve did not show any degeneration of the nerve tubes,

and the fibres of the glomus caroticum with their nerve endings on the parenchymal cells appeared normal. Therefore, the glomus was not, as previously believed, an endocrine gland but a sensory receptor.

Biscoe, Lall & Sampson (1970) performed an intracranial section of the glossopharyngeal nerve roots and concluded that this procedure induces degeneration of the nerve endings on the glomus Type I cells. Their conclusion was that these endings are *efferent* and not afferent as it has been widely accepted over the years following De Castro's (1928) observations and those of Heymans *et al.* using physiological techniques (see Heymans & Neil, 1958). More recently, however, Hess & Zapata (1972) have repeated these experiments using both light and electron microscopy. Their conclusions agree with those of De Castro, indicating that the nerve endings on the glomus Type I cells are afferent. Hess & Zapata also used physiological controls to determine the degree of sectioning of the roots. They deliberately sectioned not only the roots of the ninth nerve but also those of the tenth and eleventh nerves. These animals showed fibrillation of pharyngeal muscles (innervated by the ninth nerve), the heart did not slow down during peripheral stimulation of the vagus and the shoulder muscles were paralysed and fibrillating. It is important to notice here that De Castro and Hess & Zapata sectioned the roots at their emergence from the medulla and not close to the foramen as Biscoe *et al.* (1970) have reported doing.

Reviewing the original slides of De Castro we found some sections of the carotid body prepared nine days after either intracranial section of the ninth nerve or section of the same nerve after its emergence from the jugular foramen. Plate 10 shows photomicrographs of such tissue after glossopharyngeal nerve section either inside or outside the cranium. Intracranial section of the nerve (Plate 10*a*,*b*) shows no alterations of the nerve fibres or endings, being similar to those found in normal animals. On the contrary, Plate 10*c* and *d*, taken from a denervated carotid body, shows the absence of normal nerve endings, the presence of a few thick degenerated fibres and debris of boutons and of terminal plates (*c*). There are a few thin unmyelinated fibres well preserved (*d*). The latter may have originated in the superior cervical ganglion or from the microganglia sometimes found at the entry of the carotid nerve into the glomus (De Castro, 1926).

DISCUSSION AND SUMMARY

The results of our work reviewing De Castro's original slides have shown several interesting findings:

(1) The original slides made with methylene blue staining are extremely well preserved even after 40 years.
(2) A single nerve fibre branches to innervate several cells (more than 10) located in one or more than one glomerule.
(3) One glomus Type I cell receives very often more than one nerve fibre.
(4) The shape of the terminals ending on glomus Type I cells varies a great deal. Thus, we have boutons, plates, calices, etc. Furthermore, some endings give prolongations or finger-like processes which go to innervate another glomus cell(s). Consequently, the different types of endings do not come necessarily from different fibres.
(5) All nerve endings terminating on glomus Type I cells appear to be sensory since they degenerate after extracranial section of the glossopharyngeal nerve and are not affected by intracranial section of the same nerve.

With regard to the last point, how can we reconcile the contradictory findings of De Castro and Hess & Zapata on one hand and those of Biscoe *et al.* on the other? Two explanations may be offered:

(1) Trauma provoked by intracranial section of the ninth nerve close to the foramen jugularis, a technique employed by Biscoe *et al.*, followed by degeneration, oedema and vascular lesions.
(2) Genuine disconnection of sensory cells not located in the petrosal ganglion.

We may discard the first possibility since we may assume that Biscoe *et al.* performed a clean surgical operation. However, we cannot discard the second possibility. In fact, in most mammals (man, cat and rat) the petrosal ganglion is not always a perfectly well defined organ confined to the jugular foramen. Very often the ganglion is diffuse since enlargements of the nerve (containing ganglion cells) occur also within the intracranial trajectory of the ninth nerve roots. These enlargements are more pronounced closer to the foramen than they are near the emergence of the roots from the medulla (K. Nishi & L. J. Stensaas, unpublished; P. Zapata, unpublished.) Furthermore, for many years anatomists have described the presence of an accessory ganglion within the cranium, Ehrenritter's ganglion, which is also part of the glossopharyngeal intracranial trajectory. In fact, this ganglion has been described in the rat, in man (Cajal, 1904) and in the cat (Reighard & Jennings, 1957).

The constant presence of the Ehrenritter ganglion poses some interesting problems that are amenable to experimental investigation. Are all or most of the chemosensory neurons located in the Ehrenritter ganglion, or are they distributed evenly or unevenly between the two ganglia? Where are the baroreceptor sensory neurons located? Do they show a special preference for either of the two ganglia? These possibilities can be investigated by selective lesions of the ganglion of Ehrenritter without altering the petrosal ganglion and then studying the degeneration of chemo- and pressoreceptor endings. Another possibility is to inject radioactive aminoacids into either ganglia. This will allow the study of carotid body nerve endings with autoradiographic techniques several hours after the injections since axonal flow should transport these substances toward the nerve endings after the cells have incorporated the aminoacids into their cytoplasm.

This work was supported by grants NS 05666 and NS 07938 and from the U.S. Public Health Service and a grant from the Fundación Marques de Urquijo.

REFERENCES

Biscoe, T. J., Lall, A. & Sampson, S. R. (1970). Electron microscopic and electrophysiological studies on the carotid body following intracranial section of the glossopharyngeal nerve. *J. Physiol.*, **208**, 133–52.

Biscoe, T. J. & Stehbens, W. E. (1966). Ultrastructure of the carotid body. *J. Cell Biol.*, **30**, 563–8.

Boeke, J. (1932). Nerve endings, motor and sensory. In *Cytology and Cellular Pathology of the Nervous System*, **1**, 243–56, New York: Hoeber.

Cajal, S. R. (1904). *Textura del Sistema Nervioso del Hombre y de los Vertebrados*, Madrid: Moya.

De Castro, F. (1926). Sur la structure et l'innervation de la glande carotidienne (glomus caroticum) de l'homme et des mammifères, et sur un nouveaux système d'innervation autonome du nerf glossopharyngien. *Trab. Lab. Invest. Biol. Univ. Madrid*, **24**, 365–432.

De Castro, F. (1928). Sur la structure et l'innervation du sinus carotidien de l'homme et des mammifères. Nouveaux faits sur l'innervation et la fonction du glomus caroticum. *Trab. Lab. Invest. Biol. Univ. Madrid*, **25**, 331–80.

De Castro, F. (1940). Nuevas observaciones sobre la inervación de la región carotidea. Los quimio y preso-receptores. *Trab. Lab. Invest. Biol. Univ. Madrid*, **32**, 297–384.

De Castro, F. (1951). Sur la structure de la synapse dans les chemocepteurs: leur mécanisme d'excitation et rôle dans la circulation sanguine locale. *Acta physiol. Scand.*, **22**, 14–43.

Eyzaguirre, C., Nishi, K. & Fidone, S. (1972). Chemoreceptor synapses in the carotid body. *Fedn Proc.*, **31**, 1385–93.

Hess, A. (1968). Electron microscopic observations of normal and experimental cat carotid bodies. In *Arterial Chemoreceptors*, ed. Torrance, R. W., Oxford: Blackwell.

Hess, A. & Zapata, P. (1972). Innervation of the carotid body: normal and experimental studies. *Fedn Proc.*, **31**, 1365–82.

Heymans, C. & Neil, E. (1958). *Reflexogenic Areas of the Cardiovascular System*, London: Churchill.

Reighard, J. & Jennings, H. S. (1957). *Anatomy of the Cat*, 3rd edn enlarged by Elliott, R. New York: Holt.

EXPLANATION OF PLATES

PLATE 1

Innervation of glomus cells; (*a*), nerve fibre termination as calices on three glomus cells; (*b*), glomus cell over which a large calix is apposed. This ending gives a couple of nodose-like terminals ending on another cell; (*n*), aspect of calices when staining is incomplete. Vital staining with methylene blue (De Castro, 1940); (*c*), sinusoids; (*d*), nodose nerve ending. Glomus cell (*e*) has a calix type ending in close proximity to a sinusoid (*c'*). This and the following figures refer to preparations obtained from the cat carotid body.

PLATE 2

Serial photomicrographs of the carotid body. This and the following illustrations were taken from slides stained with methylene blue. Group of nerve fibres penetrating into a glomerule from one of its poles. Glomerules in the cat are more sausage-like than spherical. In (*f*), one may see a single fibre coming from the periglomerular plexus before penetrating into it. In this and all other photomicrographs, cal. = 20 μm.

PLATE 3

Montage showing the complete glomerule partially described in Plate 2 showing innervation from two poles and single nerve fibres entering each pole.

PLATE 4

Bundle of nerve fibres forming the periglomerular plexus; some penetrate directly into the glomerule.

PLATE 5

Nerve endings on glomus Type I cells; (*a*) and (*b*), photographs taken at two different optical planes of same preparation showing a meniscus (calix) that gives off some fine filaments, one of them showing three terminal boutons which end on the same cell innervated by the meniscus-like ending; (*c*), terminal meniscus giving off a filament which ends on a neighbouring cell; (*d*), photomicrograph taken with Nomarski optics showing nerve endings on the surface of glomus cells; (*e*) and (*f*), photographs taken from two focal planes showing nerve endings on glomus cells. One may see finely reticulated small plates, fine filaments leaving the plates and boutons during the course of fibres and terminals.

PLATE 6

Four different optical planes showing nerve endings on glomus Type I cells. There is one meniscus or calix-type ending giving some fine filaments with boutons of different size or enlargements in the form of small plates. In (*c*), the nerve fibre that ends on a glomus cell originates from the division of another larger fibre as shown in (*d*).

PLATE 7

Nerve endings photographed at two different optical planes [(*a*) and (*c*)]. (*b*) shows a drawing reconstructing the orientation of fibres and endings using photographs in (*a*) and (*c*) and other pictures not shown here.

PLATE 8

Nerve endings on glomus cells in the form of calices or terminal boutons. (*a*), (*b*) and (*c*), different optical planes; (*d*), drawing depicting a composition of the photographs shown here and others not shown.

PLATE 9

Several types of nerve endings on glomus Type I cells. (*a*) and (*b*), calix-type endings; (*c*) small boutons along a fibre that circumvents a group of glomus cells; (*d*) meniscus-like ending with a plate and a terminal bouton; (*e*) and (*f*), fibrils, plates and terminal boutons.

PLATE 10

(*a*) and (*b*), normal nerve endings on glomus Type I cells nine days after intracranial section of the glossopharyngeal nerve roots; (*c*), degenerated nerve endings in carotid body nine days after extracranial section of the glossopharyngeal nerve; (*d*) well-preserved unmyelinated fibre in the same preparation shown in (*c*). In (*b*) one nerve fibre divides inside a glomerule.

PLATE I

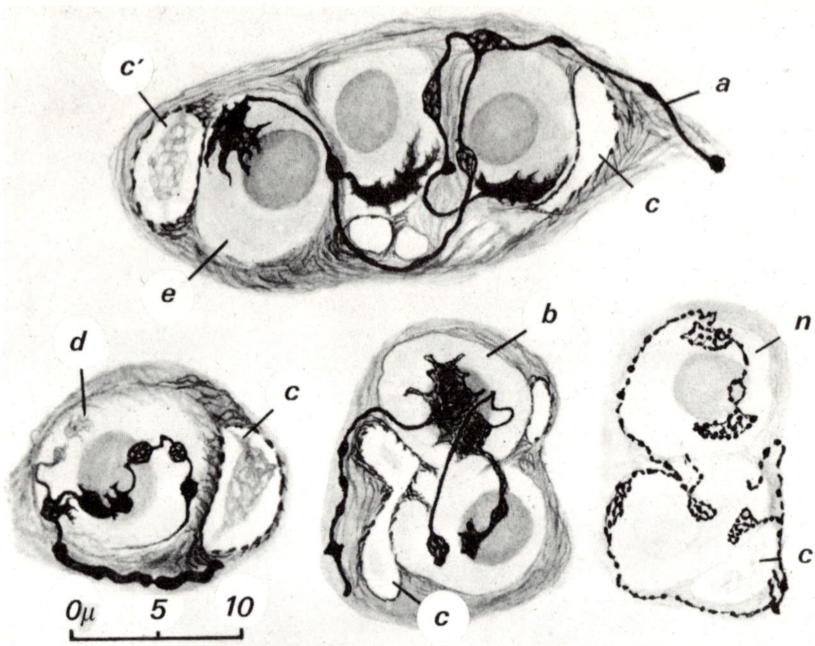

For explanation of Plate see p. 10

PLATE 2

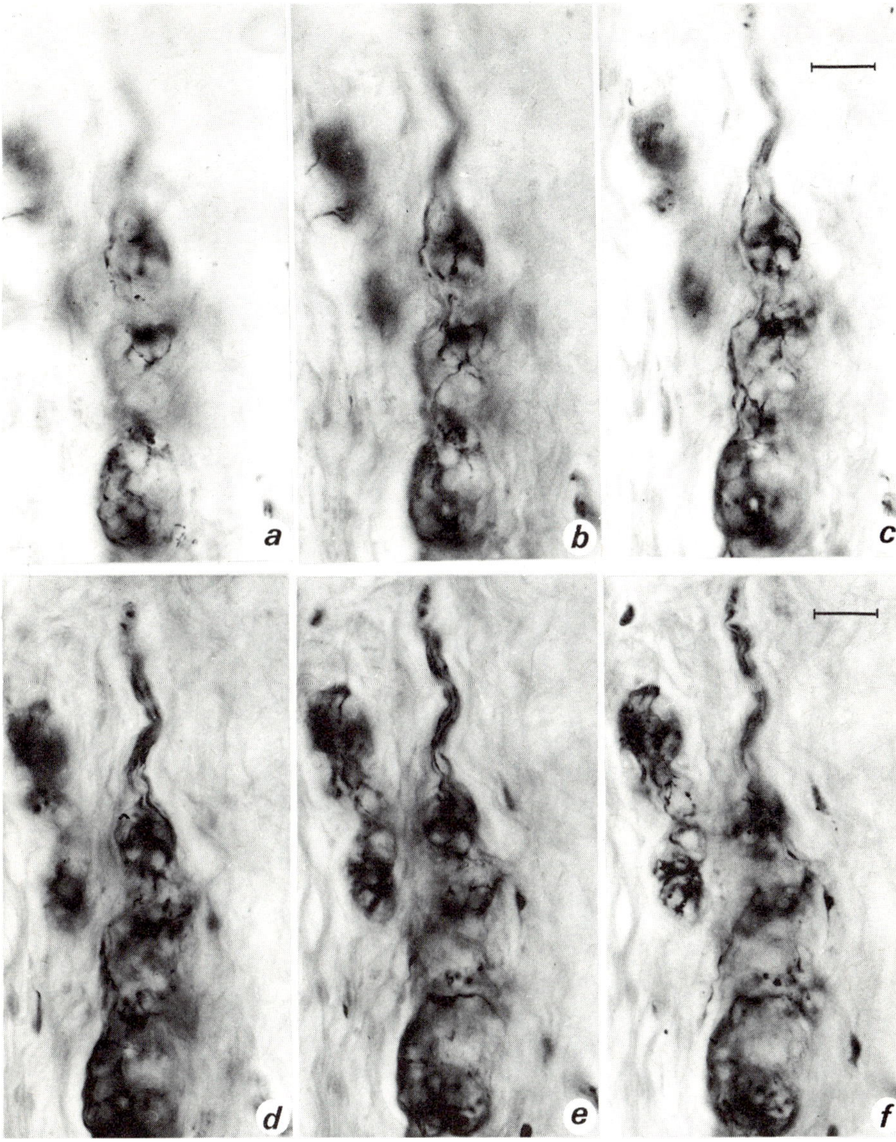

For explanation of Plate *see p. 10*

PLATE 3

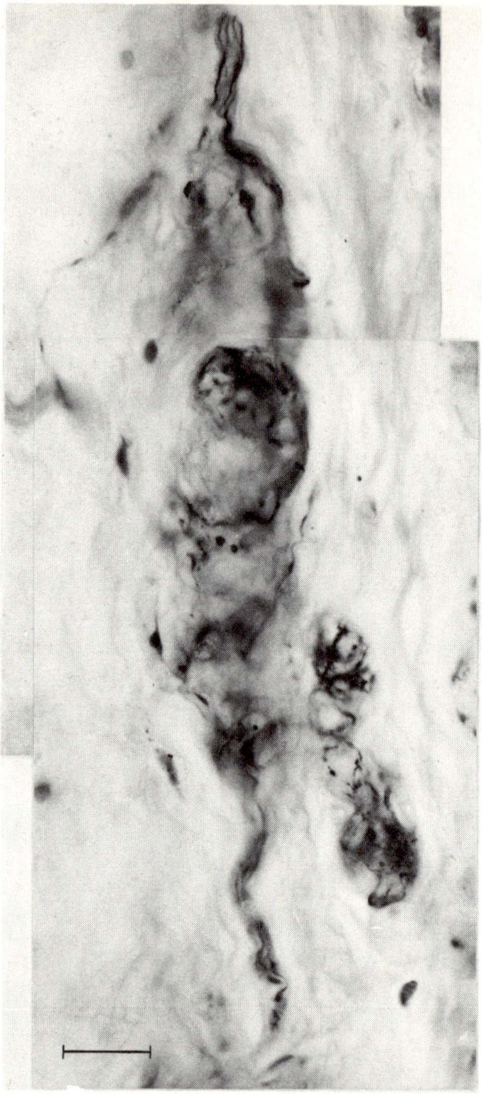

For explanation of Plate see p. 10

PLATE 4

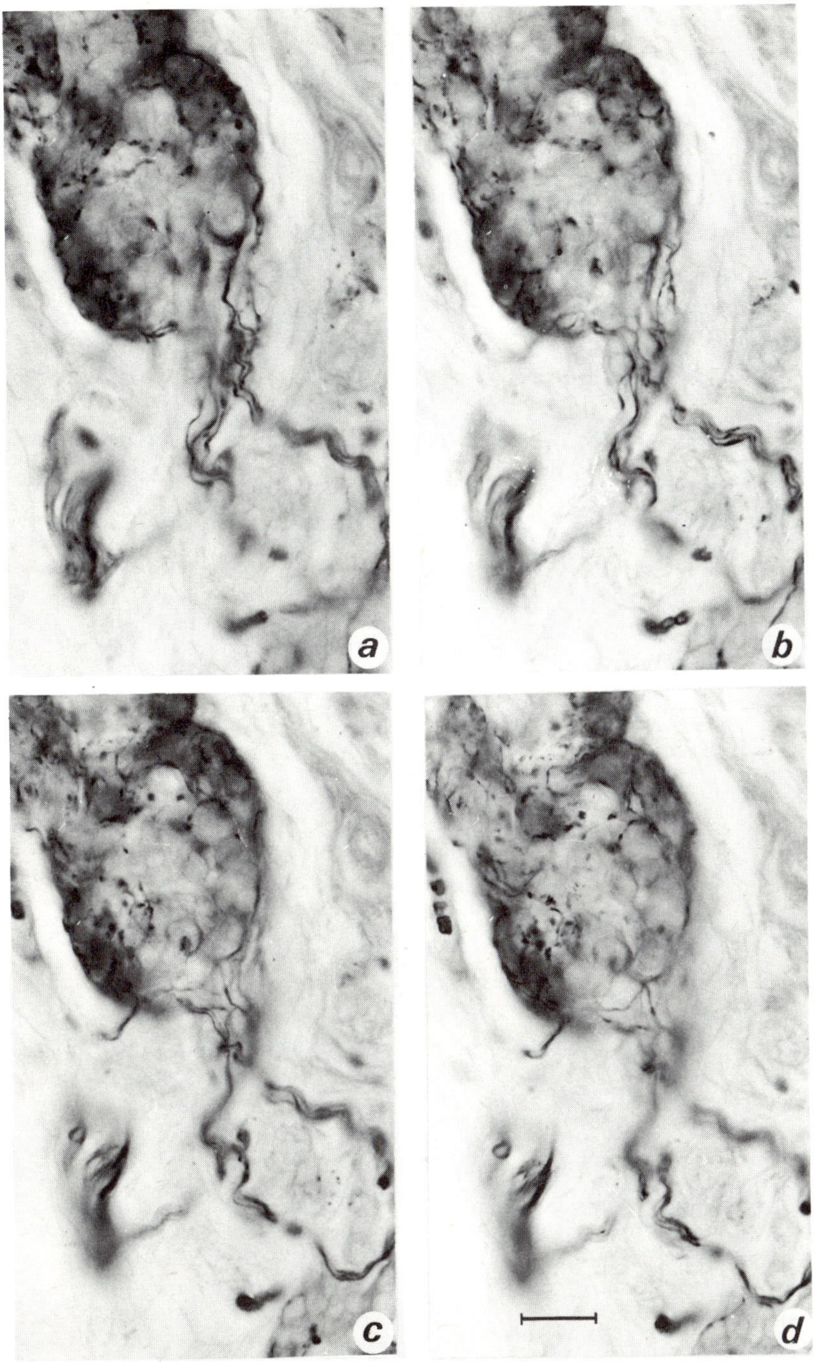

For explanation of Plate see p. 10

PLATE 5

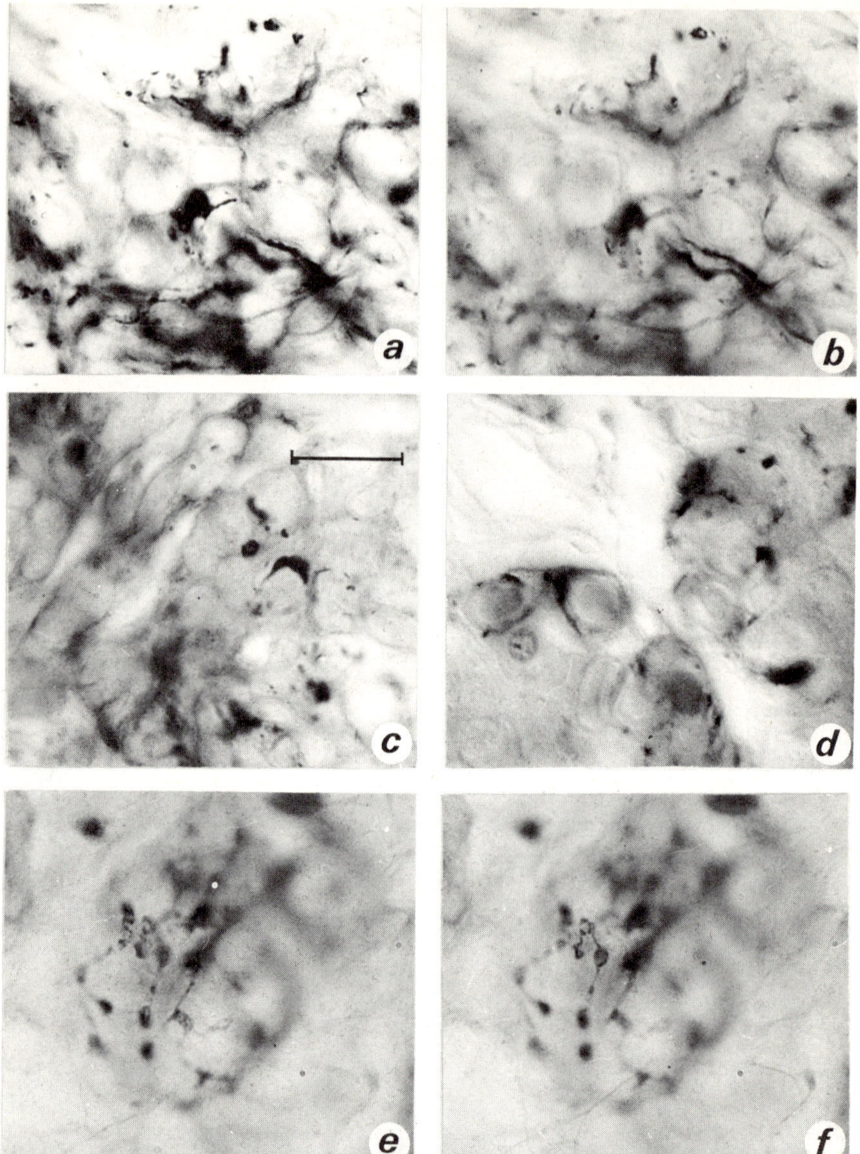

For explanation of Plate see p. 10

PLATE 6

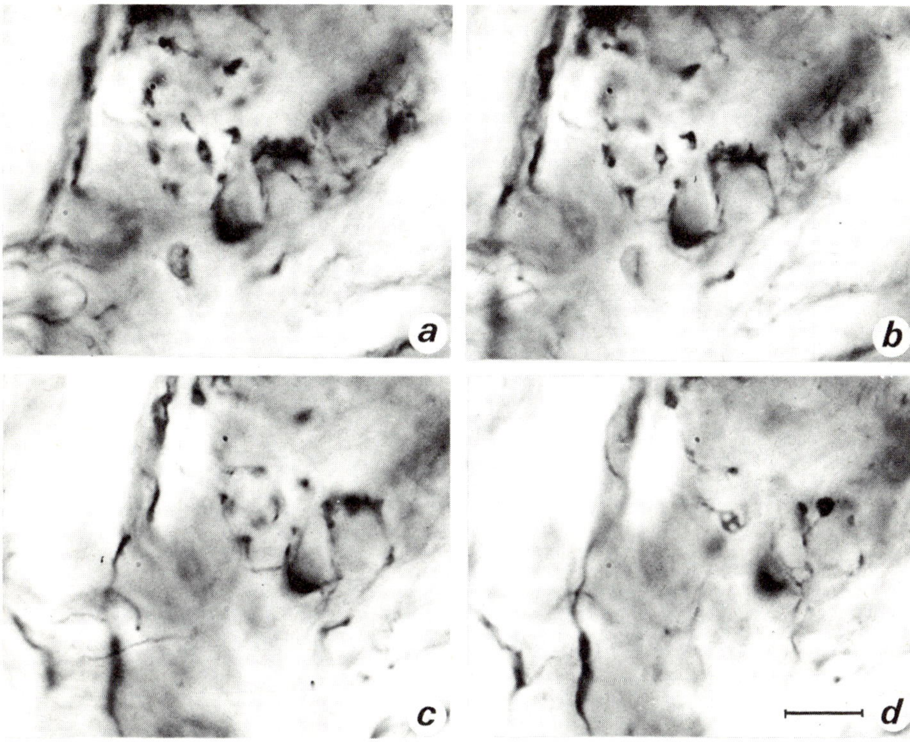

For explanation of Plate see p. 10

PLATE 7

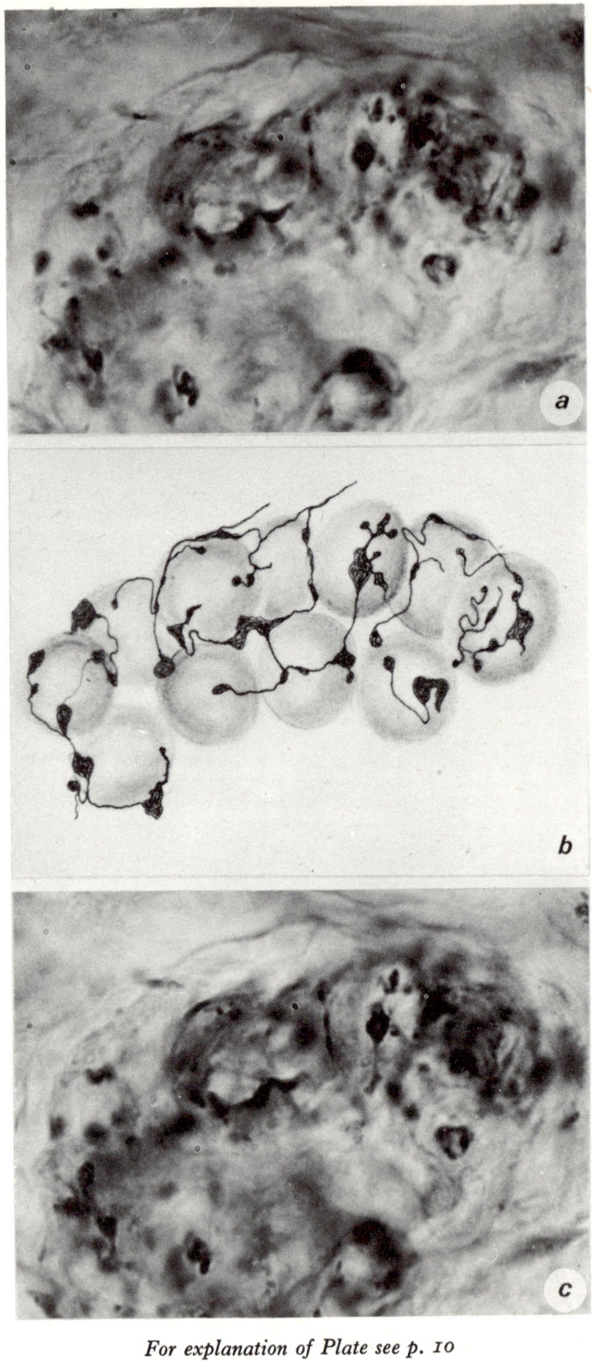

For explanation of Plate see p. 10

PLATE 8

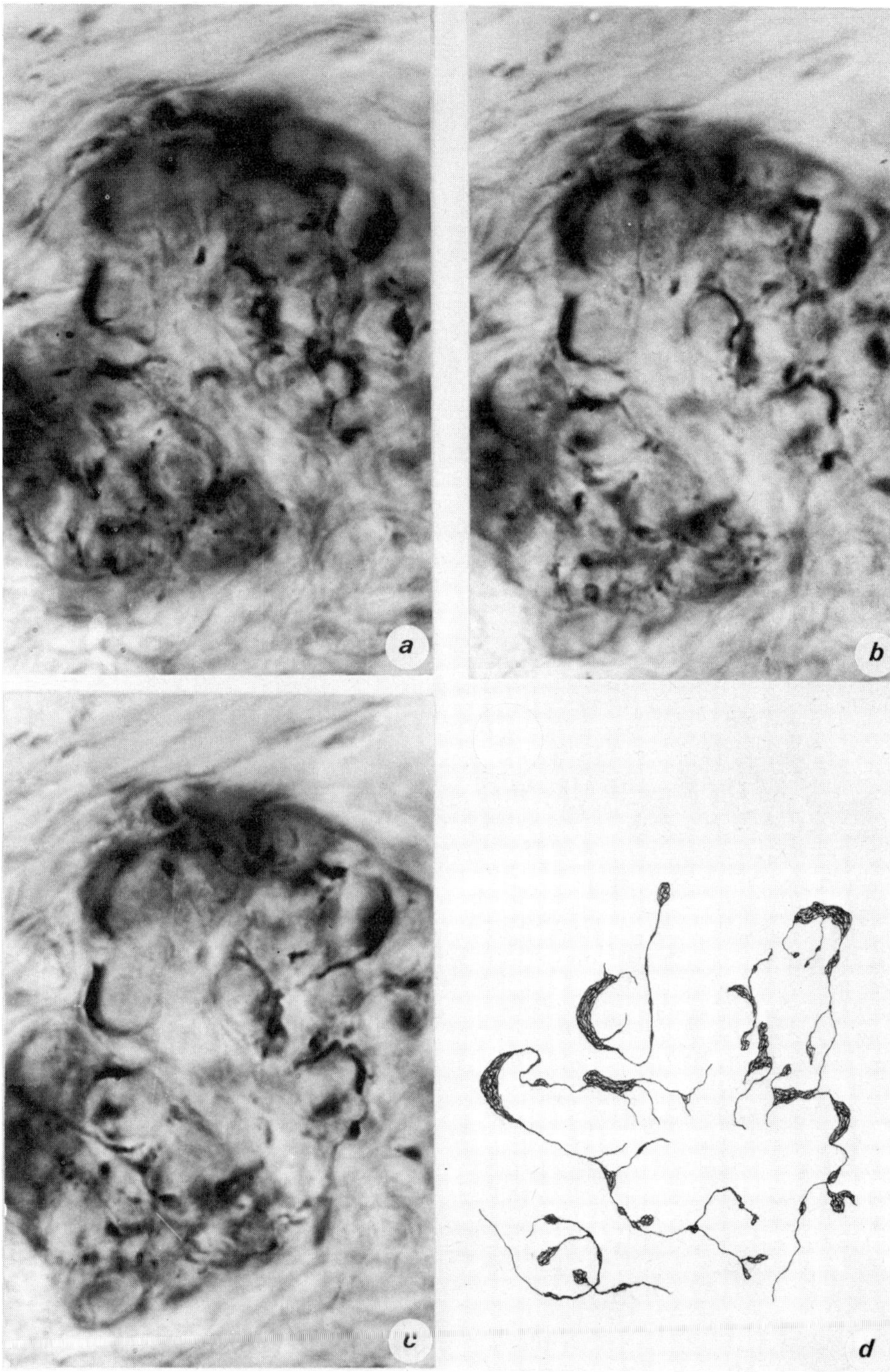

For explanation of Plate see p. 11

PLATE 9

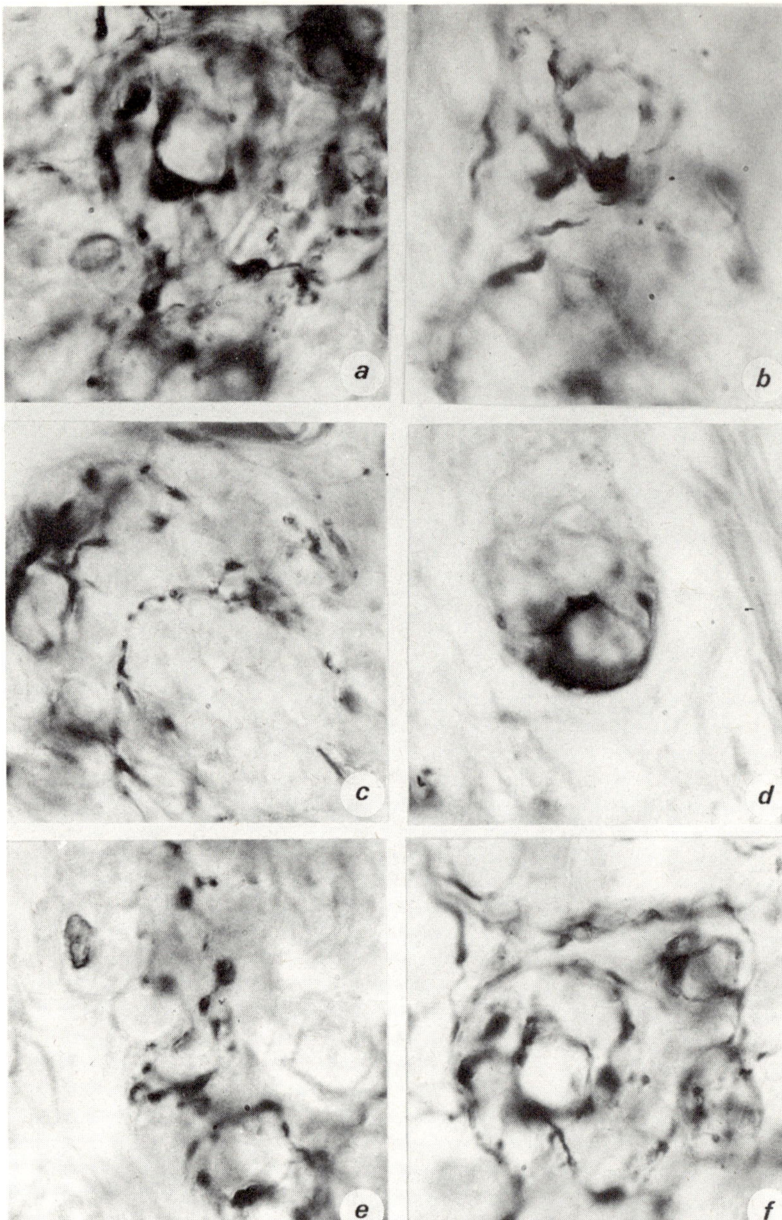

For explanation of Plate see p. 11

PLATE 10

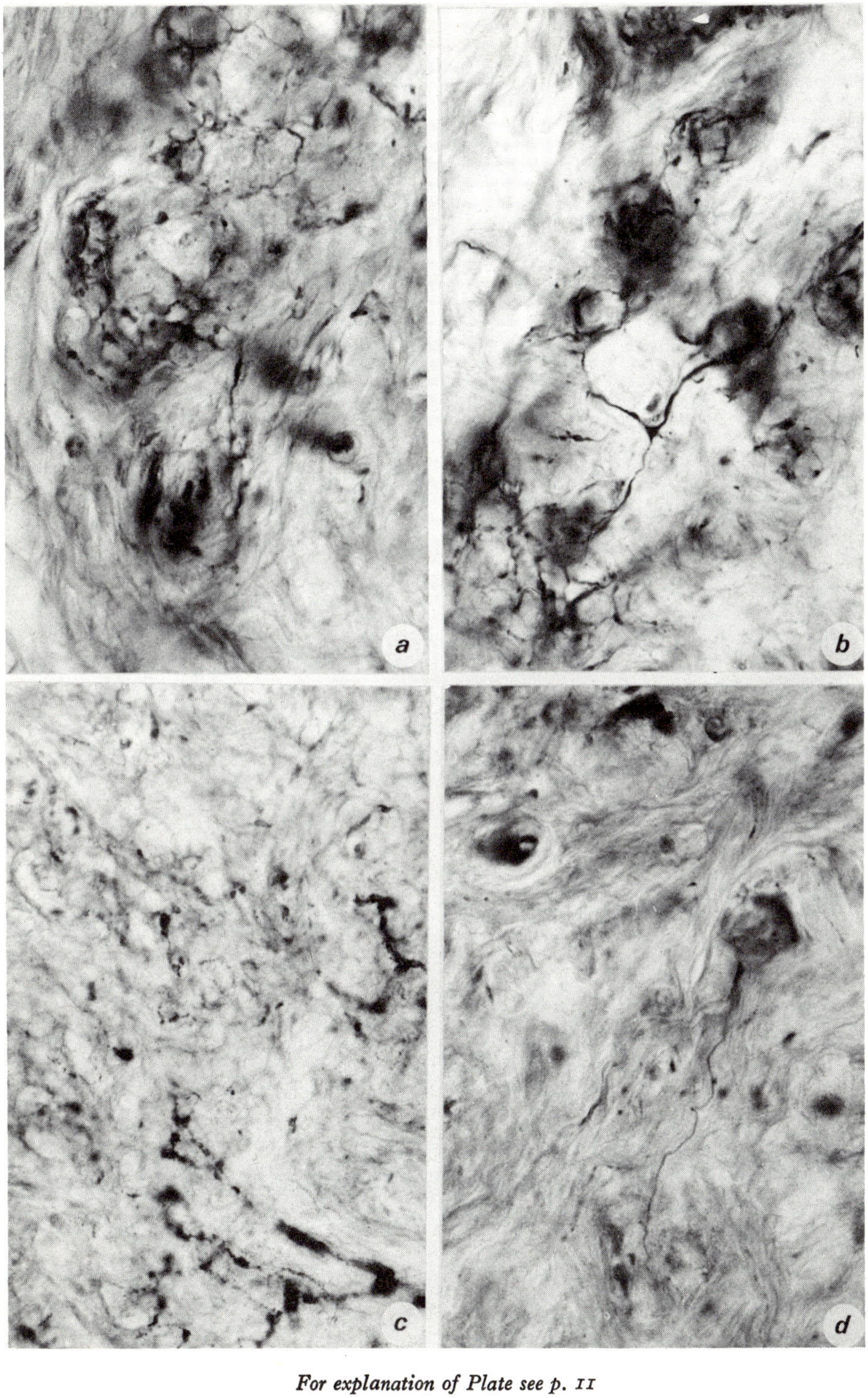

For explanation of Plate see p. 11

DISCUSSION

PURVES What were the dimensions of the small fibres which you showed and of the larger expansions?

EYZAGUIRRE I am not exactly certain on this point. The majority of the slides were viewed with oil immersion and at ×1000: so I would estimate that the fibres were about 2–3 μm and the expansions 8–10 μm.

MCDONALD Could you comment on the site of branching of the axons as they enter the carotid body. Is it close to the glomus cells?

EYZAGUIRRE This is not always easy to tell from the slides. It is certainly possible to see branching very close to the glomus cells: there is no question about that. But it may also be possible that the fibres branch further back. I seem to recall that this was seen in one or two of the lower power sections.

FILLENZ Is there any relation between fibre diameter and the type of ending or did you get all kinds of ending arising from any one fibre?

EYZAGUIRRE My impression was that there was no such correlation.

ACKER In the mouse, it has been shown by Böck (1973) that the Type I cells are frequently surrounded by a basket of nerve fibres. Did you see anything like this?

EYZAGUIRRE Yes, I did; though infrequently. I should point out of course that the vast majority of De Castro's work was done on the cat and it is possible that there is a species difference here. But I am not quite certain how often it has been reported in the mouse. Perhaps I could get back to the question of branching. It seemed to me to be clear that one fibre terminates on one cell. The same fibre, either from the terminal or before, also sends a branch across to another cell. One cell, in addition, has more than one ending – that is very common. So, the fibres branch profusely and innervate several cells.

TORRANCE Is it true that a single cell has endings from more than one fibre of the sinus nerve – that is from more than one cell in the glossopharyngeal ganglion?

EYZAGUIRRE No.

TORRANCE But you can say that there are cells which have more than one ending from the same fibre.

EYZAGUIRRE One cell, yes.

BISCOE Did you see this rosary effect with both methylene blue and with the silver staining? I ask because of the obvious possibility of artefact with silver staining.

EYZAGUIRRE I cannot be absolutely sure about this but I think it was most frequently seen with methylene blue staining.

JONES Did you ever see any nerve fibres terminating without specialized endings on them?

EYZAGUIRRE Yes. Some of the fibres petered out for no good reason at all: but I cannot be sure that this was not a staining artefact.

HESS It might be worth adding that these varicosities are very commonly seen on nerve fibres with methylene blue staining – especially non-myelinated fibres and these do not have to be nerve endings.

REFERENCE

BÖCK P. (1973). *The morphology of the carotid body of the mouse.* Dortmunder Arbeitsgesprach Arzneimittel-Forschung, **23**, 1612.

Ultrastructure of the synapses on the chief cell of the carotid body. A comparative study

By SHIGERU KOBAYASHI*

The Department of Human Morphology, University of Nottingham and Department of Anatomy, Niigata University School of Medicine, Niigata, Japan

Although the receptor theory of the carotid body was originally proposed by De Castro (1928) on the basis of observations which suggested the sensory nature of the nerve fibres which were associated with the chief (Type 1, glomus, epithelioid) cell, the appearance of the nerve endings, revealed in a number of studies with electron microscope, suggests that they subserve an efferent function since the synaptic vesicles (which elsewhere occur on the pre-synaptic side of the chemical synapse) were found only on the 'wrong' side, i.e. the nervous side.

A survey of the previous literature (Kobayashi, 1971) showed that although many electron microscopic studies have been carried out on the carotid body, most of them have been concerned with material from the cat and human. I have therefore broadened this study by examining the ultrastructure of the carotid body in several other species with special reference to the afferent or efferent nature of the nerve ending on the chief cell.

MATERIALS AND METHODS

Animals examined were Amphibia (*Hynobius nigrescens*, *Cynops pyrrhogaster pyrrhogaster*, *Bufo bufo japonicus*, *Rana nigromaculata*), birds (*Uroloncha domestica*, *Gallus domesticus*), and mammals (cat, dog, mouse, guinea-pig, human foetus). Amphibian carotid labyrinths were fixed by immersion in 1.3% osmium tetroxide in a phosphate buffer while the carotid bodies of birds and mammals were perfusion fixed with 2.5% glutaraldehyde in a phosphate buffer followed by post-fixation with osmium tetroxide. All

* Dr Kobayashi is a Wellcome Japanese Research Fellow.

fixed specimens were dehydrated through a series of ethanol, treated with propylene oxide, and embedded in Epon. Numerous, random ultra-thin sections cut on a microtome were stained doubly with uranyl acetate and Millonig's lead solution and observed in the electron microscope.

RESULTS

In all the species examined, the characteristic association of the chief cells, sustentacular cells and nerve endings which have been referred to as the chief cell unit (Grimley & Glenner, 1968), glomus, chemoreceptor unit (Feria-Velasco, Gonzalez-Angulo & Zavala, 1966) etc., formed the fundamental structure of the carotid body. There was considerable species variation as to the size and arrangement of the chief cell unit. Thus in amphibia, the unit was relatively small and contained only one or a few chief cells, and the units were to be found in the so-called trabeculae between the small blood vessels which support the rete mirabile of the carotid labyrinth. In the avian species examined, the chief cell unit was small, as in the amphibian carotid labyrinth, but it was separated from another unit by only a very thin layer of connective tissue. This might be the reason why previous authors, who studied the avian carotid body with light microscopy using thick paraffin sections, erroneously concluded that the avian carotid body consisted of a solid mass of chief cells (Watzka, 1943). Among the mammalian species, the mouse and guinea-pig had smaller chief cell units than the cat and dog and the number of chief cell units was approximately proportional to the size of the animal.

The ultrastructure of the chief cell was very similar in all the species examined and resembled that of protein-secreting endocrine cells elsewhere, such as adrenal medullary cells, pancreatic islet cells and anterior pituitary cells. Thus the nucleus of the chief cell was usually oval in shape and surrounded by a double nuclear membrane with pores. In the cytoplasm, there were abundant mitochondria, granular endoplasmic reticulum which often showed parallel arrangement, lysosomes, microtubules, fine filaments, etc. The Golgi complex was prominent and frequently contained small dense substances in its cisternae. Besides these structures, large numbers of secretory granules were dispersed in the cytoplasm. These granules consisted of a limiting membrane and contents, but the size, shape and texture of the contents were different from species to species and from cell to cell. Thus in the dog, the chief cells were easily separable into two types corresponding to noradrenaline or adrenaline storing cells of the adrenal medulla (Kobayashi, 1968). In the cat, too, it was possible to classify the chief cells into two or more types based on the ultrastructural differences of the secretory granules. The population and distribution of secretory granules varied from species to species. Thus in the mouse, guinea-pig, *Uroloncha domestica* and *Gallus domesticus*, the secretory granules were few in number, as compared with the cat and dog, and tended to gather along the cell membrane.

The ultrastructure of the sustentacular cell resembled that of the Schwann cell. It must be emphasized that the sustentacular cell only incompletely encapsulated the chief cell and associated nerve terminals so that the surface of the chief cell and nerve ending often directly faced the connective tissue space.

In all the species examined, many profiles of nerves were found on the surface of the chief cell. It was of course not always possible to distinguish clearly whether a given profile of a nerve was terminal, pre-terminal or *en passage* by an analysis of a single, two-dimensional electron micrograph; but all profiles of nerve in direct contact with the chief cell will be called nerve ending for the sake of convenience.

The nerve ending contained small agranular microvesicles of about 40 nm in diameter and occasional large-cored vesicles of about 70 nm in diameter. So-called coated vesicles, neurotubules, filaments and mitochondria were also seen. The number and distribution of these structures varied greatly from one nerve to another.

Plate 1*a*,*b* show presumed efferent, that is with nerve terminal as the pre-synaptic and the chief cell the post-synaptic element, synapses seen in the carotid bodies of *Gallus domesticus* and 15-week human foetus. The appositional membranes of the chief cell and nerve ending are separated by a narrow space and at some places they exhibit localized dense plaques associated with a widening of the intercellular gap. Agranular microvesicles are closely associated with these dense plaques to form a 'synaptic complex'. This type of synaptic contact between the nerve ending and the chief cell was recognized not only in *Gallus domesticus* and the human foetus but also in all other species examined. Some species differences were noted in the fine structure of this type of synapse. In the amphibian species, especially in *Bufo bufo japonicus* and *Rana nigromaculata*, some of these nerve endings were huge and packed with conglomerations of numerous small agranular microvesicles and/or clusters of glycogen particles as shown, for example, in Plate 2. Plate 3*a* shows a typical efferent synapse on the chief cell seen in the carotid labyrinth of *Rana nigromaculata*.

In the mouse, guinea-pig, *Uroloncha domestica* and *Hynobius nigrescens* besides the above-mentioned efferent synapses, another type of synapse was distinguishable (Plates 3*a*,*c* and 4*a*,*b*,*c*). In this type of synapse, an aggregation of microvesicles was found in the chief cell cytoplasm immediately beneath its juxtaposition to a nerve ending where the cell membrane was thickened. These microvesicles seen in the chief cell were nearly round in shape and contained electron-dense, fine granular material. Their size was variable, but most of them had a diameter of 20–40 nm (slightly smaller than the agranular synaptic vesicles in the nerve ending). These microvesicles were rigidly separated from the secretory granules and no forms intermediate between them seemed to exist.

In *Uroloncha domestica*, synapses were formed between nerve endings. The nerve ending which formed a synaptic contact with the chief cell sometimes had further synaptic contacts with other nerve terminals (Kobayashi, 1969). It was not rare to find that agranular synaptic vesicles occurred on both sides of the junction between nerve endings as shown in Plate 3*d*.

In all the species examined, in addition to these nerve endings which were recognized to form either efferent or afferent synapses with the chief cell, there were in contact with the chief cell many profiles of nerve endings which could hardly be classed as either afferent or efferent. These profiles of nerve endings contained mitochondria, glycogen particles, neurotubules, filaments, and also agranular microvesicles, as do nerve endings which form either afferent or efferent synapses with the chief cell.

In the connective tissue which separated chief cell units from one another, non-myelinated axons, enclosed within Schwann cells or sustentacular cells, were frequently found. Myelinated nerves were often seen in the mammalian and avian carotid body, but in the amphibian carotid labyrinth, they were rare. In the amphibian carotid labyrinth and avian carotid body, it was observed that the non-myelinated axons were thickened and sometimes measured up to 8 μm in diameter. As shown in Plate 5*a*,*b*, large aggregations of glycogen particles, mitochondria, fine filaments or vesicles were seen in these expanded axons. The thickness of the cytoplasmic investment by Schwann cells or sustentacular cells varied from place to place, and a part of the enlarged axon was sometimes completely devoid of cellular investment.

DISCUSSION

The chemical synapses in the central and peripheral nervous system are generally identified by the following criteria under the electron microscope: the membranes of the pre- and post-synaptic components are closely approximated without any interposed cytoplasm; electron-dense material is present closely associated with the synaptic membranes; and the pre-synaptic vesicles are present, showing a tendency to cluster against the pre-synaptic membrane (Gray & Guillery, 1966; Peters, Palay & Webster, 1970). All the structures described in the present paper as synapses on the chief cell fulfilled these criteria and two types were distinguished, based on the localization of the microvesicles.

The efferent synapse, in which the nerve terminal contained agranular microvesicles accumulated towards the specialized junctional region with the chief cell, was recognized in all the vertebrate species examined.

Another type is the afferent synapse which is provided with characteristic aggregations of dense-cored microvesicles on the 'correct' side, i.e. within the cytoplasm of the chief cell particularly in the region of the specialized membrane thickenings. This type of synapse was first described by us in the mouse and proposed to be a possible site of chemosensory transmission if the chief cell of the mouse is really a chemoreceptor (Kobayashi & Uehara, 1970). Before us, several authors including Ishii & Oosaki, 1966, Höglund, 1967, Al-Lami & Murray, 1968, and Ábrahám, 1969, suggested the presence of afferent synapses (as well as efferent synapses); however, all the structures which they proposed as chemosensory synapses did not fulfil the aforementioned criteria for chemical synapses.

Failure to detect afferent synapses in species other than the mouse, guinea-pig, *Uroloncha domestica* and *Hynobius nigrescens* does not necessarily mean the absence of afferent innervation of the chief cell. Some species difference might be assumed in the preservation of the synaptic vesicles of these afferent synapses. Or, in certain animals, the presence of numerous dense secretory granules in close relation to the plasma membrane may hinder the identification of the characteristic microvesicles, if the latter are present. At least, it might be said that some of the nerve endings which were classified neither as afferent nor as efferent type by an analysis of a given electron micrograph might exhibit either afferent or efferent synaptic contact with the chief cell in a different plane of section.

The microvesicles seen in the chief cell side of the presumed afferent synapse have small, electron-dense cores. It is known that microvesicles with dense cores are encountered in autonomic axon terminals where adrenergic nerve fibres are present (Gray & Guillery, 1966; Peters *et al.*, 1970). It is interesting that, although the so-called small-cored vesicles in the adrenergic terminal are usually slightly larger in diameter than the agranular vesicles which are presumed to contain acetylcholine, the microvesicles with dense cores seen in the chief cell of the carotid body were somewhat smaller than agranular microvesicles in the associated nerve endings.

Several authors considered that the 'secretory granules' are the synaptic vesicles and take part in the chemosensory transmission (Lever, Lewis & Boyd, 1959; Ishii & Oosaki, 1966; Grimley & Glenner, 1968). But this idea seems to be unlikely because these secretory granules are dispersed throughout the cytoplasm, while, in synapses elsewhere in the body, the occurrence of synaptic vesicles is restricted to the area subjacent to the specialized pre-synaptic membrane. Furthermore, Ω-shaped invaginations of the cell membrane embracing the granular substance, which are thought to indicate exocytotic granule release, were seen throughout the surface of the chief cell (Biscoe & Stehbens, 1966; Blümcke, Rode & Niedorf, 1967; Kobayashi, 1968). However, there is a possibility that a substance released from these granules diffuses out in the surrounding tissue and modulates the activity of various elements such as nerve endings.

With regard to the nature of the unusually swollen axons found in the amphibian and avian carotid body, several authors who were inclined to the idea that the chief cell of the carotid body is not a chemoreceptor, postulated the possibility that these enlarged axons (particularly those rich in mitochondria) correspond to the chemoreceptor (Böck, Stockinger & Vyslonzil, 1970; Kondo, 1971). On the other hand, in the mammalian carotid sinus (Knoche & Schmitt, 1964; Rees, 1967) and in the human heart (Chiba & Yamauchi, 1970), similar swelling of axons were considered as receptor nerves which were involved in baroreception. However, it should not be forgotten that expanded axons characterized by conglomerations of mitochondria, vesicles of various sizes, dense bodies, fine filaments, etc. were demonstrated in both the peripheral and central nervous systems after degeneration experiments (Hayano, 1967). Thus a conclusion concerning the nature of these swollen axons cannot yet be drawn.

Many thanks are due to the editor of *Arch. Histol. Jap.* for permission to reproduce illustrations. Plates 4a and 4b are previously published by S. Kobayashi and M. Uehara and are published here with the kind permission of Dr Uehara.

REFERENCES

ÁBRAHÁM, A. (1969). Elektronenmikroskopische Untersuchungen an menschlichen Karotiskörpern. *Z. mikrosk-anat Forsch*, **81**, 413–53.

AL-LAMI, F. & MURRAY, R. G. (1968). Fine structure of the carotid body of normal and anoxic cats. *Anat. Rec.*, **160**, 697–718.

BISCOE, T. J. & STEHBENS, W. E. (1966). Ultrastructure of the carotid body. *J. cell Biol.*, **30**, 563–78.

BLÜMCKE, S., RODE, J. & NIEDORF, H. R. (1967). The carotid body after oxygen deficiency. *Z. Zellforsch*, **80**, 52–77.

BÖCK, P., STOCKINGER, L. & VYSLONZIL, E. (1970). Die Feinstruktur des Glomus caroticum beim Menschen. *Z. Zellforsch*, **105**, 543–68.

CHIBA, T. & YAMAUCHI, A. (1970). On the fine structure of the nerve terminals in the human myocardium. *Z. Zellforsch*, **108**, 324–38.

DE CASTRO, F. (1928). Sur la structure et l'innervation du sinus carotidien de l'homme et des mammifères. Nouveaux faits sur l'innervation et la fonction du glomus caroticum. *Trab. Lab. Invest. Biol. Univ. Madrid*, **24**, 331–80.

FERIA-VELASCO, A., GONZALEZ-ANGULO, A. & ZAVALA, B. J. (1966). The ultrastructure of the normal human carotid body. 1. The chemoreceptor unit. *Bol. Inst. Estud. Med. Biol.*, **24**, 133–46.

GRAY, E. G. & GUILLERY, R. W. (1966). Synaptic morphology in the normal and degenerating nervous system. *Int. Rev. Cytol.*, **19**, 111–82.

GRIMLEY, P. M. & GLENNER, G. G. (1968). Ultrastructure of the human carotid body. A perspective on the mode of chemoreception. *Circulation*, **37**, 648–65.

HAYANO, M. (1967). Unusual axons found in experimental brain edema of cats. *Arch. Histol. Jap.*, **28**, 483–501.

HÖGLUND, R. (1967). An ultrastructural study of the carotid body of horse and dog. *Z. Zellforsch.*, **76**, 568–76.

ISHII, K. & OOSAKI, T. (1966). Electron microscopy of the chemoreceptor cells of the carotid labyrinth of the toad. *Nature, Lond.*, **212**, 1499–1500.

KNOCHE, H. & SCHMITT, G. (1964). Beitrag zur Kenntnis des Nervengewebes in der Wand des Sinus caroticus. I. Mitteilung. *Z. Zellforsch.*, **63**, 22–36.

KOBAYASHI, S. (1968). Fine structure of the carotid body of the dog. *Arch. Histol. Jap.*, **30**, 95–120.

KOBAYASHI, S. (1969). On the fine structure of the carotid body of the bird, *Uroloncha domestica. Arch. Histol. Jap.* **31**, 9–19.

KOBAYASHI, S. (1971). Comparative cytological studies of the carotid body. 2. Ultrastructure of the synapses on the chief cell. *Arch. Histol. Jap.*, **33**, 397–420.

KOBAYASHI, S. & UEHARA, M. (1970). Occurrence of afferent synaptic complexes in the carotid body of the mouse. *Arch. Histol. Jap.*, **32**, 193–201.

KONDO, H. (1971). An electron microscopic study on innervation of the carotid body. *J. Ultrastruct. Res.*, **37**, 544–62.

LEVER, J. D., LEWIS, P. R. & BOYD, J. D. (1959). Observations on the fine structure and histochemistry of the carotid body in the cat and rabbit. *J. Anat.*, **93**, 478–90.

PETERS, A., PALAY, S. L. & WEBSTER, H. F. (1970). *The fine structure of the nervous system. The cells and their processes*, New York: Hoeber.

REES, P. M. (1967). Observations on the fine structure and distribution of presumptive baroreceptor nerves at the carotid sinus. *J. comp. Neurol.*, **131**, 517–48.

WATZKA, M. (1943). Die Paraganglien. In *Möllendorffs Handbuch der mikroskopischen Anatomie des Menschen*, vol. IV, part 1, p. 262–308, Berlin: Springer.

EXPLANATION OF PLATES

PLATE 1

(*a*) An efferent synapse seen in the carotid body of *Gallus domesticus*. An aggregation of agranular microvesicles (v) is seen in the cytoplasm of the nerve ending next to the membrane thickening. Arrows indicate so-called coated vesicles; N nucleus of the chief cell; and g secretory granule of the chief cell. × 37 000 (*Arch. Histol. Jap.*, 1971, **33**, 407, Fig. 9).

(*b*) Carotid body of a 15-week human foetus. Several profiles of nerve endings are seen in the space between the sustentacular cell (Sc) and the chief cell (Cc). Two of them make efferent synaptic contacts with the chief cell at 1, 2 and 3. g indicates a secretory granule, m a mitochondrion, and N a nucleus of the chief cell. × 26 000

PLATE 2

An efferent synapse on the chief cell seen in the intervascular stroma of the capillary plexus of the carotid labyrinth of *Rana nigromaculata*. The nerve ending contains a large accumulation of agranular microvesicles (v), glycogen particles (gl) and small mitochondria. (Sc) indicates a sustentacular cell, and (c) a centriole. × 24 000 (*Arch. Histol. Jap.*, 1971, **33**, 402, Fig. 3).

PLATE 3

(*a*) *Rana nigromaculata*. An efferent synapse on the chief cell (Cc). Sc indicates a sustentacular cell process. × 37 000

(*b*) An afferent synapse seen in the carotid labyrinth of *Hynobius nigrescens*. An aggregation of microvesicles (arrows) is seen in the cytoplasm of the chief cell against the membrane thickening: m indicates a mitochondrion. × 40 000 (*Arch. Histol. Jap.*, 1971, **33**, 401, Fig. 2).

(*c*) An afferent synapse on the chief cell seen in the carotid body of *Uroloncha domestica*. A conglomeration of microvesicles containing small dense cores is located next to the membrane thickening. Af indicates an afferent nerve ending, Cc a chief cell, Cof a collagenous fibre within the connective tissue and Sc a sustentacular cell. × 48 000 (*Arch. Histol. Jap.*, 1971, **33**, 405, Fig. 6).

(*d*) A synapse between nerve endings seen in the carotid body of *Uroloncha domestica*. Note that agranular microvesicles are seen in both nerve endings. × 48 000

PLATE 4

(*a*) Afferent synapses seen in the mouse carotid body. Nerve endings contain a few mitochondria and agranular microvesicles. Aggregations of microvesicles (S) are seen in the cytoplasm of the chief cell subjacent to the specialized membrane thickenings. At the right-lower corner, a portion of a nucleus (N) is seen. n indicates a non-myelinated nerve process and p, a sustentacular cell process. × 30 000 (*Arch. Histol. Jap.*, 1970, **32**, 195, Fig. 2).

(*b*) An afferent synapse seen in the mouse carotid body. The nerve terminal contains agranular synaptic vesicles, mitochondria (m), etc. Small dense-cored vesicles are accumulated within the cytoplasm of the chief cell immediately beneath the membrane thickening. g indicates a secretory granule of the chief cell. × 80 000 (*Arch. Histol. Jap.*, 1970, **32**, 196, Fig. 3).

(*c*) An afferent synapse on the chief cell (Cc) in the mouse carotid body. Af indicates an afferent nerve ending, and g a secretory granule of the chief cell. Microvesicles in the chief cell cytoplasm are variable in size, contain fine, dense, granular substance and are definitely distinguishable from the secretory granule. × 48 000

PLATE 5

(*a*) *Rana nigromaculata*. An unusual axon swelling seen in the connective tissue surrounding the chief cell units, which is filled with numerous mitochondria, glycogen particles and neurofilaments. Sch indicates a process of Schwann cell. × 28000

(*b*) *Uroloncha domestica*. An axon swelling filled with abundant mitochondria, bundles of neurotubules, myelin figures, etc. Cc indicates a chief cell, n a non-myelinated axon of usual appearance. × 20 000 (*Arch. Histol. Jap.*, 1971, **33**, 407, Fig. 10).

PLATE I

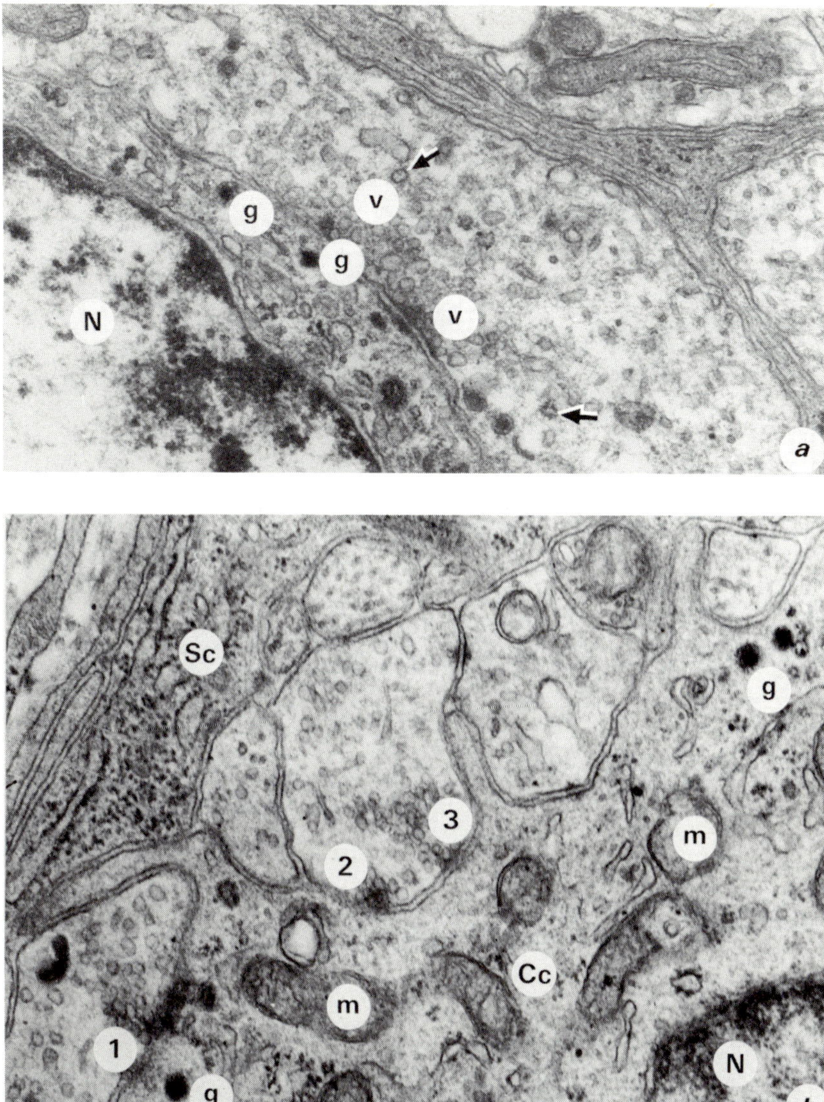

For explanation of Plate see p. 31

PLATE 2

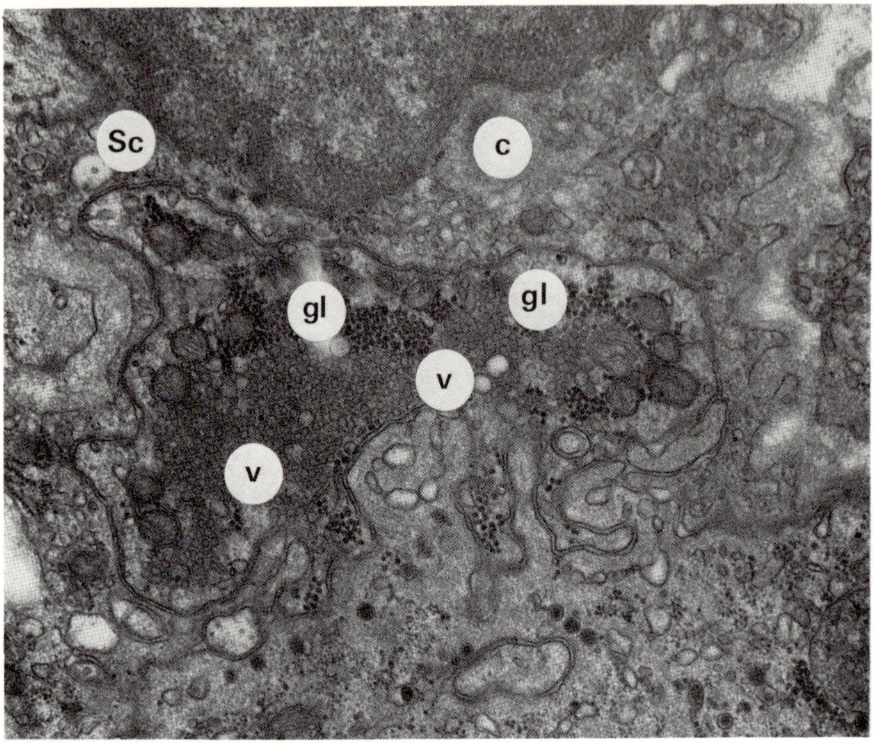

For explanation of Plate see p. 31

PLATE 3

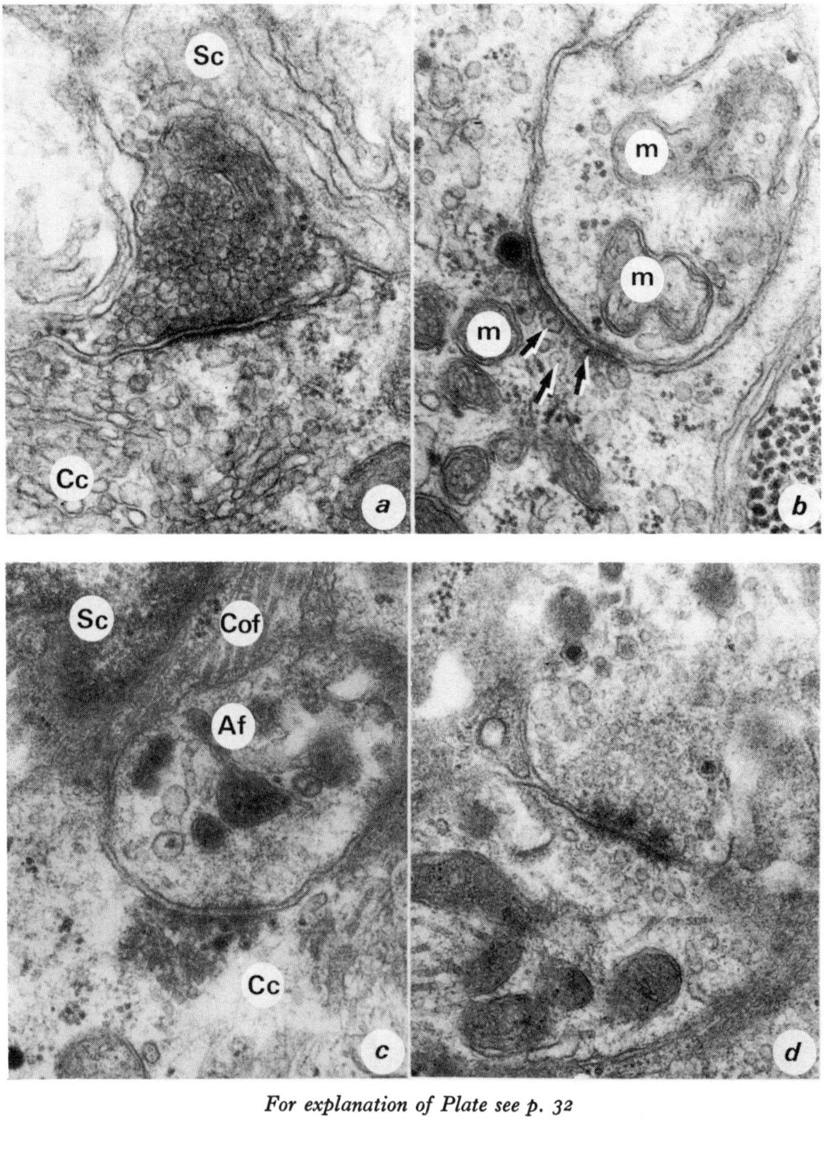

For explanation of Plate see p. 32

PLATE 4

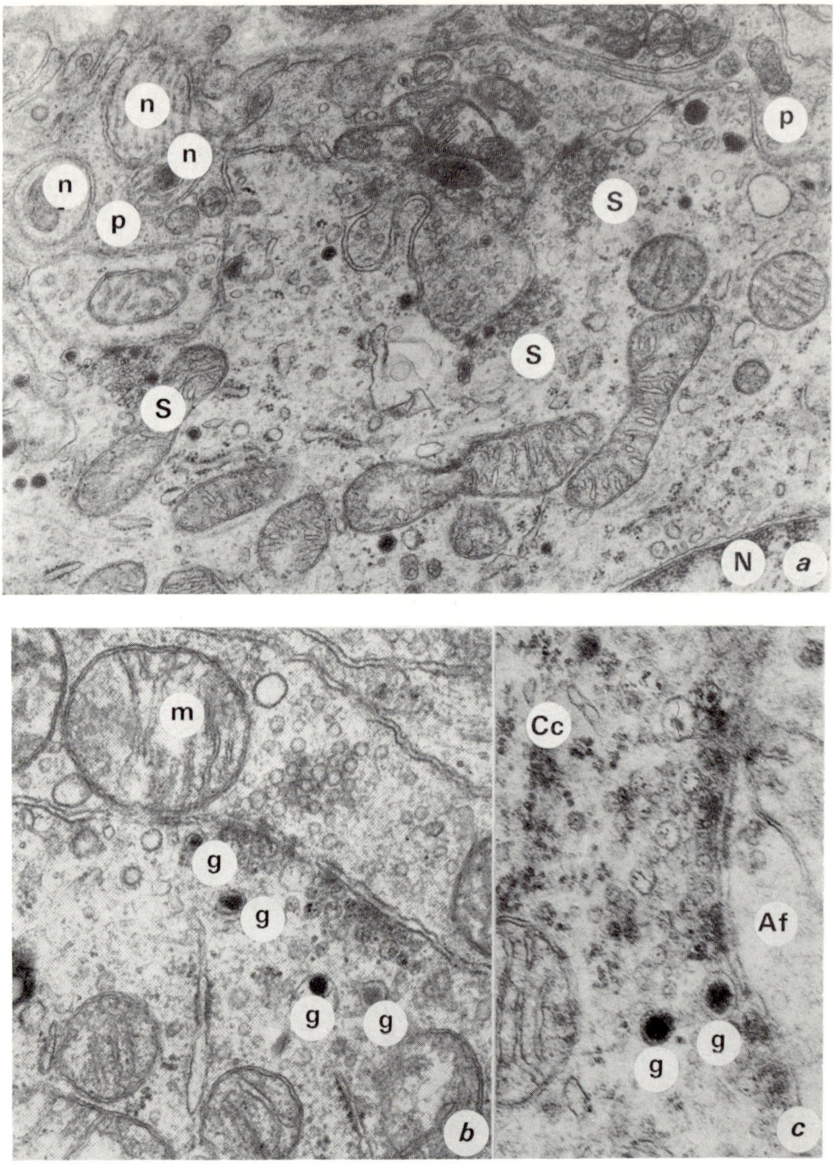

For explanation of Plate see p. 32

PLATE 5

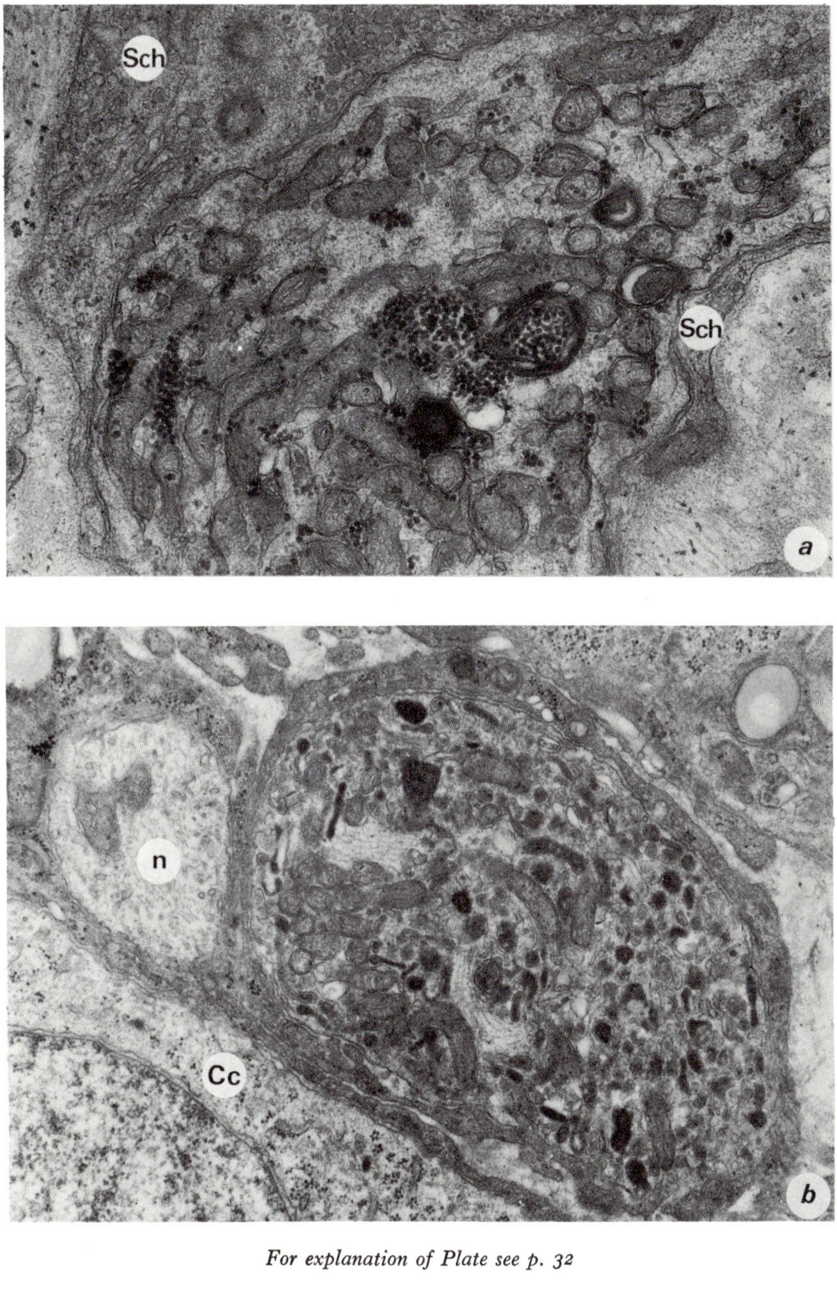

For explanation of Plate see p. 32

DISCUSSION

BANISTER Did you find no afferent-type synapses in *Rana*?

KOBAYASHI No.

BANISTER And you looked very hard for them?

KOBAYASHI Yes.

BANISTER But Dr Ishii reports such afferent synapses (Ishii & Oosaki, 1966, 1969).

KOBAYASHI Yes, I know that. But he did not demonstrate the conglomeration of small synaptic vesicles in the cytoplasm of the chief cell, and that is the reason why his synapses did not fulfil the criteria which I have suggested for afferent synapses.

BISCOE In your drawing at the end of your presentation, you show that the afferent synaptic vesicles have electron-dense cores in them; am I right?

KOBAYASHI Yes, that's right. However, their occurrence was restricted to the area immediately beneath the synaptic membrane thickening, and they are definitely distinguishable from the large-cored vesicles which were dispersed throughout the cytoplasm.

EYZAGUIRRE When you talk about afferent- or efferent-type endings on glomus cells, are they from different fibres or are they just branches from one fibre?

KOBAYASHI I cannot say with certainty from my two-dimensional pictures. My impression, however, is that they are from different fibres.

EYZAGUIRRE The reason I am asking this is that if you look at De Castro's slides and his diagrams and if you try and equate these with electron micrographs of the carotid body of the cat, the large structure encompassing the Type I cell may represent nerve endings devoid of vesicles. However, we have this prolongation which goes to a bouton and which may contain synaptic-type vesicles. So it is important to know whether these derive from the same fibre, from branches of the same fibre or from a totally different pathway.

KOBAYASHI I'm afraid that my sections cannot resolve that particular problem.

FILLENZ Could you just tell us again what are the criteria which you use for identifying nerve fibres and how you distinguish these from, say, processes of a Type I cell.

KOBAYASHI It is not easy for me to explain in words. The most important features would be the presence of microtubules, microvesicles and mitochondria, and the fibres do not contain granular endoplasmic reticulum.

FILLENZ But all these structures are to be found in Type I cells also.

KOBAYASHI Um . . . I think I can be quite certain when I view the sections but I find it difficult to describe.

WOODS Your very big endings packed with myelin figures and vesicles look very like a degenerating axon. Would you like to comment?

KOBAYASHI That's right. However, there may be another possibility. As you know, in the mammalian carotid sinus wall, similar axon swellings are considered to be baroreceptors, and, in the carotid body too, they are sometimes thought to be receptors, especially by those authors who are inclined to the idea that the chief cell is not a chemoreceptor. Therefore, the existence of similar axon swellings in the amphibian carotid labyrinth and avian carotid body would be of interest.

WOODS But I see all these appearances if I cut the sympathetic nerves.

FILLENZ How frequent are these large nerve endings in normal material?

KOBAYASHI They are quite frequently seen in the Amphibia.

MILLS In the species which you have examined, are there any consistent differences in the size, shape or cristae of the mitochondria in the nerve endings or in the chief cells?

KOBAYASHI No, I have not seen such differences.

TORRANCE When you speak of an efferent ending, where do you suppose the cell body is? That is, would you be happy to call it an efferent ending if the cell body were in an autonomic ganglion outside the central nervous system in a dorsal root ganglion, the ending being part of an axon reflex or if the cell body was within the central nervous system?

KOBAYASHI I certainly cannot make such a distinction from my electron micrographs. All I can do is to say with some certainty that it looks like an efferent-type synapse. But, where it originates – that must be determined by degeneration or other studies.

REFERENCES

ISHII, K. & OOSAKI, T. (1966). Electron microscopy of the chemoreceptor cells of the carotid labyrinth of the toad. *Nature, Lond.* **212**, 1499–1500.

ISHII, K. & OOSAKI, T. (1969). Fine structure of the chemoreceptor cell in the amphibian carotid labyrinth. *J. Anat.* **104**, 263–80.

Experiences with serial reconstruction of the glomus complex

By DAVID PALLOT
Department of Physiology, University of Bristol

I want in this contribution to consider problems about the reconstruction of a particular structure from serial electron micrographs. The first is somewhat technical and at the moment, I shall just say that there are obvious difficulties about the decision as to how many sections should be cut, what the spacing should be, how to support the sections so that they will be visible and able to be photographed satisfactorily. I think it would be best if I left this question for specific points to be raised afterwards. The second, and in many ways more difficult, problem arises when you have 600 to 700 micrographs and you have to try and communicate a picture of what a particular structure looks like with economy of illustration. It may be simplest if I illustrate this from a less complex structure than the carotid body glomus complex and consider some results from recent work on the muscle spindle in the snake. There are three or four ways in which a three-dimensional picture can be built up.

The first is shown in Fig. 1. These are tracings from serial micrographs of successive transverse sections through the sensory area of a muscle spindle. The clear areas represent nerve terminals or fibres, the black areas the satellite cells and in the middle, there is the muscle spindle proper bereft of any indication of fine structure. Each of the sections is 5 μm from its fellow and the series concentrates upon the change in shape of a nerve over this range. So, for example, we have the nerve labelled 1 which in the first section is large but over a space of 10 μm contracts down to a very small non-myelinated fibre about 0.3 μm in diameter and continues at this size over a section or two and finally expands again into a large nerve terminal. Essentially what we are doing is removing what for this particular purpose is redundant data and it seems to us that from a relatively few sections, it is possible for the reader to obtain a fair three-dimensional idea of the shape of the nerve and its relation to other immediate structures.

A second method has been to combine the presentation of a very few sections with some sort of model: drawn graphically if necessary or like the

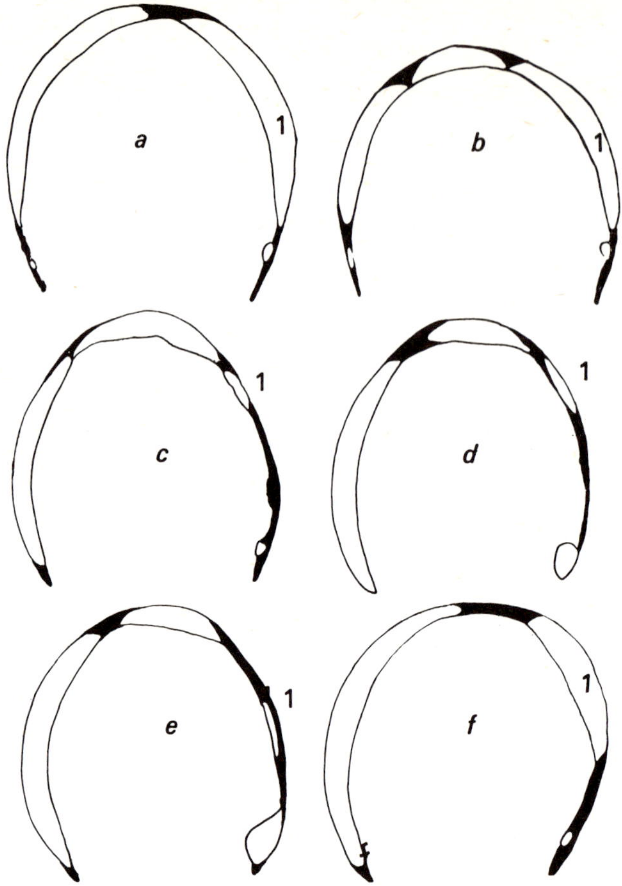

Fig. 1. Tracings of serial micrographs. Each tracing is separated from the preceding one by 5 μm. The changes in shape of ending 1 are illustrated.

one I have here, in the flesh as it were. Again, this may be satisfactory if only selected points have to be emphasized.

In Plate 1, which combines micrographs and model (Pallot & Ridge, 1972), the sections are again taken from the snake muscle spindle with intrafusal fibre showing, from left to right, a nerve bulb applied to the intrafusal fibre and covered with a satellite cell (2*a*). The next section, 20 μm away, shows that the nerve bulb has decreased in size and is now a nerve link but is still applied to the intrafusal fibre (2*b*). The next section, about 50 μm away, shows that a second bulb is now present (2*c*). From these sections, it is possible to construct a plasticine model showing these features of the nerve. Like most models, it is valuable in that it gives a graphic picture of the whole structure. On the other hand, it is derived from selected data and if the model is not accompanied by the original micrographs, the expert or sceptic or both cannot be sure either about the processes of selection or whether possibly some important information has been excluded.

A third way of presenting the data is simply to draw a diagram which illustrates the main points of importance and this is in fact what Biscoe and I (1972) did with the carotid body results, Fig. 2. I do not think that we

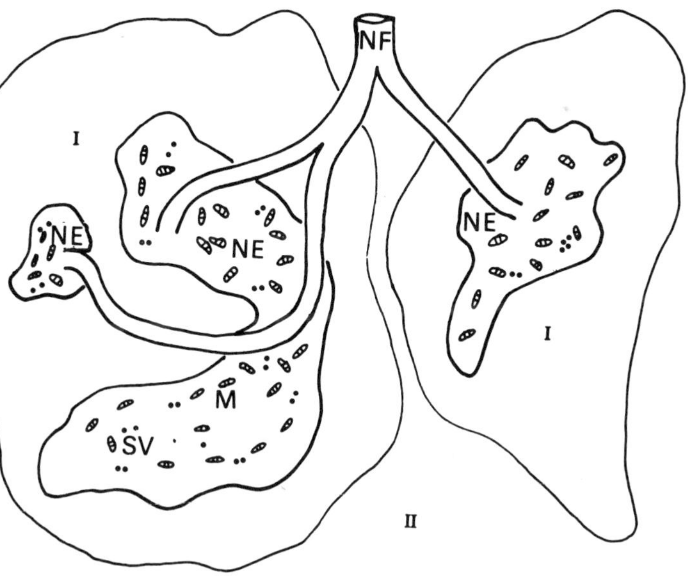

Fig. 2. A diagram illustrating the innervation of the glomus cell (I). A group of glomus cells is surrounded by a Type II cell (II). Each glomus cell is contacted by a variable number of nerve endings (NE) which are each derived from only one nerve fibre (NF). The nerve endings contain variable numbers of mitochondria (M) and synaptic vesicles (SV). (From Biscoe & Pallot (1972), *Experientia*, **28**, 222.)

need anyone to tell us the limitations of this kind of representation: but it is at least possible to draw and to illustrate some firm conclusions. First, we were unable on structural grounds to distinguish more than one class of nerve terminal which made contact with the Type I cell. Frequently it was possible to show two or even three terminals on any one Type I cell but on each occasion, it was possible to show that each terminal was derived from the same nerve fibre. The content of the terminals was variable. In some there were large numbers of mitochondria and few vesicles: in others scattered mitochondria and numerous vesicles. And it was also possible to show varying degrees of density of the membranes apposed at synaptic contacts. Similarly, it was easy to show from serial sections that the mitochondrial or vesicle content varied considerably in the same terminal. We think that this consideration alone – that is the variability of content and distribution of structures within the nerve terminal – makes it difficult from random sections to distinguish different populations of terminals on these grounds alone.

It was also possible to demonstrate the general arrangement of the Type I & II cells, which was for four or five Type I cells to be enclosed in a Type II cell. Finally, we looked in some detail at the course and possible termination of the small non-myelinated fibres of about 0.2 μm diameter which have been observed by a number of workers in random EM sections. We were not able to demonstrate any specific termination of these fibres, e.g. on or near a Type I cell: indeed many fibres passed through the whole series of sections representing a nest of glomus and Type II cells.

I hope that in emphasizing the difficulties inherent in serial reconstruction from electron micrographs, I have not suggested that the method is of limited value: clearly with many structures it can be a very powerful analytical tool. However, when the subject of study is as complex as the carotid body, it may be that the best strategy is to make a preliminary assault and then retire in good order, reformulate the questions and come back another day.

REFERENCES

BISCOE, T. J. & PALLOT, D. J. (1972). Serial reconstruction with the electron microscope of carotid body tissue. The Type I cell nerve supply. *Experientia*, **28**, 222.

PALLOT, D. J. & RIDGE, R. M. A. P. (1972). The fine structure of the long-capsule muscle spindles in the snake *Natrix* sp. *J. Anat.*, **113**, 61–74.

PLATE I

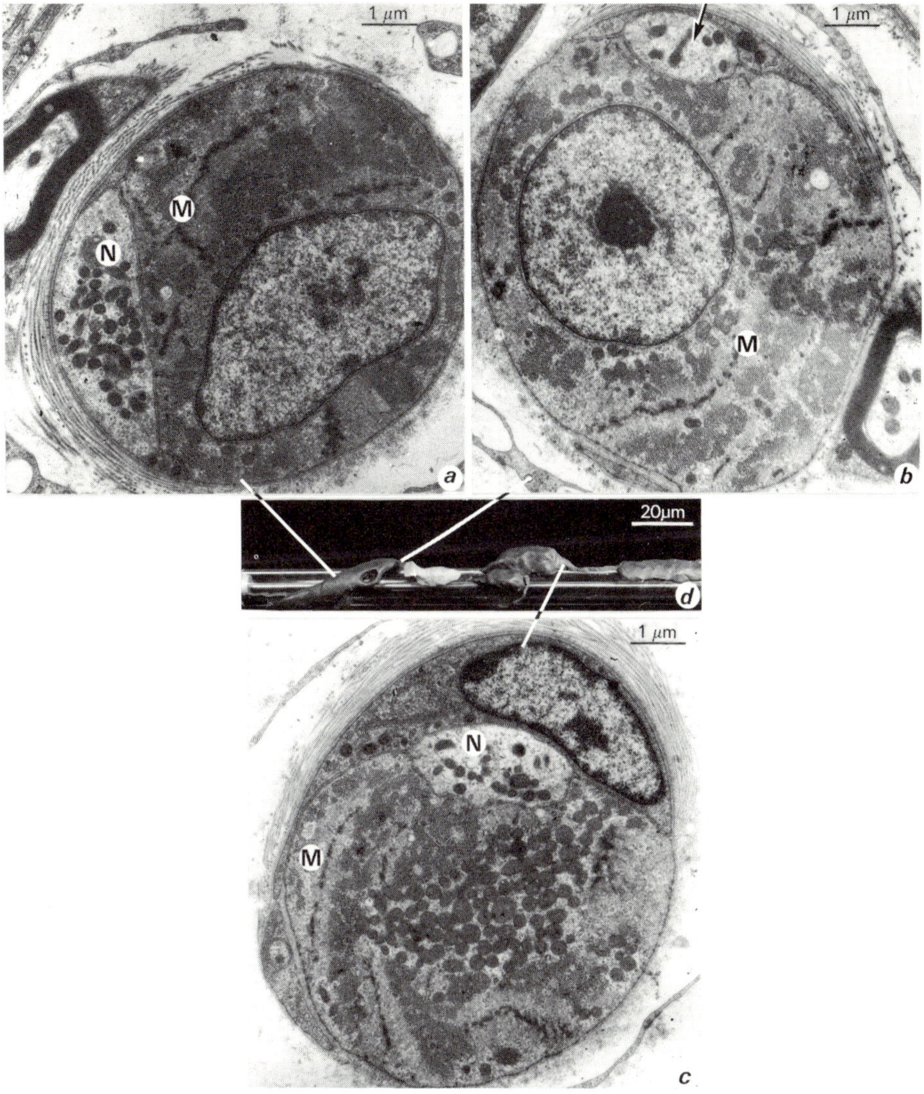

This illustrates the combination of electron micrographs and a model. The lines from each micrograph (*a*), (*b*), (*c*), show the position of each section on the model. M, intrafusal fibre; N, nerve terminal. (From Pallot & Ridge (1972). *J. Anat.*, **113,** 61–74.)

DISCUSSION

EYZAGUIRRE How do you think that this picture agrees with De Castro?
PALLOT I think that it fits in quite well with the slides which you showed in your communication earlier. I think that we are agreed that there is more than one ending on each Type I cell but whereas, as I recall, you could not be certain – or rather De Castro could not – whether these nerve endings were derived from the same nerve fibre, we can be rather more certain from the number of cells which we have studied.
BISCOE Yes. What we have done is largely to confirm De Castro's results but with a lot more effort and probably not as well because, by the nature of his method, he had a bigger sample.
PURVES Can I come back to the dimensions. What sort of size are the pre-terminal portions of the nerves in your sections?
PALLOT They are about 0.2 μm in diameter.
BISCOE I might add that one of the most important reasons for this study was to determine where and how these fibres terminated. But we were frustrated in this because the fibres traversed the reconstructed portion, that is, they were visible in 200 consecutive pictures.
PURVES Do you think there is correspondence between these fibres and those shown by De Castro.
EYZAGUIRRE I think that is difficult to answer because we are dealing with fibres that are almost at the limit of resolution of the light microscope: and another point is that the optics which I used may have been helping me to visualize only the fibres with better staining.
PURVES But it seems to me to be important before we leave this point whether you think it has been established that these small fibres of 0.2 μm or so form an entirely separate population or whether as Dr Hess (Hess & Zapata, 1972) has suggested, these small fibres are really the pre-terminal portions of the nerves which make synaptic contact with the Type I cells.
PALLOT I think that there are too many small fibres for them to be concerned with the Type I innervation: but obviously this is simply an impression. On the other hand, this impression is fortified by the observation that the same nerve fibre supplies two Type I cells and in two reconstructions which were admittedly incomplete, it looked as though a nerve fibre supplied three or four cells in the nest. I suppose that one nerve fibre could even supply all the cells in one complex. If this is so, then it makes the disparity between the number of these small fibres and the likely number of nerve terminals even more marked.

SAMPSON You are quite certain about the termination or lack of it in these small nerve fibres?

PALLOT Alas, yes. It was rather disappointing that we were not able to resolve this particular question. They could of course terminate more specifically, for example, on the next glomus complex: but we did not see this.

TORRANCE Ábrahám (1968) showed pictures in a number of species in which the small nerve fibres went round and round the complex. Do you conceive that that is what is happening?

PALLOT We have not seen anything like that in serial section. It certainly is a possibility but in order to answer that question, we should have to reconstruct a large section and we have not done it.

TORRANCE How far from the ending do you get the branching?

PALLOT The branches which we saw were very close to the Type I cell – about 3–4 μm: certainly inside the Type II cell envelope.

FILLENZ As a matter of interest, has anyone counted the number of fibres in the sinus nerve.

EYZAGUIRRE Yes, we have. The figure is 500 to 600 unmyelinated fibres. I do not know how many glomus cells there are and I would be inclined to leave that task to someone else.

MCDONALD I may be that someone else! In the rat, there are approximately 500 axons in the sinus nerve of which about 10% are myelinated. There are approximately 10 000 glomus cells in the carotid body of which about 20% are Type II and 80% Type I. There are thus about four times as many granule-containing cells as Schwann-type sheath cells. This means that there are about 16 cells to each fibre. But, of course, we do not know where all the sinus nerve axons go to. But if we assume that all the glomus cells are innervated – and that is a very big 'if' – this suggests that there must be substantially more glomus cells than there are axons and presumably there must be branching closer to the glomus cells.

BISCOE What was the level of your section of the sinus nerve?

MCDONALD We counted the myelinated axons at the junction of the glossopharyngeal nerve and the sinus nerve and also counted them in sections at various levels over the 1.5 mm length of the sinus nerve. There was no appreciable difference in the number of myelinated axons among these levels; however, the count could be affected by a contribution from the pharyngeal branch of the vagus which joins the sinus nerve at about its midpoint.

BISCOE Which way does that contribution go – towards the carotid body or the CNS?

MCDONALD Usually this nerve travels rostrally with the carotid sinus nerve toward the glossopharyngeal nerve.

O'REGAN Do your results suggest that the nest rather than the single glomus cell is the functional unit?

PALLOT We think that that is a reasonable assumption though I must stress that we have not reconstructed a whole nest.

O'REGAN And you only get one fibre supplying that?

PALLOT Possibly, we have seen one fibre supplying up to three cells.

WILLSHAW Going back a bit to your technique, I presume that one of your greatest difficulties was in building up a three-dimensional model from stacks of sections and I take it that over a period of time, you built up a three-dimensional model in your head and then had a further difficulty in projecting this as a two-dimensional drawing. Is it possible that you could trace the outline of the relevant structures in your sections and store these in some form of data retrieval system and use that to make the logical connections between the various parts. That would take the human element out of it and possibly speed the whole process up considerably?

PALLOT It is partly a question of cash and partly that when we did this study, we were still fumbling with the basic methods.

HESS What are the dimensions of your intrafusal fibre?

PALLOT The region in which we are interested is about 150 μm long and 5 μm in diameter.

HESS Then wouldn't it save money and effort to cut longitudinal sections?

PALLOT I have actually tried to do this but technically it is very much more difficult to keep the plane of section exactly horizontal and if you do not, then it makes interpretation of the sections more difficult. But with transverse sections you know where you are even although it is more tedious and requires more piecing together of the sections.

McDONALD Before we leave this subject I wonder if you would summarize the similarities and differences between your results and those of De Castro?

PALLOT Basically, the results are similar. More than one nerve ending per Type I cell and these endings are derived from the same nerve fibre.

BISCOE You might ask why we actually did the study and my reply would be that in addition to sorting out the question which Dr Pallot has discussed, we were trying to see whether there was one or more types of nerve ending on the Type I cell. Our findings in the cells we studied were that there was only one type and that you could get these quite different structural appearances in the same nerve ending. This, of course, De Castro could not ascertain with the light microscope.

TORRANCE On the afferent/efferent point, have you seen in the branching of the same nerve fibre, endings which can be called afferent and also others which can be called efferent?

PALLOT It is possible in the same nerve ending to show pictures which suggest on morphological grounds that this ending could be afferent or efferent within the same ending but in different sections.

TORRANCE But have you seen endings which are purely afferent or purely efferent.

PALLOT No.

BISCOE Nor, as I recall did we observe Kobayashi's criterion for afferent endings i.e. synaptic vesicles on the Type I cell side of the synaptic junction.

FILLENZ So what are the criteria which you use for saying whether a nerve ending is afferent or efferent?

PALLOT A nerve ending that is packed full of synaptic vesicles might be regarded as being efferent.

FILLENZ May I then ask you what criteria you use for identifying a nerve fibre? My point is that with the exception of the large dense-cored vesicles, I cannot think of a single structure which you have discussed which one does not see in what are called nerves. There are neurotubules, mitochondria, coated vesicles and agranular vesicles in both Type I cells and nerves.

PALLOT I am sure you will not think me flippant, though others might, if I say that they looked like nerve fibres.

FILLENZ Yes, that is rather what Dr Kobayashi said. But my difficulty is that since Type I cells have processes and since they also contain agranular vesicles, I wonder whether you can be really certain.

HESS You don't have ribosomes in the nerve cell?

FILLENZ Yes, I agree: but the absence of ribosomes is scarcely a sufficient criterion because we have seen numerous sections of the Type I cell without ribosomes and it therefore follows that the Type I cell processes may not have them.

BISCOE You don't see that many neurotubules in a Type I cell.

FILLENZ On the other hand, you do see many nerve fibres without neurotubules.

TORRANCE But from the point of view of this paper, you are in a strong position because you have traced back from the nerve ending to the nerve fibre and that seems to me to be absolutely sound; whereas if you had used single sections, you would be susceptible to some of the arguments raised by Dr Fillenz.

PALLOT Yes, I agree.

FILLENZ May I say that I understand the criteria for efferent fibres but what about the criteria for afferent fibres?

PALLOT It has been suggested that there are different types of endings on the Type I cell with respect to the number and arrangement of mitochondria and synaptic vesicles. What I would like to say is that we have shown that in random sections through a Type I cell with a nerve ending, you can show on the one hand that the nerve ending has very few vesicles and very few mitochondria and yet, 2 μm away, it will have clumps of vesicles and many mitochondria. So unless you undertake serial reconstruction, it is very hazardous to conclude that there is more than one type of nerve ending on the Type I cell.

FILLENZ Yes, but the difficulty is that the possibility has been raised that you may have two different functions going on in the same cell, i.e. it could be afferent at one point and efferent at another.

PALLOT I don't think it is possible even with reconstruction to answer that possibility.

REFERENCES

ÁBRAHÁM, A. (1968). Special characteristics in the structure of the nervous system in the carotid body. In *Arterial Chemoreceptors*, pp. 57–64, ed. TORRANCE, R. W., Oxford: Blackwell.

HESS, A. & ZAPATA, P. (1972). Innervation of the cat carotid body: normal and experimental studies. *Fedn Proc.*, **31**, 1365–82.

The significance of the ultrastructure of the rat carotid body in structure and function of chemoreceptors

By ARTHUR HESS

Department of Anatomy, Rutgers Medical School-CMDNJ, Piscataway, New Jersey, U.S.A.

Various interpretations and thoughts about chemoreceptor function have evolved from studies of the ultrastructure of the carotid body. Suggestions have been made as to the afferent or efferent nature of nerve terminals on glomus cells or the significance of the granulated vesicles in the glomus cells based only upon the structure of these elements as seen in the electron microscope in normal material. Thus, for example, the suggestion that motor endings occur on the chemoreceptive cells of carotid bodies is based on the appearance of the vesicles in identified nerve terminals in the electron microscope. Similarly, afferent terminals have been placed in other locations, on capsule cells or even as free endings, rather than on the chemoreceptive cells. These conclusions are in doubt. Any determination of the afferent or efferent nature of a nerve terminal based on its content or appearance or location of synaptic vesicles cannot be accepted because of the variety of appearances such structures can present in different animals and in different locations. Even the identification of nerve terminals in several instances is questionable, and distinguishing nerve terminals from other cells, such as sectioned pieces of capsule cell cytoplasm, can present a problem. In one kind of experiment involving the cat carotid body with performance of the same supposedly definitive operations to determine the afferent or efferent nature of the nerve terminals on the glomus cells, the two sets of authors disagree and have come to diametrically opposite conclusions, one saying the terminals are efferent (Biscoe, Lall & Sampson, 1970); the other, afferent (Hess & Zapata, 1972). It was hoped that, by studying the carotid body of other animals and relating the findings to the results of others, a more meaningful interpretation could be gained. Hence, not only will a simple description of the carotid body of the rat be presented, but also a comparison of the ultrastructure of the carotid body of this animal with others, especially the cat, will be made, and it will be attempted to glean structural and functional implications from this comparison.

MATERIAL AND METHODS

The rat carotid body was fixed *in situ* with glutaraldehyde, dissected, post-fixed in osmium tetroxide, dehydrated in increasing concentrations of ethyl alcohol, and embedded in Epon. Thin sections were stained with lead citrate and uranyl acetate and examined in the electron microscope. Thick plastic sections were stained by a methylene blue–azure II mixture and studied in the light microscope. Paraffin cross and longitudinal sections were stained by haematoxylin–eosin. Cryostat frozen sections were stained by a dilute methylene blue–azure II mixture.

150–300 mg kg^{-1} of 6-hydroxydopamine hydrobromide were injected intravenously or intraperitoneally, and the carotid bodies studied one to three hours later or nine days after two injections of the drug. Histofluorescence studies were performed on cryostat sections after exposure to formaldehyde vapour (see Fuxe, Hökfelt, Jonsson & Ungerstedt, 1970). Others of these treated carotid bodies were fixed in osmium tetroxide and prepared for electron microscopy, along with normal carotid bodies initially fixed in osmium tetroxide (see Results below). Some carotid bodies after 6-hydroxydopamine treatment and fixation in osmium tetroxide yielded especially satisfactory preparations and are used below for portions of the description of the carotid body.

RESULTS

The rat carotid body is located very close to or on the superior cervical sympathetic ganglion about 1 mm above the bifurcation of the common carotid into external and internal carotid arteries. It is elongated and, measured in section excluding connective tissue ensheathment, about $\frac{3}{4}$–1 mm long and $\frac{1}{3}$–$\frac{1}{2}$ mm wide. Its capsule is quite dense and its borders definite, so that it probably should be considered as a compact carotid body.

The organ is organized into units or islets of collections of glomus cells called glomera (Plate 5*e*). Each unit apparently has a hilus through which nerves and connective tissue enter.

The glomus cells of the carotid body are about 8–14 μm in diameter. They contain the usual organelles found in glomus cells (Plates 1, 3*a*) of other orders of mammals: a centrally located nucleus, ribosomes, Golgi apparatus, scattered double membranes and mitochondria, microtubules, and a cilium. In a corner of the cell, lamellar double membranes occur upon which and between which, ribosomes occur; this structure thus appears similar to a Nissl body. Most significantly, the cells contain membrane-bound vesicles of 60–90 nm in diameter with a dense central core, varying in density and in amount, sometimes filling the vesicle. These will be called dense-core or granulated vesicles. The number of granulated vesicles in each cell varies: some cells are loaded, others nearly empty. However, some granulated vesicles can be found in all cells, and a cell devoid of any granulated vesicles is an extreme rarity. The great majority of vesicles are scattered in the cytoplasm of the glomus cell. However, there are loci of accumulations or groups of vesicles localized near the membrane of a glomus cell adjacent to another glomus cell or capsule cell or under nerve terminals; these relations will be described in more detail below.

The glomus cells are apposed to each other or to a capsule cell which ramifies on and between them. At the hilus, the capsule cell cytoplasm is extensive in amount (Plate 3*b*), but it thins out as one moves away from the hilus, until the glomus cells approximate each other (Plate 1). The capsule cell cytoplasm (Plate 3*b*) is not usually dense, contains scattered ribosomes and double membranes to which ribosomes are attached, and has a Golgi apparatus.

The glomus cells are in general ovoid (Plate 3*a*,*b*), but have short processes extending from them (Plate 1). At times, an elongated process of one cell wraps around another glomus cell (Plate 2*c*). At other times the intertwining short processes of adjacent glomus cells form an elaborately complex meshwork (Plate 2*a*). Membrane thickenings of adjacent glomus cells occur frequently (Plates 1, 2*b*), so that it appears that all the cells of a unit are linked to each other. Accumulations or alignment of granulated vesicles near the membrane of a glomus cell adjacent to another glomus cell (Plate 4*a*,*c*), or, indeed more rarely, adjacent to a capsule cell (Plate 4*b*), have been found. These accumulations of granulated vesicles often, but not necessarily, occur at the level of membrane thickenings. The membranes of adjacent glomus cells or glomus cell and capsule cell are usually separated by a distance of about 30 nm (Plates 1, 3*b*). However, in one instance, an apparent tight junction was formed between the membrane of one glomus cell and that of the process of another wrapping around it (Plate 2*c*).

The dominant nerve terminal (Plates 5*a*–*d*, 6*a*,*d*, 8*a*,*c*) encountered is large with scattered vesicles. At the hilus, the nerve fibres enter and form conspicuous terminal and preterminal fibres around the glomus cells so that some glomus cells appear almost ensheathed by nerve fibre (Plate 3*a*). The nerve fibres enter the unit and run in the capsule cell. As the nerve fibres pass away from the hilus, they run in the capsule cell cytoplasm, which ensheaths them individually in a mesaxon (Plate 8*b*). Hence, the capsule cell replaces the terminal Schwann cell and ensheaths the nerve fibre in a manner entirely analogous to that of the Schwann cell. Microtubules are conspicuous in these preterminal fibres (Plate 8*b*). The terminal nerve fibres are characterized and identified as nerve fibres by their absence of ribosomes, double membranes and Golgi apparatus, which capsule and glomus cells regularly contain. The vesicles in the nerve terminal are scattered and variable in size and density; mitochondria are abundant.

The nerve fibres also present small short inward projections or invaginations of the membrane facing glomus cells (Plates 5*a*,*c*,*d*, 6*a*,*d*, 8*a*,*c*). Random thickenings occur at the same level on the apposed membranes of nerve terminal and glomus cell (Plates 5*a*–*d*, 6*d*, 8*a*,*c*); these thickenings appear to be of about equal density and extent on both sides of the respective membranes. The glomus cell frequently extends thin fingers of cytoplasm to embrace and partially enclose the nerve terminal (Plates 5*c*, 6*a*, 8*a*). These extensions of glomus cell are usually so thin that they do not contain any organelles and appear only slightly dense.

The granulated vesicles in the glomus cell very frequently are seen grouped or accumulated and apposed to the membrane of the glomus cell under the nerve terminal (Plates 5*b*, 6*a*, 8*a*), often, but not always, at the level of mutual membrane thickening of nerve terminal and glomus cell. There may be more than one locus of accumulated vesicles in the glomus cell under a single nerve terminal (Plate 5*c*,*d*). The vesicles in these accumulations are not exclusive in appearance, but appear similar to others in the glomus cell and are variable in size and density and extent of content of dense material. Some of these vesicles are smaller in diameter than the granulated vesicles and are devoid of dense-core material; such vesicles are also seen in the glomus cell, but are not as conspicuous as their dense-cored counterparts. Frequently the same nerve terminal is apposed to two or more glomus cells (Plates 5*b*, 6*d*).

A very small nerve terminal is also seen on the glomus cell (Plate 4*f*, *g*). This terminal is more dense than the conspicuous terminals described above. The accumulated vesicles in it are clear, devoid of dense cores, and fairly uniform in size. The portions of membrane of nerve terminal and that of apposed glomus cell are very dense and smooth and regular. The glomus cell membrane presents a density under the nerve terminal which is usually thicker than that of the nerve terminal.

Nerve terminals are found sharing the above morphological characteristics. The granulated vesicles in the glomus cell are accumulated under a nerve terminal, which in turn presents groups of clear vesicles of uniform size accumulated near thickenings of the membrane of the same or an adjacent glomus cell (Plates 4*d*,*e*, 7*a*,*b*). Also found are terminals with small vesicles pushing into or apposed to large, clear terminals with varying-sized vesicles (Plate 6*b*,*c*).

6-*Hydroxydopamine experiments*

6-Hydroxydopamine is a monoamine which can replace catecholamines in peripheral nerves (Tranzer & Richards, 1971). Catecholamines are rendered fluorescent by formaldehyde; 6-hydroxydopamine does not fluoresce after such treatment. Hence, if 6-hydroxydopamine replaces the catecholamine, the structure involved is diminished in intensity of fluorescence in histofluorescence studies.

The carotid body fluoresces fairly intensely after formaldehyde treatment (Plate 9*c*); the fluorescence colour indicates that catecholamines are involved; it is apparently the glomus cells which fluoresce. After administration of 6-hydroxydopamine, the carotid body is consistently less intense in fluorescence (Plate 9*d*), indicating that the 6-hydroxydopamine has replaced the catecholamines.

In preliminary experiments, we have attempted to localize the 6-hydroxydopamine, and thus the catecholamines, in the carotid body. After initial fixation in osmium tetroxide, the dense-cored vesicles in peripheral nerve appear clear, which indicates that the dense-core material is dissolved by osmium tetroxide. After 6-hydroxydopamine treatment and fixation in osmium tetroxide, the vesicles again appear dense. This indicates that the vesicles have taken up the 6-hydroxydopamine, which is not dissolved by osmium tetroxide fixation (Tranzer & Richards, 1971).

Initial fixation of the carotid body in osmium tetroxide results in the appearance of numerous clear vesicles in the glomus cells and only very few granulated vesicles are seen (Plate 9*a*). After administration of 6-hydroxydopamine and fixation in osmium tetroxide, the vesicles in the glomus cells are again dense (Plate 9*b*). This indicates that the 6-hydroxydopamine is taken up by the vesicles in the glomus cells. It would thus appear that catecholamines endogenous in the glomus cell are located within the vesicles, are dissolved by initial fixation in osmium tetroxide, and are replaced by 6-hydroxydopamine.

DISCUSSION

Comparison of the structure of the rat carotid body especially with that of the cat, on which experimental studies have been performed, can prove enlightening. The rat carotid body conforms morphologically to the generally accepted scheme of the glomus cell as the presynaptic receptor cell stimulating the postsynaptic afferent terminal. The large conspicuous nerve terminals on glomus cells of the rat appear to have the structure and to contain the organelles characteristic of sensory fibres elsewhere. Their vesicles are scattered and variable in size and density in contrast to the orientation and accumulation, uniformity in size, and clear interiors of most vesicles in motor nerve terminals. The short invaginations of the nerve fibre membrane is indicative of pinocytosis and a morphological correlate of the occurrence of this process in cells. 'While pinocytosis also occurs at presynaptic sites, it occurs more commonly in postsynaptic areas' (Pappas & Waxman, 1972). On the contrary, as is well known, the nerve terminals on glomus cells of the cat present accumulations of uniform-sized clear vesicles and can indeed appear efferent in nature. The glomus cells of the rat present accumulations of granulated vesicles under the apposed nerve terminal. These groups of vesicles are thus in the 'right' place, the presynaptic cell. The cat glomus cell only rarely, if at all, has such accumulations of dense-cored vesicles. It may thus be suggested that if the rat had been studied initially as intensely as the cat, it is doubtful if an efferent innervation of receptor glomus cells would ever have been sought.

The rat glomus cells contain organelles similar to those of the cat. However, the rat glomus cells appear more irregular in shape with more processes. Perhaps most significantly, the rat glomus cells are apposed to each other very frequently with membrane thickenings between adjacent cells, whereas, although this can occur in the cat, the cat glomus cell is most frequently ensheathed by capsule cell cytoplasm, even though at times very thin. Hence, the rat glomus cells are generally much more intimately related to each other than those of the cat. Whether such frequent cell junctions are physiological and allow stimulation or inhibition of one cell by another glomus cell can, at this time, be only a matter of speculation.

The significance of what appears to be a motor innervation of rat glomus cells is not known. The larger thickenings of the membrane of the glomus cell under these endings are similar to postsynaptic densities (Akert, Pfenninger, Sandri & Moor, 1972), which might indicate that these nerve terminals are indeed efferent in nature. It has been suggested that the sensory nerve terminals of the cat with their accumulations of synaptic vesicles are 'presynaptic dendrites' (Hess & Zapata, 1972), afferent fibres which reciprocally influence the cells which stimulate them and upon which they terminate. Accumulations of granulated vesicles in rat glomus cells under nerve terminals which in turn have accumulations of synaptic vesicles apposed to the glomus cell certainly appear morphologically as if a reciprocal physiological relation may occur. The intimate relation of rat glomus cells to each other may be influential, and one cell may stimulate or inhibit another. Accumulations of granulated vesicles in adjacent cells may account for this. Motor-type terminals ending on apparent sensory terminals (see Verna, this volume) is another mechanism which has been said to mediate inhibitory influences in the nervous system and which can be found in the rat carotid body. It is entirely possible that the rat carotid body, necessarily performing the same function as that of the cat, uses slightly different morphological arrangements to achieve the same goals. Guesses as to the actual cellular functional mechanisms in the carotid body based on morphological data alone will have to await more studies, both anatomical and physiological, before they can be proven right or wrong. Nevertheless, such speculations may give insight into truly meaningful experiments, which are yet to be performed.

The motor terminals on the glomus cells of the rat may be autonomic in nature. Neuronal cell bodies are consistently found in the rat carotid body and could be the parent ganglion cells of these terminal fibres. Chronic denervation experiments with survival or degeneration of these terminals would perhaps answer this question.

If the catecholamines are located in the vesicles of the glomus cells as suggested here, then it would perhaps provide further evidence that these substances are transmitter agents accumulated in the presynaptic cell under sensory nerve terminals, a most obvious and advantageous location for a transmitter substance to induce stimulation of a postsynaptic nerve fibre. Catecholamines have been shown to inhibit and also to stimulate brain structures. Their precise location in the glomus cells perhaps can be more definitively determined by the use of other false transmitter substances and replacement and depletion by other drugs.

This investigation was supported by Research Grant #NS-07662 from the National Institute of Neurological Diseases and Stroke of the US Public Health Service. Ms Helen Ferguson, Mr P. J. Adamo and Ms Irene Cassady provided technical efficiency and co-operation throughout all aspects of this investigation.

REFERENCES

Akert, K., Pfenninger, K., Sandri, C. & Moor, H. (1972). Freeze etching and cytochemistry of vesicles and membrane complexes in synapses of the central nervous system. In *Structure and Function of Synapses*, pp. 67–86, eds Pappas, G. D. and Purpura, D. P. Raven Press: New York.

Biscoe, T. J., Lall, A. & Sampson, S. R. (1970). Electron microscopic and electrophysiological studies on the carotid body following intracranial section of the glossopharyngeal nerve. *J. Physiol., Lond.*, **208**, 133–52.

FUXE, K., HÖKFELT, T., JONSSON, G. & UNGERSTEDT, U. (1970). Fluorescence microscopy in neuroanatomy. In *Contemporary Research Methods in Neuroanatomy*, pp. 275–314, eds NAUTA, W. J. H. and EBBESSON, S. O. E. Springer: New York.

HESS, A. & ZAPATA, P. (1972). Innervation of the cat carotid body: normal and experimental studies. *Fedn Proc.*, **31**, 1365–82.

PAPPAS, G. D. & WAXMAN, S. G. (1972). Synaptic fine structure–morphological correlates of chemical and electrotonic transmission. In *Structure and Function of Synapses*, pp. 1–43, eds PAPPAS, G. D. and PURPURA, D. P. Raven Press: New York.

TRANZER, J. P. & RICHARDS, J. G. (1971). Fine structural aspects of the effect of 6-hydroxydopamine on peripheral adrenergic neurons. In *6-Hydroxydopamine and Catecholamine Neurons*, pp. 15–31, eds MALMFORS, T. and THOENEN H., American Elsevier: New York.

EXPLANATION OF PLATES

All photographs are electron micrographs of the rat carotid body, except for Plates 5*e* and 9*c*,*d*, which are light micrographs.

PLATE 1

Survey picture showing the glomus cells, their organelles, the short processes coming from them, and the membrane thickenings (arrows) between adjacent cells. ×13 000

PLATE 2

(*a*) The complex arrangement of the intertwining processes of glomus cells. ×13 000
(*b*) A membrane thickening between two adjacent glomus cells. ×25 000
(*c*) An elongated process of a glomus cell wrapping around another glomus cell and forming an apparent tight junction at the arrow. ×13 000

PLATE 3

(*a*) Glomus cells ensheathed by nerve fibres. ×7000
(*b*) Extensive capsule cell cytoplasm, light in density, ramifying between glomus cells. ×13 000

PLATE 4

(*a*) Accumulation of granulated vesicles (arrow) in a glomus cell adjacent to another glomus cell. ×13 000
(*b*) Accumulation of granulated vesicles (arrow) in a glomus cell adjacent to a capsule cell. ×25 000
(*c*) Granulated vesicles aligned along the membrane of a glomus cell adjacent to another glomus cell. ×25 000
(*d*) The glomus cell process on the left has an accumulation of granulated vesicles apposed to a nerve terminal on the right, which, in turn, has synaptic vesicles grouped at the junction with the glomus cell. An apparent reciprocal arrangement. ×25 000
(*e*) Accumulation of granulated vesicles occurs in the small glomus cell process on the top at its junction with the nerve fibre, which, reciprocally, has synaptic vesicles massed at its area of apposition with the glomus cell. ×25 000
(*f*), (*g*) Apparent small motor terminals with synaptic vesicles, very regular dense membranes on nerve terminal and glomus cell, and an apparent postsynaptic density attached to the glomus cell membrane. ×25 000

PLATE 5

(*a*) An apparent sensory nerve terminal on a glomus cell. Membrane thickenings on nerve

fibre and glomus cell membranes at the long arrows. The short arrow indicates the invaginating membrane of the nerve fibre (pinocytosis). × 13 000
(*b*) A sensory terminal ending on two different glomus cells. × 13 000
(*c*) A sensory terminal containing an apparent pinocytotic vesicle (arrow) and embraced by the process of a glomus cell (G). × 25 000
(*d*) Accumulations (arrows) of granulated vesicles in the glomus cell under a sensory terminal. × 13 000
(*e*) Light micrograph. The glomus cells are organized as units or islets of cells. × 200

PLATE 6

(*a*) Sensory terminal embraced by glomus cell process at the top and containing a pinocytotic vesicle (arrow). × 25 000
(*b*), (*c*) Apparent small motor terminals ending on sensory nerve fibres. Membrane thickenings occur. Synaptic vesicles accumulate in motor terminal. × 25 000
(*d*) Sensory terminal with variable sizes and densities of vesicles. Membranes of nerve fibre and glomus cell thickened at the arrows. The nerve fibre is apposed to two different glomus cells. × 25 000

PLATE 7

(*a*) Apparent motor terminal ending on a glomus cell, which has an accumulation of granulated vesicles at the junction with the nerve fibre (top arrow). Accumulations of synaptic vesicles (bottom arrow) occur in the nerve terminal at its junction with another glomus cell. × 25 000
(*b*) Apparent motor terminal ending on a glomus cell which exhibits postsynaptic densities (single arrows at the top). Another glomus cell has an accumulation of granulated vesicles (left arrow) at its junction with the nerve fibre. × 25 000

PLATE 8

(*a*) Fingers of glomus cell cytoplasm embrace the sensory terminal at the top. Accumulation of granulated vesicles occurs in the glomus cell under the nerve terminal at the bottom, which has an apparent pinocytotic vesicle in it (arrow). × 25 000
(*b*) The capsule cell forms mesaxons around the preterminal nerve fibres with prominent microtubules. × 25 000
(*c*) Vesicular and mitochondrial content of an apparent sensory nerve terminal. Membrane thickenings at the arrows. × 25 000

PLATE 9

(*a*), (*b*) 6-Hydroxydopamine experiment.
(*a*) Glomus cell fixed in osmium tetroxide only. The dense core of the granulated vesicles is apparently dissolved and very few dense-cored vesicles remain. × 25 000
(*b*) Glomus cell treated with 6-hydroxydopamine and then fixed in osmium tetroxide only. The dense cores of the granulated vesicles are now restored, and many granulated vesicles are present. × 25 000
(*c*), (*d*) 6-Hydroxydopamine experiment, histofluorescence of catecholamines, light micrographs.
(*c*) The usual fluorescence seen in the untreated carotid body. × 150
(*d*) The reduced intensity of fluorescence after 6-hydroxydopamine treatment. × 150

PLATE I

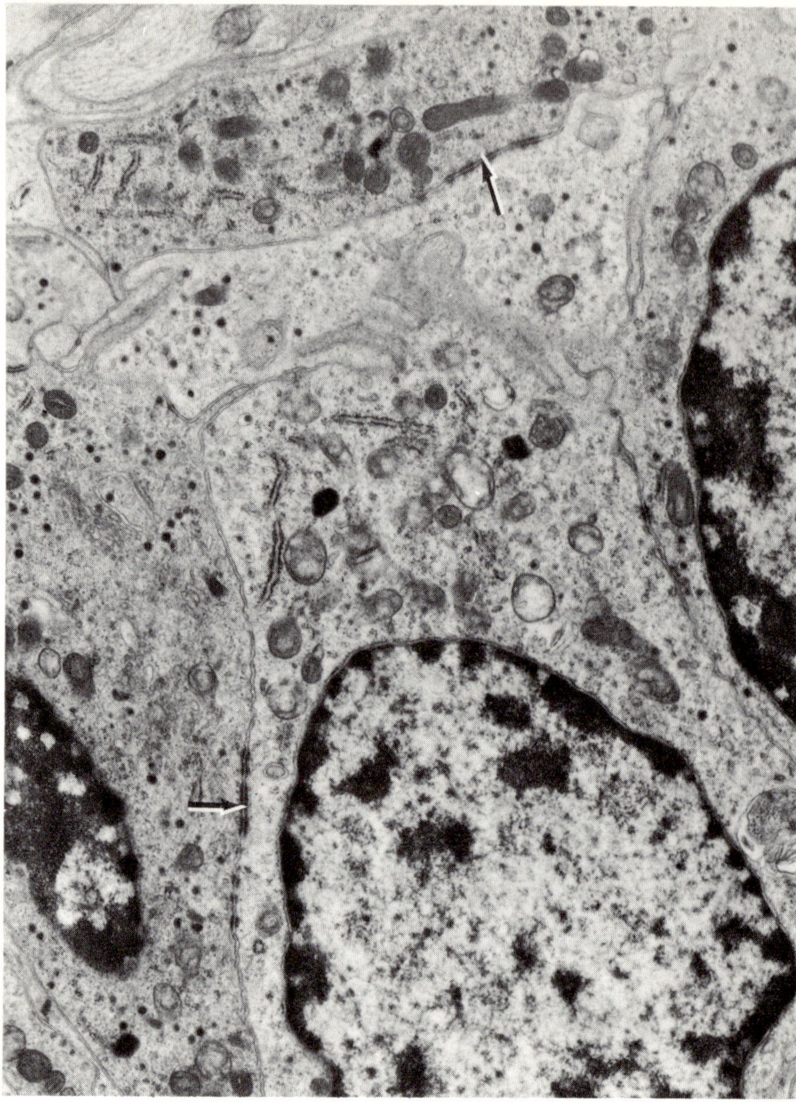

For explanation of Plate see p. 58

PLATE 2

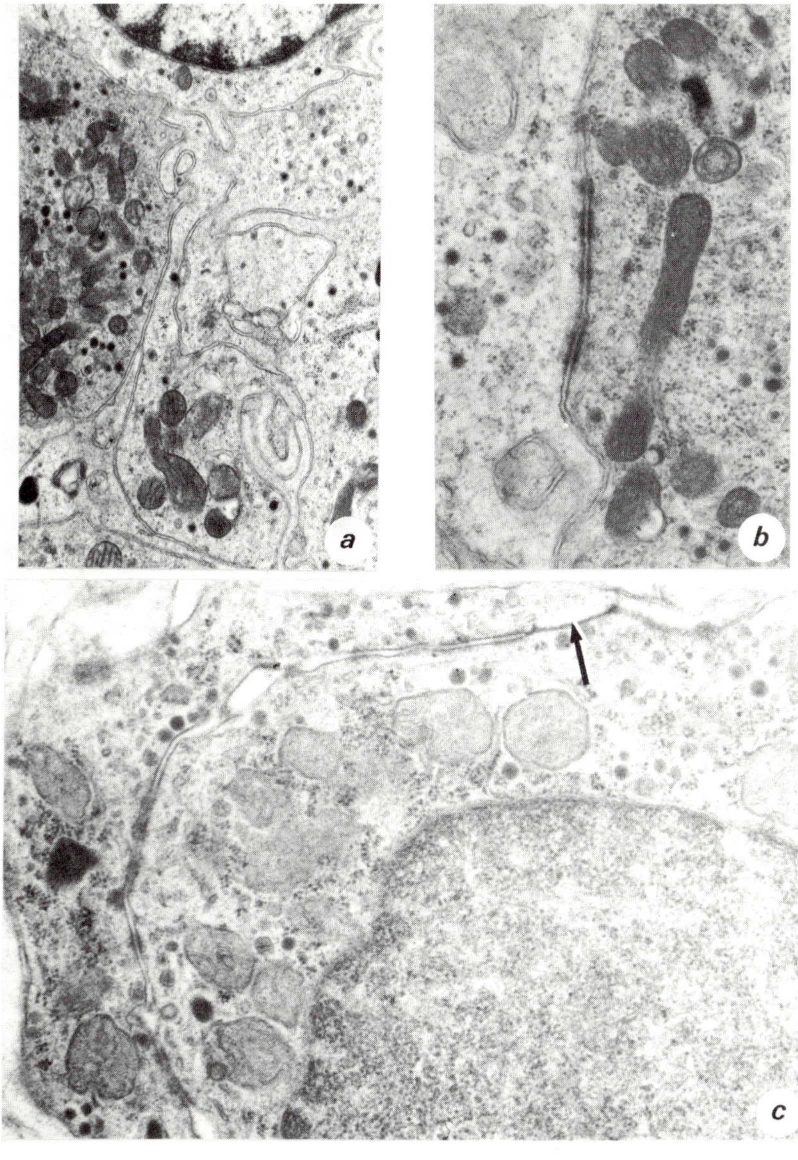

For explanation of Plate see p. 58

PLATE 3

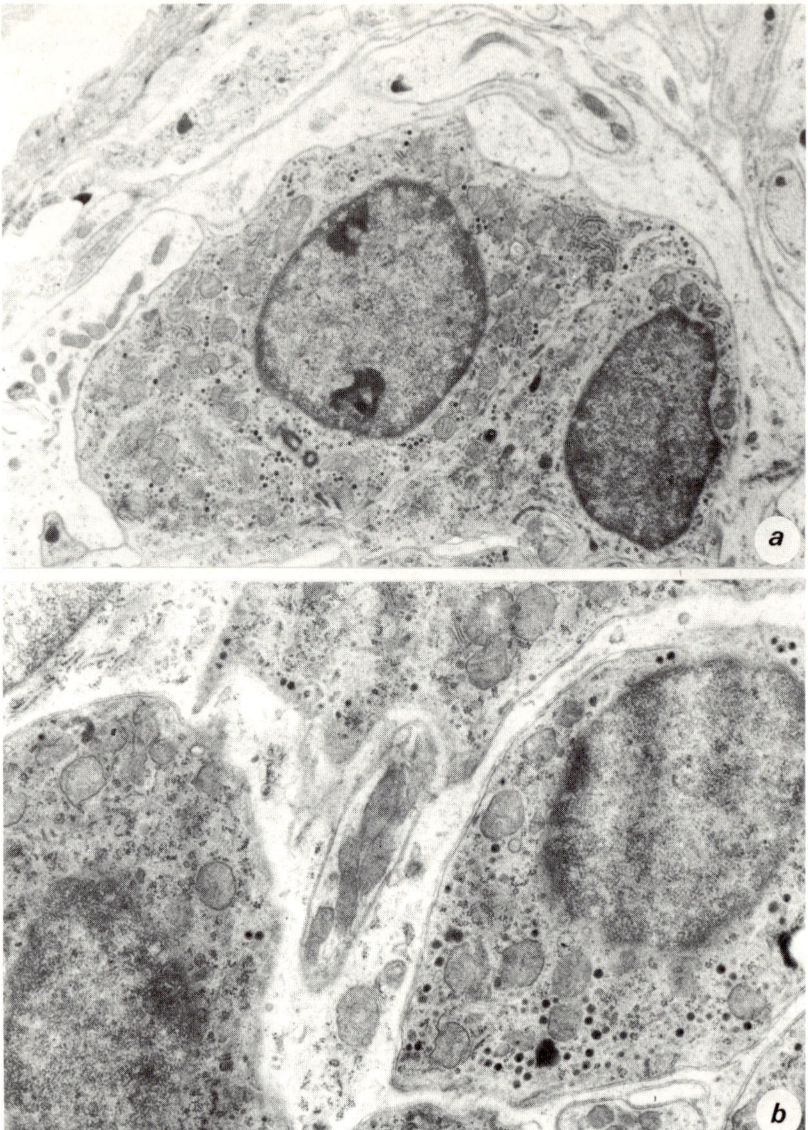

For explanation of Plate see p. 58

PLATE 4

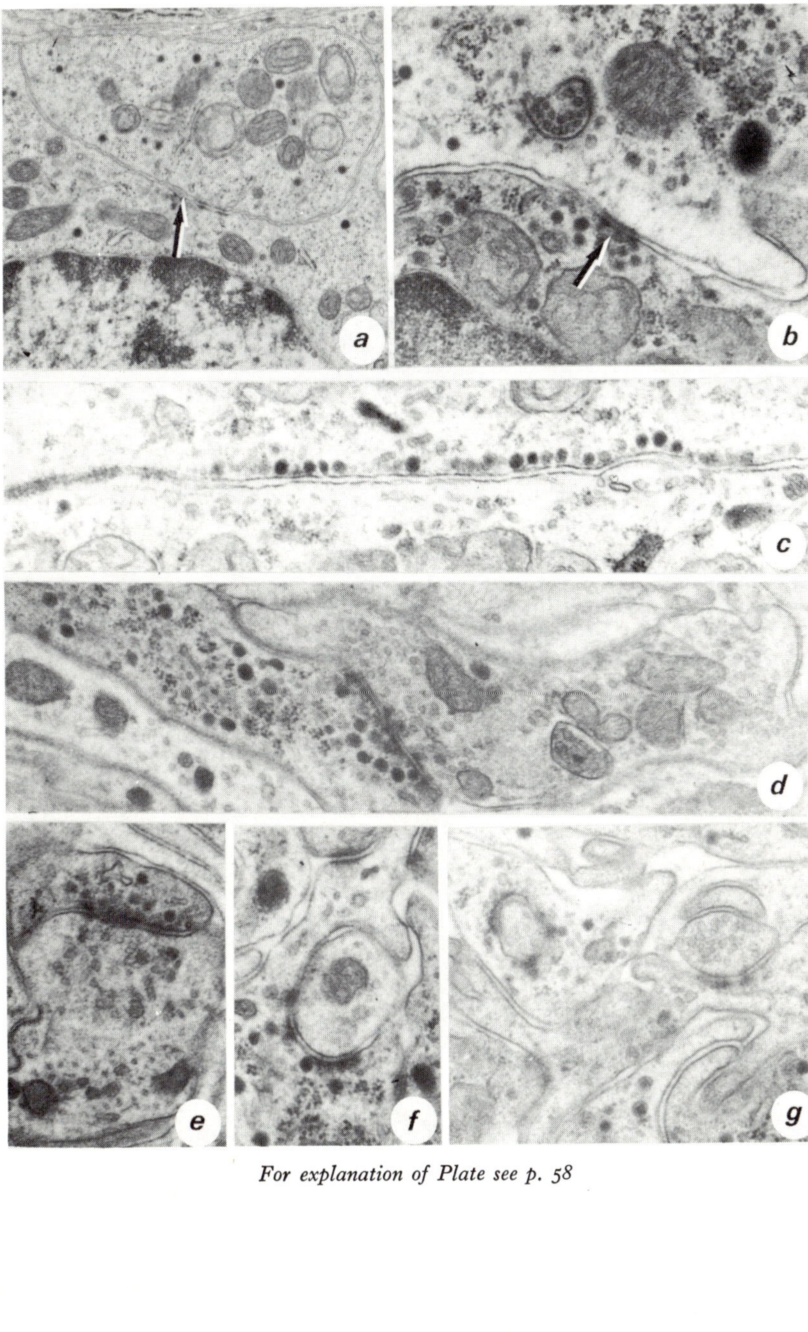

For explanation of Plate see p. 58

PLATE 5

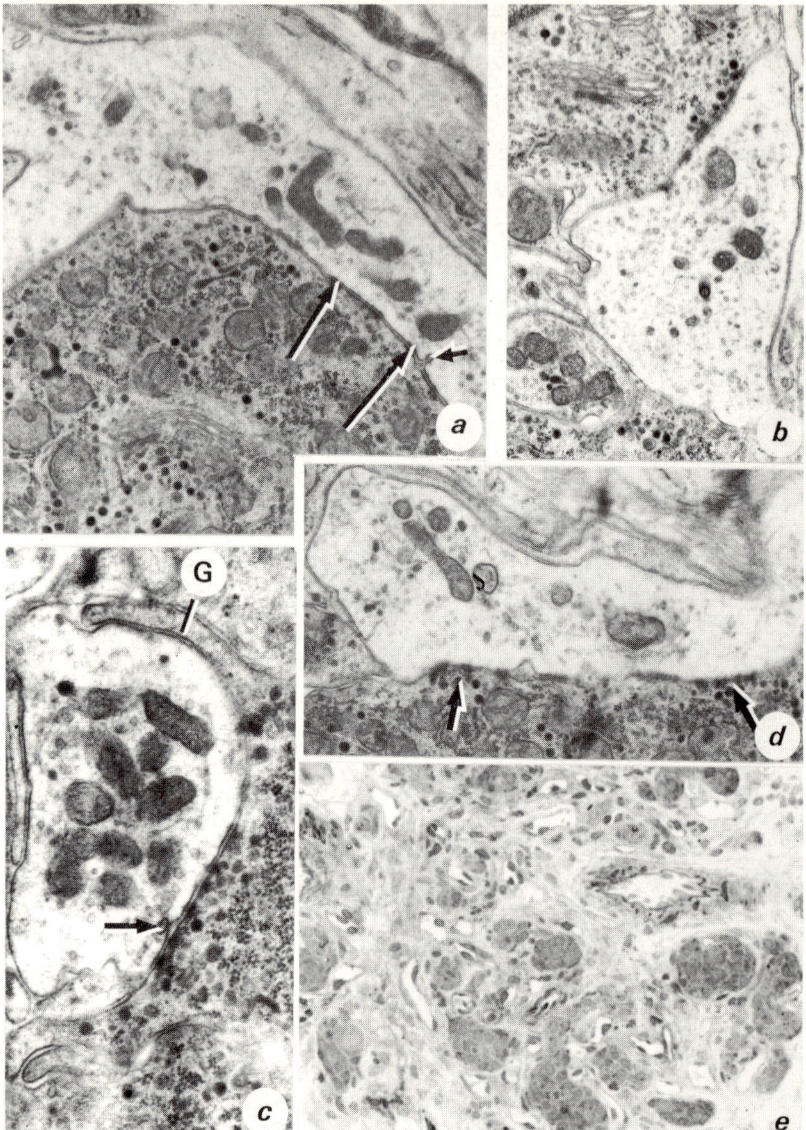

For explanation of Plate see pp. 58–9

PLATE 6

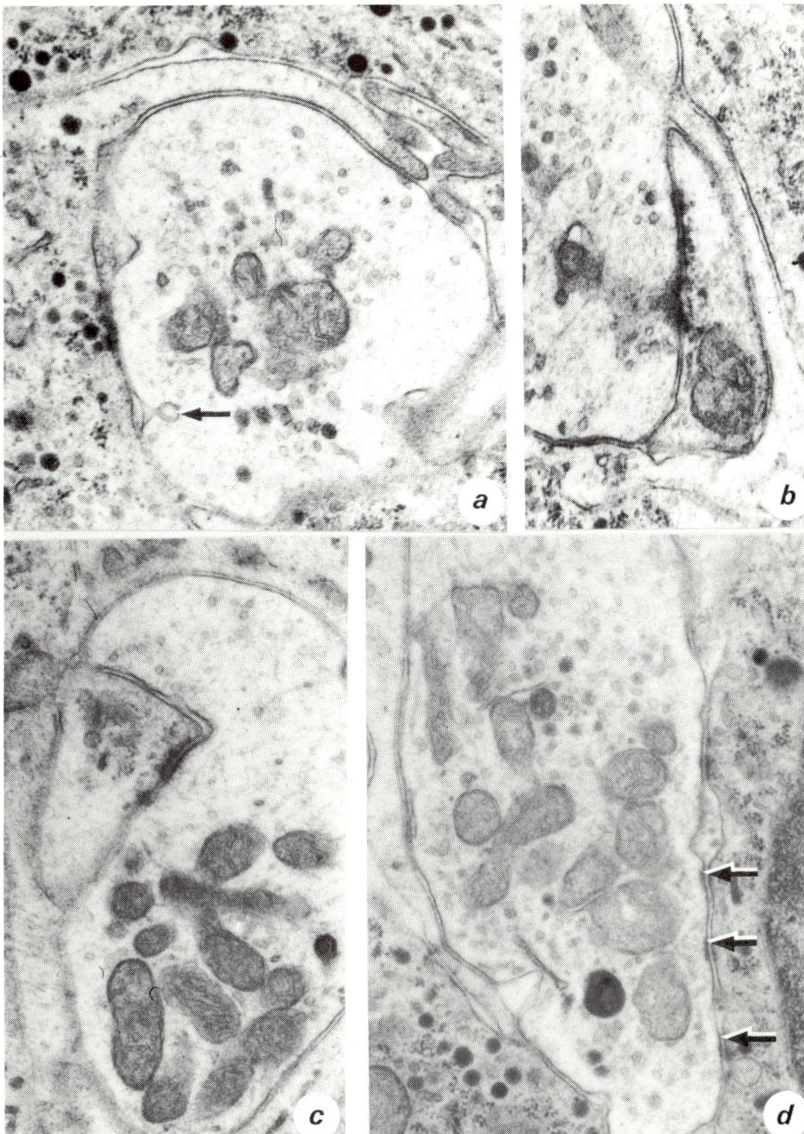

For explanation of Plate see p. 59

PLATE 7

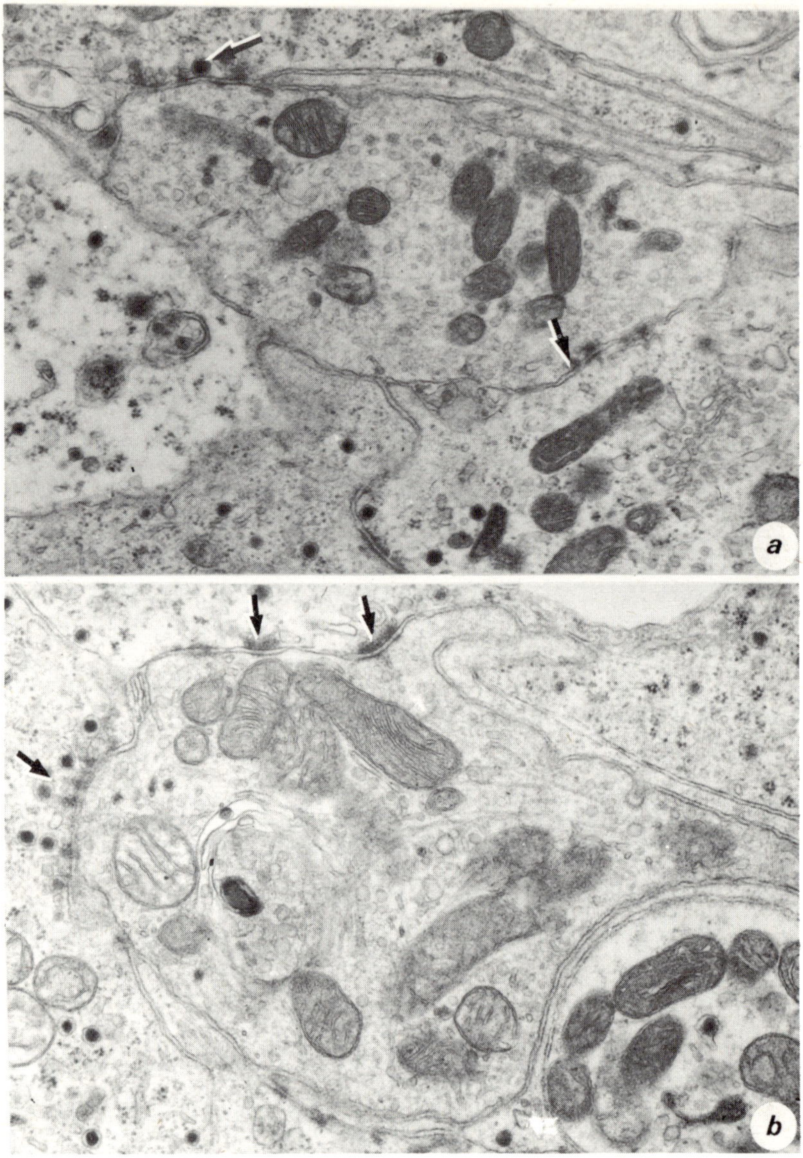

For explanation of Plate see p. 59

PLATE 8

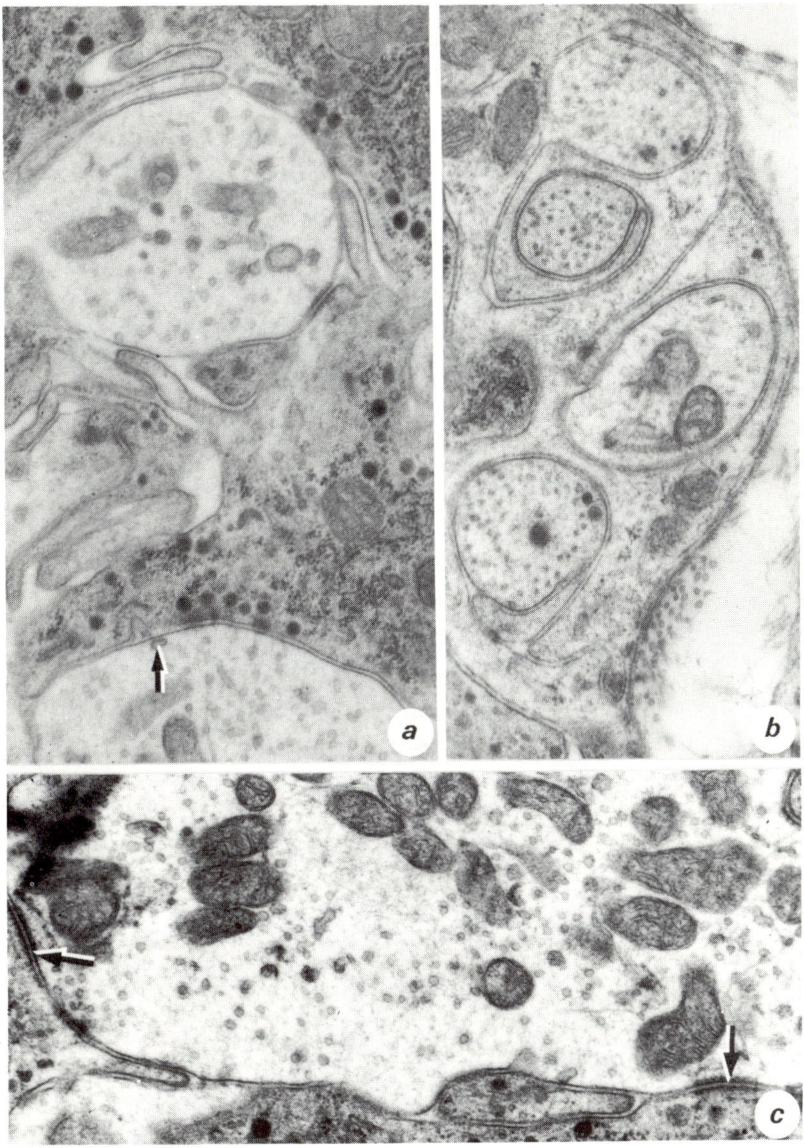

For explanation of Plate see p. 59

PLATE 9

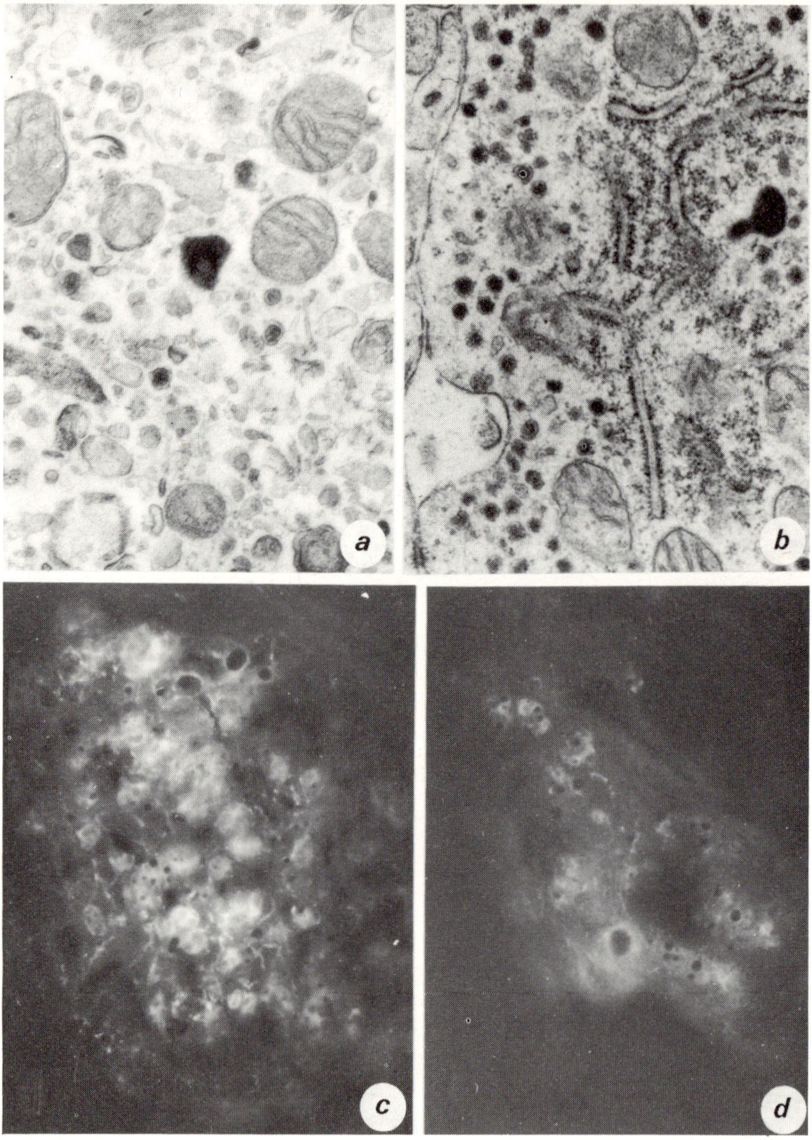

For explanation of Plate see p. 59

DISCUSSION

WOODS You mentioned seeing desmosomes between Type I and Type II cells. Could this possibly have been between two Type I cells because I have never seen the sustentacular cell forming a desmosome against the Type I cell?

HESS I have seen this type of membrane thickening between Type I and Type II cells, but admittedly not frequently. And I identify the Type II cell by the fact that I do not see granulated vesicles, the cytoplasm is lighter and the disposition of ribosomes and mitochondria is quite different from the Type I cell.

WOODS I have seen endings which are quite large and with different appearances in the same ending. For example, on one ending against a Type I cell, you see strong thickening and darkening in the Type I cell and then slightly further along, you get even thickening in both nerve ending and cell, and yet further along, you get darkening only in the nerve ending.

HESS Was the vesicular content different?

WOODS Yes, the vesicle contents were different but not sufficiently distinct to call them sensory or motor, as compared, for example, to the olfactory bulb where you can say that one synapse goes a particular way and a bit further along, it goes back the other way.

HESS Yes, I would certainly agree that there are marked differences between the types of thickening seen between Type I cell and sensory fibre.

WOODS The large ending which you showed in one of the last slides looked to me very much like a Type I cell because there were not many vesicles visible, and those which were looked like Golgi vesicles.

HESS No, I think I would stand by my interpretation.

WOODS I thought it was relatively easy to make this distinction between nerve endings and Type I cells until I found a long process ending in a swelling and making synaptic contact with another profile. I thought the long process was a nerve fibre as it contained many micro-tubules but then I saw that the swelling at its end had ribosomes in it (see Plate 7, Woods, this Symposium, p. 204).

HESS Oh yes: but I don't show that kind of picture!

SAMPSON Without arguing about whether or not we should call a certain structure an afferent terminal, there are two points in your paper I take issue with. The first is that if you had looked at the rat before you looked at the cat, you would look at a Type I cell and say, obviously, that is a receptor cell. Don't you think it possible, especially if you are showing efferent terminals, that the Type I cell is not the receptor but it could be part of a modulating system which affects the afferent terminal?

HESS What I am saying is that if I look at the rat carotid body and I see a glomus cell with accumulated granular vesicles underneath a clear-looking nerve terminal which had varying sized vesicles, then I would say there is no problem; there is the sensory ending and there is the receptor cell.

SAMPSON The receptor ending, certainly. But the cell could be part of a modulating system and be a secretory cell.

HESS I quite agree: all or most sensory systems are capable of being modulated.

SAMPSON And the other point which I wanted to make was about the pinocytotic vesicles. Surely they also occur in motor nerves as incorporated fragments of membrane. I am not of course saying that the evidence indicates that it is a motor terminal but that equally it does not prove that it is an afferent terminal.

HESS I agree.

BISCOE I'm a little confused. Are you saying that you see electron-dense cored vesicles at the post-synaptic side of the efferent terminals. Did I hear you say that?

HESS I do see accumulations of dense-cored vesicles next to those small nerve endings which have accumulations of synaptic vesicles. So that by analogy with Kobayashi's criterion of an afferent terminal which has an accumulation of granulated vesicles in the Type I cell under the nerve terminal – and he could not illustrate such a thing in his efferent terminals – so I am saying under terminals which look efferent, I have also seen and I think shown quite clearly, granulated vesicles in the Type I cell.

BISCOE So it is conceivable that the endings which you are calling afferent could in fact be efferent if you did the serial reconstruction and found synaptic vesicles a few microns away.

HESS They might be all of the same terminal – I don't deny that.

FILLENZ I just wanted to say that Dr Kobayashi's dense-cored vesicles in the Type I cells were finely granular and different in size and appearance from the main body of dense-cored vesicles whereas yours seemed to be more uniform.

HESS This may be a species problem. I am of course looking at the rat and Dr Kobayashi, the mouse.

KOBAYASHI It may be useful if I reiterate what I believe to be the most important morphological feature in the identification of the 'afferent' synapse. In Plate 4*b*, (p. 36) you can see a nerve terminal apposed with a Type I cell and I think that there is an obvious contrast between the clear vesicles in the nerve terminal and the small dense-cored vesicles in the Type I cell which are accumulated close to the synaptic cleft. I have had the opportunity of looking at comparable sections prepared by Drs McDonald and Verna: they both find these vesicles in comparable accumulations and in situations where the synapse appears to be 'afferent'. So there seems to be some consensus on this point and therefore I cannot immediately understand why Dr Hess has not found them unless it is on some technical ground. This figure should also make it clear that these small dense-cored vesicles can be distinguished fairly easily from the large-cored vesicle in the Type I cell. I believe that the former, from their disposition – have a part to play in synaptic transmission while the latter which are found scattered thoughout the cytoplasm of the Type I cell are secretory since there is also some evidence that they can be released at any part of the cell surface.

ACKER I would like to ask whether the fixation methods are really so good that you can make these distinctions between the size and appearance of vesicles and secondly, don't you think that it is necessary to control important variables, for example hypoxia, prior to fixing?

HESS I usually fix *in situ*: I simply pour the fixative over the carotid body.

ACKER But don't you think that it is possible that these appearances which have caused so much discussion may be due to differences in blood gas tensions at the time of fixation?

HESS That may be so: but I don't think that the question of clear or dense-cored vesicles would be answered by differences in such variables.

HOWE I'd be grateful for some clarification about the dimensions of these vesicles because it became increasingly unclear to me as you showed your slides whether we were talking about vesicles involved in some transmission process or the more characteristic vesicles in the Type I cell.

HESS I don't really make such a distinction for it would seem to me that it was the same dense-cored vesicles found scattered in the Type I cell which were accumulated next to the membrane. They range between 60 and 90 nm. This makes them very different from the vesicles seen in adrenergic nerve terminals which are usually smaller and with a smaller dense core. But they are certainly smaller but similar in their granularity with the catecholamine-containing granules in the adrenal medulla.

BISCOE Did you say that you saw electron-dense cored vesicles along the cell membrane and at other sites than the nerve ending?

HESS Yes, cell to cell and Type I to Type II cell.

HOWE But are these in any way different from the granules which you see elsewhere in the Type I cytoplasm?

HESS No.

SPECKMANN I must press you about Dr Acker's question. Is it possible that the different synaptic appearances which you have described could be due to differences in synaptic activity?

HESS I think that it is possible but I still think that you can argue by analogy with other structures – for example, the hair cell or muscle spindle which was discussed earlier – and say, from the arrangement of the synaptic vesicles that this terminal is sensory and this one is motor.

SPECKMANN But that is rather a static approach, that is, you have not considered experiments with nerve stimulation.

HESS I agree, but it is only very recently that changes in the arrangement of synaptic vesicles in motor nerve fibres have been observed with stimulation of the nerve. For years this was tried but no changes seen.

MCDONALD Could you summarize how you think the structures which we have been discussing are hooked up.

HESS Type I cell to a sensory nerve terminal; Type I cell to Type I cell; Type I to Type II cell; Type I cell to an ending with an accumulation of synaptic vesicles which same ending has an accumulation of synaptic vesicles making contact with another Type I cell.

WILLSHAW How do you visualize the granules forming within the Type I cell? Do you have 'old' and 'young' vesicles?

HESS I know of no way in which you can distinguish the maturity of such granules unless you are lucky enough to get pictures of early-formed vesicles – such as I believe has been shown in adrenergic fibres.

ACKER But are you not assuming that the catecholamines which you showed in your fluorescent slide – I think it was low powered showing the whole carotid body at ×200 – are solely in the granules?

HESS Yes, they contain catecholamine.

FILLENZ Surely not. Dopamine is known to be in the cells but it has not been definitely located in the vesicles.

HESS I am arguing by analogy with adrenergic nerve endings where the vesicles are supposed to contain catecholamines.

EYZAGUIRRE Yes, but the analogy may not be sound. Have you thought of the possibility of sorting this question out with ultra-centrifugal separation?

HESS Yes, but I have not done it.

MCDONALD Is there anything else which could account for the increase in density after treatment with 6-hydroxydopamine? Did you conclude that after treatment with 6-OHDA, the 6-OHDA was within vesicles?

HESS Yes, that is the same conclusion that others have drawn with respect to the adrenergic nerve fibre so treated. There is one other point. If you fix with osmium tetroxide, the vesicles of the Type I cell appear as extremely light. If you pre-treat the animal with 6-OHDA and fix with osmium, the vesicles appear dense. This is similar to the results obtained by

Tranzer & Richards (1971), who made adrenergic vesicles dense in peripheral nerves after 6-hydroxydopamine and osmium tetroxide fixation.

REFERENCE

TRANZER, J. P. & RICHARDS, J. G. (1971). Fine structural aspects of the effect of 6-hydroxydopamine on peripheral adrenergic neurons. In *6-Hydroxydopamine and catecholamine neurons*, pp. 15–32, eds MALMFORS, P. & THOENEN, H., Amsterdam: Elsevier.

Observations on the innervation of the carotid body of the rabbit

BY ALAIN VERNA

Laboratoire de Cytologie, Université de Bordeaux II,
Avenue des Facultés, 33-Talence, France

The histological structure of arterial chemoreceptors is now well known thanks to considerable research. The comparative study of the carotid labyrinth in amphibia and of the carotid body in birds, reptiles and mammals shows moreover a remarkable unity in structure (Kobayashi, 1971*a*); but, if these observations are relatively concordant, the same cannot be said for their interpretation. Thus, the function of glomic cells has not yet been established, and the afferent or efferent nature of their innervation remains a subject of controversy (Kobayashi, 1971*b*). These divergences of opinion can be easily explained by the complexity of the carotid body, which is supposed to contain receptors and their afferent innervation, a vaso-motor efferent innervation, and, at least for certain authors, an inhibitory, efferent innervation which is non vaso-motor (Fidone & Sato, 1970; Biscoe, 1971; Neil & O'Regan, 1971).

It is clearly impossible to tell by a morphological study alone whether a cell is sensory or not, whether a nervous fibre is afferent or efferent. Nonetheless, the observations presented in this paper seem likely to bring some elements of an answer to some of the numerous problems still posed by the carotid body.

The carotid body of the rabbit, relatively little studied up to now, does not present any notable histological particularities. It is made up of cell clusters disseminated in an abundant connective tissue. These clusters are formed by the association of two habitual types of cell: Type I cells, characterized by abundant dense-cored vesicles, and Type II cells (devoid of vesicles) surrounding the Type I cells with a thin cytoplasmic sheath (Plate 1). The cell clusters are closely associated with the capillaries, the endothelium of which is occasionally fenestrated. One may also note that the Type II cells do not completely cover the Type I cells (Plate 2*a*); there is therefore no anatomical obstacle to the exchanges between blood and Type I cells.

In the connective tissue surrounding the clusters of glomic cells, one can observe numerous nerve fibres. The myelinated fibres are abundant at the periphery of the organ, more scarce in the centre. Some fibres lose their myelin sheath in the immediate proximity of the cell clusters (Plate 2*b*). The non-myelinated fibres are very numerous both at the periphery and in the centre of the carotid body, where they run around the capillaries and the cell clusters.

Among the non-myelinated nerve fibres running around the capillaries, some are of extremely small diameter, in the region of 0.1 μm (Plate 3*a*), others show dilatations containing agranular and granular vesicles of about 50 nm in diameter (Plate 3*b*). These aspects are characteristic of the sympathetic endings, the presence of which has already been established in the carotid body (Biscoe & Stehbens, 1966). The extent of this peri-vascular innervation and its aminergic nature can easily be demonstrated by Falck's method: a section of the carotid body treated according to this method shows in effect numerous varicose nerve fibres highly fluorescent, around the blood vessels and the cell clusters (Plate 4*a*). Twenty-four hours after excision of the superior cervical ganglion, these fibres disappear, which clearly demonstrates the sympathetic origin of this aminergic innervation (Plate 4*b*). Histo-autoradiography after administration of tritiated noradrenalin finally makes it possible to demonstrate that the peri-vascular sympathetic innervation is noradrenergic (Plate 5*a*). This same technique, furthermore, makes it possible to note that rare noradrenergic endings (probably therefore, of sympathetic origin) come into contact with the clusters of glomic cells where they are more or less completely enveloped by the Type II cells (Plate 5*b*). Nevertheless, we have never noted any contact between noradrenergic endings and Type I cells.

The endings which come into contact with the Type I cells are recognizable by their mitochondria which are smaller and more electron-dense than those of the glomic cells, and by the presence of clear vesicles of 35 to 50 nm in diameter. The size and shape of these endings are very variable, as is their content of organelles. One can thus note *en passant* endings of small diameter (Plate 6*a*), 'synaptic knobs' of greater size, rich to a greater or lesser extent in clear vesicles and in mitochondria (Plate 6*b*), and endings of considerable size (up to 10 μm) particularly rich in mitochondria (Plate 7). The existence of intermediary aspects, nevertheless, makes this classification debatable, and makes it possible to think that these various 'types' of endings result from the direction of the plane of the sections. In this respect, one must recall that the polymorphism of the endings innervating the Type I cells, already noted by De Castro (1926), has recently been underlined by Biscoe & Pallot (1972) after study of serial sections.

This point of view is confirmed by Plate 8: this particularly favourable longitudinal section shows successive aspects of the same, single ending, which are very different. Initially, this has a small diameter (0.6 μm) and is rich in small, clear vesicles. But this ending is prolonged by a considerable dilatation, and is rich in mitochondria and in glycogen, but devoid of clear vesicles.

It seems difficult, therefore, in these conditions, to admit the existence of a double innervation (afferent and efferent) of the Type I cells on the basis of purely morphological criteria (shape, presence or absence of clear vesicles) used up to now by various authors (Höglund, 1967; Al Lami & Murray, 1968; Ábrahám, 1969; Kobayashi, 1971*b*; Eyzaguirre, Nishi & Fidone, 1972).

The presence of clear vesicles of the synaptic type, on the 'wrong side' of the junctions, in other words, in the endings innervating the Type I cells, seems to have been at the bottom of the doubts expressed by several authors (Biscoe, 1971; Sampson, 1972) concerning the interpretations of De Castro (De Castro, 1926; De Castro & Rubio, 1968).

In fact, apparently identical vesicles can be seen also in the Type I cells in proximity to nerve endings (Plate 9*a*). In a few cases, the clear vesicles build up on both sides of the junction, around a same zone of membrane thickening (Plate 9*b*). The significance of these vesicles is therefore obscure, all the more so since they are only rarely associated with the membrane differentiations, whether it be a question of clear vesicles of Type I cells or of nerve endings.

On the other hand, other structures appear to be more significant: namely, a mass of dense-cored vesicles of 70 to 110 nm in diameter, exclusively situated in the Type I cells and closely associated with the membrane thickenings by an electron-dense material. Certain vesicles appear to be more or less completely 'emptied' of their dense content, and the dense-cored vesicles are occasionally accompanied by clear vesicles of 35 to 45 nm in diameter (Plates 10–11*a*). These characteristic associations (dense-cored vesicles, electron-opaque material, membrane thickening) therefore seem to constitute a true synaptic complex (Fig. 1).

This type of structure is visible at the level of nerve endings of very different aspects: rich or not rich in clear vesicles (Plate 10*a*, 10*b*), rich both in mitochondria and in clear vesicles (Plate 10*c*) or of very small diameter (Plate 11*a*). This observation therefore confirms that these various endings are merely different aspects of the same, highly polymorphous innervation, and moreover means that we may consider this innervation as being the post-synaptic element of the junction between the Type I cell and the nerve.

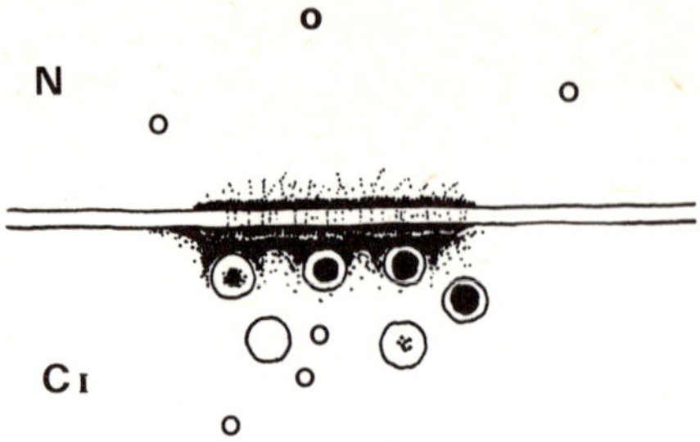

Fig. 1. Synaptic complex between the Type I cell and the nerve; N, nerve ending; C I, Type I cell. (From *J. Microsc. Paris*, 1973, **16**, 299–308, Fig 1.)

The synaptic complexes described above are not peculiar to the carotid body of the rabbit: comparable structures have been described in the mouse and in the guinea-pig (Kobayashi, 1971*b*) and in various amphibia (Kobayashi, 1971*b*; Poullet-Krieger, 1973). One may finally add that the abundance of mitochondria is generally considered as being characteristic of sensory fibres or of dendrites (Sotelo & Palay, 1968; Böck, Stockinger & Vyslonzil, 1970; Kobayashi, 1971*b*; Kondo, 1971; Chiba, 1972).

These various arguments are therefore concordant, and allow one to envisage the hypothesis of the afferent nature of the innervation of Type I cells.

This point of view is not shared by Biscoe (1971). According to this author, on the contrary, the innervation of Type I cells appears to be efferent, and the afferent fibres which must necessarily exist in the carotid body appear to be represented by the non-myelinated endings of approximately 0.1 μm in diameter, previously described by De Kock & Dunn (1966) in the folds of Type II cells. Several objections have already been posed to this interpretation: in particular, the possibility of these fibres, despite their small size, being at the origin of the endings innervating Type I cells (Eyzaguirre, *et al.*, 1972; Hess & Zapata, 1972). Our own observations (Verna, 1971) have, moreover, shown that an ending of 6 μm in diameter can have as origin a fibre of 0.3 μm (Plate 11*b*) and that it is perfectly conceivable that an ending of even smaller size can have as origin a fibre of a diameter of around 0.1 μm. What is more, the study of sympathetic innervation following administration of tritiated noradrenalin shows us a further source of error: the existence of contacts between sympathetic fibres and Type II cells (Plate 5*b*). Among the fibres presumed by Biscoe to be afferent, some of them are perhaps in fact sympathetic efferent fibres; one must recall in this respect that the fibres which circulate in proximity of the capillaries are sometimes of extremely small diameter (Plate 3*a*).

A further type of junction can be observed in the carotid body of the rabbit: namely, synapses, in the strict sense of the term, between nerve fibres. The asymmetry of the area of junction makes it possible to identify without difficulty the pre- and post-synaptic compartments by comparison with the well-known synapses of the central nervous system (Peters, Palay & Webster, 1970).

The pre-synaptic cytoplasm is characterized by the presence of clear vesicles of 40 to 50 nm in diameter, grouped in compact masses always associated with a slightly thickened zone of the membrane. Moreover, it contains a few mitochondria and rare dense-cored vesicles of 130 to 180 nm in diameter.

On the post-synaptic side, the membrane thickening is more marked and the mitochondria generally more numerous. This cytoplasm may also contain clear vesicles of 35 to 50 nm in diameter, but these vesicles are scattered and have no relation with the membrane thickenings. One may finally observe in the cytoplasm a few rare dense-cored vesicles (smaller than those of the pre-synaptic element), and possibly a few grains of glycogen. These observations are summed up in Fig. 2.

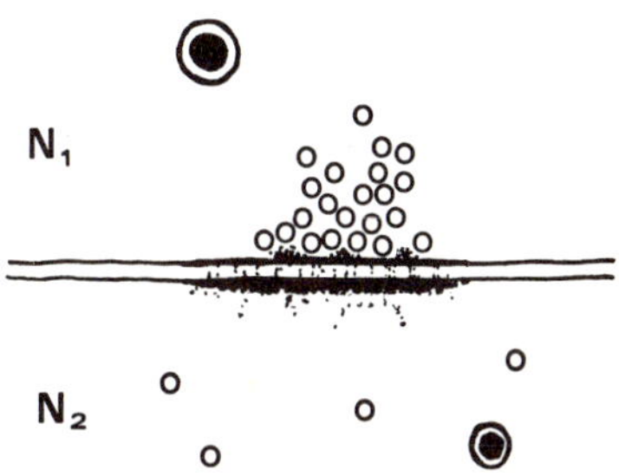

Fig. 2. Area of junction between nerve endings. N_1, pre-synaptic element; N_2, post-synaptic element.

Such synapses are visible in the connective tissue surrounding the clusters of glomic cells (Plates 12, 13*a*) or, much more rarely, in these clusters themselves (Plate 13*b*).

The identification of the nerve fibres involved in these synapses is obviously extremely difficult without the help of degeneration experiments or serial reconstruction. Nevertheless, one must point out that the post-synaptic elements often present cytological features which remind one of the endings innervating the Type I cells: numerous mitochondria, scattered clear vesicles, the possible presence of grains of glycogen. One may therefore presume that the synapses described above are situated at the pre-terminal level of the fibres innervating the Type I cells. This interpretation is confirmed by Plate 13*b*, in which the post-synaptic element is indeed in contact with the glomic cells. If we admit the afferent nature of the innervation of the Type I cells, the post-synaptic element of the junctions described here is a dendrite, and the pre-synaptic element may be either an axon, or another dendrite. The dendro-dendritic hypothesis, nonetheless, appears unlikely (although it cannot be dismissed altogether) because of the constant cytological differences presented by the pre- and post-synaptic elements; it seems more plausible to consider these synapses as being axo-dendritic. Synaptic contacts between nerve endings have also been observed

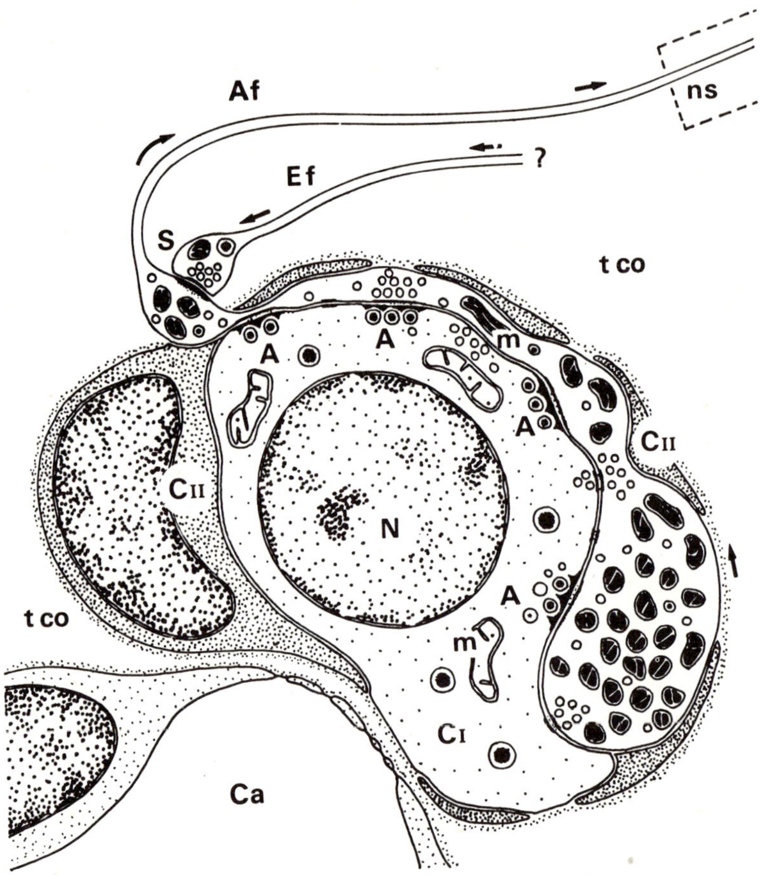

Fig. 3. Presumptive diagram of the innervation of glomic cells. A Type I cell (CI) receives a highly polymorphous, afferent nerve ending (Af). One can observe the afferent synaptic complexes (A) at the level of the various aspects of this ending. Clear vesicles are present both in the ending and in the Type I cell. An efferent fibre (Ef) controls the activity of the afferent ending at the level of an axo-dendritic synapse (S). t co, connective tissue; Ca, capillary; N, nucleus; m, mitochondria; ns, sinus nerve. The arrows show the direction of the circulation of the nervous influxes. (The Schwann cells have not been shown.) (From *J. Microsc. Paris*, 1973, **16**, 299–308, Fig. 2.)

in the swallow by Kobayashi (1969) but this author, having admitted the efferent nature of the innervation of Type I cells (because of the abundance of clear vesicles) concluded from this that it was a question of axo-axonic synapses.

This interpretation is to be linked with recent research showing an efferent innervation that is chemo-inhibitory in the sinus nerve (Fidone & Sato, 1970; Neil & O'Regan, 1971): the pre-synaptic element of the junctions presumed to be axo-dendritic described here may very well correspond to the efferent inhibitory fibres observed by electro-physiologists.

To sum up, the innervation of glomic cells may be conceived according to the following schema:

(i) Chemo-afferent innervation on Type I cell.
(ii) Efferent innervation controlling the activity of the fibres at the level of the axo-dendritic synapses which are presumed to be inhibitory.

This diagram is obviously merely a working hypothesis, and needs to be confirmed by a considerable amount of further research.

In particular, the origin of the pre-synaptic element of the inter-neuronal connections described here needs to be determined (sinus nerve? ganglion cells?). Moreover, this diagram has no intention of prejudging the identity of the chemoreceptor proper (Type I cell?; the nerve ending itself?).

In the case of the carotid body, the importance of confronting morphological observation with physiological observations is particularly obvious: thus, for example, certain recent publications deny the existence of a direct inhibitory innervation, and attribute the effects observed to a simple vasodilatation (Goodman, 1973). Should this interpretation be confirmed, it will then be necessary to find another explanation for the connections between nerve fibres described in this paper.

In conclusion, let us hope that the observations and the hypotheses presented in this paper contribute, if not to enlightening the problems posed by the carotid body, at least to stimulating fresh experiments.

REFERENCES

Ábrahám, A. (1969). Elektronenmikroskopische Untersuchungen an menschliche Karotiskörpern. *Z. mikrosk. Anat. Forsch.*, **81**, 413–53.

Al Lami, F. & Murray, R. G. (1968). Fine structure of the carotid body of normal and anoxic cats. *Anat. Rec.*, **160**, 697–718.

Biscoe, T. J. (1971). Carotid body: structure and function. *Physiol. Rev.*, **51**, 437–95.

Biscoe, T. J. & Pallot, D. (1972). Serial reconstruction with the electron microscope of carotid body tissue. The Type I cell nerve supply. *Experientia*, **28**, 33–4.

Biscoe, T. J. & Stehbens, W. E. (1966). Ultrastructure of the carotid body. *J. Cell. Biol.*, **30**, 563–8.

Böck, P., Stockinger, L. & Vyslonzil, E. (1970). Die Feinstruktur des Glomus caroticum beim Menschen. *Z. Zellforsch.*, **105**, 543–68.

CHIBA, T. (1972). Fine structure of the baroreceptor nerve terminals in the carotid sinus of the dog. *J. Electron Microsc.*, **21**, 139–48.

DE CASTRO, F. (1926). Sur la structure et l'innervation de la glande intercarotidienne (glomus caroticum) de l'homme et des mammifères, et sur un nouveau système d'innervation autonome du nerf glossopharyngien. Etudes anatomiques et expérimentales. *Tra.b Lab· Invest. Biol. Univ. Madr.*, **24**, 365–432.

DE CASTRO, F. & RUBIO, M. (1968). The anatomy and innervation of the blood vessels of the carotid body and the role of chemoreceptive reactions in the autoregulation of the blood flow. In *Arterial chemoreceptors*, pp. 267–76, ed. TORRANCE, R. W., Blackwell: Oxford and Edinburgh.

DE KOCK, L. L. & DUNN, A. E. G. (1966). An electron microscope study of the carotid body. *Acta. Anat.*, **64**, 163–78.

EYZAGUIRRE, C., NISHI, K. & FIDONE, S. (1972) Chemoreceptor synapses in the carotid body. *Fedn Proc.*, **31**, 1385–93.

FIDONE, S. J. & SATO, A. (1970). Efferent inhibition and antidromic depression of chemoreceptor A-fibers from the cat carotid body. *Brain Research*, **22**, 181–93.

GOODMAN, N. W. (1973). Efferent control of arterial chemoreceptors mediated by glossopharyngeal fibres and artifacts introduced by stimulation techniques. *J. Physiol.*, **230**, 295–311.

HESS, A. & ZAPATA, P. (1972). Innervation of the cat carotid body: normal and experimental studies. *Fedn Proc.*, **31**, 1365–83.

HÖGLUND, R. (1977). An ultrastructural study of the carotid body of horse and dog. *Z. Zellforsch.*, **76**, 568–76.

KOBAYASHI, S. (1969). On the fine structure of the carotid body of the bird, *Uroloncha domestica. Arch. Histol. Jap.*, **31**, 9–19.

KOBAYASHI, S. (1971a). Comparative cytological studies of the carotid body. 1. Demonstration of monoamine-storing cells by correlated chromaffin reaction and fluorescence histochemistry. *Arch. Histol. Jap.*, **33**, 319–39.

KOBAYASHI, S. (1971b). Comparative cytological studies of the carotid body. 2. Ultrastructure of the synapses on the chief cell. *Arch. Histol. Jap.*, **33**, 397–420.

KONDO, H. (1971). An electron microscopic study on innervation of the carotid body of guinea pig. *J. Ultrastruct. Res.*, **37**, 544–62.

NEIL, E. & O'REGAN R. G. (1971). Efferent and afferent impulse activity recorded from few-fibre preparations of otherwise intact sinus and aortic nerves. *J. Phyiol. Lond.*, **215**, 33–47.

PETERS, A., PALAY, S. L. & WEBSTER, H. DE F. (1970). *The fine structure of nervous system: the cells and their processes.* New York: Hoeber.

POULLET-KRIEGER, M. (1973). Innervation du labyrinthe carotidien du Crapaud *Bufo bufo*: étude ultrastructurale et cytochimique. *J. Microsc. Paris*, **18**, 55–64.

SAMPSON, S. R. (1972). Innervation of the carotid body: another point of view. *Fedn Proc.*, **31**, 1383–4.

SOTELO, C. & PALAY, S, L. (1968). The fine structure of the lateral vestibular nucleus in the rat. I. Neurons and neuroglial cells. *J. Cell Biol.*, **36**, 151–79.

VERNA, A. (1971). Infrastructure des divers types de terminaisons nerveuses dans le glomus carotidien du Lapin. *J. Microsc. Paris*, **10**, 59–66.

EXPLANATION OF PLATES

All the electron micrographs are of the carotid body of the rabbit, fixed by perfusion (glutaraldehyde alone or glutaraldehyde and formaldehyde in a cacodylate buffer) postfixed by immersion (OsO_4 1% in the same buffer) and embedded in araldite.

PLATE I

Low magnification showing the organization in clusters of glomic cells: several Type I cells (CI) are enveloped by the cytoplasm, considerably thinned, of a Type II cell (CII). Two nerve endings (N) are visible on the periphery of the cluster. A non-myelinated nerve fibre (n) is circulating between the glomic cells and the capillaries (Ca); ×6250.

PLATE 2

(*a*) A Type I cell (CI) is separated from a capillary (Ca) by only an extremely small space occupied by a few collagenous fibriles. Note the fenestration of the endothelium (arrows); CII, Type II cell; × 18 600.
(*b*) Myelinated nerve fibre (Mn) losing its myelin sheath in the proximity of a cluster of glomic cells; CI, Type I cell; CII, Type II cell: Ca, capillary; N, nerve ending (thought to be of sympathetic origin) containing small granular vesicles; × 9600.

PLATE 3

(*a*) Non-myelinated nerve fibre of extremely small diameter (arrow), situated in proximity to a capillary (Ca); × 37 000.
(*b*) Nerve ending thought to be of sympathetic origin (N) because of the presence of agranular and granular vesicles of approximately 500 Å in diameter. This ending is in proximity to a capillary. E, endothelium; × 27 000.

PLATE 4

(*a*) Normal glomus treated according to Falck's method. Note the high fluorescence of the cell clusters – (in inset, the detail of a cluster; × 900) and the abundance of varicose nerve fibres, circulating around the vessels and the cell clusters; × 100.
(*b*) Glomus treated according to Falck's method 24 hours after excision of the superior cervical ganglion: the abundant aminergic innervation visible in Plate 4*a* has disappeared; × 100.

PLATE 5

(*a*) Histo-autoradiograph following administration of tritiated noradrenalin. Two non-myelinated fibres situated in a little bundle circulating in the connective tissue are considerably marked by the radioactive noradrenalin. Ca, capillary; × 16 400.
(*b*) Histo-autoradiograph following administration of tritiated noradrenalin. A noradrenergic ending is visible in the cytoplasmic folds of a Type II cell (CII); CI, Type I cell; × 17 600.

PLATE 6

(*a*) *En passant* ending (N) on a Type I cell (CI). Note the succession of membrane thickenings and the presence of clear vesicles in the ending; CII, Type II cell; × 21 200.
(*b*) 'Synaptic knob' (N) on a Type I cell (CI). Note the small size of the mitochondria and the presence of clear vesicles, either associated (arrow) or non-associated (asterisk) to the membrane thickenings. CII, Type II cell; × 41 000. (From *J. Microsc., Paris*, 1971, **10,** 59–66, Plate 3*a*)

PLATE 7

Nerve ending (N) of large size, particularly rich in mitochondria. CI, Type I cell; CII, Type II cell; Ca, capillary; × 22 200. (From *J. Microsc., Paris*, 1973, **16,** 299–308, Plate 1.)

PLATE 8

Longitudinal section in a nerve ending (N). Initially of small diameter and rich in clear vesicles, this ending is at this stage comparable to that of Plate 6*a*; it is then prolonged by a considerable dilatation, rich in mitochondria, and becomes comparable to the ending in Plate 7; × 10 300. In inset, detail at higher magnification (× 31 000) of the narrowed part of this ending (rich in clear vesicles). CI, Type I cell; CII, Type II cell; Ca, capillary. (From *J. Microsc., Paris*, 1973, **16,** 299–308, Plate 2.)

PLATE 9

(*a*) Nerve ending (N) on a Type I cell (CI). Note the presence of a considerable mass of clear vesicles (arrow) in the Type I cell. CII, Type II cell; ×31 000. (From *J. Microsc., Paris*, 1973, **16**, 299–308, Plate 3*b*.)
(*b*) Nerve ending (N) on a Type I cell (CI). In this case, one can observe the clear vesicles on both sides of a same zone of membrane thickening; ×24 300. (From *J. Microsc., Paris*, 1971, **10**, 59–66, Plate 3*b*.)

PLATE 10

(*a*) Synaptic complex, characterized by the presence in the Type I cell (CI) of dense-cored vesicles associated with membrane thickenings. The nerve ending (N) contains a relatively considerable number of clear vesicles; ×39 400. (From *J. Microsc., Paris*, 1973, **16**, 299–308, Plate 4*a*.)
(*b*) Synaptic complex (arrow) comparable to the preceding one. In this example, the nerve ending (N) is practically devoid of clear vesicles; CI, Type I cell; CII, Type II cell; ×28 000. (From *J. Microsc., Paris*, 1973, **16**, 299–308, Plate 4*b*.)
(*c*) Synaptic complex (arrow) comparable to the preceding ones, but situated at the level of an ending (N) rich both in mitochondria and in clear vesicles; CI, Type I cell. Note the presence of clear vesicles of the same diameter as the dense-cored vesicles ('emptied' dense-cored vesicles?) and of vesicles of smaller diameter; ×44 500. (From *J. Microsc., Paris*, 1971, **10**, 59–66, Plate 5*a*.)

PLATE 11

(*a*) Synaptic complex (arrow) at the level of a nerve ending (N) of small diameter. Ca, capillary; CI, Type I cell; CII, Type II cell; n, pre-terminal fibre; ×38 000. (From *J. Microsc., Paris*, 1973, **16**, 299–308, Plate 5*b*.)
(*b*) Nerve ending of large size, rich in mitochondria. Note that this ending, however, has as origin a fibre of very small diameter (arrow), in this particular spot at least. Ca, capillary, Sch, Schwann cell; ×10 000. (From *J. Microsc., Paris*, 1971, **10**, 59–66, Plate 4*b*.)

PLATE 12

Synaptic contact between nerve endings. The pre-synaptic element (N_1) is characterized by a considerable mass of clear vesicles (arrows). In the post-synaptic element (N_2) the membrane thickening is more marked, and the mitochondria more numerous. Sch, Schwann cell; ×22 400. (From *J. Microsc., Paris*, 1973, **16**, 299–308, Plate 6.)

PLATE 13

(*a*) Synaptic contact between nerve endings. The pre-synaptic element (N_1) reveals a considerable mass of clear vesicles (arrow). The post-synaptic element (N_2) is comparable to certain aspects of the innervation of Type I cells by the abundance of mitochondria, and the scattering of clear vesicles. Sch, Schwann cell; ×29 400. (From *J. Microsc., Paris*, 1973, **16**, 299–308, Plate 7*a*.)
(*b*) Synaptic contact between nerve endings. N_1, pre-synaptic element; N_2, post-synaptic element. The junction is characterized by a considerable mass of clear vesicles (arrow) on the pre-synaptic side, and by a membrane thickening that is more marked on the post-synaptic side. The element N_2 is itself in contact (non-synaptic on this section plane) with a Type I cell (CI). CII, Type II cell; ×27 500. (From *J. Microsc., Paris*, 1973, **16**, 299–308, Plate 7*b*.)

PLATE 1

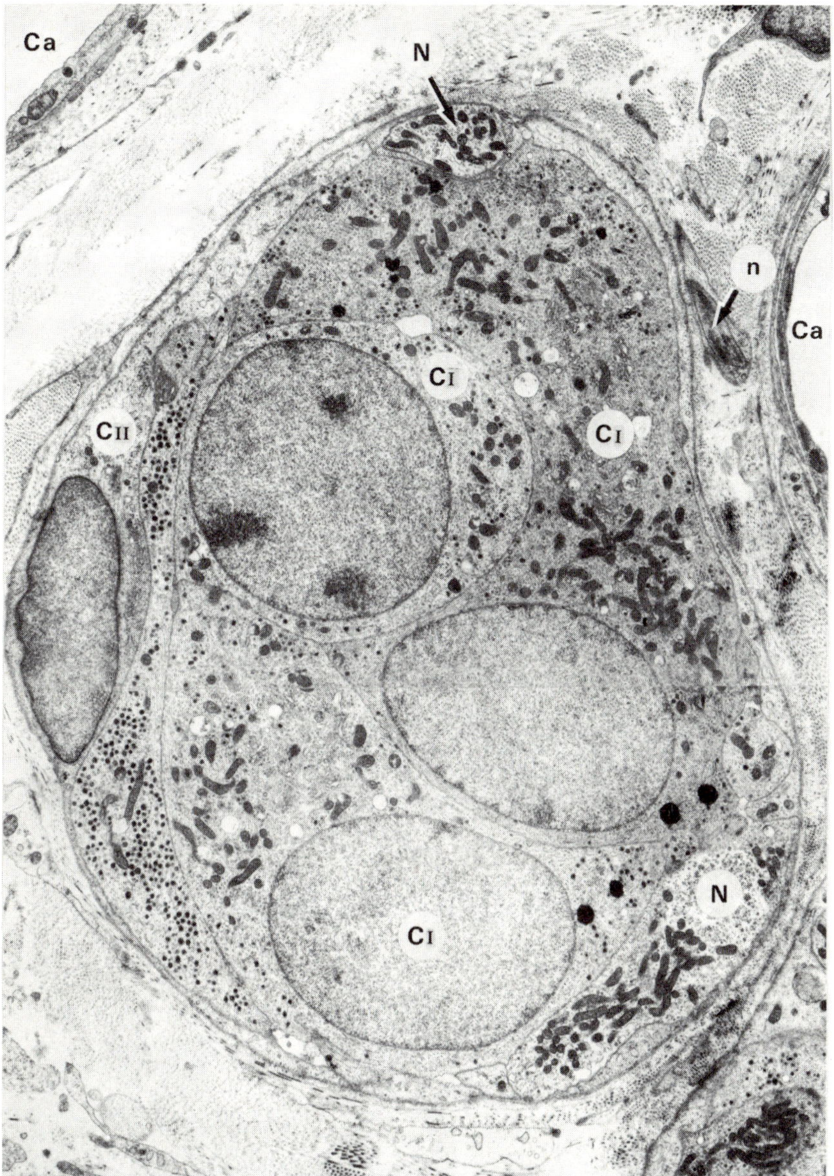

For explanation of Plate see p. 82

PLATE 2

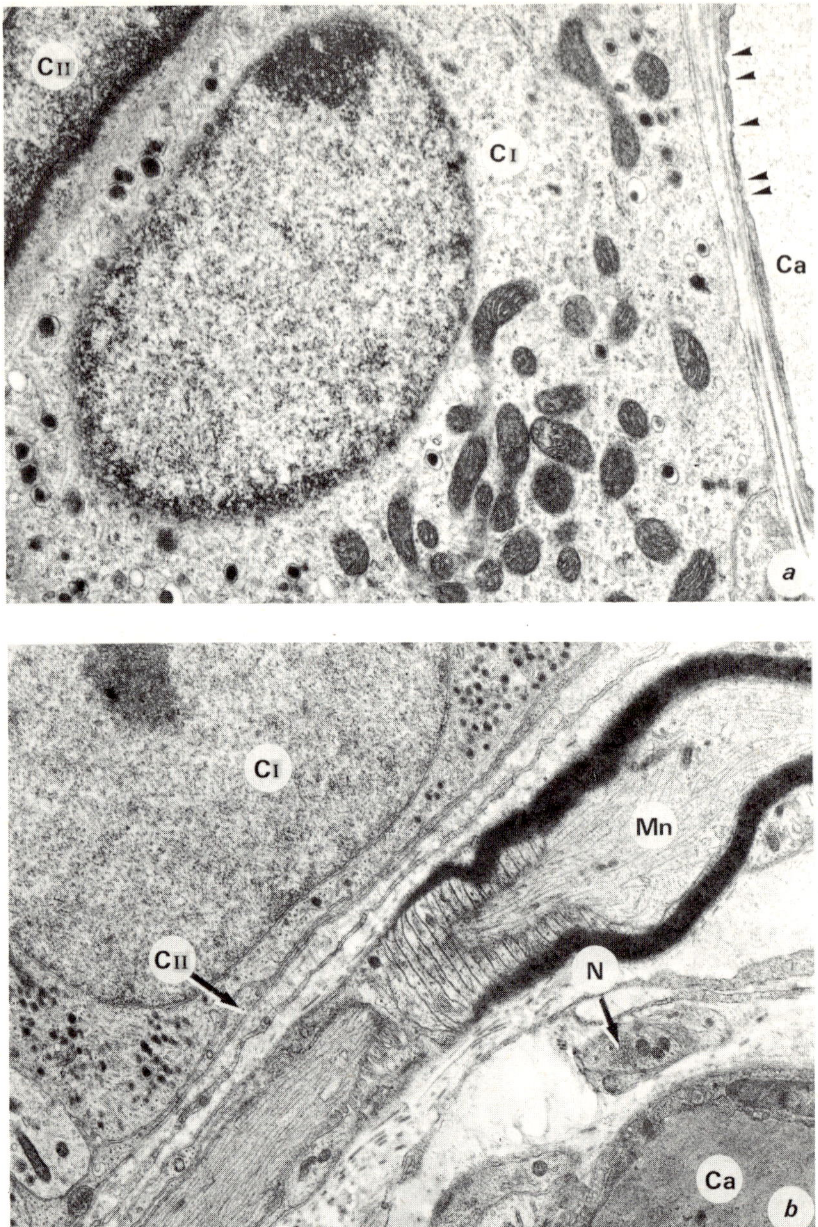

For explanation of Plate see p. 83

PLATE 3

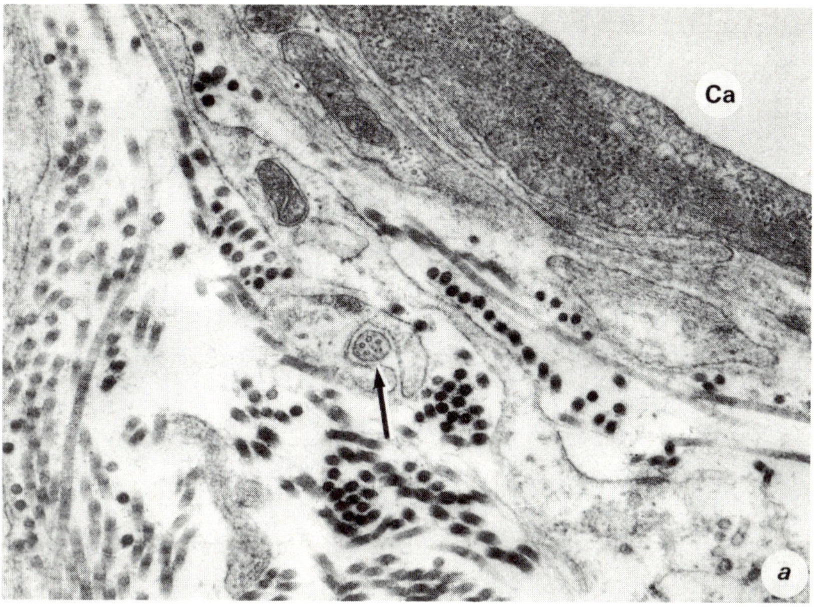

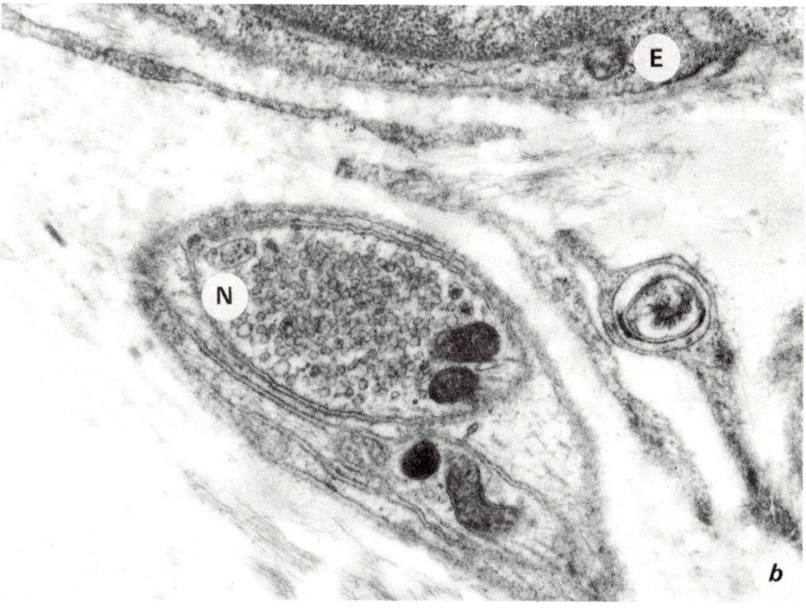

For explanation of Plate see p. 83

PLATE 4

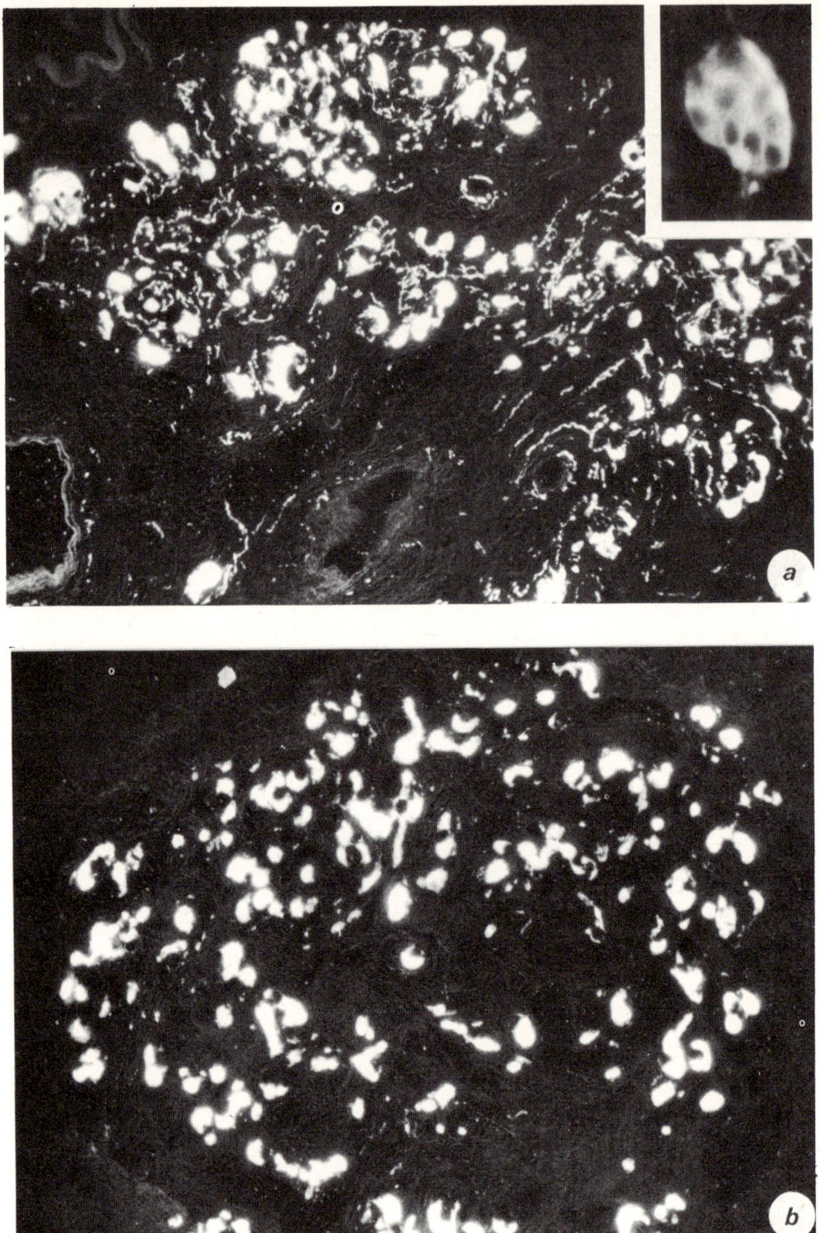

For explanation of Plate see p. 83

PLATE 5

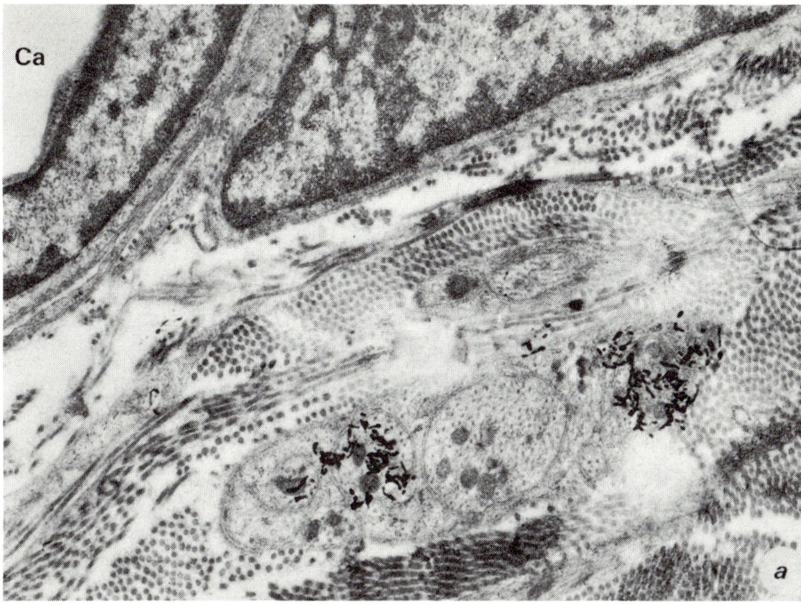

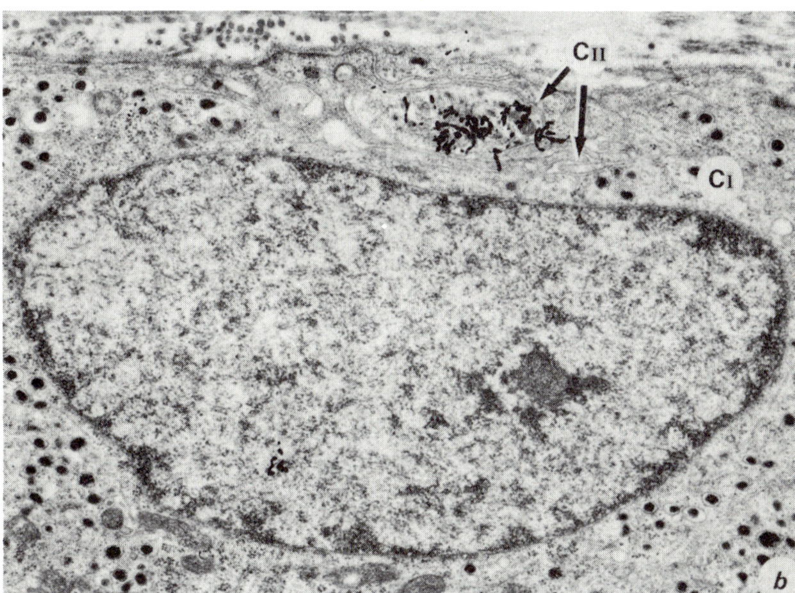

For explanation of Plate see p. 83

PLATE 6

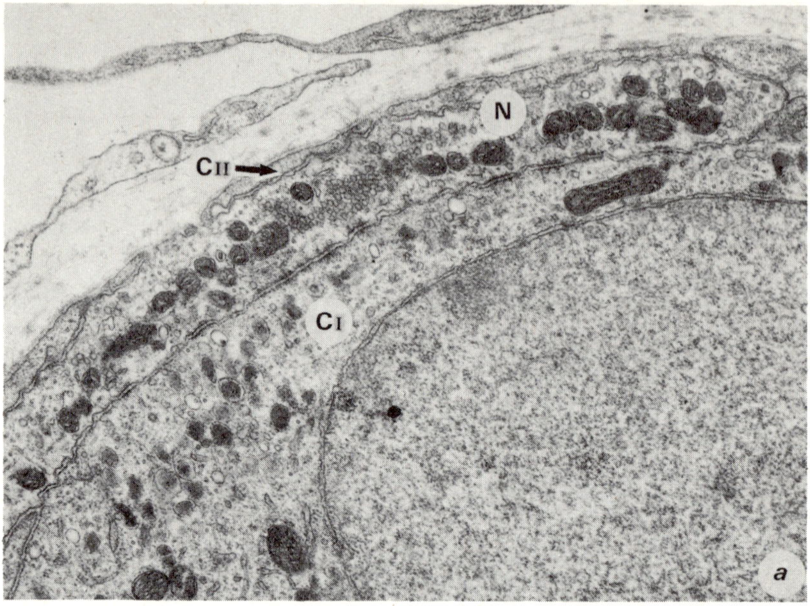

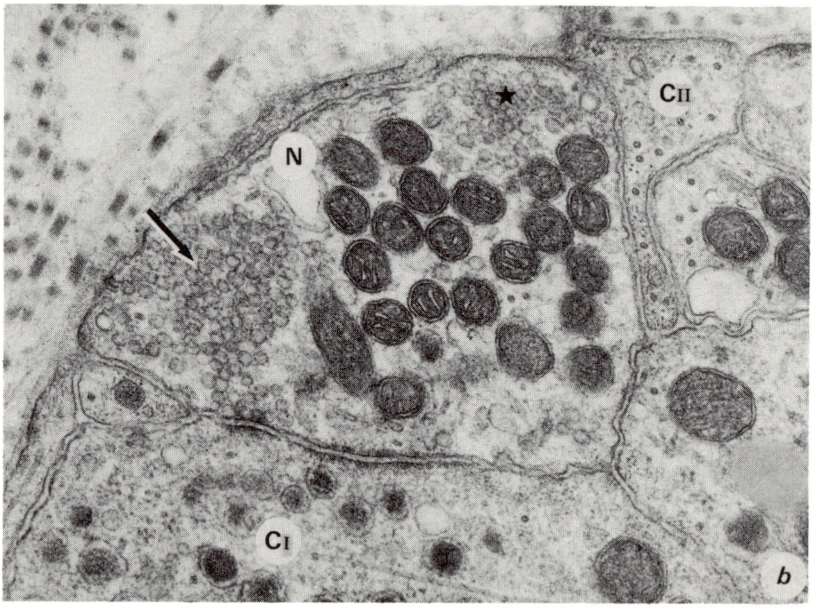

For explanation of Plate see p. 83

PLATE 7

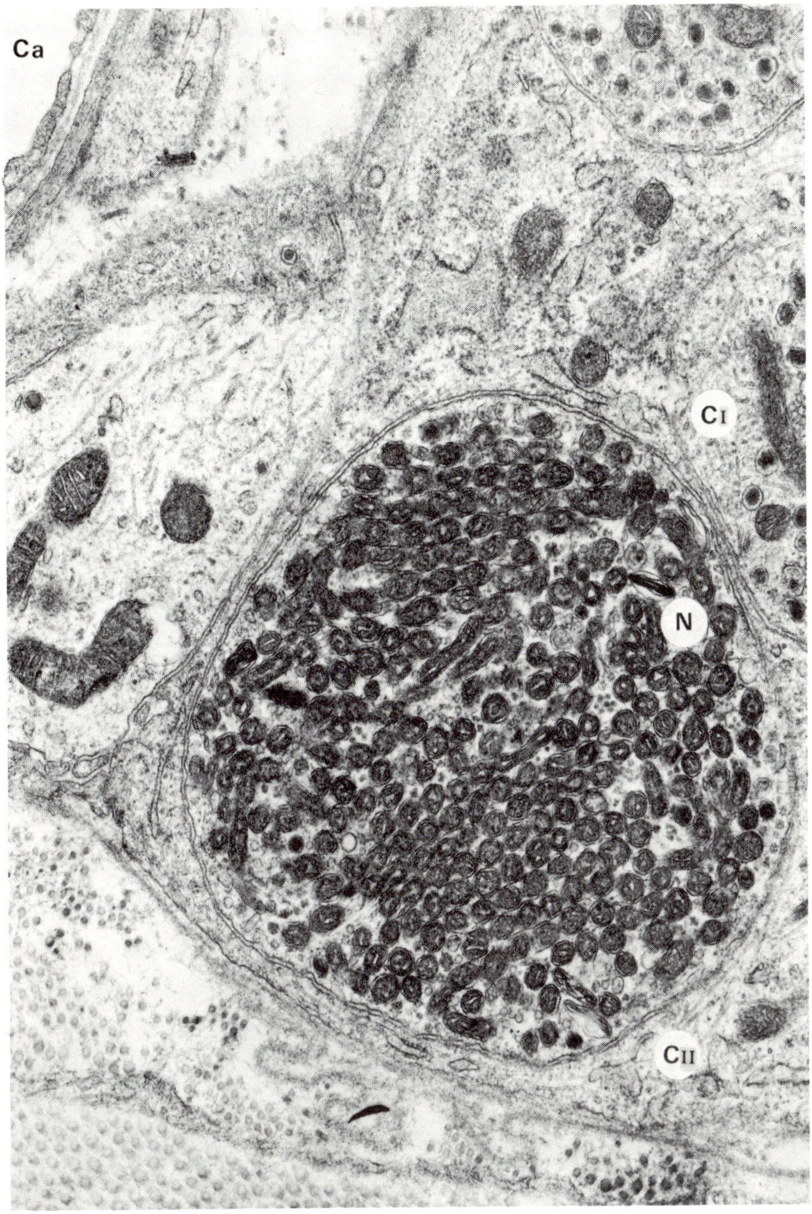

For explanation of Plate see p. 83

PLATE 8

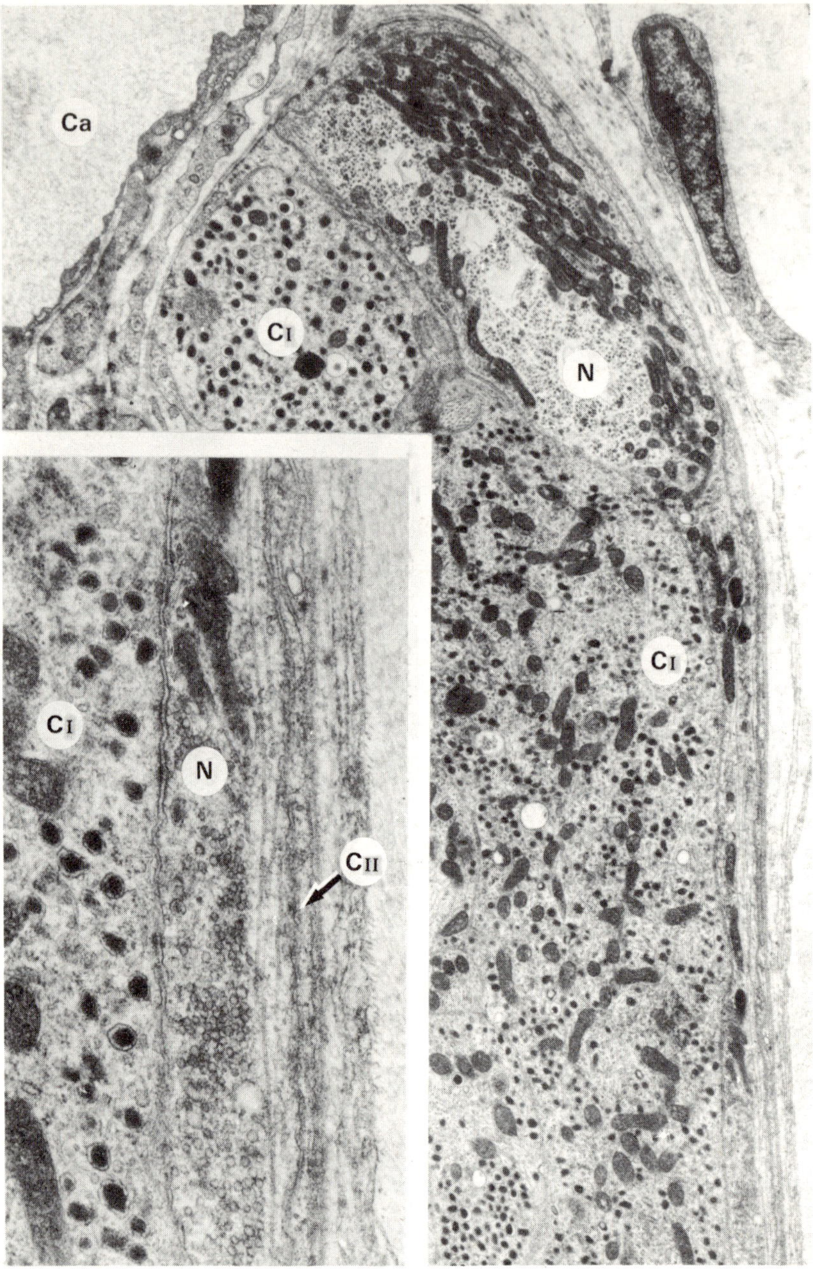

For explanation of Plate see p. 83

PLATE 9

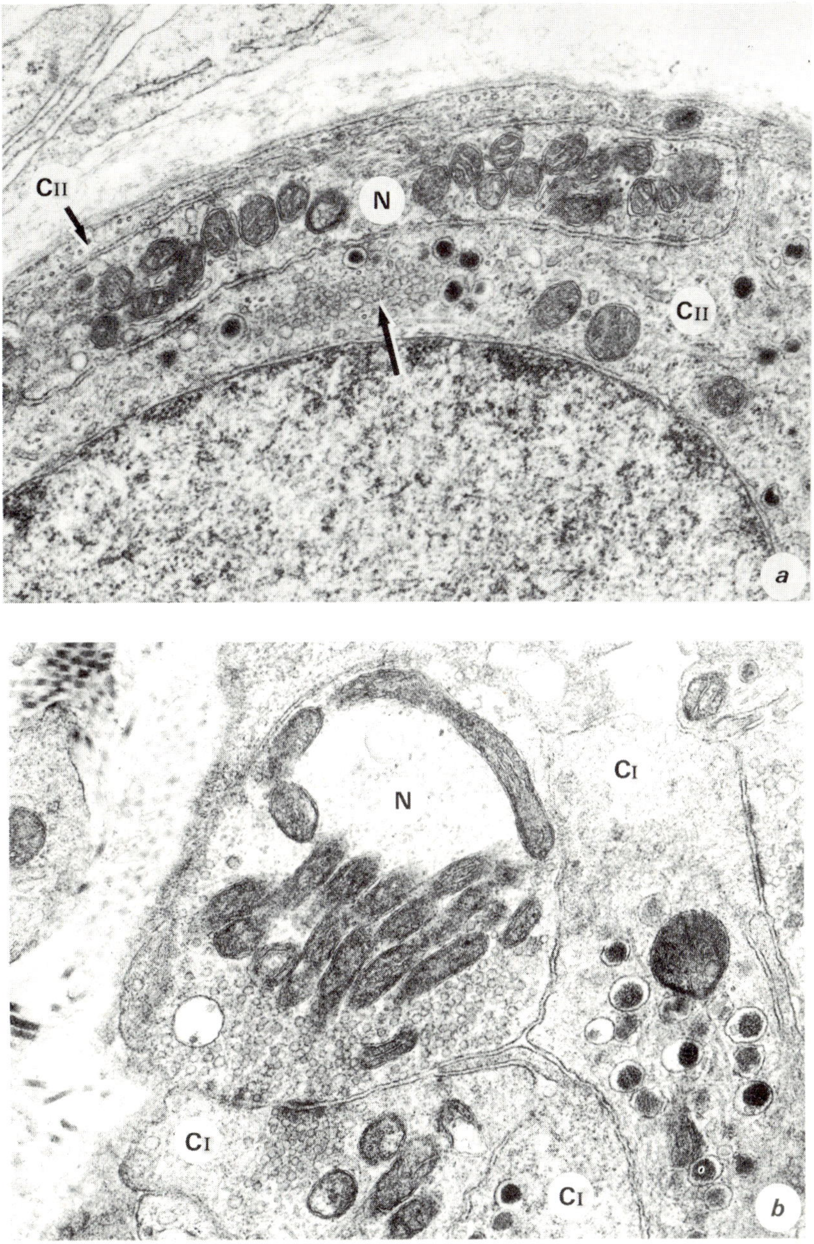

For explanation of Plate see p. 84

PLATE 10

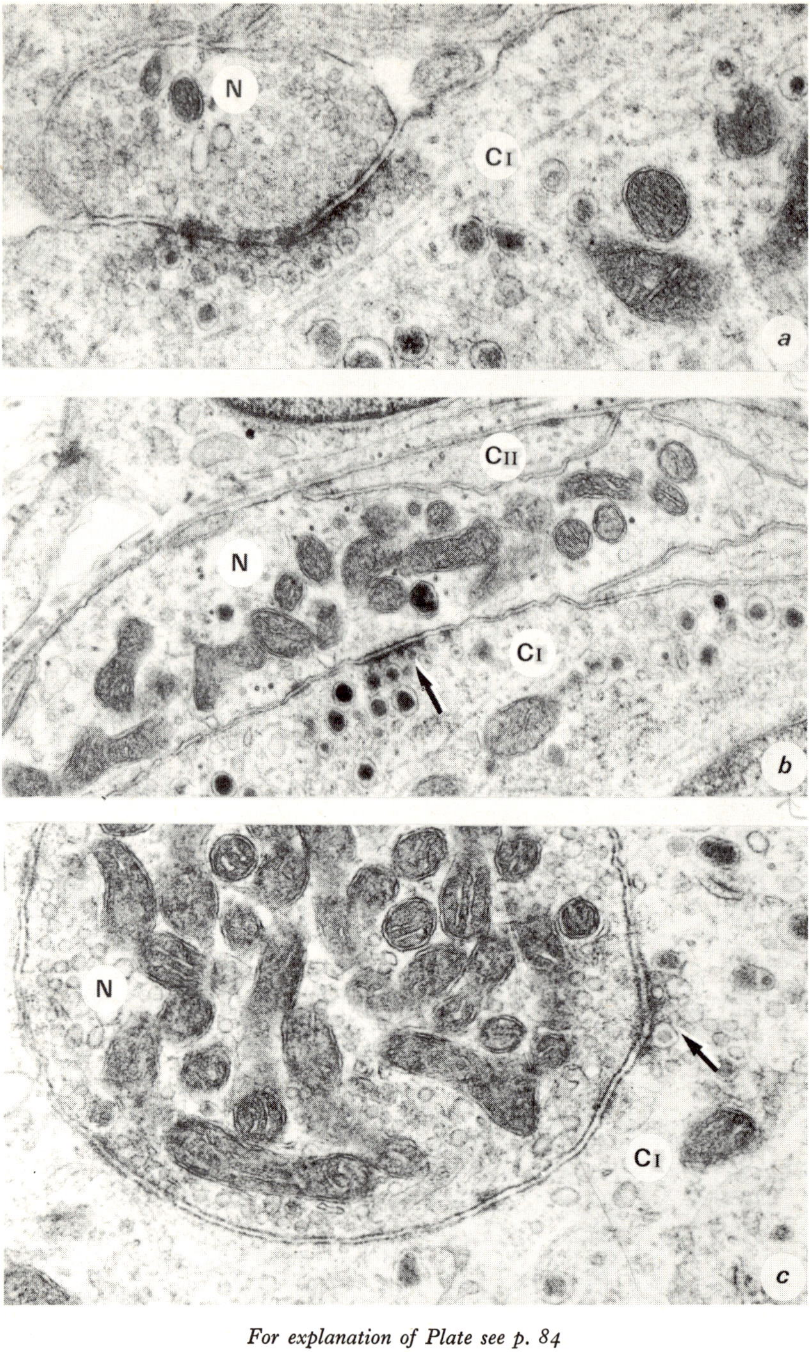

For explanation of Plate see p. 84

PLATE II

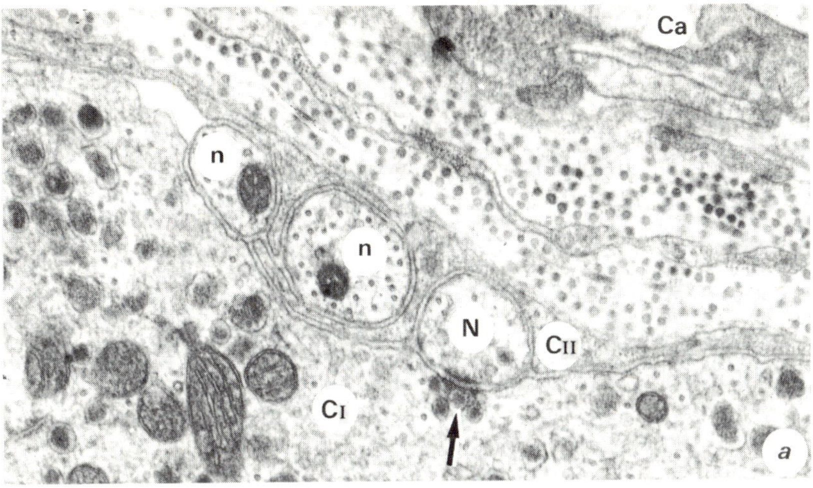

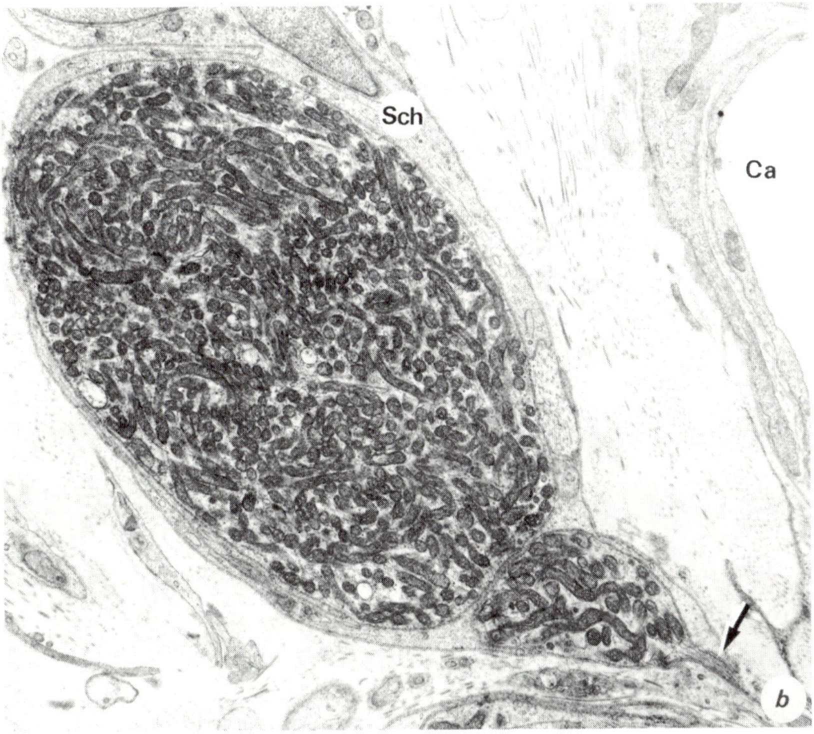

For explanation of Plate see p. 84

PLATE 12

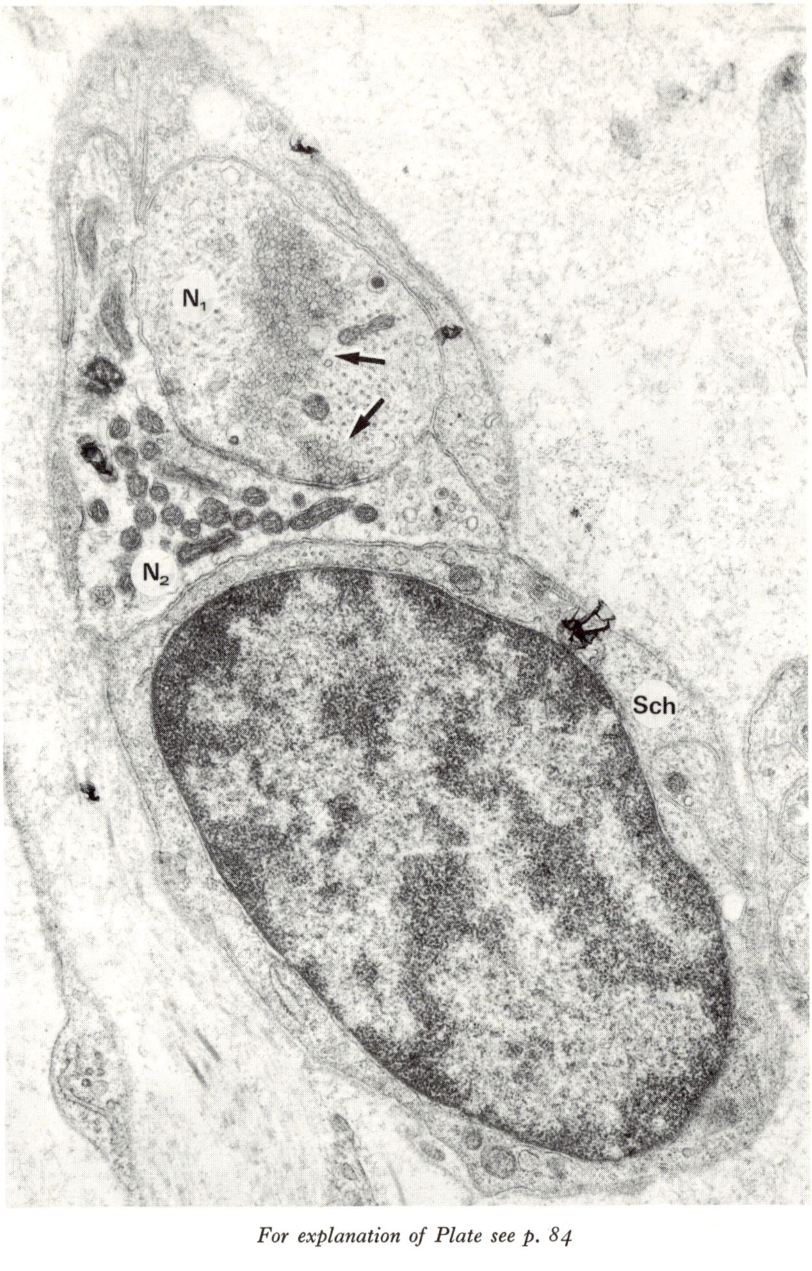

For explanation of Plate see p. 84

PLATE 13

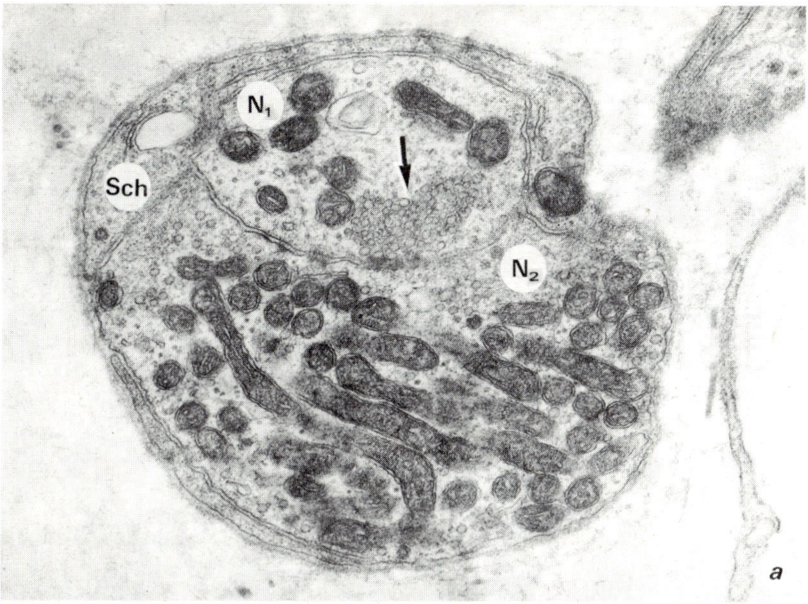

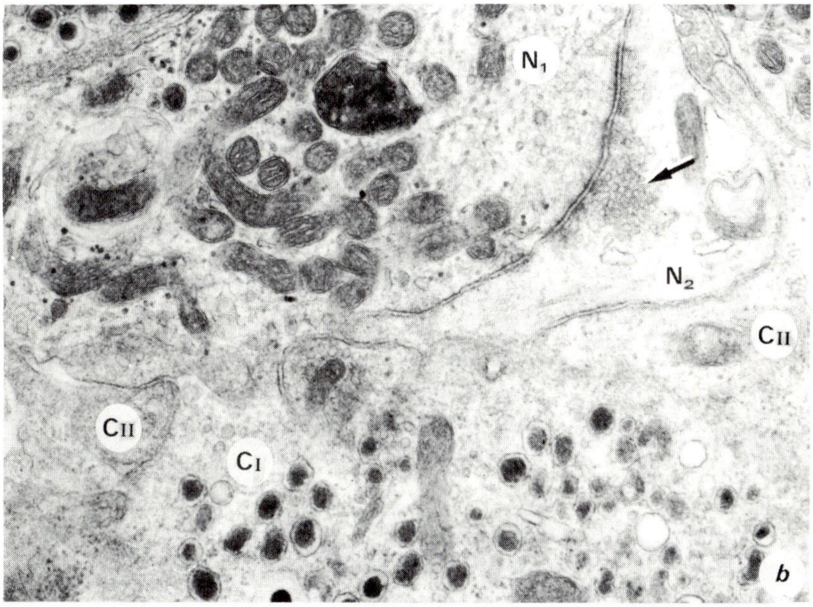

For explanation of Plate see p. 84

DISCUSSION

FILLENZ I got the impression that the dense-cored synaptic vesicles were smaller than those in the general population.

VERNA You are correct; the smaller ones are about 70 to 100 nm; those vesicles scattered in the cytoplasm of the Type I cell are 150 to 280 nm.

ACKER Dr H. Starlinger in our Institute, has measured the catecholamine content of carotid body tissue using fluorescent techniques. We have shown that this may differ from animal to animal by 500%: but in the same animal, the content differs by only 10–20%. My point is that you cannot really compare fluorescence or other features quantitatively except in the same animal.

VERNA My two pictures were of the same rabbit. Normal carotid body on the right; on the left, after extirpation of the superior cervical ganglion.

EYZAGUIRRE You find clear vesicles as well as dense-cored vesicles in the Type I cell?

VERNA Yes. But not always close to the synaptic membrane.

SAMPSON Have you seen whether hypoxic gas mixtures cause any changes in mitochondria?

VERNA Hypoxia no: but after NaCN the shape but not the number of mitochondria are affected.

ACKER We have tried antimycin or amytal. The mean oxygen consumption remains within control levels but changes with time.

EYZAGUIRRE Have you tried fixing the carotid body in high magnesium like Birks (1971) for the neuro-muscular junction?

VERNA No, I do not think anyone has done this.

PURVES How many of these axo-dendritic junctions have you seen and have you seen any nerve endings, in addition to those you describe as being sensory, which appear to be purely motor?

VERNA I have seen only a few axo-dendritic junctions and in many carotid bodies, none. I think that this may be because they are situated some distance from the Type I cell – probably before the branching of the nerve terminal. I have never seen any motor-like nerve endings upon Type I cells.

TORRANCE Have these junctions or anything like them been seen with light microscopy, for example by De Castro?

EYZAGUIRRE There are what appear to be junctions between nerves in some of De Castro's sections – particularly in the silver stained sections: but I cannot be sure of their composition. Certainly, De Castro did not comment upon them.

HOWE In your final slide, summarizing your findings, you describe the terminals making contact with the Type I cell as afferent. Could you justify this label for us?

VERNA First, the abundance of mitochondria which is present, is also present in most sensory terminals. Secondly, the arrangement of structures, particularly of the dense-cored vesicles in Type I cell suggests transmission in the direction indicated, that is, from Type I cell to nerve terminal. Finally, there is the presence of filamentous material in the nerve ending, which is usually considered as being post-synaptic ('synaptic web').

REFERENCE

BIRKS, R. I. (1971). Effects of stimulation on synaptic vesicles in sympathetic ganglia, as shown by fixation in the presence of Mg^{2+}. *J. Physiol.*, **216**, 26–28P.

A quantitative analysis of synaptic connections in the rat carotid body

DONALD M. McDONALD AND ROBERT A. MITCHELL

The Cardiovascular Research Institute and the Departments of Anatomy, Anesthesia, and Physiology, University of California, San Francisco, California 94143

De Castro (1928) postulated that the carotid body's glomus cells (Type I cells) are chemoreceptors which are innervated by afferent (sensory) nerves. Biscoe, Lall & Sampson (1970) questioned this hypothesis when they inferred from their data that glomus cells are innervated only by efferent nerves. They proposed that chemoreceptors are afferent nerves not ending on glomus cells, and that glomus cells are part of a feedback mechanism which regulates the sensitivity of these nerves. Questions raised by these two conflicting hypotheses led to our study of the carotid body. We examined: (1) ultrastructural characteristics and synaptic connections of glomus cells; (2) the source, proportion, ultrastructural characteristics, and synaptic connections of each type of nerve ending on glomus cells; and (3) ultrastructural changes produced by hypoxia in glomus cells and nerve endings.

We found that glomus cells are innervated by afferent axons from the glossopharyngeal nerve and a few (preganglionic) efferent axons from the cervical sympathetic trunk, but not by efferent axons from the glosso-pharyngeal nerve. Afferent nerves and glomus cells are interconnected by reciprocal synapses; efferent nerves are presynaptic to glomus cells. Hypoxia produces ultrastructural changes rapidly in afferent nerves but not in efferent nerves or glomus cells. These data, together with those from neurophysiological studies in which we examined the function of afferent nerves, glomus cells, and reciprocal synapses between them (Mitchell & McDonald, this Symposium), support our hypothesis that afferent nerves are chemoreceptors sensitive to hypoxia and glomus cells are dopaminergic interneurons. Reciprocal synapses complete an inhibitory feedback loop through which glomus cells adjust the sensitivity of the chemoreceptive nerves.

METHODS

Procedure for preparing carotid bodies for electron microscopy

Anaesthetized adult rats of the Long-Evans strain were connected by a tracheal cannula to a Harvard rodent respirator and ventilated with oxygen. We used two other gas mixtures (5 % O_2-6 % CO_2-89 % N_2 and 100 % N_2) when examining the effects of hypoxia. After a period of controlled ventilation, we inserted a cannula into the left ventricle and perfused a mixture of glutaraldehyde and hydrogen peroxide (Peracchia & Mittler, 1972) to fix the tissue. Less than 10 seconds elapsed between incising the ventricle and beginning the perfusion. After the perfusion, we removed the carotid bodies, treated them with osmic acid and then uranyl acetate, dehydrated them in methanol, and embedded them in epoxy resin. Sections of the carotid bodies on grids were stained with lead citrate and examined with a Siemens Elmiskop I electron microscope.

Quantitative methods

We used the diameter and concentration of synaptic vesicles and dense-cored vesicles to characterize glomus cells and nerve endings. We measured the diameter and number of vesicles in electron micrographs having a magnification of ×100 000 with a Zeiss Particle Size Analyser and measured areas of nerve endings and glomus cells on prints with a compensating polar planimeter or calculated them from the weight of the paper that they occupied.

RESULTS

Structural characteristics of glomus cells

The carotid body of the rat is composed of an anastomosing network of blood vessels which surround collections of glomus cells (Type I cells) and sheath cells (Type II cells). We found two types of glomus cells, both having a globular cell body 8–15 microns in diameter, and large vesicles with electron-dense cores. Some cells, which we call Type A, have dense-cored vesicles nearly 30 % larger and 150 % more numerous than those in the remainder, which we call Type B. The difference in size and concentration of dense-cored vesicles in Type A and B glomus cells is statistically significant (Table 1, Fig. 1). Type B cells typically have cytoplasmic processes which are more numerous, longer, thinner, and more branched than those of Type A cells. Type A and B cells and their respective dense-cored vesicles are illustrated in Plate 1*a–c*.

Morphological differences between Type A and B cells are sufficiently

Table 1. *Diameter and concentration of vesicles in sections of Type A and B glomus cells*

Type of glomus cell	Vesicle diameter (nm) A	Vesicle diameter (nm) B	Vesicle concentration* A	Vesicle concentration* B
Mean ± SD	116 ± 20	90 ± 11	3.4 ± 1.3	1.4 ± 0.6
P†	< 0.001		< 0.001	
Number of glomus cells counted	6	10	50	26
Number of vesicles measured	530	334	5254	1159

* Vesicles per μm^2 of section (glomus cell cytoplasm, excluding the nucleus).
† Unpaired t-test.

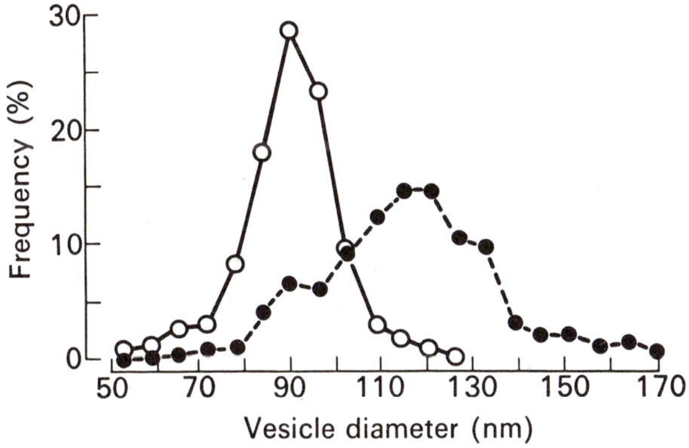

Fig. 1. This graph illustrates the difference in size of large dense-cored vesicles in Type A glomus cells (mean diameter 116 nm, broken line) and type B cells (mean diameter 90 nm, solid line). (From McDonald & Mitchell, 1974.)

prominent that the two types can be distinguished by electron microscopy without measurements of their dense-cored vesicles. In 19 carotid bodies which we examined, Type A cells comprised 51 ± 10 % (mean ± SD) of 3708 glomus cells. The remainder were Type B cells.

Source of nerves ending on glomus cells

To determine the source of nerves ending on glomus cells we cut nerves which enter the carotid body. We measured the effect of each lesion on glomus cell innervation by comparing the number of nerve endings on glomus cells in experimental carotid bodies with the number in controls. We counted all nerve endings on glomus cells visible by electron microscopy in a single section from each carotid body and counted all glomus cells whose nucleus appeared in the section (mean = 200 glomus cells per carotid body). The data for each carotid body are expressed as a ratio: the number of nerve endings to the number of glomus cells. To describe nerves ending on glomus cells, we use the term *afferent* for axons that conduct from the carotid body toward the central nervous system and *efferent* for axons that conduct in the opposite direction.

In 10 control carotid bodies the ratio of nerve endings to glomus cells was 1.0 ± 0.3 (mean $\pm$ SD). In carotid bodies from five rats whose glossopharyngeal nerve was cut central to its sensory (petrosal) ganglion with 6 hours to 25 days allowed for degeneration, the ratio did not deviate significantly from the control value (Fig. 2).* We found no degenerating nerve

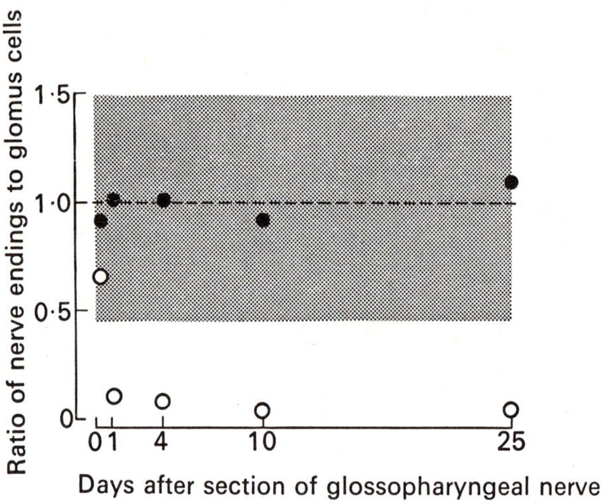

Fig. 2. This graph illustrates the difference in the effect on glomus cell innervation of cutting the glossopharyngeal nerve central to the petrosal ganglion (suprapetrosal, solid circles) and cutting the nerve peripheral to the ganglion (infrapetrosal, open circles). The broken line indicates the mean ratio of nerve endings to glomus cells in single sections from control carotid bodies. The stippled region indicates the 95 % confidence interval for the mean. (From McDonald & Mitchell, 1974.)

endings on glomus cells in these carotid bodies. However, the ratio was less than 10 % of control one day after the glossopharyngeal nerve was cut peripheral to the sensory ganglion, and 25 days after surgery, the ratio was less than 5 % of the control value (Fig. 2). Many nerve endings on glomus cells were degenerating 6 hours after the nerve was cut peripherally, and from 4 until 25 days after surgery the few nerve endings that remained on glomus cells appeared to be normal.

We conclude from these experiments that more than 95% of nerve

* In this group of rats we also cut the vagus nerve rostral to its superior ganglion and the rostral rootlets of the spinal accessory nerve.

endings on glomus cells are from afferent axons which have their cell bodies in the sensory ganglion of the glossopharyngeal nerve. The remainder are from axons that are not in the glossopharyngeal nerve; thus no nerve endings on glomus cells are from efferent axons in the glossopharyngeal nerve whose cell bodies are in the brain.

To determine if sympathetic axons are the source of nerve endings that remain after the glossopharyngeal nerve is cut, we removed a superior cervical sympathetic ganglion and cut the ipsilateral carotid sinus nerve in two rats. One day after the operation, the ratio of nerve endings to glomus cells was only 0.01 and at 20 days it was zero. We cut the superior cervical sympathetic trunk in another two rats and examined the carotid bodies 10 days later. We found no nerve endings on glomus cells which resembled those remaining intact after the glossopharyngeal nerve was cut peripheral to its sensory ganglion (the structure of such endings is described in the next section). We conclude from these experiments that the few nerve endings on glomus cells which do not come from axons in the glossopharyngeal nerve arise from (preganglionic) efferent axons in the cervical sympathetic trunk.

Structural characteristics of afferent and efferent nerve endings

We use the term *afferent nerve ending* for terminals on glomus cells of afferent axons which degenerate after the glossopharyngeal nerve is cut and *efferent nerve ending* for terminals on glomus cells of preganglionic sympa-

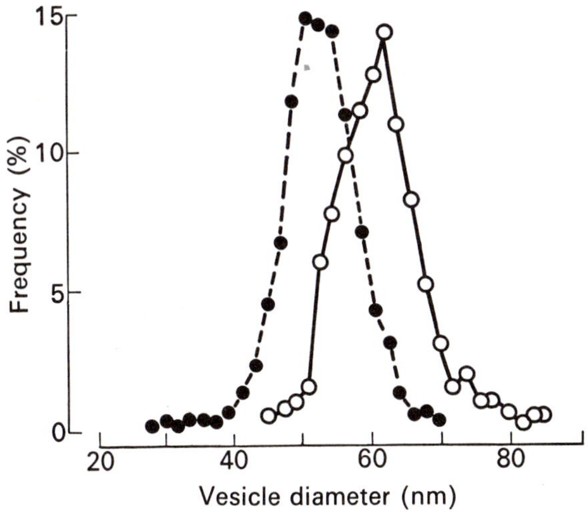

Fig. 3. This graph illustrates the difference in size of synaptic vesicles in afferent nerve endings (mean diameter 61 nm, solid line) and efferent nerve endings (mean diameter 53 nm, broken line). (From McDonald & Mitchell, 1974.)

thetic axons which do not degenerate after the glossopharyngeal nerve is cut.

Afferent nerve endings vary in size and shape: some are small boutons; others are large calices. Efferent nerve endings are also small boutons, and consequently cannot be distinguished by their size and shape alone. Although both afferent and efferent nerve endings contain synaptic vesicles which lack dense cores and large vesicles which have dense cores, vesicles in afferent and efferent nerve endings differ in size and concentration. Synaptic vesicles in afferent nerve endings are 15 % larger and 60 % less numerous than those in efferent nerve endings (Fig. 3, Plate 3*a*,*b*). Large dense-cored vesicles in afferent nerve endings are similar in size but 80 % less numerous than those in efferent nerve endings. The differences in size and concentration of vesicles in afferent and efferent nerve endings are

Table 2. *Diameter* of vesicles in sections of nerve endings on glomus cells*

Type of ending	Afferent	Efferent	*P*†
Synaptic vesicles	61 ± 7	53 ± 6	<0.001
Large dense-cored vesicles	95 ± 14	97 ± 11	NS
Number of endings	28	19	
Number of synaptic vesicles	1647	2539	
Number of large dense-cored vesicles	88	309	

* Outside diameter in nm (mean ± SD).
† Unpaired t-test for differences between afferent and efferent nerve endings; NS = not significantly different.

Table 3. *Concentration* of vesicles in sections of nerve endings on glomus cells*

Type of ending	Afferent	Efferent	*P*†
Synaptic vesicles	29 ± 14	76 ± 32	<0.001
Large dense-cored vesicles	2 ± 1	9 ± 3	<0.001
Number of endings	28	19	
Area per ending (μm^2)	2 ± 1	2 ± 1	NS

* Vesicles per μm^2 of section (mean ± SD).
† Unpaired t-test for differences between afferent and efferent endings; NS = not significantly different.

statistically significant (Tables 2, 3) and have a magnitude sufficient that the two types of endings can be distinguished in electron micrographs without quantitative determinations (Plate 2*a*,*b*).

Synaptic connections of nerve endings and glomus cells

Investigators have used various criteria to identify synapses in the carotid body. Some have called synapses regions in which synaptic vesicles accumulate near cytoplasmic densities attached symmetrically to apposing membranes (zonulae adherentia, Plate 4*b*,*d*,*f*) (Abbott, Daly & Howe, 1972; Hess & Zapata, 1972). In contrast, we identified synapses in the carotid body by the criteria used to identify chemical synapses in the central nervous system, i.e. regions where synaptic vesicles accumulate at asymmetrical junctions involving the pre- and post-synaptic membranes (Plate 4*a*,*c*,*e*) (Peters, Palay & Webster, 1970). Cytoplasmic densities divided into cone-shaped subunits (pre-synaptic dense projections) are attached to the pre-synaptic membrane (Plate 4*a*), and cytoplasmic dense material in the form of a band (post-synaptic web) lines the post-synaptic membrane.

The following data apply to synapses between nerve endings and Type A glomus cells. Of nearly 1400 nerve ending-glomus cell synapses which we identified, more than 99 % involved Type A cells; less than 1 % involved Type B cells. Consequently, we assume that few Type B cells are innervated by either afferent or efferent nerves.

Afferent nerve-glomus cell synapses

Afferent nerve endings are both pre- and post-synaptic to glomus cells. Where they are pre-synaptic, synaptic vesicles accumulate in endings near pre-synaptic dense projections (Plate 3*c*). Where they are post-synaptic, two types of vesicles associate with pre-synaptic dense projections in glomus cells (Plates 3*c*, 4*c*). One type is a large dense-cored vesicle (mean diameter = 116 nm) identical to those present throughout the glomus cell cytoplasm; the other type is a synaptic vesicle (mean diameter = 53 nm) which contains little or no electron-dense material; few of these accumulate at sites other than synapses. Where glomus cell bodies are pre-synaptic to afferent nerves, synaptic vesicles and large dense-cored vesicles are present in about equal concentration in the glomus cell at the synaptic junction (Plate 4*c*); but where glomus cell processes are pre-synaptic to afferent nerves, synaptic vesicles predominate over dense-cored vesicles. This feature makes glomus cell processes resemble nerve endings which are pre-synaptic to afferent nerves. However, we found no structures other than glomus cells which are pre-synaptic to afferent nerve endings. Consequently, our data do not support the observation of Verna (1973) that efferent nerves are pre-synaptic to afferent nerve endings.

Some afferent nerve endings and glomus cells are interconnected by reciprocal synapses. These are paired synaptic contacts: at one synapse an afferent nerve ending is pre-synaptic to a glomus cell, and at an adjacent synapse the same ending is post-synaptic to the same glomus cell (Plate 3*c*).

In single sections about 20% of afferent nerve endings are pre-synaptic to glomus cells, 35% are post-synaptic, and 5% form reciprocal synapses. The remainder have no synapses in the portion of the ending appearing in the section. We could not determine what fraction of individual synapses are part of reciprocal synapses because we did not examine serial sections.

Large dense-cored vesicles in glomus cells may store catecholamines (Chen & Yates, 1969), but the nature of the contents of synaptic vesicles in these cells is unknown. To test the ability of the synaptic vesicles to take up catecholamines, we used an analogue of dopamine, *5-hydroxydopamine* (5-OHDA). This substance has been used to mark noradrenergic nerves, which take up 5-OHDA in preference to noradrenaline (Thoenen & Tranzer, 1971). In animals treated with 5-OHDA, synaptic vesicles in noradrenergic nerves become filled with an electron-dense precipitate easily distinguished from the small dense cores which they have in the absence of the drug (Chiba, 1973). Synaptic vesicles in cholinergic nerves do not have dense cores either before or after 5-OHDA treatment (Chiba, 1973).

We infused 5-OHDA hydrobromide (200 mg kg^{-1}) (Regis Chemical Company, Chicago, Illinois) intravenously over 30 minutes into anaesthetized rats, then prepared the carotid bodies for electron microscopy using our standard procedure. These experiments yielded striking results. Nearly 40% of synaptic vesicles in glomus cells (at 24 synapses which we analysed) contained a dense precipitate in rats receiving 5-OHDA (Plate 5*b*,*d*). The amount of precipitate varied from a small patch adjacent to the inner leaflet of the vesicle membrane to a mass filling the vesicle. Some small vesicles near the Golgi apparatus in glomus cells also contained a precipitate. In rats not receiving 5-OHDA less than 1% of synaptic vesicles in glomus cells contained a dense core (Plate 5*a*,*c*). Large dense-cored vesicles in rats receiving the drug did not appear more dense than those in controls. Synaptic vesicles in afferent and efferent nerve endings were devoid of precipitate.

Efferent nerve–glomus cell synapses

Efferent nerve endings from pre-ganglionic sympathetic axons are pre-synaptic but not post-synaptic to glomus cells (Plate 2*b*). Although some glomus cells are connected synaptically to both afferent and efferent nerves, most are probably innervated only by afferent nerves. Efferent nerve endings do not have direct synaptic connections with afferent nerve endings.

Glomus cell–glomus cell synapses

Some Type A and Type B glomus cells are pre-synaptic or post-synaptic to other glomus cells. We found synapses between adjacent Type A cells, between Type B cells, and between Type A and B cells. Synapses where Type A cells are pre-synaptic to other glomus cells resemble those between A cells and afferent nerve endings: both large dense-cored vesicles and synaptic vesicles accumulate at synaptic junctions (Plate 4, compare *c* with *e*). Although few Type B cells are connected by synapses to afferent or efferent nerve endings, many (perhaps more Type B than Type A cells) are pre-synaptic to other glomus cells. At these synapses dense-cored vesicles typical of those present throughout the cytoplasm of Type B cells are concentrated near synaptic junctions. Some synapses interconnecting glomus cells are reciprocal.

Structural changes produced by hypoxia in glomus cells and nerve endings

We examined the effects of hypoxia on the ultrastructure of glomus cells and nerve endings in anaesthetized rats, breathing either 100 % N_2 or 5 % O_2-6 % CO_2-89 % N_2. We examined the effects of brief but severe hypoxia (100 % N_2 for 1 minute, producing a P_{a,O_2} of less than 15 torr) and prolonged but less severe hypoxia (5 % O_2 for 30 minutes, producing a P_{a,O_2} of about 35 torr). We compared carotid bodies of rats made hypoxic with those of rats breathing oxygen.

Changes in afferent nerve endings

Conspicuous changes appear in the ultrastructure of afferent nerve endings coincident with their increased activity produced by brief, severe hypoxia (Plate 6, compare *a* with *b*). Synaptic vesicles are distributed rather evenly in the cytoplasm of afferent nerve endings in severely hypoxic rats, while they are often aggregated in rats breathing oxygen. Mitochondria are markedly swollen and have a dispersed matrix after 1 minute severe hypoxia; in contrast, they are small and have a dense, compact matrix in hyperoxic rats (Plate 6, compare *a* with *b*).

Prolonged but less severe hypoxia produces ultrastructural changes in afferent nerve endings qualitatively similar but quantitatively different from those produced by brief, severe hypoxia (compare Plate 6 with 7). After 30 minutes of hypoxia, the number of synaptic vesicles in afferent nerve endings is reduced to less than half of the control value (Fig. 4). Mitochondria have a size and matrix-density between that of rats breathing oxygen and that of rats breathing nitrogen for 1 minute (Plate 7). These changes are reversible. Synaptic vesicles and mitochondria in afferent nerve endings of rats made hypoxic for 30 minutes, and then ventilated with oxygen for 10 minutes, are indistinguishable from controls. The glossopharyngeal nerve need not be intact for hypoxia to produce changes in afferent nerve endings (before they begin to degenerate following axonal transection) (Fig. 4). Hypoxia does not change the concentration of large dense-cored vesicles or mitochondria in afferent nerve endings.

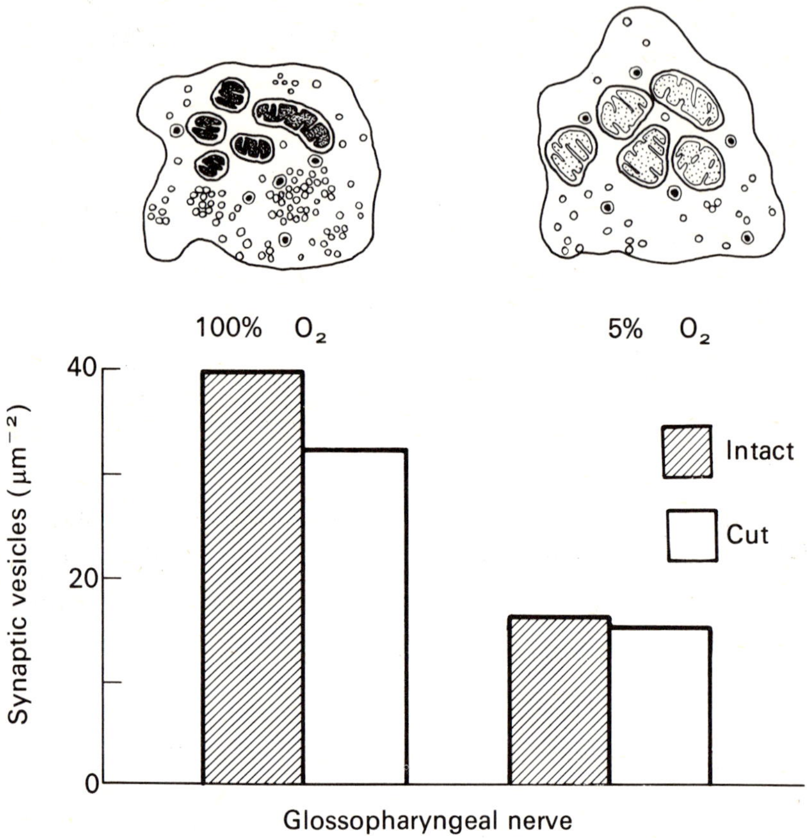

Fig. 4. This graph illustrates that the concentration of synaptic vesicles in afferent nerve endings in rats breathing 5 % O_2 is less than half of that in rats breathing 100 % O_2. This reduction in synaptic vesicles is not prevented by cutting the glossopharyngeal nerve and therefore must result from the direct action of hypoxia on the carotid body.

Changes in efferent nerve endings

In sharp contrast to afferent nerve endings, efferent nerve endings show neither a decrease in the concentration of synaptic vesicles nor mitochondrial swelling in response to hypoxia.

Changes in glomus cells

Glomus cells also differ from afferent nerve endings in their response to hypoxia. Brief, severe hypoxia does not change the ultrastructure of glomus cells detectably (Plate 6*b*); prolonged but less severe hypoxia produces slight enlargement and decreased matrix-density of mitochondria (Plate 7). The concentration of dense-cored vesicles in Type A glomus cells does not change in either state.*

DISCUSSION

Our studies show that most nerve endings on glomus cells in the rat carotid body are from afferent axons in the glossopharyngeal nerve. The remainder are from (pre-ganglionic) efferent axons in the cervical sympathetic trunk. We obtained no evidence in support of the hypothesis of Biscoe *et al.* (1970) that glomus cells are innervated by efferent axons from the glossopharyngeal nerve. The gradual disappearance over many months of nerve endings from glomus cells which Biscoe's group observed after suprapetrosal section of the glossopharyngeal nerve probably resulted from a process different from the one we studied.

Some investigators have concluded that nerve endings on glomus cells are from efferent rather than afferent axons because they contain synaptic vesicles and are pre-synaptic to glomus cells (Ábrahám, 1968; Biscoe & Stehbens, 1966; Höglund, 1967; Kobayashi, 1971). Our data indicate, however, that their cell bodies are in the sensory ganglion of the glossopharyngeal nerve. Consequently, we regard them as *dendrites* of sensory neurons (using the nomenclature of Bodian, 1962), analogous to presynaptic dendrites in other parts of the nervous system (Reese & Shepherd, 1972).

We found that afferent nerve endings are post-synaptic to glomus cells, in addition to being pre-synaptic to them. Other investigators have found that nerve endings (thought by them to be afferent) are post-synaptic to glomus cells in the mouse (Kobayashi & Uehara, 1970), guinea-pig (Kobayashi, 1971; Kondo, 1971), and rabbit (Verna, 1973). In addition, we have morphological evidence of similar synapses in the cat, although they occur with substantially less frequency than in the rat. Apparently, junctional specializations of glomus cell–afferent nerve synapses are not well-developed in some species. However, some synapses in other parts of the peripheral nervous system, e.g. autonomic nerve endings on vascular or epididymal smooth muscle, also lack junctional specializations (Grillo, 1966).

Our finding that afferent nerve endings and glomus cells are interconnected by reciprocal synapses integrates the observations that these

* Prolonged but less severe hypoxia was produced in these rats by ventilating them with 9 % O_2-6 % CO_2-85 % N_2 for 30 minutes. The mean concentration of dense-cored vesicles in micrographs of 35 Type A glomus cells from two hypoxic rats was compared with that of vesicles in the same number of glomus cells from two control rats.

structures are both pre- and post-synaptic to one another. Although reciprocal synapses have not been reported previously to exist in the peripheral nervous system, they are present in several parts of the central nervous system (Reese & Shepherd, 1972).

We are intrigued by the similarity of reciprocal synapses in the olfactory bulb and retina to those in the carotid body. In each, reciprocal synapses interconnect dendrites or perikarya of two types of cells, one a neuron carrying sensory information (mitral cell, bipolar cell, or afferent nerve), and the other an axonless interneuron (granule cell, amacrine cell, or glomus cell). Support for the concept that glomus cells are interneurons comes not only from their synaptic connections (their synaptic input coming from afferent and efferent nerves and their synaptic output going to afferent nerves) but also from their similarities to interneurons (small intensely fluorescent (SIF) cells) in sympathetic ganglia. SIF cells are post-synaptic to incoming pre-ganglionic axons and pre-synaptic to sympathetic ganglion cells (Matthews & Raisman, 1969) and are considered by Libet (1970) to be dopaminergic interneurons which inhibit ganglion cells. Glomus cells and SIF cells contain large dense-cored vesicles of the same size, and some glomus cells, like SIF cells, contain dopamine (Chiocchio, King, Carballo & Angelakos, 1971) which may be released at synapses on afferent nerve endings.

Both glomus cell–afferent nerve and SIF cell–ganglion cell synapses characteristically have large dense-cored vesicles and synaptic vesicles near synaptic junctions. The large dense-cored vesicles may contain catecholamines (Chen & Yates, 1969), and our studies of the uptake of 5-OHDA suggest that the synaptic vesicles also store catecholamines. Synaptic vesicles in glomus cells do not have a dense core in the absence of 5-OHDA; thus they resemble those in dopaminergic neurons in the brain Hökfelt, 1971) and differ from those in noradrenergic neurons in the peripheral nervous system (Fillenz, 1971).

Extending the concept that afferent nerves are connected by reciprocal synapses to interneurons, we postulate that afferent nerves are chemoreceptors sensitive to hypoxia, and that glomus cells modulate the sensitivity of these nerves through reciprocal synapses. We consider the ultrastructural changes which hypoxia produces selectively in afferent nerve endings as evidence that they are stimulated by hypoxia. We do not believe that these changes are toxic effects of hypoxia because they occur also in endings of nerves stimulated electrically in the absence of hypoxia (Friesen & Khatter, 1971; Heuser & Reese, 1973). The mitochondrial swelling we observed may result from a change in energy state associated with increased neural activity (Hackenbrock, 1972). Reduction in the number of synaptic vesicles presumably occurs during neural activity when the rate of transmitter release (by exocytosis of synaptic vesicles) exceeds the rate of formation of new vesicles.

There have not been previous reports of effects of hypoxia on afferent nerve endings. Blümcke, Rode, & Niedorf (1967) reported that hypoxia depletes glomus cells of their large dense-cored vesicles, but to obtain depletion they put rats into an atmosphere of 2.5 % oxygen for 10 minutes. They found that 5 % O_2 for up to 20 minutes did not produce depletion. These data together with those from our own studies indicate that the ultrastructure of afferent nerve endings is more sensitive to hypoxia-induced change than is the ultrastructure of glomus cells.

Although we infer from our data that hypoxia stimulates afferent nerve endings, the data do not show whether it does so directly or indirectly via glomus cells. However, the hypothesis that afferent nerve endings are themselves chemoreceptors is supported by our finding that regenerated axons from the carotid sinus nerve are stimulated by hypoxia in the absence of glomus cells (Mitchell, Sinha & McDonald, 1972).

If afferent nerves are chemoreceptors, what do glomus cells do? Several lines of evidence from neurophysiological experiments support the hypothesis that glomus cells regulate the activity of chemoreceptive nerves by releasing dopamine (reviewed by Mitchell & McDonald, this Symposium). Several investigators have shown that antidromic electrical stimulation of most of the carotid sinus nerve decreases chemoreceptor activity recorded from a small slip of the nerve (Fidone & Sato, 1970; Sampson & Biscoe, 1970; Neil & O'Regan, 1971; Goodman, 1973). Inhibition of chemoreceptors produced in this manner ('efferent inhibition') is mimicked by dopamine injected intra-arterially and blocked by drugs which abolish the inhibitory action of dopamine (Sampson, 1972, and this Symposium). Drugs which block electrically-evoked inhibition also increase the sensitivity of chemoreceptors to hypoxia (Mitchell & McDonald, this Symposium).

Electrically-evoked inhibition of chemoreceptors, although itself an artificial phenomenon, is probably based on the activation of a feedback circuit which is a normal component of the carotid body. We postulate that the following events are initiated by the electrical stimulus: (1) afferent nerves fire antidromically; (2) afferent nerve endings release a transmitter at synapses which depolarizes glomus cells; (3) glomus cells release dopamine at synapses; and (4) dopamine decreases the sensitivity of afferent nerve endings to hypoxia. Fig. 5 summarizes neural circuits which involve glomus cells and nerves which end on them. The feedback circuit between afferent nerve endings and glomus cells may be responsible for the

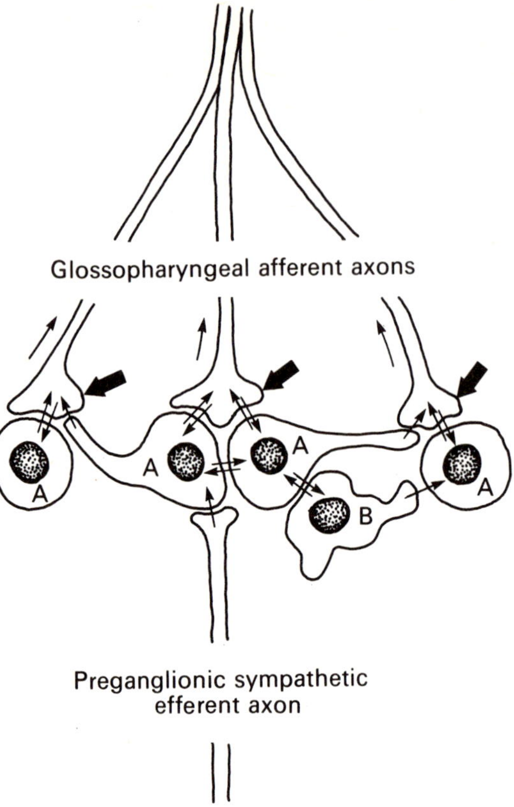

Fig. 5. This diagram summarizes neural circuits involving glomus cells, afferent nerves, and efferent nerves in the rat carotid body. The direction of nerve conduction and synaptic transmission are indicated by thin arrows. Reciprocal synapses, indicated by two arrows pointing in opposite directions, interconnect afferent nerve endings and glomus cells and also adjacent glomus cells (A, Type A; B, Type B). We postulate that hypoxia stimulates afferent nerve endings (thick arrows) and that glomus cells are interneurons which regulate the sensitivity of these nerve endings.

non-linear sensitivity of chemoreceptors to hypoxia (Mitchell & McDonald, this Symposium).

SUMMARY

1. In an analysis of the ultrastructure of the rat carotid body we found two types of glomus cells, which are present in about equal proportion. Large dense-cored vesicles in Type A glomus cells have a mean diameter of 116 nm, while those in Type B have a mean diameter of 90 nm.
2. Two types of nerves end on Type A glomus cells; few if any nerves end

on Type B. More than 95 % of nerve endings on glomus cells come from afferent axons in the glossopharyngeal nerve. The remainder come from (pre-ganglionic) efferent axons in the sympathetic trunk. None comes from efferent axons in the glossopharyngeal nerve.

3. Afferent nerve endings are bouton-like or caliceal in shape and have synaptic vesicles with a mean diameter of 61 nm. They also contain a few large dense-cored vesicles. Some regions of afferent nerve endings are pre-synaptic to glomus cells, some are post-synaptic, and some form reciprocal synapses.
4. Efferent nerve endings have the shape of boutons and contain synaptic vesicles with a mean diameter of 53 nm and frequent large dense-cored vesicles.
5. Both Type A and B glomus cells are interconnected by synapses.
6. At synapses involving Type A cells both large dense-cored vesicles and clear-cored synaptic vesicles accumulate near pre-synaptic junctional specializations. Because the synaptic vesicles acquire dense cores in rats treated with 5-hydroxydopamine, they can probably take up and store catecholamines.
7. Hypoxia produces changes in the ultrastructure of afferent nerve endings before it does in efferent nerve endings or glomus cells. These changes may reflect the increased activity of afferent nerves produced by hypoxia.
8. We postulate that afferent nerve endings are chemoreceptors stimulated by hypoxia. Glomus cells are dopaminergic interneurons which are interconnected with afferent nerve endings by reciprocal synapses. Through an inhibitory feedback loop based on reciprocal synapses, glomus cells modulate the sensitivity of chemoreceptive nerves.

We thank Mr Richard W. Blewett and Ms Dorothy Herbert for their technical assistance in the laboratory, and Mr Blewett for his editorial assistance on the manuscript.

This work was supported in part by NIH Grants GM 15571, 1-FO-3-HD-45487, HL-06285, and Pulmonary SCOR HL-14201 and 5-K3-1319411 from the US Public Health Service.

REFERENCES

Abbott, C. P., Daly, M. de B. & Howe, A. (1972). Early ultrastructural changes in the carotid body after degenerative section of the carotid sinus nerve in the cat. *Acta anatomica*, **83**, 161–85.

Ábrahám, A. (1968). Electron microscopic investigations on the human carotid body (preliminary communication). *Z. Zellforsch. mikroskop. Anat.*, **79**, 309–15.

Biscoe, T. J., Lall, A. & Sampson, S. R. (1970). Electron microscopic and electrophysiological studies on the carotid body following intracranial section of the glossopharyngeal nerve. *J. Physiol., Lond.*, **208**, 133–52.

Biscoe, T. J. & Stehbens, W. E. (1966). Ultrastructure of the carotid body. *J. Cell Biol.*, **30**, 563–78.

BLÜMCKE, S., RODE, J. & NIEDORF, H. R. (1967). The carotid body after oxygen deficiency. *Z. Zellforsch. mikroskop. Anat.*, **80**, 52–77.

BODIAN, D. (1962). The generalized vertebrate neuron. *Science*, **137**, 323–26.

CHEN, I-L. & YATES, R. D. (1969). Electron microscopic radioautographic studies of the carotid body following injections of labeled biogenic amine precursors. *J. Cell Biol.*, **42**, 794–803.

CHIBA, T. (1973). Electron microscopic and histochemical studies on the synaptic vesicles in mouse vas deferens and atrium after 5-hydroxydopamine administration. *Anat. Rec.*, **176**, 35–48.

CHIOCCHIO, S. R., KING, M. P., CARBALLO, L. & ANGELAKOS, E. T. (1971). Monoamines in the carotid body cells of the cat. *J. Histochem. Cytochem.*, **19**, 621–6.

DE CASTRO, F. (1928). Sur la structure et l'innervation du sinus carotidien de l'homme et des mammiferès. Nouveaux faits sur l'innervation et la fonction du glomus caroticum. *Trav. Lab. Rech. Biol., Université de Madrid*, **25**, 330–80.

FIDONE, S. J. & SATO, A. (1970). Efferent inhibition and antidromic depression of chemoreceptor A-fibers from the cat carotid body. *Brain Research*, **22**, 181–93.

FILLENZ, M. (1971). Fine structure of noradrenaline storage vesicles in nerve terminals of the rat vas deferens. *Phil. Trans. R. Soc., Ser. B*, **261**, 319–23.

FRIESEN, A. J. D. & KHATTER, J. C. (1971). Effect of stimulation on synaptic vesicles in the superior cervical ganglion of the cat. *Experientia*, **27**, 285–7.

GOODMAN, N. W. (1973). Efferent control of arterial chemoreceptors mediated by glossopharyngeal fibres and artifacts introduced by stimulation techniques. *J. Physiol., Lond.*, **230**, 295–311.

GRILLO, M. A. (1966). Electron microscopy of sympathetic tissues. *Pharmacol. Rev.*, **18**, 387–99.

HACKENBROCK, C. R. (1972). States of activity and structure in mitochondrial membranes. *Ann. N. Y. Acad. Sci.*, **195**, 492–505.

HESS, A. & ZAPATA, P. (1972). Innervation of the cat carotid body: normal and experimental studies. *Fedn Proc.*, **31**, 1365–82.

HEUSER, J. E. & REESE, T. S. (1973). Evidence for recycling of synaptic vesicle membrane during transmitter release at the frog neuromuscular junction. *J. Cell Biol.*, **57**, 315–44.

HÖGLUND, R. (1967). An ultrastructural study of the carotid body of horse and dog. *Z. Zellforsch. mikroskop. Anat.*, **76**, 568–76.

HÖKFELT, T. (1971). Ultrastructural localization of intraneuronal monoamines; some aspects on methodology. In *Histochemistry of Nervous Transmission*, ed. O. ERÄNKÖ, *Progress in Brain Research*, **34**, 213–22, Amsterdam: Elsevier.

KOBAYASHI, S. (1971). Comparative cytological studies of the carotid body. 2. Ultrastructure of the synapses on the chief cell. *Arch. Histol. Jap.*, **33**, 397–420.

KOBAYASHI, S. & UEHARA, M. (1970). Occurrence of afferent synaptic complexes in the carotid body of the mouse. *Arch. Histol. Jap.*, **32**, 193–201.

KONDO, H. (1971). An electron microscopic study on innervation of the carotid body of guinea pig. *J. Ultrastruct. Res.*, **37**, 544–62.

LIBET, B. (1970). Generation of slow inhibitory and excitatory postsynaptic potentials. *Fedn Proc.*, **29**, 1945–56.

MATTHEWS, M. R. & RAISMAN, G. (1969). The ultrastructure and somatic efferent synapses of small granule-containing cells in the superior cervical ganglion. *J. Anat.*, **105**, 255–82.

MCDONALD, D. M. & MITCHELL, R. A. (1974). The innervation of glomus cells, ganglion cells, and blood vessels in the rat carotid body: a quantitative ultrastructural analysis. *J. Neurocytol.* (in press).

MITCHELL, R. A., SINHA, A. K. & MCDONALD, D. M. (1972). Chemoreceptive properties of of regenerated endings of the carotid sinus nerve. *Brain Research*, **43**, 681–5.

NEIL, E. & O'REGAN, R. G. (1971). The effects of electrical stimulation of the distal end of the cut sinus and aortic nerves on peripheral arterial chemoreceptor activity in the cat. *J. Physiol., Lond.*, **215**, 15–32.

PERACCHIA, C. & MITTLER, B. S. (1972). Fixation by means of glutaraldehyde-hydrogen peroxide reaction products. *J. Cell Biol.*, **53**, 234–8.

PETERS, A., PALAY, S. L. & WEBSTER, H. DE F. (1970). *The Fine Structure of the Nervous*

System, New York, Evanston & London: Hoeber Medical Division, Harper & Row.

REESE, T. S. & SHEPHERD, G. M. (1972). Dendro-dendritic synapses in the central nervous system. In *Structure and Function of Synapses*, pp. 121–36 eds. G. D. PAPPAS & D. P. PURPURA, New York: Raven Press.

SAMPSON, S. R. (1972). Mechanism of efferent inhibition of carotid body chemoreceptors in the cat. *Brain Research*, **45**, 266–70.

SAMPSON, S. R. & BISCOE, T. J. (1970). Efferent control of the carotid body chemoreceptor. *Experientia*, **26**, 261–2.

THOENEN, H. & TRANZER, J. P. (1971). Functional importance of subcellular distribution of false adrenergic transmitters. In *Histochemistry of Nervous Transmission*, ed. O. ERÄNKÖ, *Progress in Brain Research*, **34**, 223–36, Amsterdam: Elsevier.

VERNA, A. (1973). Terminaisons nerveuses afférentes et efférentes dans le glomus carotidien du lapin. *J. Microsc., Paris*, **16**, 299–308.

EXPLANATION OF PLATES

PLATE 1

(*a*) This electron micrograph illustrates at low magnification two types of glomus cells in the rat carotid body. The Type A glomus cell (below) contains many prominent large dense-cored vesicles. The Type B cell (above) contains fewer large dense-cored vesicles, and they are less prominent than those in Type A cells. Scale = 1 μm.
(*b*) Large dense-cored vesicles (mean diameter 90 nm) from a Type B cell. Scale = 0.1 μm.
(*c*) Large dense-cored vesicles (mean diameter 116 nm) and a few synaptic vesicles (mean diameter 53 nm) from a Type A cell. Scale = 0.1 μm. (From McDonald & Mitchell, 1974.)

PLATE 2

(*a*) An afferent nerve ending (A) with scattered synaptic vesicles is shown next to a portion of a Type A glomus cell (G). Large dense-cored vesicles and a few synaptic vesicles accumulate in the glomus cell near the nerve ending. S, nucleus of a sheath cell. Scale = 0.5 μm.
(*b*) An efferent nerve ending (E) with densely-packed synaptic vesicles is pre-synaptic to a Type A glomus cell (G). S, nucleus of a sheath cell. Scale = 0.5 μm. (From McDonald & Mitchell, 1974.)

PLATE 3

(*a*) This electron micrograph illustrates at high magnification synaptic vesicles (mean diameter 61 nm) in an afferent nerve ending. Scale = 0.1 μm.
(*b*) This micrograph shows at high magnification the densely-packed synaptic vesicles (mean diameter 53 nm) and frequent large dense-cored vesicles (mean diameter 97 nm) in an efferent nerve ending. Scale = 0.1 μm.
(*c*) This afferent nerve ending (A) is interconnected with a Type A glomus cell (G) by a reciprocal synapse (arrows). Scale = 0.1 μm. (From McDonald & Mitchell, 1974.)

PLATE 4

(*a*) Large dense-cored vesicles and synaptic vesicles accumulate in a Type A glomus cell near pre-synaptic dense projections which are part of a synaptic junction.
(*b*) Such symmetrical junctions as these (zonulae adherentia) join glomus cells and afferent nerve endings and also join adjacent glomus cells. We do not consider them synaptic junctions.
(*c*) A synapse between a Type A glomus cell (above) and an afferent nerve ending.
(*d*) Zonulae adherentia (arrows) join a Type A glomus cell (above) and an afferent nerve ending.

(*e*) A synapse between a Type A glomus cell (above) and a Type B cell.
(*f*) Zonulae adherentia (arrows) join two Type A glomus cells.
Scale for (*a*)–(*f*) (located in *f*) = 0.1 μm.

PLATE 5

(*a*) Synaptic vesicles without dense cores accumulate at a synapse between a Type A glomus cell (above) and an afferent nerve ending in a control rat; compare with (*b*).
(*b*) Synaptic vesicles containing a dense precipitate (arrows) are in a Type A glomus cell near a synaptic junction in a rat which received 5-hydroxydopamine.
(*c*) A group of clear-cored synaptic vesicles and large dense-cored vesicles in a Type A glomus cell of a control rat; compare with (*d*).
(*d*) A group of synaptic vesicles, many of which contain a dense precipitate, and large dense-cored vesicles in a Type A glomus cell of a rat treated with 5-hydroxydopamine.
Scale for (*a*)–(*d*) (located in (*d*)) = 0.1 μm.

PLATE 6

(*a*) This micrograph shows an afferent nerve ending (A) from a rat ventilated with 100 % O_2 for 30 minutes before fixation. Some synaptic vesicles in the endings are clustered. Mitochondria are small and have a dense matrix. Portions of Type A glomus cells are next to the ending. Scale = 0.5 μm.
(*b*) This micrograph shows an afferent nerve ending (A) from a rat ventilated with 100 % N_2 for 1 min before fixation. Synaptic vesicles in the ending are dispersed. Mitochondria are swollen and have a pale matrix. Mitochondria in portions of Type A glomus cells which are adjacent to the ending are similar to those in the control (*a*). Scale = 0.5 μm.

PLATE 7

The afferent nerve ending (A) in this micrograph is from a rat which was ventilated with 5 % O_2-6 % CO_2-89 % N_2 for 30 minutes before fixation. In the ending synaptic vesicles are reduced in number and mitochondria are slightly enlarged and have a pale matrix (compare with control, Plate 6*a*). Mitochondria in portions of Type A glomus cells which surround the ending also have a pale matrix. Scale = 0.5 μm.

PLATE I

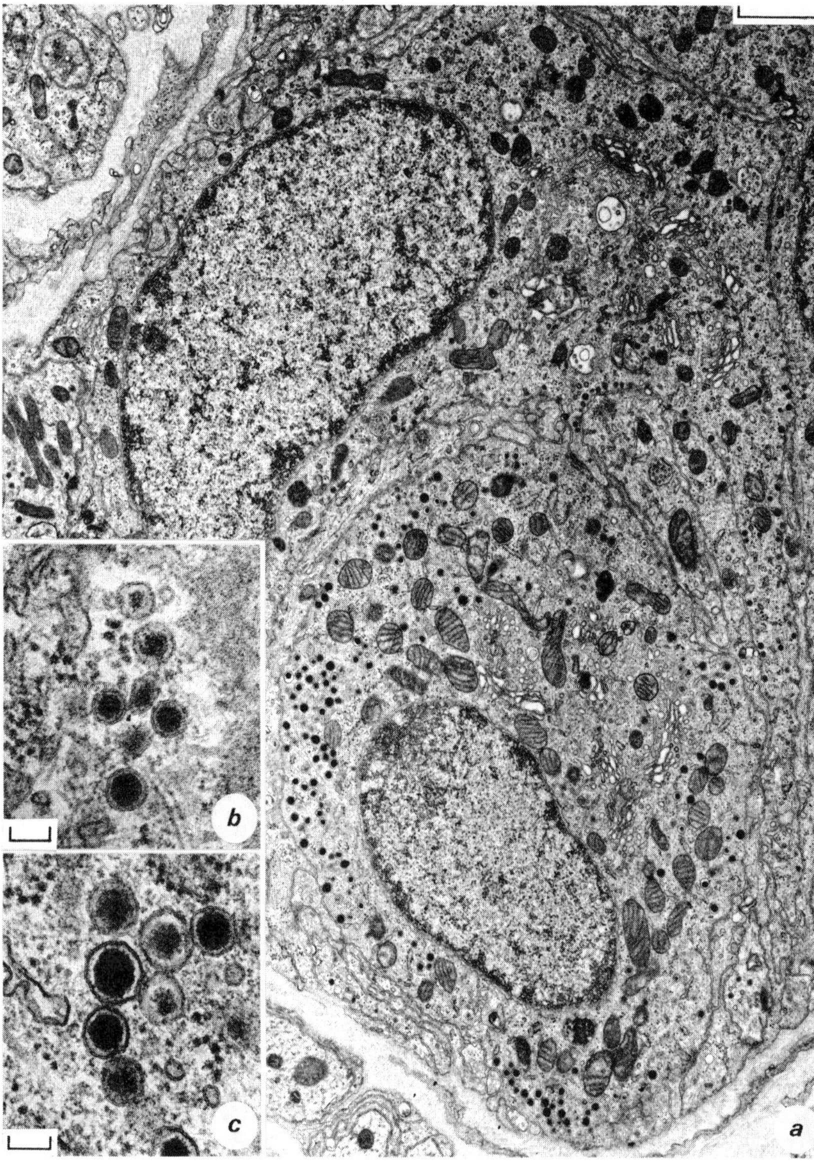

For explanation of Plate see p. 117

PLATE 2

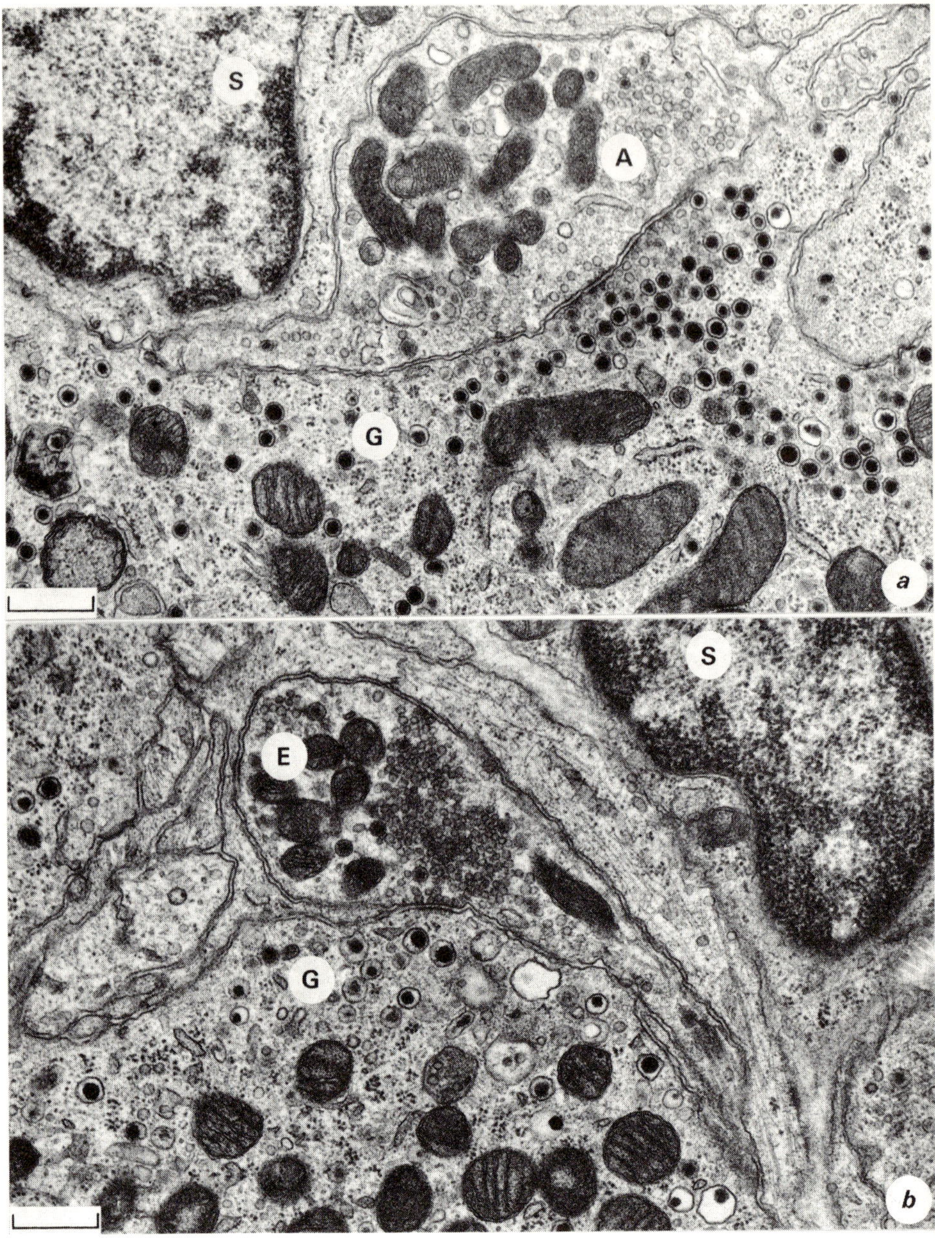

For explanation of Plate see p. 117

PLATE 3

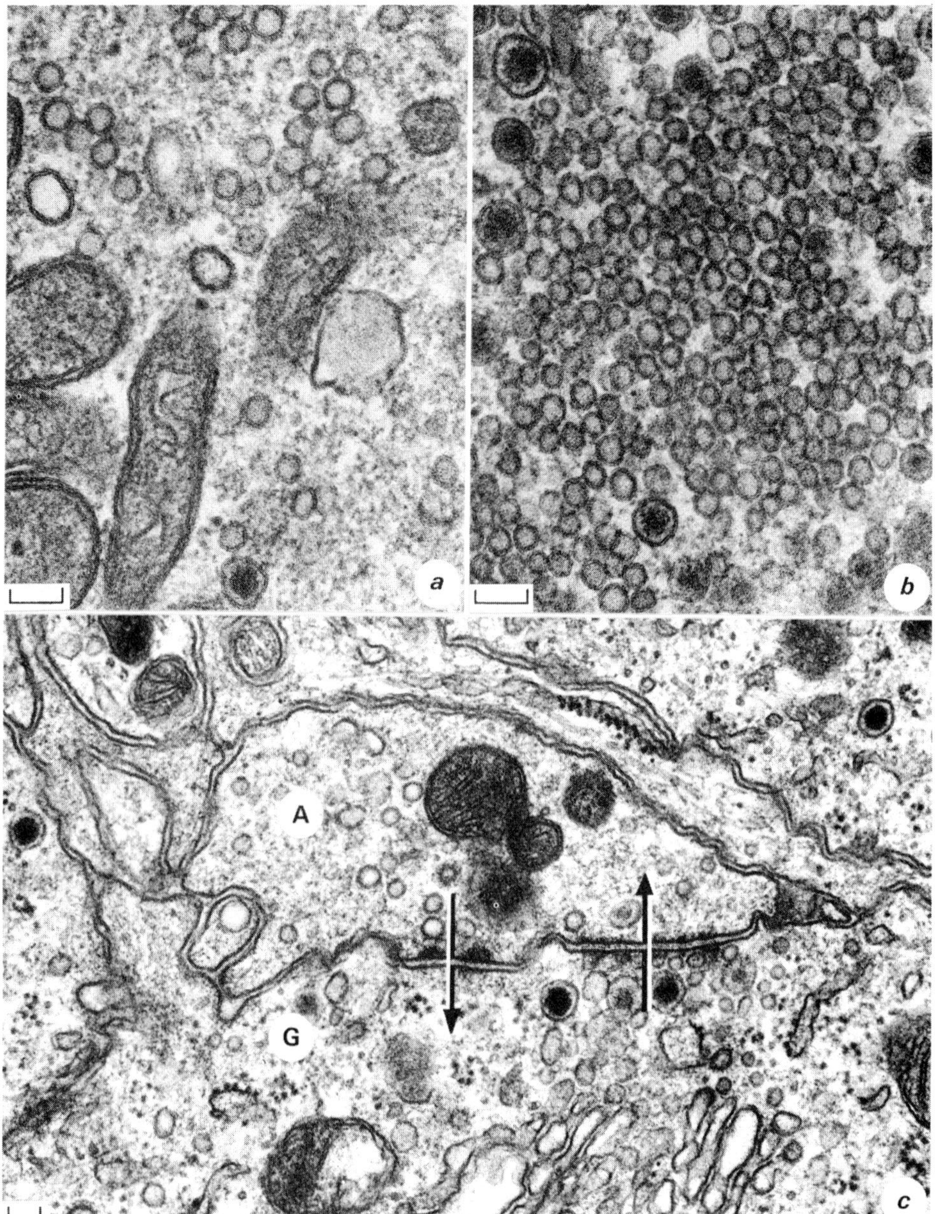

For explanation of Plate see p. 117

PLATE 4

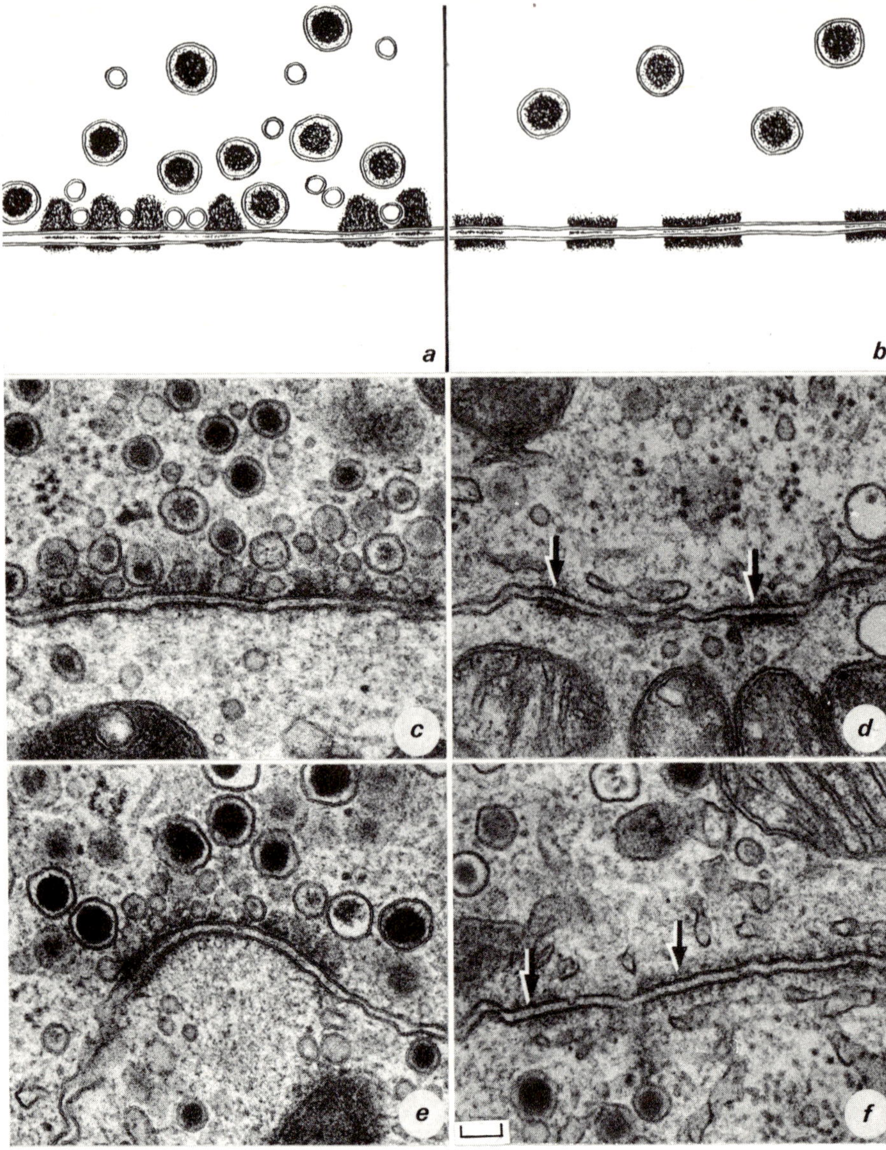

For explanation of Plate see pp. 117–18

PLATE 5

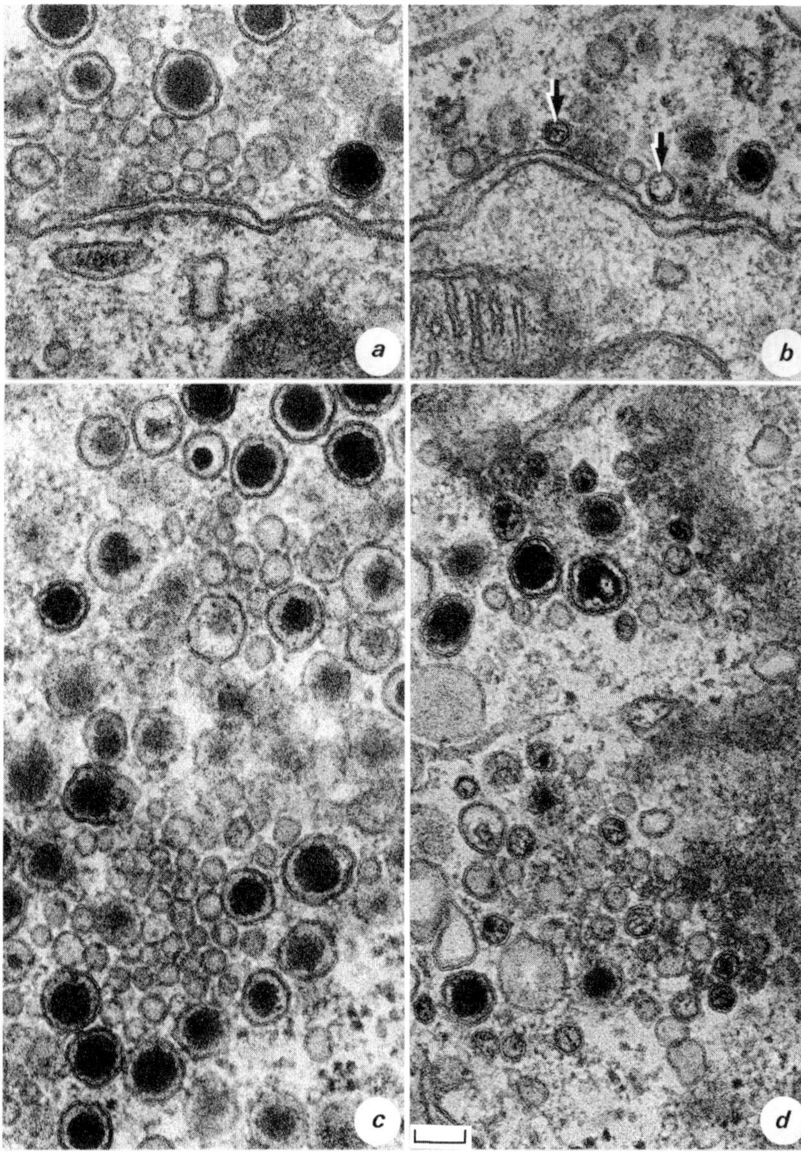

For explanation of Plate see p. 117

PLATE 6

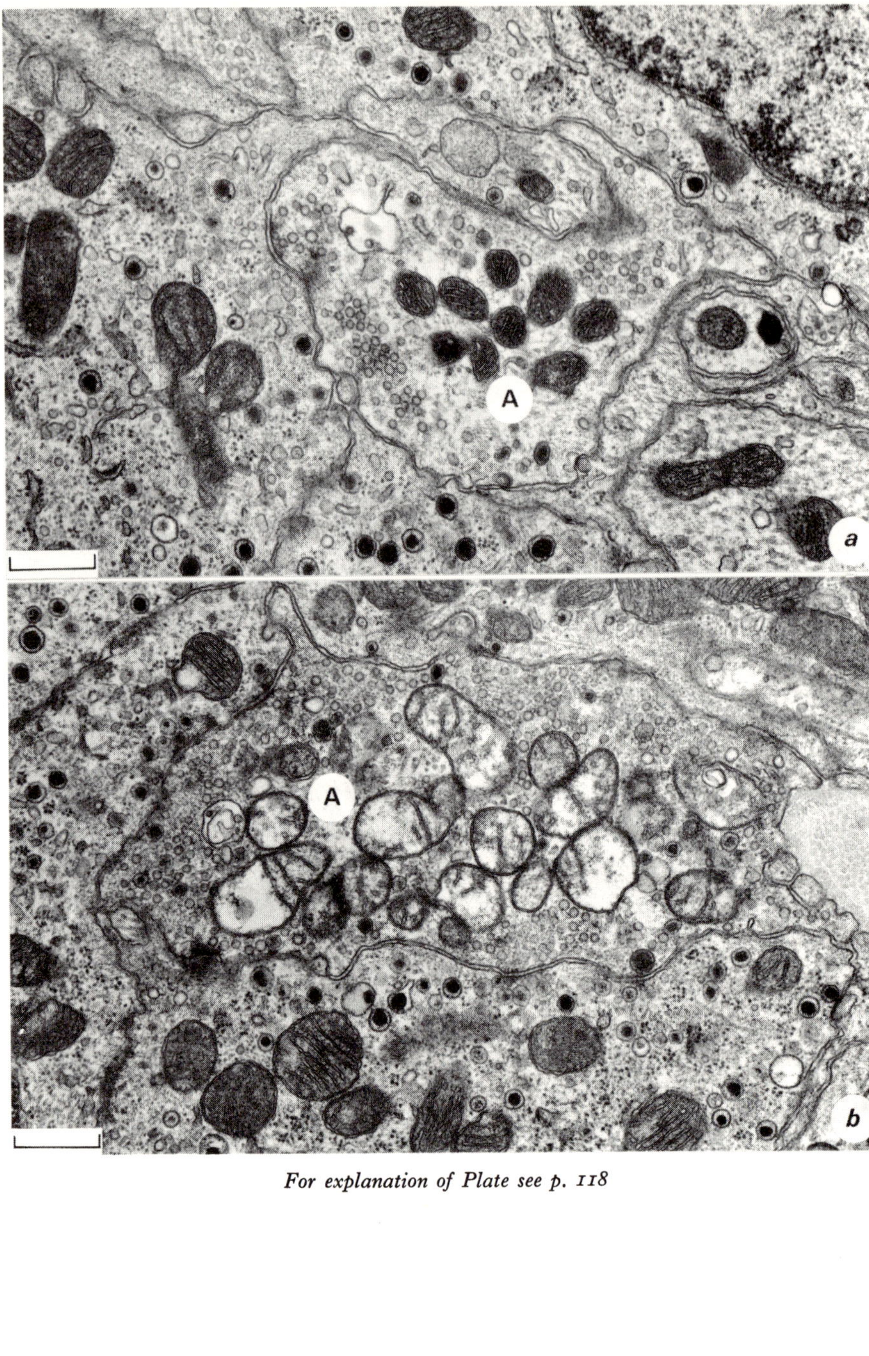

For explanation of Plate see p. 118

PLATE 7

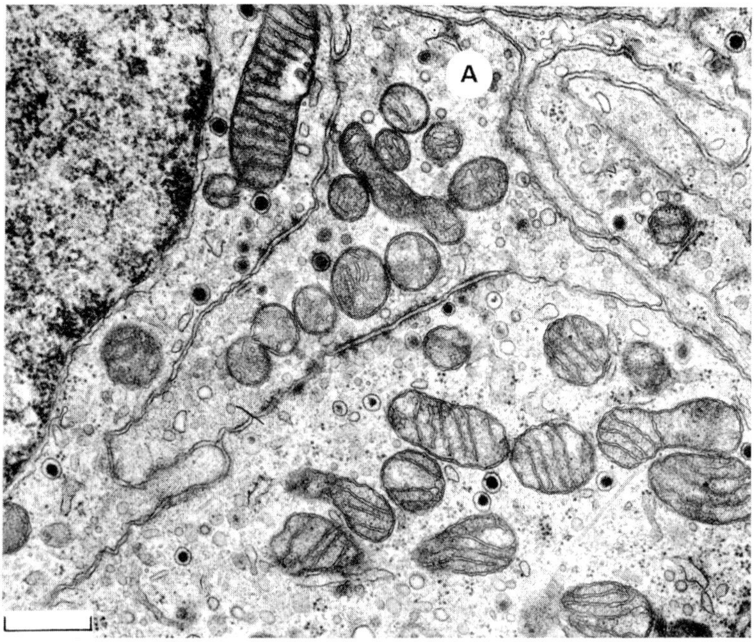

For explanation of Plate see p. 118

DISCUSSION

HESS Have you ever seen any nerve endings in the rat which do not have accumulations of synaptic vesicles of uniform size?

McDONALD I was going to go into this later because it has to do with the oxygen tension of the animal at the time of fixation. The number of synaptic vesicles in the nerve ending varies with the oxygen tension of the animal. If the animal is ventilated with 100% oxygen before fixation, the number of vesicles, as I have mentioned, is of the order of 30–40 per μm^2 – range 15 to 60 per μm^2. This is quite outside the mean observed in the efferent terminals where these vesicles are much more numerous. After the animal has been made hypoxic by ventilating it with 5% oxygen in nitrogen for 30 minutes, the number of vesicles is approximately halved and the mitochondria are swollen.

Because of this variability, we have characterized afferent endings as a class rather than trying to make sub-types because we have found, as Dr Verna has shown, the same nerve ending in different places can have different appearances.

HESS When I find an ending that I would call motor or efferent with accumulated vesicles, I can find a cell or a group of carotid body cells from the same rat, from the same glomus, from the same cluster of cells, which has numerous clear vesicles. Do you imply one group represents hypoxic endings and one normal?

McDONALD No. I want to stress that afferent endings have a variety of appearances in any carotid body. There may be sub-types; but the number and size of their synaptic vesicles suggests that the endings form one population.

WILLSHAW Of the endings which you have seen on the Type I cell approximately 95% appear to degenerate when you cut the sinus nerve below the petrosal ganglion. Can I make the point that this is 95% of the endings and not 95% of the fibres running in the sinus nerve.

McDONALD We have not examined the sinus nerve with respect to the unmyelinated fibres; however, all the myelinated axons in the sinus nerve degenerate within four days in the rat. In addition to counting the nerve endings around the glomus cells, we counted axons. The normal ratio of the number of axons to the number of cells is about 1.0. After we cut the glossopharyngeal nerve below the petrosal ganglion, this ratio fell to about 0.05. There were few axons near glomus cells – so you can see that these axons behave in the same way that the nerve endings do. Let me emphasize that the number of these axons does not change significantly if the glossopharyngeal nerve is cut intracranially.

WILLSHAW Yes, but Biscoe & Pallot (1972) showed that one axon branches to give many endings supplying many Type I cells. So if 95 % of the endings are affected, these endings are derived from comparatively few axons. Would you therefore allow that other axons which account for much more than 5 % of those travelling in the sinus nerve end somewhere other than on the Type I cell?

McDONALD Yes. We have shown, for example, axons which terminate upon ganglion cells. We have not studied these quantitatively.

O'REGAN Can you compare these findings in the rat with those in the cat in which most of the experimental work has been done?

McDONALD Yes. They are virtually identical with those obtained in the rat. We have done similar studies in cats, but in cats we have not studied degeneration sequentially. We chose to examine the cat carotid body 10 days after cutting rostral or caudal to the petrosal ganglion because at this time we had found in the rat that the difference was marked. Under control conditions in the cat there are approximately 1.7 nerve endings per glomus cell; 10 days after we sectioned the glossopharyngeal nerve intracranially, we found the ratio unchanged. However, 10 days after we sectioned the glossopharyngeal nerve caudal to the petrosal ganglion, this ratio fell to 0.10 endings per glomus cell.

O'REGAN Where do you think these endings come from and where are their cell bodies?

McDONALD Their cell bodies are in the petrosal ganglion, we presume, because the endings degenerate when the glossopharyngeal nerve is cut caudal to the petrosal ganglion, but not when the nerve is cut rostral to the ganglion.

O'REGAN What about the nerves which did not degenerate?

McDONALD These nerve endings degenerate when the superior cervical sympathetic ganglion is removed or when the cervical sympathetic trunk is cut proximal to the cervical ganglion in both rat and cat. These nerves represent only a small percentage of the total, and are pre-synaptic, not post-synaptic, to the glomus cells. We found no evidence of the existence of reciprocal synapses between glomus cells and efferent endings. We presume that the cell bodies of these efferent nerves are in the spinal cord.

O'REGAN What about other sources for the nerves which do not degenerate? Could there be a parasympathetic supply to the glomus cells?

McDONALD No. Efferents to glomus cells appear to be sympathetic for the reasons I have given. If some are parasympathetic they do not appear to travel in the sinus nerve. It is interesting that pre-ganglionic parasympathetic endings on ganglion cells in the carotid body degenerate rapidly – within a day or so of cutting their axons. These parasympathetic endings follow about the same time course in their degeneration as the afferent endings do.

O'REGAN Yes, but post-ganglionic parasympathetic fibres, do they supply glomus cells?

MCDONALD We have no evidence that they do, but we cannot exclude the possibility.

EYZAGUIRRE I just want to mention that De Castro described contamination of the sinus nerve with vagal fibres and he assumed that most of these were destined for ganglion cells in the carotid body. The destination of other such fibres he did not know.

TORRANCE McDonald did make the point that there are some endings which do not degenerate after extra-cranial section of the sinus nerve.

EYZAGUIRRE Exactly. If these are parasympathetic fibres they may not come through the ninth nerve foramen but may join the sinus nerve later on.

SHEPHERD Did you study the effect of hypoxia and high P_{O_2} on the efferent endings as well as on the afferent fibres?

MCDONALD Yes. There is a clear difference; efferent endings do not show the structural changes we have seen in afferent endings. In afferent endings 1 minute of hypoxia causes mitochondria to swell and synaptic vesicles to change their distribution. In rats breathing 100 % oxygen, synaptic vesicles in afferents often are aggregated in clumps. In rats breathing 100 % nitrogen for 1 minute synaptic vesicles are scattered. These changes do not occur in efferent endings, so we have another criterion for distinguishing between afferent and efferent terminals.

CASPERS Did you compare the vesicle content of synapses in the central part of the carotid body with those in the periphery which might correlate with the systematic distribution of tissue P_{O_2} shown by Acker, Lübbers & Purves (1971)? This might be of interest in testing the relation between the vesicle count and P_{O_2}.

MCDONALD No. We have not made this comparison.

HESS Do you think that all the afferent terminals have reciprocal synapses with glomus cells?

MCDONALD Perhaps. While in single sections we have found that only about 5 % of afferent endings are connected to glomus cells by reciprocal synapses, the synaptic junction occupies only a very small amount of the total area of the interface of afferent endings and glomus cells.

ACKER You said that the number of vesicles is reduced with hypoxia. Are they consumed or what?

MCDONALD We presume that there is a continuous exchange between synaptic vesicle membrane and the axoplasmic membrane. As transmitter is released by exocytosis of synaptic vesicles, the vesicle membrane is temporarily incorporated into the axoplasmic membrane. Ultimately this membrane is recycled and new vesicles are formed. We presume that the number of synaptic vesicles falls during stimulation when more vesicles are lost than formed.

ACKER Does that mean that the axoplasmic membrane becomes greater with hypoxia?

McDONALD Perhaps, but we have not determined this in our study of single sections. The variability in the size of afferent endings makes such a measurement difficult.

ABBOUD Did you see any change in the polarity of the synapses with hypoxia or suprapetrosal section?

McDONALD No.

BISCOE I was asked about my experimental results. Insofar as they are controversial, I would just like to say that I would fully accept McDonald's results: first, because the actual pictures are better than those I've seen anyone else take and second because the work has obviously been thoroughly and completely done.

HOWE Dr McDonald, I take it you would agree that your results are based on single sections taken largely, as I understand, from the centre of the carotid body and this does assume a certain uniformity of distribution of synapses, etc. throughout the structure.

McDONALD Yes, we did realize that there could be variability and therefore we studied single sections from many carotid bodies. However, we did note that there was not an even distribution of the two types of glomus cells, i.e. Type A and Type B. Because Type B glomus cells rarely are innervated, the ratio of the number of nerve endings to glomus cells varies from one part of the carotid body to another. But although we found that this ratio varied between 0.4 and 1.7 between different groups of 50 cells, the mean ratio for 2000 glomus cells from 10 carotid bodies was 1.0.

PURVES If you don't do serial sections of individual glomus complexes, how can you be so sure that this particular percentage – 50 % – of glomus cells is not innervated?

McDONALD First we distinguish between Type A and Type B glomus cells by the size and concentration of their dense-cored vesicles. The mean size of the vesicles in Type A glomus cells was 120 nm; the mean diameter in Type B glomus cells was 90 nm. The mean concentration of vesicles in Type A glomus cells was 3.4 per μm^2 of cytoplasm and the mean concentration in Type B glomus cells was $1.4 \mu m^2$. Then we determined the frequency of innervation of each type of glomus cell. On the average, in single sections there are about two endings per Type A glomus cell but nearly zero per Type B glomus cell.

PURVES Yes, but whatever index you use, you still must convince me that from single sections, you can distinguish two populations of glomus cells, innervated and non-innervated. I mean, can you really see all the way round the glomus cell, as it were?

MCDONALD No, I admit I could obtain more information by studying serial sections, but I believe that we have answered the question about the relative frequency of each type of glomus cell by analysing statistically our data from a large number of random sections.

TORRANCE You say that half of the glomus cells in a section have no nerve endings on them in that section?

MCDONALD Right.

TORRANCE And so you say that if you cannot see endings on half of the cells, half of the cells have no endings at all. Is that right?

BULLER Surely you can consider the statistics of this quite simply, by considering a number of partially marked balls in a container. If you take a section or sections through the mass, you will get a proportion which are not marked. And if you are saying that 50% are not marked, or insignificantly so, if you raise that assumption, I don't think you are doing it on the evidence you have provided. Thus you haven't given any evidence about the volume of tissue which you are dealing with which would be necessary for a dimensional analysis.

TORRANCE There is a further point, that you must consider, namely the length of contact of the nerve ending with the cell.

MCDONALD If Type A glomus cells were innervated by large endings and Type B glomus cells by small endings, I admit that more endings may appear on Type A cells than B cells in single sections. But we find 100–200 times more endings on Type A cells than B cells in single sections.

TORRANCE From only one section of that cell? And are you really justified in suggesting that there are two types of cell on the basis of their vesicle content in one section? For example, it could well be that the appearance and size of vesicles varies depending on whether they are close to or far away from the synapse.

MCDONALD No. Let me make our procedure clear. We measured the size of all dense-cored vesicles in the glomus cell cytoplasm irrespective of their relation to synapses; then we determined the mean diameter of the whole population. Furthermore, the difference between dense-cored vesicles in Type A and B glomus cells is not only related to the mean vesicle diameter but also to their concentration in the cytoplasm. There are twice as many dense-cored vesicles in Type A than in Type B glomus cells.

BULLER But I think that Purves' point still remains. Even if you identify two types of glomus cells on the vesicle size and content, you are still not able to make a clear correlation between appearance of the cell and its innervation because as I see it, you cannot be absolutely sure from single sections whether the cell is innervated or not.

McDONALD I'm wondering how else to explain these data. We have a population representing 50 % of the cells on which less than 1 % of the nerves end. Assuming a relatively large population, how else can we explain these data?

BULLER My comment is that I don't think you have a unique solution for the data you have presented. This does not mean that you are wrong but that if we had an hour or so, I am certain that we could come up with an alternative explanation.

McDONALD That may be so. But I think that ours is the most likely explanation.

HESS One point about the distribution. I have noticed, as I'm sure you must have, that if you study the point at which the nerve enters the carotid body, you can see very many more nerve endings than farther away in the body. Also, I have noted in the rat, there is a very different organization of nerves and cells: a high ratio near the nerve entry and a much lower ratio farther away. I think that De Castro also observed this in the cat.

PURVES Can I come back to the question of orientation of sections. I could imagine that if you assume that all contacts between nerve and cell involve reciprocal synapses, a slight variation in the plane of section would show what appears to be a purely motor or purely sensory arrangement of vesicles, etc. Can you tell us what proportion of synaptic contacts observed showed them to be unequivocally reciprocal and how this compares with the proportion of sensory or motor synapses?

McDONALD About 35 % of afferent endings in any given section are post-synaptic to Type A glomus cells, about 20 % of afferent endings are pre-synaptic to Type A glomus cells, and about 5 % are connected by reciprocal synapses to Type A glomus cells.

REFERENCES

ACKER, H., LÜBBERS, D. W. & PURVES, M. J. (1971). Local oxygen tension field in the glomus caroticum of the cat and its change at changing arterial P_{O_2}. *Pflügers Arch. ges. Physiol.*, **329**, 136–55.

BISCOE, T. J. & PALLOT, D. (1972). Serial reconstruction with the electron microscope of carotid body tissue. The Type 1 cell nerve supply. *Experientia*, **28**, 222.

HISTOCHEMICAL AND NEUROPHYSIOLOGICAL ASPECTS OF THE CAROTID BODY

The function of the Type I cell of the carotid body

By MARIANNE FILLENZ
University Laboratory of Physiology, Oxford

The problem with the carotid body is not so much to discover the mechanism underlying the chemoreceptor function, as to find functions for all the structures and mechanisms available in this organ. There are at least two distinct cell types, a rich supply of nerve fibres with a wide range of diameters and a rich vascular supply with many fenestrated capillaries. The organ contains a variety of biologically active substances and enzyme systems concerned with their synthesis and degradation.

The following questions need to be examined:

1. Are all these structures in the carotid body concerned with monitoring the P_{O_2} and P_{CO_2} of arterial blood? Or:
2. Does the carotid body subserve a number of different functions?
3. If so, to what extent are these functions interrelated?

In the present paper the Type I cells of the carotid body will be examined in the light of these questions, and an attempt will be made to see what conclusions can be drawn from a comparison with other cell types with which they share important properties.

The most firmly established fact about the Type I cell is that it contains a high concentration of catecholamines. This is most clearly demonstrated with the aid of the Hillarp and Falck histochemical fluorescence method for catecholamines (Fillenz & Woods, 1966). The use of this method has confirmed the presence of catecholamines in a wide variety of cells. Many of these catecholamine-storing cells are still very incompletely understood, but two cell types have been very thoroughly studied and much is known both about their function and mode of operation: one is the adrenal medullary cell and the other is the post-ganglionic sympathetic neurone. Although they have a common embryological origin, there are important differences in their function and their characteristics. The adrenal medulla is an endocrine gland, whose secretion is controlled by impulses in pre-ganglionic fibres, although it is also subject to hormonal influences (Axelrod, 1971). The post-ganglionic sympathetic neurone modifies the activity of a variety of effector cells by means of its neurotransmitter, noradrenaline. The activity of post-ganglionic sympathetic neurones has until recently been thought to be controlled entirely by impulses in pre-ganglionic fibres: this view has to be modified in the light of recent findings (Libet, 1970).

Recently a third catecholamine containing cell type has been intensively investigated: this was called the small granular cell by earlier histologists and is now known as the small, intensely fluorescent (SIF) cell (Eränkö & Härkönen, 1965). This cell type has many characteristics which make it intermediate between adrenal medullary cells and postganglionic sympathetic neurones. There is evidence which suggests that it plays the role of an interneurone in ganglionic transmission and therefore can modulate the effect of pre-ganglionic impulses on post-ganglionic neurones; however it does not look like a typical neurone and some of its features suggest that it may have an endocrine function. There are many striking similarities between SIF cells and the Type 1 cells of the carotid body which may provide some clues to the functions of the latter.

Table 1 presents a comparison of adrenal medullary cells, SIF cells and ganglion cells. The adrenal medullary cells produce the highest fluorescence intensity with the Hillarp and Falck technique and calculations based on

Table 1. *A comparison of some properties of adrenal medullary cells, SIF cells and sympathetic ganglion cells*

	Adrenal medullary cell	SIF cell	Sympathetic ganglion cell
Morphology	Oval cell, no processes	Several, short processes	axon
Fluorescence intensity	+++	++	+
Catecholamines	NA,A	DA	NA
Vesicle diameter (nm)	200–300	100–200	50–100
Catecholamine concentration (μg g^{-1})	12 000–16 000	7000	1–10, 1000–3000
Half-life	14 days	?	12 hours
6-OH-DA	0	0	– –
Guanethidine	0	++	– –

the extraction and assay of catecholamines give a figure of 12 000–16 000 $\mu g\ g^{-1}$ for the concentration of catecholamines in the adrenal medullary cell (Helpap & Hempel, 1969). There is evidence for the presence of two catecholamines – adrenaline and noradrenaline – which are stored in different cells (Coupland, 1965). The catecholamines are stored in a special subcellular particle, the chromaffin granule, which is 200–300 nm in diameter. It is not known whether there is an extravesicular pool of catecholamines and if so what proportion of the total store it constitutes.

In the ganglion cell the distribution of catecholamines is uneven (Geffen & Livett, 1971) as is shown by the distribution of fluorescence intensity. The highest concentration of catecholamines is found in the axon varicosities (1000–3000 $\mu g\ g^{-1}$), which are brilliantly fluorescent. Differential and density gradient centrifugation experiments have shown that there are two populations of subcellular particles which store catecholamines, the small dense-cored vesicles 50 nm in diameter and the large dense-cored vesicles 85–100 nm in diameter (Bisby & Fillenz, 1971). The relationship between these two kinds of vesicles and their respective functions are at present not understood.

SIF cells are small oval cells with no axons but a number of short processes; they are intensely and evenly fluorescent, thus resembling adrenal medullary cells rather than ganglion cells (Eränkö & Eränkö, 1971). Electron microscopy reveals the presence of dense-cored vesicles (Matthews & Raisman, 1969; Williams & Palay, 1969) 100–200 nm in diameter. These are thought to be the storage organelles for the catecholamines, by analogy with the dense-cored vesicles in other catecholaminergic cells. However, no subcellular fractionation studies have been carried out to show that catecholamines in SIF cells are particle bound. Moreover, Eränkö & Eränkö (1972*b*) have reported an increase in fluorescence intensity of SIF cells in newborn rats after hydrocortisone administration, but on electron-microscopic examination no dense-cored vesicles were seen in these cells.

No figures are available for the catecholamine concentration in SIF cells. However, approximate calculations can be made as follows: the cat superior cervical ganglion contains 5.25 $\mu g\ g^{-1}$ noradrenaline and 0.8 $\mu g\ g^{-1}$ of dopamine (Laverty & Sharman, 1965); there is evidence (Bjorklund *et al.*, 1970) that the dopamine is contained in SIF cells. The ratio of the number of ganglion to SIF cells is about 1000:1 and the ratio of their volumes is 9:1. Hence the approximate concentration of dopamine in SIF cells works out about 7000 $\mu g\ g^{-1}$.

The turnover time for catecholamines in the adrenal medulla is normally slow (7–33 days) and in nerve terminals, which have a much lower content, it is fast (12–24 hours). No figures are available for the turnover time of catecholamines in SIF cells.

The most striking differences between the three cell types are pharmacological: 6-hydroxydopamine in newborn rats causes degeneration of nine-tenths of the ganglion cells (Eränkö & Eränkö, 1972*a*) and in adult animals causes degeneration of the axon terminals of ganglion cells (Tranzer & Thoenen, 1967). It has no effect on adrenal medullary cells or on SIF cells in either new-born or adult animals.

Guanethidine in young rats causes destruction of ganglion cells, but a three–five-fold increase in the number of SIF cells (Eränkö & Eränkö, 1971); in adult animals the destruction of ganglion cells is less marked and the SIF cells show either no change or a slight decrease in fluorescence intensity. Hydrocortisone injections lead to a ten-fold increase in the number of SIF cells when given to newborn rats and have a less marked effect on SIF cells when given to adult rats; hydrocortisone has no effect on ganglion cells in either newborn or adult animals (Eränkö & Eränkö, 1972*b*).

Although in many of their characteristics listed in Table 1 SIF cells appear to be intermediate between ganglion cells and adrenal medullary cells, pharmacologically they resemble the latter. However, there is independent evidence for their role as interneurones: there is evidence from electron microscopy (Matthews & Raisman, 1969; Williams & Palay, 1969) that they receive pre-ganglionic cholinergic terminals with which they form synapses where SIF cells constitute the post-synaptic element; they also form synapses with cell bodies and dendrites of ganglion cells, where SIF cell processes have the characteristics of the pre-synaptic element. There is in addition electrophysiological evidence (Libet, 1970) for inhibition of ganglion cells involving the action of two transmitters: the muscarinic action of acetylcholine and dopamine. The SIF cells are the most likely candidates for this interneuronal role. This inhibition has a latency of 50–100 msec and a duration of 10–30 seconds, both very much longer than for synaptic events in ordinary ganglionic transmission. The fact that SIF cells function as inhibitory interneurones, of an unusual kind, does not exclude the possibility that they also have an endocrine role.

Table 2. *A comparison of some properties of the SIF and Type I cell*

	SIF cell	Type 1 cell
Catecholamines	DA,NA	DA,NA,5-HT
Capillaries	fenestrated	fenestrated
Effect of cortisone	++	++
Vesicle diameter (nm)	100–200	100–200

Table 2 presents a comparison of SIF cells and Type I carotid body cells. With the fluorescence histochemical method the two cell types are indistinguishable. Electron microscopy shows that both cell types have a number of short processes, often in close relation to fenestrated capillaries. Both cells have cilia and contain dense-cored vesicles 100–200 nm in diameter. The catecholamine in SIF cells is dopamine according to Bjorklund *et al.* (1970), but is noradrenaline in the superior cervical ganglion of the rat according to Eränkö & Eränkö (1971); in the carotid body, although dopamine-containing Type I cells are commonest, some Type I cells contain dopamine and noradrenaline and some contain 5-hydroxytryptamine (Morita, Chiocchio & Tramezzani, 1969). Pharmacologically SIF cells and Type I cells are also very similar: both show an increase in fluorescence intensity after cortisone treatment (Eränkö & Eränkö, 1971) and are relatively insensitive to 6-hydroxydopamine: SIF cells are unaffected whereas Type I cells show some decrease in their fluorescence intensity and the number of dense-cored vesicles (Lassmann & Böck, 1972). The similarity between SIF cells and Type I cells suggests that the latter could also be acting as interneurones: the confirmation of this hypothesis depends on the elucidation of the nature of the synapses on Type I cells. However, it is also possible that both types of cells release catecholamines into the blood stream: this is suggested by their close association with fenestrated capillaries. In addition, Helpap & Hempel (1969) have shown a turnover rate of 14 hours for the catecholamines in the carotid body. This, taken together with the high catecholamine content of the carotid body, implies the release of very large quantities of catecholamines. Leduc (1961) measured the excretion of dopamine in the urine in cold-treated rats and found that it was ten times greater than noradrenaline and 20 times greater than adrenaline; the source of this dopamine is not known, but it rose in response to cold just as did noradrenaline, which came from sympathetic nerve terminals.

In conclusion, although there is as yet no direct evidence for release of dopamine from Type I cells of the carotid body, it seems likely that they release it at a high rate. It seems possible that the release is not restricted to the synaptic contacts with nerve fibres and the rate of release, as indicated by the turnover rate, seems too high for it merely to mediate chemoreceptor impulses. Finally, there seems to be little correlation between storage and release of catecholamines and chemoreceptor function.

REFERENCE

Axelrod, J. (1971). Noradrenaline: fate and control of its biosynthesis. *Science*, **173**, 598–606.

BISBY, M. A. & FILLENZ, M. (1971). The storage of endogenous noradrenaline in sympathetic nerve terminals. *J. Physiol.*, **215**, 163–79.

BJORKLUND, A., CEGRELL, L., FALCK, B., RITZEN, M. & ROSENGREN, E. (1970). Dopamine-containing cells in sympathetic ganglia. *Acta physiol. Scand.*, **78**, 334–8.

COUPLAND, R. E. (1965). *The Natural history of the chromaffin cell*, London: Longmans.

ERÄNKÖ, L. & ERÄNKÖ, O. (1971). Effect of guanethidine on nerve cells and SIF cells in sympathetic ganglia of newborn and adult rats. *Acta Pharmacol. Toxicol., Kbh*, **30**, 403–16.

ERÄNKÖ, L. & ERÄNKÖ, O. (1972*a*). Effect of 6-hydroxydopamine on the ganglion cells and the SIF cells in the superior cervical ganglion of the rat. *Acta physiol. Scand.*, **84**, 115–24.

ERÄNKÖ, L. & ERÄNKÖ, O. (1972*b*). Effect of hydrocortisone on histochemically demonstrable catecholamines in the sympathetic ganglia and extraadrenal chromaffin tissue of the rat. *Acta physiol. Scand.*, **84**, 125–33.

ERÄNKÖ, O. & ERÄNKÖ, L. (1971). Small intensely fluorescent granule containing cells in the sympathetic ganglion of the rat. *Progr. Brain Res.*, **34**, 39–51.

ERÄNKÖ, O. & HÄRKÖNEN, M. (1965). Monoamine containing small cells in the superior cervical ganglion of the rat and an organ composed of them. *Acta physiol. Scand.*, **63**, 511–12.

FILLENZ, M. & WOODS, R. I. (1966). Some observations on the rabbit carotid body. *J. Physiol.*, **186**, 39–40P.

GEFFEN, L. B. & LIVETT, B. G. (1971). Synaptic vesicles in sympathetic neurons. *Physiol. Rev.*, **51**, 98–157.

HELPAP, B. & HEMPEL, K. (1969). Uber den Katecholamin Stoffwechsel des Caroticuskorperchens der Ratte. *Virchows Arch. Abt.*, **3**, 270–81.

LASSMANN, H. & BÖCK, P. (1972). Die Wirkung von 6-Hydroxydopamin auf den Katecholamingehalt des Glomus caroticum. *Z. Zellforsch. mikrosk. Anat.*, **127**, 220–9.

LAVERTY, R. & SHARMAN, D. F. (1965). The estimation of small quantities of 3,4-dihydroxyphenylethylamine in tissues. *Brit. J. Pharmacol.*, **24**, 538–48.

LEDUC, J. (1961). Catecholamine production and release in exposure and acclimation to cold. *Acta physiol. Scand.*, **53**, *Suppl.*, **183**, 1–101.

LIBET, B. (1970). Generation of slow inhibitory and excitatory post-synaptic potentials. *Fedn Proc.*, **29**, 1945–56.

MATTHEWS, M. & RAISMAN, J. (1969). The ultrastructure and somatic efferent synapses of small granule containing cells in the superior cervical ganglion. *J. Anat.*, **105**, 255–82.

MORITA, E., CHIOCCHIO, S. R. & TRAMEZZANI, J. H. (1969). Four types of main cells in the carotid body of the cat. *J. Ultrastr. Res.*, **28**, 399–410.

TRANZER, J.-P. & THOENEN, H. (1968). An electron microscopic study of selective, acute degeneration of sympathetic nerve terminals after administration of 6-hydroxydopamine. *Experientia*, **24**, 155–6.

WILLIAMS, T. H. & PALAY, S. L. (1969). Ultrastructure of the small neurons in the superior cervical ganglion. *Brain Res.*, **15**, 17–35.

DISCUSSION

GOODMAN You used the terms pre- and post-synaptic to describe the connections. In view of previous difficulties about defining these terms, would you give us your definition.

FILLENZ I was careful not to use these terms in connection with the carotid body glomus complex but only in connection with the SIF cells. In the SIF cells, the distinction is made partly on vesicle arrangement: but of course one can identify the pre-ganglionic cholinergic terminals most frequently on ganglion and partly on SIF cells. The junction between SIF and ganglion cells, is based partly on the distribution of vesicles but also on assymetric membrane thickness.

GOODMAN Secondly, was any microelectrode work done in the Libet studies? What did the results show?

FILLENZ Yes. There is hyperpolarization with a latency of 2–300 msec and duration of 7 seconds recorded from the post-ganglionic sympathetic nerves.

MCDONALD I'd like to comment on another similarity between SIF cells and glomus cells. Both contain two types of vesicles: dense-cored vesicles, that are so prominent, and smaller vesicles, which look more like synaptic vesicles, and are restricted to the pre-synaptic region. Small vesicles accumulate both where the SIF cell is pre-synaptic to ganglion cells and where the glomus cell is pre-synaptic to sensory nerve terminals.

FILLENZ If one regards the synaptic region as being one of release, one might say that the small vesicles are concerned with release and could represent recapture of membranous material.

MCDONALD In the rat, these vesicles normally contain little or no electron dense material but after 5-OHDA has been administered, they do.

FILLENZ That might almost suggest that they correspond to the small vesicles in the sympathetic nerve terminal.

MCDONALD Yes, I agree. Both sympathetic nerve endings and glomus cells contain the two populations of vesicles, but the unusual feature of the small vesicles in glomus cells in the rat is that they do not have an electron-dense core. What intrigues me about this fact is that glomus cells may contain dopamine. After all, there exist dopamine-containing neurones in the central nervous system which have synaptic vesicles without dense cores.

FILLENZ At least, they are very difficult to demonstrate. What fixative do you use?

MCDONALD A combination of glutaraldehyde and hydrogen peroxide developed by Peracchia & Mittler (1972): I am not certain what the active agent is.

ABBOUD Your suggestion that the carotid body can function as a secretory organ is interesting. Do you think that sufficient material can be secreted, for example, to have systemic effects?

FILLENZ No: what I was suggesting is that catecholamines may enter into the blood stream and exert their effect locally rather than remotely. For example, in the SIF cell it is quite possible that the very long duration of inhibition of the post-ganglionic neurone is due to a blood-borne effect of the transmitter.

SPECKMANN Is there electrophysiological evidence as to the mechanisms underlying the inhibition of the ganglionic cells induced by the two types of transmitters which you mentioned?

FILLENZ Yes. If atropine is given, the hyperpolarization disappears. It can be mimicked by iontophoretic application of dopamine. It is blocked by α-blockers: in the presence of α-blockers, acetylcholine does not produce hyperpolarization.

SPECKMANN Is anything known about the ion fluxes during this period of inhibition?

FILLENZ There is an increase in the resistance of the membrane.

SPECKMANN An increase?

FILLENZ In other words it is not due to an increased conductance. In fact there is also evidence that there is a decrease in potassium permeability which is responsible for the slow excitatory post-synaptic potential.

MILLS Following on from Dr Abboud's question, if a substance is secreted by the carotid body, it would have to have a higher activity than catecholamines because if they were totally extruded, this would be of the order of 30–40 ng.

FILLENZ No, I did not mean to imply a remote action: simply that it was conceivable that catecholamines could have a local action which may be wider than a purely synaptic action.

BISCOE Is anything known about the action of these drugs upon blood vessels. Could they affect the vasculature of the carotid body, and the veins?

FILLENZ Apart from the fact that dopamine is thought to be a vasodilator, I don't think much is known.

SAMPSON Dopamine is not strictly a vasodilator: it may have vasodilator effects in various vascular beds which become more prominent after blockade of α-receptors.

TORRANCE De Castro described aberrant carotid bodies up and down the carotid sheath. Could these be collections of SIF cells? That is, there would be interneurones but no efferent neurones.

FILLENZ They would be difficult to tell apart.

McDONALD We have studied the distribution of glomus-like bodies – non-chromaffin paraganglia – in the rat. In each rat we found these bodies, which contain SIF cells, in the superior cervical sympathetic ganglion, the nodose ganglion, and along the superior laryngeal nerve. These bodies invariably are associated with the superior laryngeal nerve near the larynx. The ultrastructure of these bodies suggests that these SIF cells are very similar to the glomus cells of the carotid body. Superior laryngeal nerve paraganglia are innervated by two types of nerves. One type probably is afferent. We found collections of SIF cells in the sympathetic trunks and vagus nerves, around the arch of the aorta, where they might be considered aortic bodies, along the recurrent laryngeal nerves, and on the posterior wall of the trachea. There are hundreds of these nests of cells in each rat, and the structure of many is remarkably like that of the carotid body.

EYZAGUIRRE In the cat this is also the case and there has been a recent paper by Matsuura (1973) showing this widespread dissemination of small islands of cells: and the afferent nerves respond similarly to the carotid body nerve.

REFERENCES

MATSUURA, S. (1973). Chemoreceptor properties of glomus tissue found in the carotid region of the cat. *J. Physiol.*, **235**, 57–74.

PERACCHIA, C. & MITTLER, B. S. (1972). Fixation by means of gluteraldehyde-hydrogen peroxide reaction products. *J. Cell Biol.*, **53**, 234–8.

Localization and quantitation of carotid body enzymes: their relevance to the cholinergic transmitter hypothesis

BY J. V. JONES

University Department of Medical Cardiology, Glasgow Royal Infirmary, Castle Street, Glasgow

Although the localization of acetylcholinesterase has often been interpreted as an indication of sites at which cholinergic mechanisms may act, this assumption, in the absence of a method of localizing acetylcholine directly cannot be proved. However, where there is a high level of acetylcholinesterase (as opposed to butyrylcholinesterase) activity this assumption is more likely to be true (Koelle, 1955; Garrett, Howard & Nixon, 1969).

In the carotid body the mechanism is more complex and the interpretation of cholinesterase distribution more difficult if the transmitter, which is possibly acetylcholine, is not wholly destroyed by enzymic hydrolysis but is instead partially washed away in the blood stream (Neil, 1951). With such a transmitter hypothesis the finding of low levels of acetylcholinesterase by Hollinshead & Sawyer (1945) does not exclude the possibility of acetylcholine being the transmitter. With this in mind, a number of workers have studied the distribution of cholinesterases in carotid bodies of a variety of animals (Koelle, 1950, 1951; Ross, 1957; Pryse-Davies, Dawson & Westbury, 1964; Palkama, 1965; Rogers, 1965; Fillenz & Woods, 1966; Becker, Drukker & Meijer, 1967; Serafini-Fracassini & Frasson, 1966; see Table 1).

The process of investigation was carried one stage further by Biscoe & Silver (1966) when they studied the distribution of cholinesterases in the carotid body of the cat in normal animals and in those after section of either the glossopharyngeal or sympathetic nerves. Both butyrylcholinesterase and acetylcholinesterase were found to form strands and plexuses which were extracellular although enzyme activity was found in a few cells which were presumed to be ganglion cells. They concluded that a cholinergic pathway controlling the blood vessels of the carotid body ran in the post-ganglionic sympathetic pathway.

It is clear, therefore, from these studies with the light microscope, that some disagreement existed about this aspect of carotid body structure. In order to try and shed some light on the problem, a study, using the electron microscope, of the localization of cholinesterases in the carotid body was undertaken (Jones & Ballard, 1971; Ballard & Jones 1971). The enzyme was found in two main sites, butyrylcholinesterase being localized (by inference) to the Type II cell membrane (Plate 1) while acetylcholinesterase was found in groups of peripheral axons (Plate 2). No intracellular reaction product was seen and cholinesterase was absent where Type I cell and nerve fibres were adjacent (Plate 3).

Table 1. *Localization by light microscope of cholinesterases in carotid body*

Author	Species	Enzyme	Site
Koelle, 1950; 1951	Cat	AChE	Type I cell (nuclear, perinuclear and intracytoplasmic); variable intensity, some cells negative.
		BuChE	Type I cell (nuclear, perinuclear and intracytoplasmic); variable intensity, some cells negative
Ross, 1957	Cat	AChE	Type I cell: intracellular.
		BuChE	Type I cell: intracellular.
Becker *et al.*, 1967	Rabbit	AChE	Type I cell: weakly positive. Type II cell and nerve fibres: strongly positive.
		BuChE	Type II cell and nerve fibres: strongly positive.
Serafini-Fracassini & Frasson, 1966	Dog	AChE	Nerve fibres.
		BuChE	Type I cells.
Pryse-Davies *et al.*, 1964	Human	AChE	Type I cells: weakly positive.
		BuChE	Type I cells and surrounding structures: strongly positive.

However, with the histochemical method employed in this study several problems are encountered. Spontaneous break-down of the incubation medium may occur with incubation periods of longer than 50 minutes and results in non-specific staining (Robinson, 1969). This can be overcome by transferring sections to fresh incubation medium at shorter intervals. Furthermore, if low concentrations of anticholinesterases are incorporated in the incubation medium and inhibit the reaction then any staining that is abolished can be assumed to be due to cholinesterase activity and not to spontaneous break-down.

Another problem that has recently been recognized with Karnovsky's (1964) method is that the ferricyanide ions present in the incubation medium may be incapable of penetrating cell membranes. If this proves to be true, no intracellular reaction product would be detected with this method. Work is in progress to clarify this issue in the carotid body.

Lee (1968) has pointed out that a more true reflection of whether cells are cholinergic or not would be obtained by studying the localization of the enzyme choline acetyltransferase. Hebb (1968) has assayed carotid bodies biochemically for their choline acetyltransferase activity. Moderately high levels were found and for a variety of reasons Hebb concluded that the enzyme is present in the Type I cells themselves. In our study, a total of five carotid bodies from five cats (one from each animal) were assayed for their choline acetyltransferase content. The mean weight of the carotid bodies assayed was 2.06 mg while the mean choline acetyltransferase activity was 179 μg acetylcholine formed per gram tissue per hour (Ballard & Jones, 1972).

The possibility of more precise localization of sites of enzymic activity came with the advent of histochemical methods for demonstrating choline acetyltransferase activity. Burt (1969, 1970) first described a method for the light microscope and almost simultaneously Kasa, Mann & Hebb (1970*a*, *b*) described methods for light and electron microscopy. Using this latter method we found choline acetyltransferase activity, which seemed to be associated with vesicles, in the cytoplasm of Type I cells (Plates 4, 5 and 6). No enzyme was found associated with the large nerve endings synapsing with the Type I cells.

With such new histochemical methods the results must be viewed as only preliminary. Reactions other than those mediated by choline acetyltransferase can cause the release of free CoA from acetyl CoA and hence give false positive results. These non-specific reactions can be greatly diminished by incorporation of an anticholinesterase in the incubation medium. It now seems that difluorophosphate (DFP) has a higher specificity in its ability to abolish non-specific reactions than have other anticholinesterases (Burt, 1973, personal communication). Iso-OMPA* as used in our studies would seem to be much less efficient than DFP in this respect. The specificity of the histochemical method can be improved, however, by the incorporation of a specific inhibitor of choline acetyltransferase in the incubation medium with consequent abolition of reaction product, or by omission of choline from the incubation medium (Kasa & Morris, 1972). In our experiments choloroacetylcholine perchlorate was used as the inhibitor. Various authors disagree about the specificity of this inhibitor but in our hands it seemed to abolish what we regarded as the specific staining in the Type I cells. We feel that the sum of the evidence, both histochemical and biochemical, would indicate that our localization of choline acetyltransferase to vesicles of the Type I cell is valid. This, of course, clearly requires confirmation.

* Iso-OMPA = tetramonoisopropylpyrophosphotetramide

Choline acetyltransferase is also very sensitive to fixation so that sites of low enzymic activity will not be detected by the histochemical method employed. It would seem likely, therefore, that there are other sites of enzymic activity not yet described.

The carotid body was found to contain large amounts of acetylcholine (20–30 μg g^{-1} tissue) when assayed by Eyzaguirre, Koyano & Taylor (1965); acetylcholine was identified chromatographically. We obtained similar results when a total of twelve carotid bodies from eight cats were

Table 2. *Whole carotid body assay for acetylcholine*

No.	Weight (mg)	g ACh
1	4.3 (2)	4.6
2	3.6 (2)	11.1
3	1.6 (1)	37.5
4	2.9 (1)	69.9
5	1.6 (1)	12.5
6	2.2 (1)	36.0
7	5.2 (2)	4.0
8	6.1 (2)	9.8

No. of carotid bodies given in brackets.

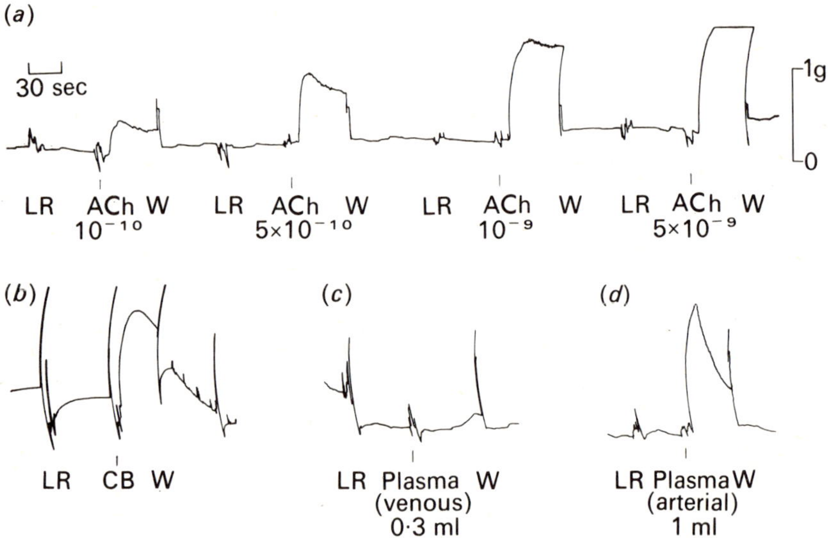

Fig. 1. Effect on the leech muscle strip of (*a*) standard concentrations of acetylcholine; (*b*) extract of whole carotid body; (*c*) carotid body venous plasma and (*d*) plasma taken direct from a cat femoral artery. ACh, acetylcholine; CB, carotid body; W, wash solution; LR, leech ringer solution.

assayed on leech muscle (Forrester, 1966) for their acetylcholine content (Table 2; Fig. 1). The mean weight of the carotid bodies assayed was 2.3 mg and the mean acetylcholine content was 23.1 $\mu g\ g^{-1}$ tissue.

Subsequently it was shown by Metz (1969) that acetylcholine could be measured in the venous blood from the carotid body of the dog and that it increased progressively in concentration from perfusion with oxygenated blood to perfusion with hypoxic blood to perfusion with asphyxial blood. We endeavoured to repeat these experiments using the cat as the experimental animal. Altogether 15 experiments were performed with essentially negative results (Fig. 1). Initial experiments employing blood as the carotid body perfusate indicated that acetylcholine was possibly being released into the venous blood as shown by a small rise in the tension of the assay leech muscle strip. The results were neither constant nor consistent and the degree of tension increase in the muscle did not seem to vary with the different gaseous tensions of the perfusing blood. Indeed it seemed that the response was related to the size of the sample obtained rather than its alleged acetylcholine content. Further curare which should have blocked acetylcholine effects on the leech muscle strip did not affect the responses obtained.

This prompted a control test using blood taken directly from a femoral artery. A 2 ml sample of whole blood was centrifuged and the plasma decanted and made up to 2 ml again with eserinized leech Locke solution. This caused a brisk rise in tension of the leech muscle strip when injected into the bath. In view of these observations the fluid used to perfuse the carotid bodies was changed from blood to Krebs–Henseleit solution. Consistently negative results were now obtained.

The distribution of the enzyme myosin ATPase was studied with the light microscope using two histochemical methods (Guth & Samaha, 1969, 1970; Meijer, 1970). The reaction product was virtually confined to the groups of glomus cells and at high magnification was seen to form fine strands that seemed to surround presumed Type I cells rather than to be in their cytoplasm. It was concluded, therefore, that the myosin ATPase was present in or on the Type II cell (Plates 7 and 8).

Hebb (1968) claimed that the low concentrations of choline acetyltransferase (5 % of that of the superior cervical ganglion) in the carotid body were compatible with the high concentrations of acetylcholine if the acetylcholine present were destroyed at a slow rate. Our finding of butyrylcholinesterase, and not acetylcholinesterase, on the membrane of the Type II cell supports this idea. Butyrylcholinesterase is known to hydrolyse acetylcholine at a much slower rate than does acetylcholinesterase (Gerebtzoff, 1959).

This slow rate of destruction fits the observation of Metz that acetylcholine could be recovered from the carotid body venous blood (in dogs) in conditions of hypoxia and asphyxia, when presumably relatively large amounts of transmitter are released, but not when the animal breathes room air and the carotid body is relatively inactive; under these latter conditions butyrylcholinesterase might well be able to hydrolyse all released acetylcholine. There are objections to Metz's work, however, e.g. the venous blood from the carotid body was allowed to drip on to adjacent muscle from where it was collected. Also adequate control experiments do not seem to have been done. MacIntosh & Perry (1950) state that in order to identify a substance as acetylcholine at least two different assay methods should be used; Metz used only the one. However, in our experiments, in a smaller animal, it is possible that acetylcholine was being released but was below the threshold for the assay methods used.

In an attempt to clarify some of the previous anomalous pharmacological findings (Heymans, 1955; Sampson, 1971), Nishi & Eyzaguirre (1971) performed dose–response studies with ganglionic blocking agents. They argued that due to the presence of Type II cells and their processes such drugs might not be able to gain adequate access to the afferent nerve endings and in this way natural stimuli would still be effective while applied acetylcholine was blocked. They proceeded to show that large doses of ganglionic blocking drugs could, after all, block the effect of natural stimuli like hypoxia on the carotid body discharge.

The demonstration in the present study of cholinesterase around the membranes of the Type II cells and of choline acetyltransferase within Type I cells has led us to a similar conclusion. Endogenously released acetylcholine, possibly from the Type I cell in response to hypoxia, is unlikely to be blocked by perfused or superfused anticholinergic agents but would have to diffuse through the Type II cells before reaching the postulated site of action of the endogenously released acetylcholine at the inner surface of the Type II cell. Exogenous acetylcholine, on the other hand, if it acts at the outer membrane of the Type II cell, is more likely to be blocked by anticholinergic agents.

Such considerations may explain the findings of Moe, Capo & Peralta (1948) and Douglas (1952) that the respiratory responses to injections of acetylcholine could be blocked by anticholinergic agents while the responses to hypoxia were left unaffected, and those of Sampson (1971) who showed that, *in vivo*, the response of the carotid body to hypoxia and to some stimulating agents is not blocked by mecamylamine while the response to injected acetylcholine is reduced by this drug.

The most determined antagonist of the acetylcholine transmitter hypothesis in recent times has been Paintal (1967, 1969, 1971*a*,*b*) who has shown that, in the aortic bodies at least, cooling of the chemoreceptors results in loss of the responses to applied acetylcholine while the responses to hypoxia are still present. However, the argument to exogenously administered drugs may still hold in this situation. Because of the very high metabolic rate of glomus cells (Daly, Lambertsen & Schweitzer, 1954), considerable heat must be generated within the glomus cells so that it is possible that cold perfusing blood would cool the outer aspect of the glomus cell groups while the centre would still be at a considerably higher temperature, owing to the heat generated by the metabolically very active Type I cells, and still be able to respond. In this case there would be an expected reduction although not abolition of the response to natural stimuli: this is indeed the case (Paintal, 1971*a*).

Paintal (1967) has advocated a similar system. He suggests that the Type I cell is metabolically very active and accounts for the high metabolic rate of the carotid body, but it is the Type II cell of low metabolic rate that is the oxygen sensor cell. Slight alteration of Paintal's theory gives the following hypothesis. The Type I cell is metabolically active and also acts as the sensor mechanism. Upon stimulation by natural stimuli the Type I cell releases a substance, possibly acetylcholine, to act on the Type II cell. The Type II cell could, in turn, initiate the afferent impulses by contracting and stimulating the naked nerve fibres shown to be enclosed by the Type II cell (De Kock & Dunn, 1966, 1968).

The demonstration of heavy staining for myosin ATPase in the carotid body and its probable localization in the Type II cell would indicate the presence of a contractile mechanism in these cells. Further support comes from the demonstration of large numbers of microfilaments in the Type II cell (Biscoe & Stehbens, 1966). Such microfilaments in other cells were formerly thought to have a structural role but current thoughts (Wessels, 1971) ascribe a much more active role in cell movement, changes in cell shape and cell contraction to these microfilaments.

Circumstantial support for this contractile hypothesis comes from a variety of sources. Large desmosomes are seen to connect Type I cells to one another and to connect Type I cells to nerve endings. Significantly no desmosomes are seen to connect Type I cells to Type II cells. The carotid body is thought to change in size during hypoxia (De Castro, 1951) while signs of compression of Type I cells have been reported at the electron microscopical level (Kondo, 1971).

The present results are compatible with the hypothesis that acetylcholine is a transmitter in the carotid body and that it is destroyed, at least in part, by enzymic hydrolysis while not excluding the possibility that some of the transmitter may be washed away in the blood stream. A hypothesis on the mode of chemoreception has been suggested, that the Type I cell responds to natural stimulation by releasing acetylcholine, the acetylcholine then causes the Type II cell to contract thereby stimulating the naked nerve endings enclosed by these Type II cells.

REFERENCES

BALLARD, KATHLEEN, J. & JONES, J. V. (1971). The fine structural localization of cholinesterases in the carotid body of the cat. *J. Physiol.*, **219**, 747–53.

BALLARD, KATHLEEN J. & JONES, J. V. (1972). Demonstration of choline acetyltransferase activity in the carotid body of the cat. *J. Physiol.*, **227**, 87–94.

BECKER, A. E., DRUKKER, J. & MEIJER, A. E. F. H. (1967). Histochemical characteristics of chemoreceptor organs (glomera). *Histochemie*, **11**, 195–204.

BISCOE, T. J. & SILVER, A. (1966). The distribution of cholinesterases in the cat carotid body. *J. Physiol.*, **183**, 501–12.

BISCOE, T. J. & STEHBENS, W. E. (1966). Ultrastructure of the carotid body. *J. Cell Biol.*, **30**, 563–78.

BURT, A. M. (1969). The histochemical demonstration of choline acetyltransferase activity in the spinal cord of the rat. *Anat. Rec.*, **163**, 162.

BURT, A. M. (1970). A histochemical procedure for the localisation of choline acetyltransferase activity. *J. Histochem. Cytochem.*, **18**, 408–15.

DALY, M. DE B., LAMBERTSEN, C. J. & SCHWEITZER, A. (1954). Observations on the volume of blood flow and oxygen utilisation of the carotid body in the cat. *J. Physiol.*, **125**, 67–89.

DE CASTRO, F. (1951). Sur la structure de la synapse dans les chémorécepteurs: leur mécanisme d'excitation et rôle dans la circulation sanguine local. *Acta physiol. Scand.*, **22**, 14–43.

DE KOCK, L. L. & DUNN, A. E. G. (1966). An electron microscope study of the carotid body. *Acta anat.*, **64**, 163–78.

DE KOCK, L. L. & DUNN, A. E. G. (1968). Electron microscopic investigation of the nerve endings in carotid body. In *Arterial Chemoreceptors*, pp. 179–88, ed. TORRANCE, R. W., Oxford: Blackwell.

DOUGLAS, W. W. (1952). The effect of a ganglionic blocking drug, hexamethonium, on the response of the cat's carotid body to various stimuli. *J. Physiol.*, **118**, 373–83.

EYZAGUIRRE, C., KOYANO, H. & TAYLOR, J. R. (1965). Presence of acetylcholine and transmitter release from carotid body chemoreceptors. *J. Physiol.*, **178**, 463–76.

FILLENZ, M. & WOODS, R. I. (1966). Some observations on the rabbit carotid body. *J. Physiol.*, **186**, 39–40P.

FORRESTER, T. (1966). An isometric leech muscle preparation with stable sensitivity for the assay of acetylcholine. *J. Physiol.*, **187**, 12–14P.

GARRETT, J. R., HOWARD, E. R. & NIXON, N. H. (1969). Autonomic nerves in rectum and colon in Hirschsprung's disease. *Arch. Dis. of Childh.*, **44**, 404–17.

GEREBTZOFF, M. A. (1959). *Cholinesterases. Histochemical Contribution to the Solution of Some Functional Problems.* London: Pergamon.

GUTH, L. & SAMAHA, F. L. (1969). Qualitative differences between actomyosin ATP-ase of slow and fast mammalian muscle. *Expl. Neurol.*, **25**, 138–52.

GUTH, L. & SAMAHA, F. J. (1970). Procedure for the histochemical demonstration of actomyosin ATP-ase. *Expl. Neurol.*, **28**, 365–7.

HEBB, C. (1968). Discussion of paper by K. D. Lee. In *Arterial Chemoreceptors*, p. 139, ed. TORRANCE, R. W., Oxford: Blackwell.

HEYMANS, C. (1955). Action of drugs on carotid body and sinus. *Pharmac. Rev.*, **7**, 119–42.

HOLLINSHEAD, W. H. & SAWYER, C. H. (1945). Mechanisms of carotid body stimulation. *Am. J. Physiol.*, **144**, 79–86.

JONES, J. V. & BALLARD, KATHLEEN J. (1971). Cholinesterases in the carotid body of the cat as seen with the electron microscope. *Nature New Biol., Lond.*, **233**, 146–8.

KARNOVSKY, M. J. (1964). The localization of cholinesterase activity in rat cardiac muscle by electron microscopy. *J. Cell Biol.*, **23**, 217–32.

KASA, P., MANN, S. P. & HEBB, CATHERINE (1970*a*). Localization of choline acetyltransferase. *Nature, Lond.*, **266**, 812–14.

KASA, P., MANN, S. P. & HEBB, CATHERINE (1970*b*). Ultrastructural localization of choline acetyltransferase in spinal neurones. *Nature, Lond.*, **226**, 814–16.

KASA, P. & MORRIS, D. (1972). Inhibition of choline acetyltransferase and its histochemical localization. *J. Neurochem.*, **19**, 1299–1304.

KOELLE, G. B. (1950). The histochemical differentiation of types of cholinesterases and their localization in tissues of the cat. *J. Pharmac. exp. Ther.*, **100**, 158–79.

KOELLE, G. B. (1951). The elimination of enzymic diffusion artifacts in the histochemical localization of cholinesterases and a survey of their cellular distributions. *J. Pharmac. exp. Ther.*, **103**, 153–71.

KOELLE, G. B. (1955). The histochemical identification of acetylcholinesterase in cholinergic, adrenergic and sensory neurones. *J. Pharmac. Exp. Ther.*, **114**, 167–84.

KONDO, H. (1971). An electron microscopic study on innervation of the carotid body of the guinea pig. *J. Ultrastruc. Res.*, **37**, 544–62.

LEE, K. D. (1968). Enzyme activity of the carotid body. In *Arterial Chemoreceptors*, pp. 133–8, ed. TORRANCE, R. W., Oxford: Blackwell.

MACINTOSH, F. C. & PERRY, W. L. M. (1950). Biological estimation of acetylcholine. *Meth. Med. Res.*, **3**, 78–92.

MEIJER, A. E. F. H. (1970). Histochemical method for the demonstration of myosin ATPase in muscle. *Histochemie*, **22**, 51–8.

METZ, B. (1969). Release of acetylcholine from carotid body by hypoxia and hypoxia plus hypercapnia. *Respir. Physiol.*, **6**, 386–94.

MOE, G. K., CAPO, L. R. & PERALTA, B. (1948). Action of tetraethylammonium on chemoreceptor and stretch receptor mechanisms. *Am. J. Physiol.*, **153**, 601–5.

NEIL, E. (1951). Chemoreceptor areas and chemoreceptor circulatory reflexes. *Acta physiol. scand.*, **22**, 54–65.

NISHI, K. & EYZAGUIRRE, C. (1971). The action of some cholinergic blockers on carotid body chemoreceptors *in vivo*. *Brain Res.*, **33**, 37–56.

PAINTAL, A. S. (1967). Mechanism of stimulation of aortic chemoreceptors by natural stimuli and chemical substances. *J. Physiol.*, **189**, 63–84.

PAINTAL, A. S. (1969). Further evidence that acetylcholine is not a transmitter at chemoreceptors. *J. Physiol.*, **204**, 94–95P.

PAINTAL, A. S. (1971*a*). The responses of chemoreceptors at reduced temperatures. *J. Physiol.*, **217**, 1–18.

PAINTAL, A. S. (1971*b*). Action of drugs on sensory nerve endings. *Ann. Rev. Pharmac.*, **11**, 231–40.

PALKAMA, A. (1965). Histochemistry and electron microscopy of the carotid body. *Annls. med. exp. biol. Fenn.*, **43**, 260–6.

PRYSE-DAVIES, J., DAWSON, I. M. P. & WESTBURY, G. (1964). Some morphologic, histochemical and chemical observations on chemodectomas and the normal carotid body, including a study of the chromaffin reaction and possible ganglion cell elements. *Cancer*, **17**, 185–202.

ROBINSON, P. M. (1969). A cholinergic component in the innervation of the longitudinal smooth muscle of the guinea pig vas deferens. *J. Cell Biol.*, **41**, 462–76.

ROGERS, D. C. (1965). The development of the rat carotid body. *J. Anat.*, **99**, 89–101.

ROSS, L. (1957). A cytological and histochemical study of the carotid body of the cat. *Anat. Rec.*, **129**, 433–56.

SAMPSON, S. R. (1971). Effects of mecamylamine on response of carotid body chemoreceptors *in vivo* to physiological and pharmacological stimuli. *J. Physiol.*, **212**, 655–66.

SERAFINI-FRACASSINI, A. & FRASSON, P. (1966). Histochemical observations on the carotid body of the dog. *Acta anat.*, **63**, 240–8.

WESSELLS, N. K. (1971). How living cells change shape. *Scient. Am.*, **225**, 76–85.

PLATE 1

Two Type I (I) cells are seen separated by a Type II (II) cell process. Reaction product (ChE) is seen between the two types of cell: CT, connective tissue.

PLATE 2

Group of axons (nf) found outwith the glomus and staining positively for cholinesterase (ChE).

PLATE 3

Shows a large nerve ending (NE) synapsing with a Type I (I) cell. Reaction product is seen between the Type I and Type II (II) cells, between Type II cells and the outer aspect of the nerve ending, between adjacent Type II cells and between Type II cell and the surrounding connective tissue. No reaction product (ChE) is seen at the point of synapse between the Type I cell and the nerve ending.

PLATE 4

Carotid body incubated for the localization of choline acetyltransferase activity. Two Type I (I) cells are seen partially separated by a thin Type II (II) cell process. Reaction product (ChAc) is seen within the cytoplasm of the Type I cell. This section has been ightly stained with lead citrate and uranyl acetate: N, nucleus.

PLATE 5

Carotid body incubated for the localization of choline acetyltransferase activity and showing, at greater magnification typical sites of reaction product (ChAc). Section not stained with lead citrate and uranyl acetate: I, Type I cell; II, Type II cell.

PLATE 6

Carotid body incubated for the localization of choline acetyltransferase activity and showing three typical sites of reaction product (A, B, C). The reaction product at site A seems to be associated with a vesicle and probably at B also. This section has been counterstained with lead citrate and uranyl acetate.

PLATE 7

Whole carotid body incubated for the detection of myosin ATPase by the method of Meijer. The deposit of reaction product is present mostly in the form of clumps.

PLATE 8

Higher power view of one of the clumps seen in a carotid body incubated for the detection of myosin ATPase activity by the method of Meijer (1970). Reaction product seems to surround the presumed Type I cells in the form of fine strands which may represent Type II cells, rather than be intracellular.

PLATE 1

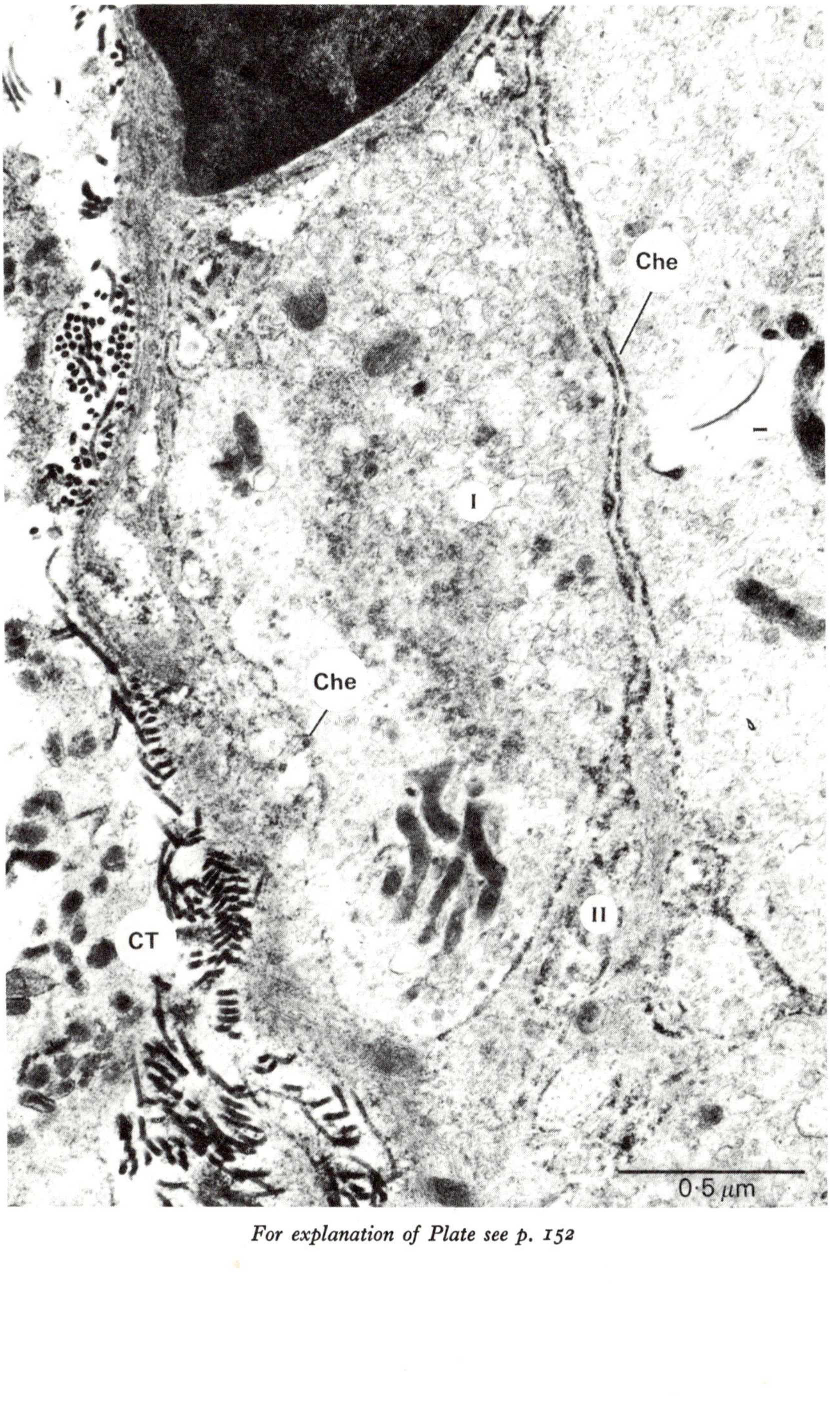

For explanation of Plate see p. 152

PLATE 2

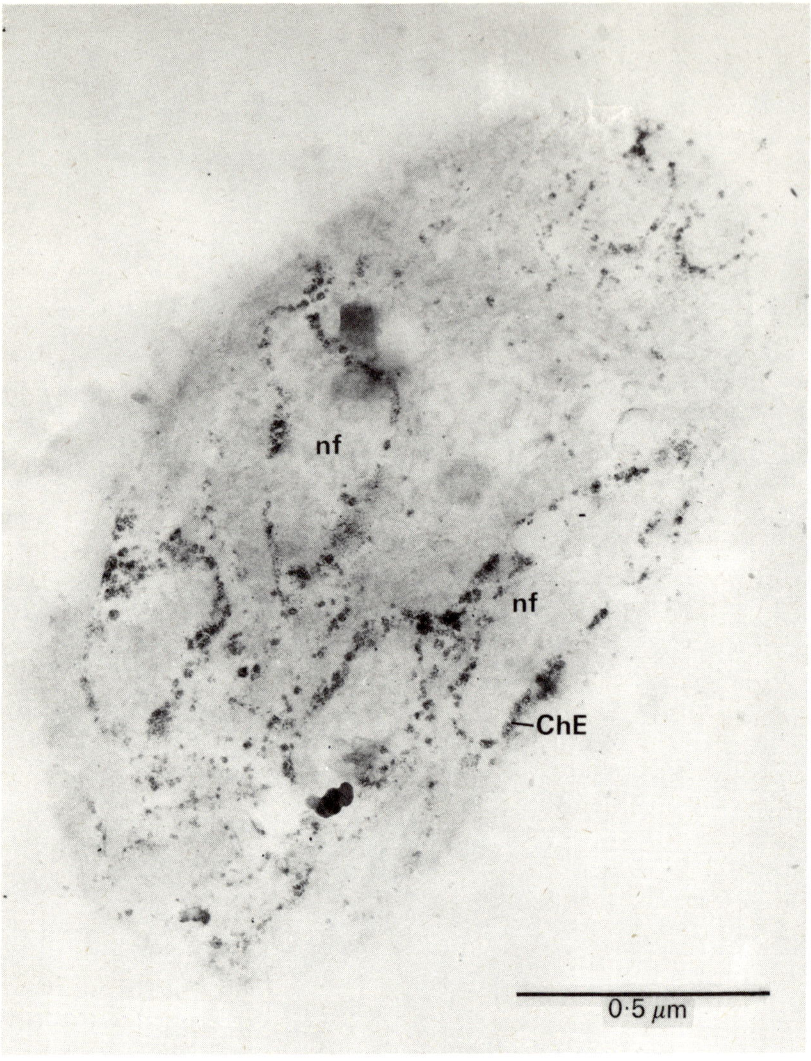

For explanation of Plate see p. 152

PLATE 3

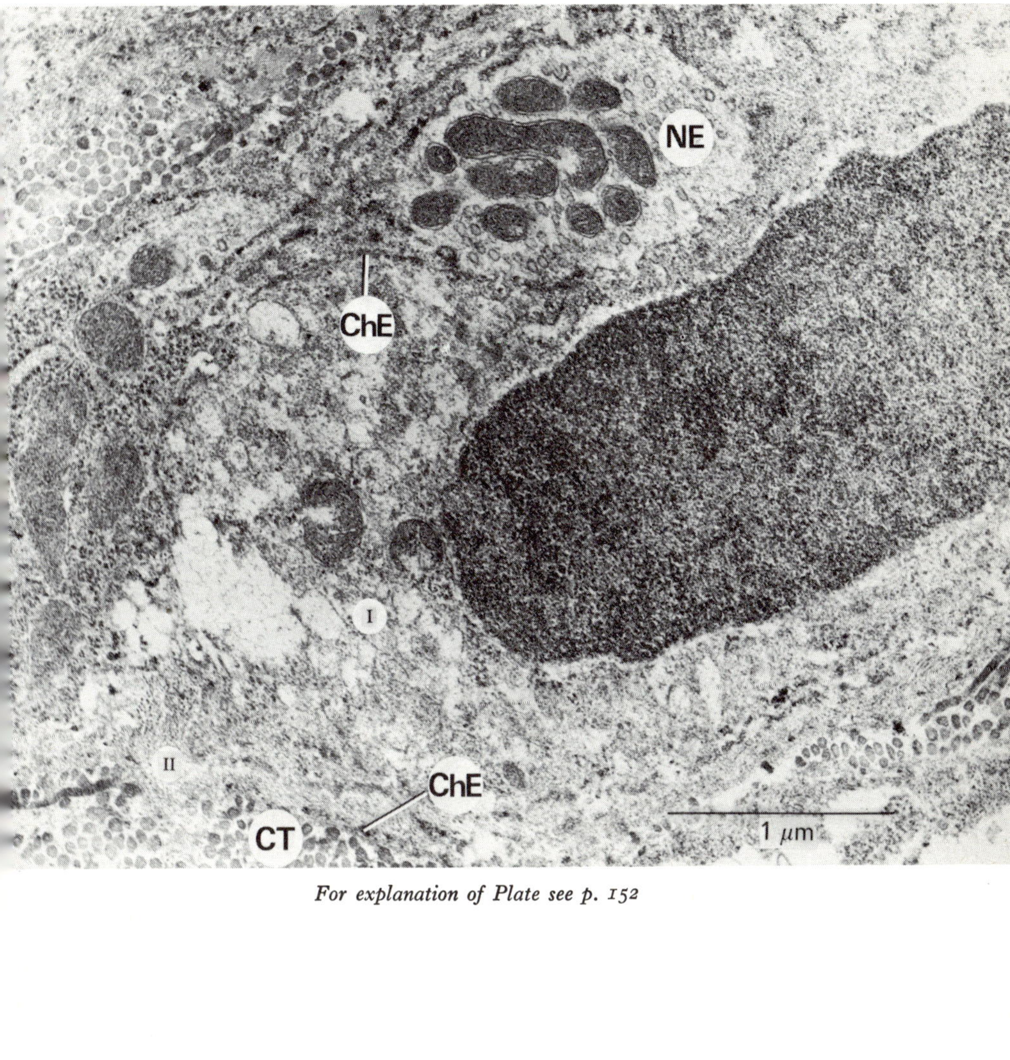

For explanation of Plate see p. 152

PLATE 4

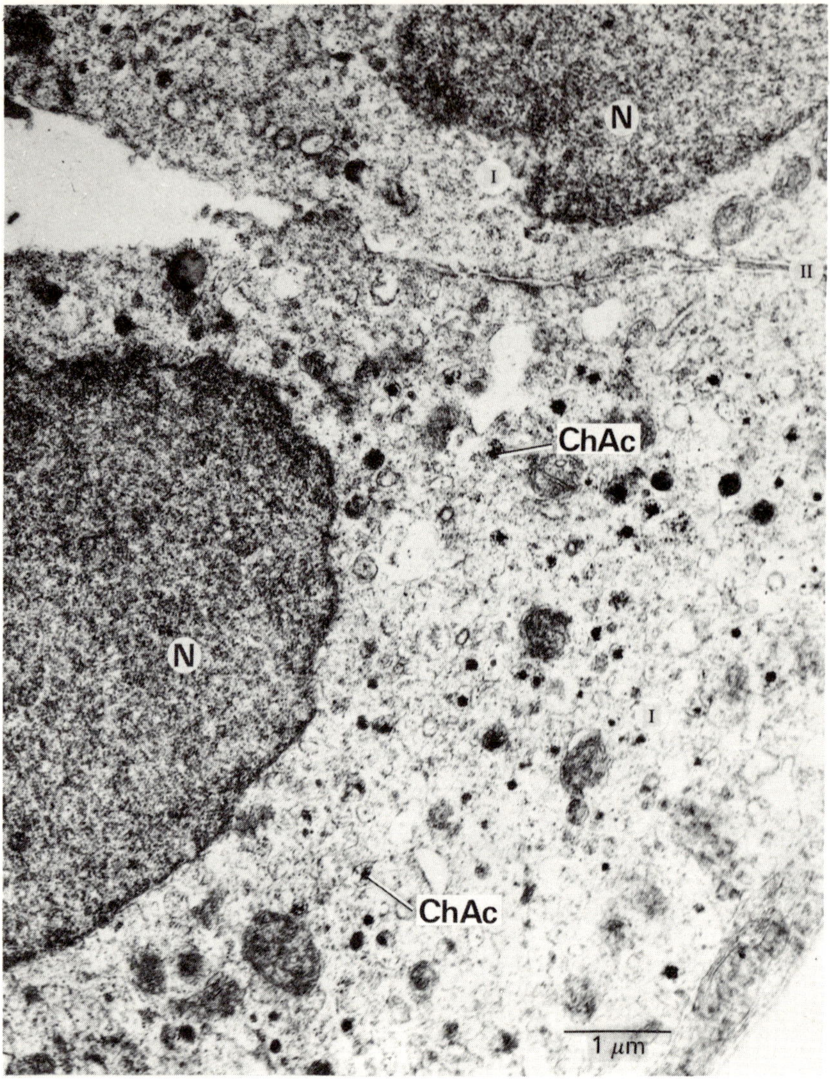

For explanation of Plate see p. 152

PLATE 5

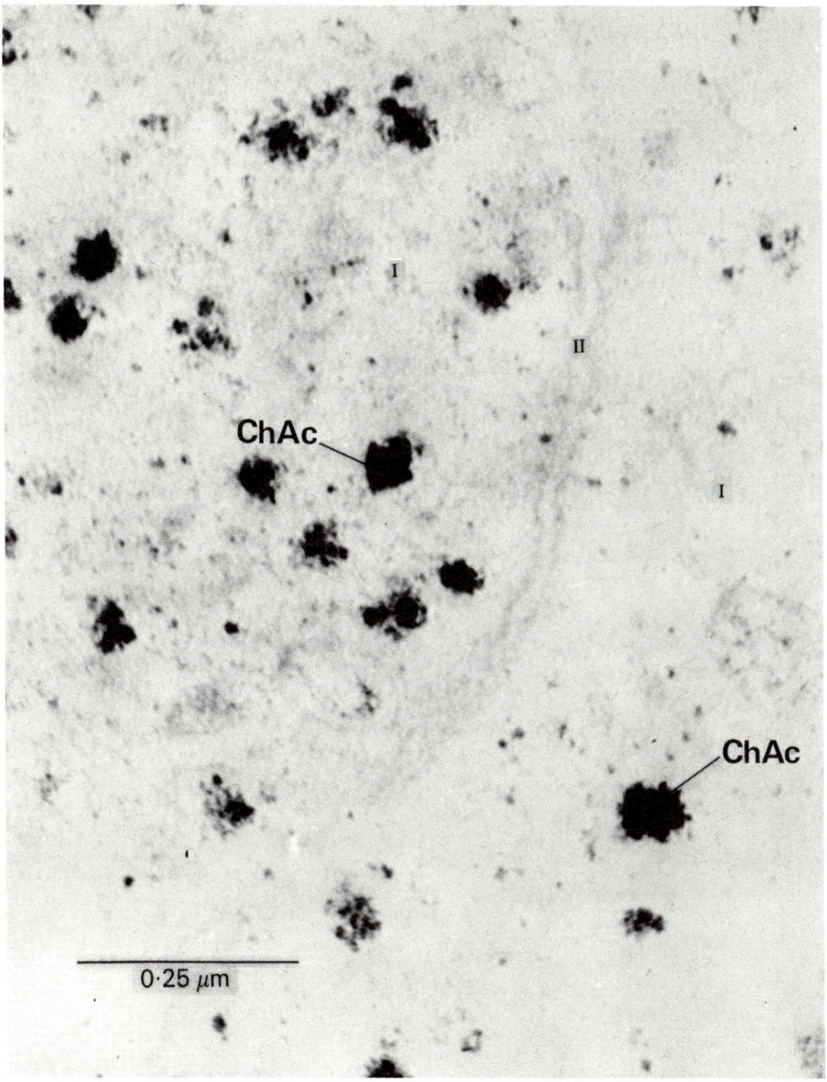

For explanation of Plate see p. 152

PLATE 6

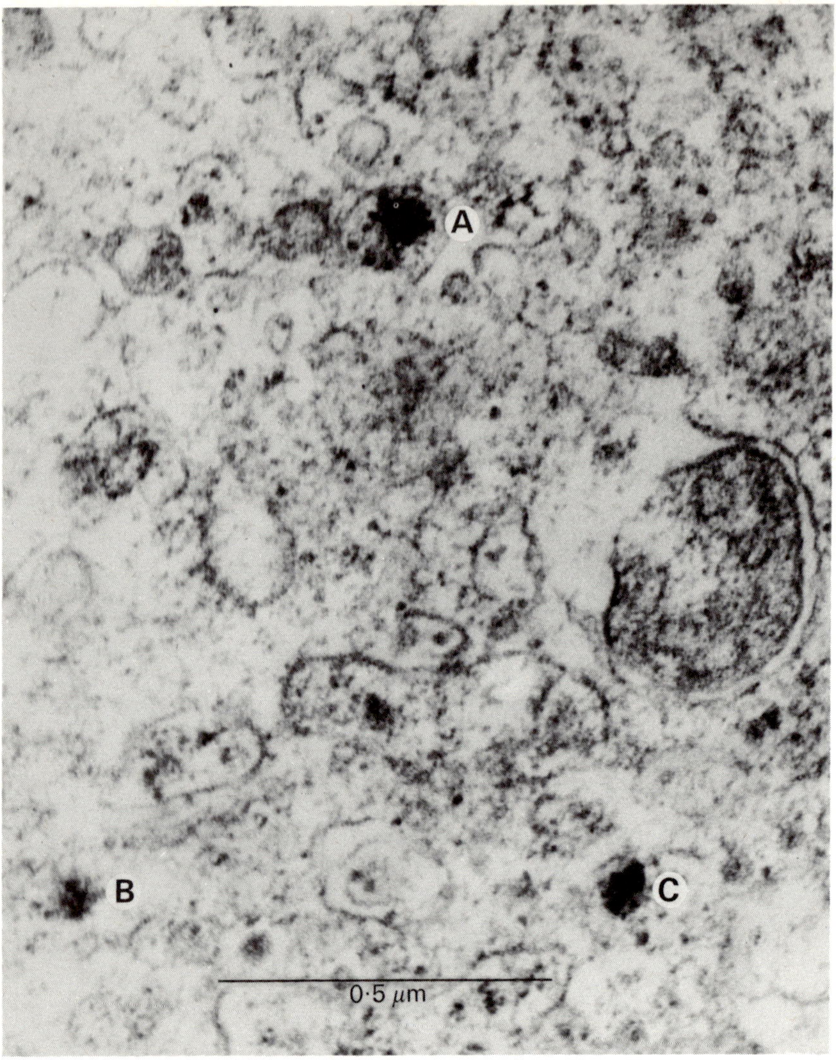

For explanation of Plate see p. 152

PLATE 7

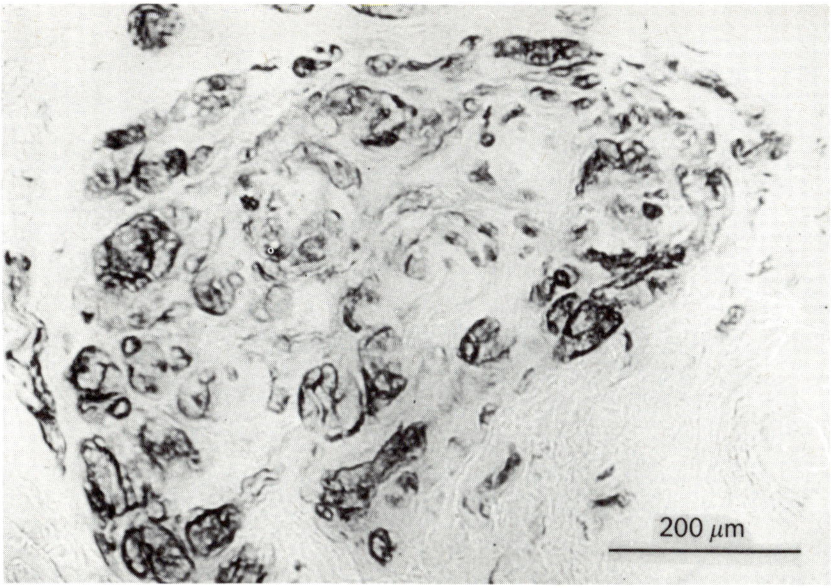

For explanation of Plate see p. 152

PLATE 8

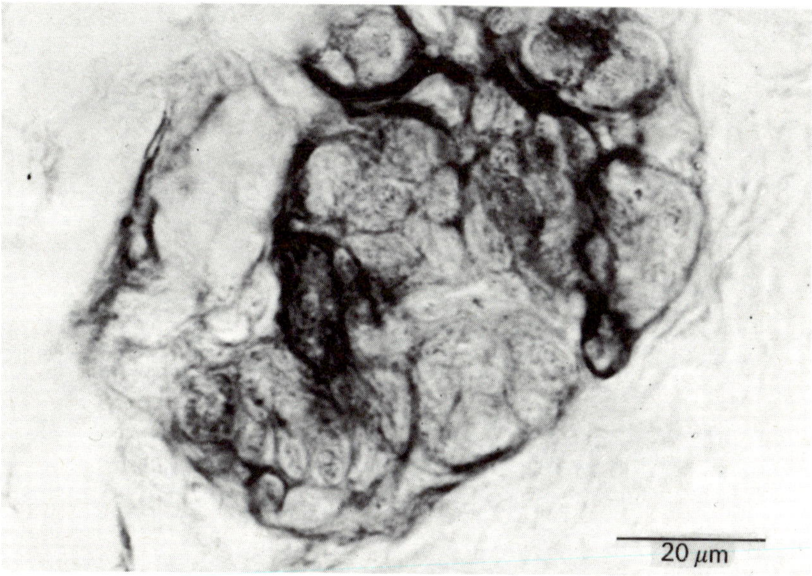

For explanation of Plate see p. 152

DISCUSSION

EYZAGUIRRE I understand that you have not been able to recover acetylcholine from the venous effluent – is that right? Could that be because the bioassay method is not sensitive enough.

JONES It certainly is one possibility. We are certainly not excluding the fact that acetylcholine may be there.

EYZAGUIRRE Fidone has been developing radiochemical methods for detecting the release of ACh. Preliminary experiments have shown that there is release of this substance in small amounts during natural carotid body stimulation *in vitro*. But the main problem is to equate receptor activity with the amount released. Everyone who has obtained release, including ourselves, has had to hit the carotid body very hard with electric current or severe asphyxia; and I think this is true of any transmitter proposed. I can illustrate the method which we have been using in an attempt to analyse the problem using radioactive choline and autoradiography, Plate 9. We plan to study this question quantitatively and it should be possible to see whether there is any loss of this material with appropriate stimulation.

BISCOE How are these sections prepared?

EYZAGUIRRE They are prepared by rapidly feeze drying the carotid bodies in liquid propane and cut on a microtome after osmium fixation and vacuum embedding in epon or araldite.

McDONALD If these data are presented as evidence for the acetylcholine hypothesis, I ask whether you have used other tissues which have membrane-bound vesicles as controls – for example, the adrenal medulla?

EYZAGUIRRE No. About 80% of the choline is bound to the phospholipids of the membrane and only a fraction is converted to acetylcholine. The way to separate these fractions is to use high-voltage electrophoresis. If you compare this with the superior cervical and nodose ganglia, the sympathetic ganglia being cholinergic has approximately the same peaks as the carotid body: the nodose ganglion being sensory does not have the acetylcholine peak but it has the phospholipid and choline peaks.

BANISTER Have you compared your estimates of ACh with those in denervated sympathetic ganglion cells and in the adrenal medulla to compare those with the carotid body?

JONES No, this has been done before by Dr Catherine Hebb.

BANISTER I know: but the values are very low for a cholinergic mechanism.

JONES But if it is not being destroyed at a high rate, does the initial content need to be high? Our findings are similar to those of Dr Hebb.

ACKER This business of the myosin filaments in the Type II cell. Do they also occur near or around the capillary?

JONES Possibly; it is possible to see myosin ATPase near the capillaries. But this is by no means certain proof.

BISCOE What fixative did you use?

JONES Formaldehyde with phosphate buffer.

BISCOE Because it seems to me that if you are going to make claims about the localization of these materials, for me to be persuaded, I would need to see better fixation. Is there some reason why you couldn't improve this?

JONES Well: the difficulty is that you will lose enzyme – particularly choline acetylase. I agree, formaldehyde 1 % is an atrocious fixative. The cholinesterase fixative was 4 % formaldehyde: again there was almost certainly loss of enzyme and I think that we have the dilemma that with proper fixation, you lose enzymes. I cannot see a way round it.

BISCOE Do you not think it would be worth using a cryostat – that is, a freezing attachment to your microtome?

JONES It may be.

PLATE 9

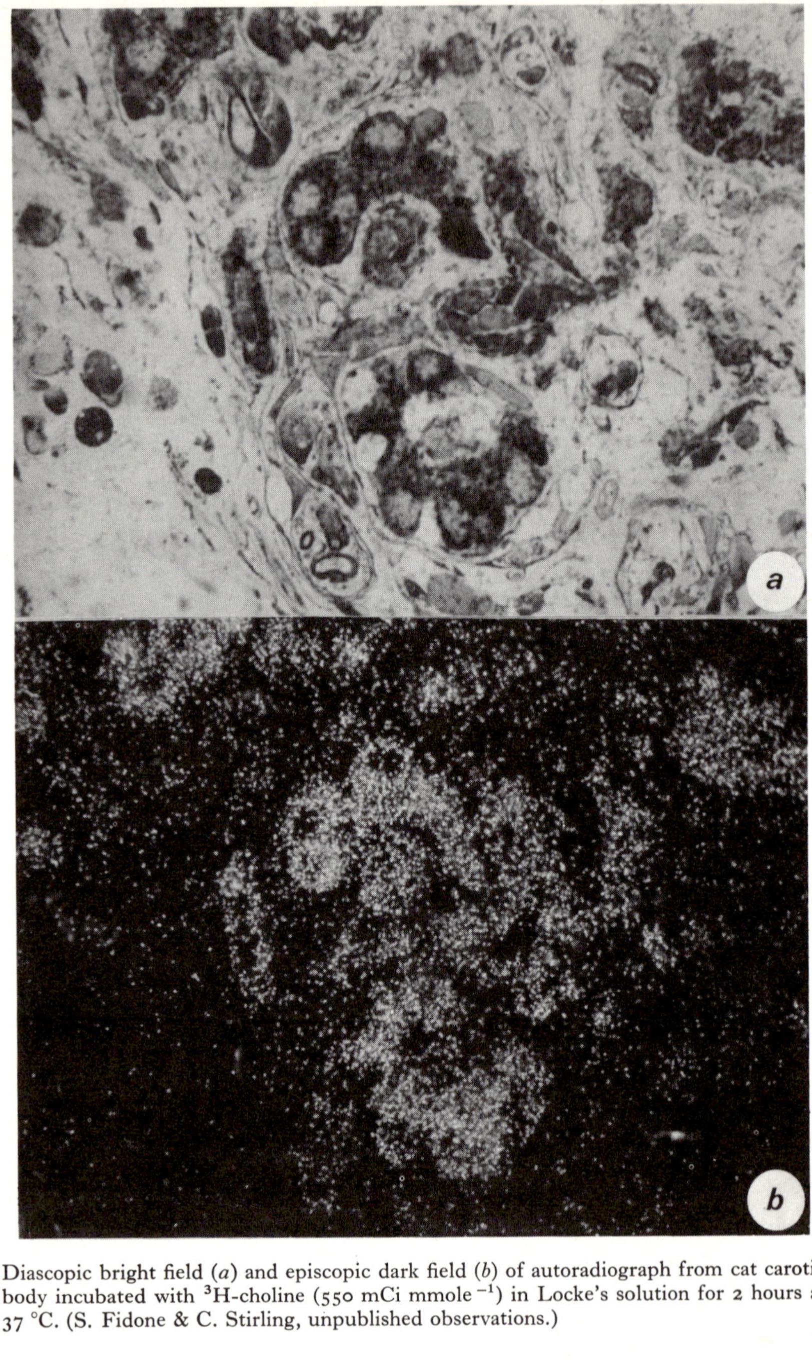

Diascopic bright field (*a*) and episcopic dark field (*b*) of autoradiograph from cat carotid body incubated with ^{3}H-choline (550 mCi mmole^{-1}) in Locke's solution for 2 hours at 37 °C. (S. Fidone & C. Stirling, unpublished observations.)

Responses of spinal and cortical neurons to changes of P_{O_2} and P_{CO_2} in blood and tissue

By E.-J. SPECKMANN and H. CASPERS
Physiologisches Institut der Universität Münster/Westf., Fed. Rep. Germany

In the context of research on chemoreceptor function it may be valuable to compare the responses of specialized neural tissue in the carotid body with the reactions of other neural tissue in the mammalian CNS to changes of gas pressures in blood and tissue. It will be shown that a reduction of P_{O_2} and an increase in P_{CO_2} have, in part, opposite effects on most cortical and spinal neurons. Therefore we shall start, in this contribution, by describing separately the neural responses to *isolated hypoxia* and *hypercapnia* and then consider the interactions between these agents during primary *asphyxia* in a final section.

The investigations reported here were carried out on spinal and cortical neurons of anaesthetized and artificially ventilated cats and rats. In order to test the influence of supra-spinal structures on the response of spinal neurons, the experiments were performed both with intact cerebro-spinal connections and after sectioning the cord at C4. Significant differences of results obtained in the two preparations were not detectable. The techniques applied for the recording of unit activity and for measurements of gas pressures in blood and in tissue have been described in detail elsewhere (Caspers & Speckmann, 1972; Speckmann & Caspers, 1967, 1969; Speckmann, Caspers & Sokolov, 1970).

EFFECTS OF HYPOXIA

An isolated hypoxia, i.e. at nearly constant P_{CO_2}, was achieved by a reduction of the oxygen content in the inspired gas mixture so that a critical disturbance of the circulation could be avoided. Under these conditions, lowering of oxygen tension in tissue was accompanied by an initial decrease in membrane potential of the lumbar neurons (Fig. 1*c*,*d*). The depolari-

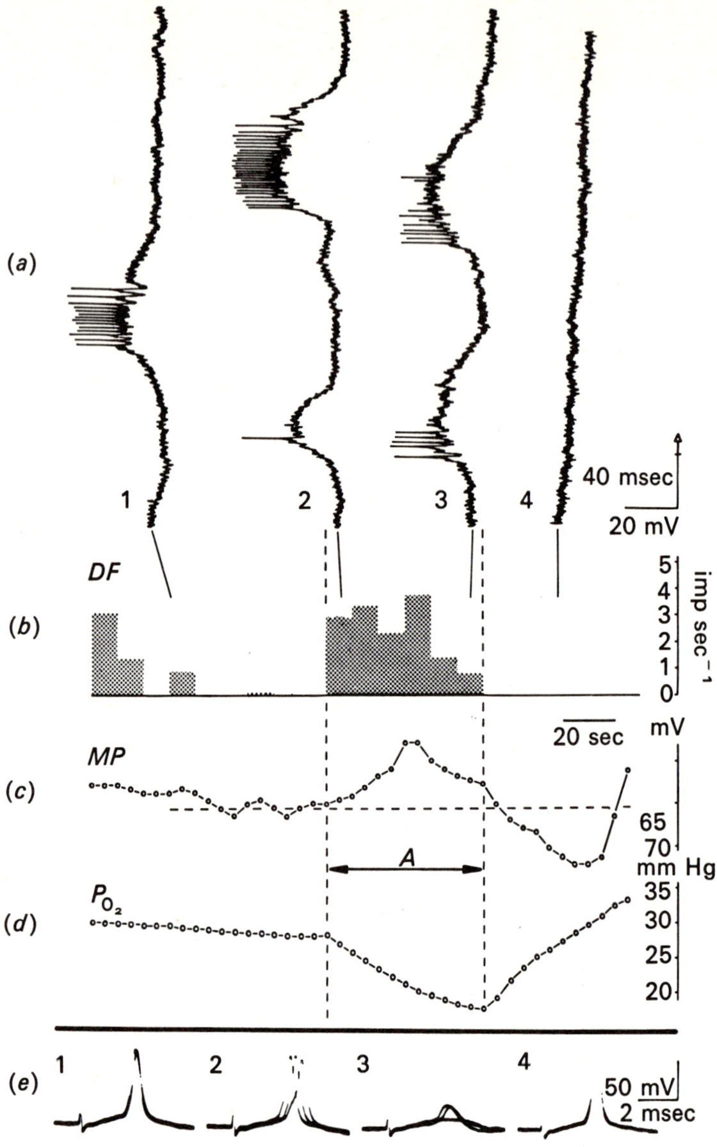

Fig. 1. Changes in membrane potential (*MP*), in discharge frequency (*DF*) and in post-synaptic and action potentials [(*a*) 1–4] of a rat lumbar interneuron during a decrease of tissue P_{O_2} in the immediate surrounding of the impaled unit. The fall in P_{O_2} was caused by a short ventilatory arrest (*A*). The selected oscilloscopic tracings of the MP [(*a*) 1–4] are related to the graph (*c*) by black lines. (*e*) Hypoxic effects on post-synaptic and action potentials elicited by dorsal root stimulation in the rat lumbar cord. (*e*) 1: Control recordings, three sweeps superimposed; (*e*) 2 and 3: during artificial ventilation with 8 % O_2 in N_2; (*e*) 4: re-ventilation with air. (From Caspers & Speckmann (1972). *Epilepsia*, **13**, 699–725.)

zation was associated with an enhancement of spontaneously occurring excitatory post-synaptic potentials (e.p.s.p.) in respect to frequency, amplitude and duration (Fig. 1*a*). The increase in synaptic activity was most evident in interneurons. As is shown by the graph in Fig. 1*b*, these neural reactions led to the initial increase of discharge frequency which is well-known to occur from many previous studies (Caspers & Speckmann, 1971, 1972; Collewijn & van Harreveld, 1966; Creutzfeldt, Kasamatsu & Vaz-Ferreira, 1957; Eccles, Løyning & Oshima, 1966; Glötzner, 1967; Kolmodin & Skoglund, 1959; Niechaj & van Harreveld, 1968; Speckmann *et al.*, 1970). With progressive hypoxia, however, the spike amplitudes became fractionated (Fig. 1*a*3). A more informative display of this phenomenon is given in part *e* of Fig. 1. The curves of Fig. 1*e*1 represent control action potentials elicited by weak dorsal root stimulation. With persisting hypoxia (Fig. 1*e*2,3), spike potentials at first showed a reduced slope of rise, then the B-spike was blocked and finally all somatic spike generation failed. In this phase of oxygen deficiency, e.p.s.p. could still be evoked by pre-synaptic stimulation. These changes were reversed when the animal was re-ventilated with air (Fig. 1*e*4).

In summary, pure hypoxia causes a biphasic response of neuronal structure in the mammalian CNS. With slight or initial progressive hypoxia, excitability and excitation increase; with more severe or prolonged oxygen deficiency, a depression takes place with a rapid decline of spike activity.

EFFECTS OF HYPERCAPNIA

In order to study the neural activity when P_{CO_2} was raised, we applied the so-called apnoea technique (Speckmann & Caspers, 1967; Speckmann *et al.*, 1970). This method allows P_{CO_2} to be raised without causing hypoxia in cerebral or spinal tissue. Prior to the period of apnoea caused by turning off the ventilation pump, the paralysed animal had been ventilated for about 1 hour with 100% oxygen. During the arrest of artificial ventilation, the oxygen container was left open to the animal's trachea. When the pump was switched off, arterial P_{O_2} fell only slowly from its initial high values while the P_{CO_2} progressively increased. After the onset and termination of an apnoeic period, tissue P_{O_2} increased rapidly and this was followed by a slow decay (Figs. 2, 3*d*, 4*b*). These fluctuations in tissue P_{O_2} can be interpreted as being due to changes in local blood flow, in arterial P_{O_2}, and in oxygen consumption.

Examination of the discharge frequency of spontaneously active neurons during hypercapnia induced by this method showed that two neural reaction types can be distinguished. In a small population of cells, the dis-

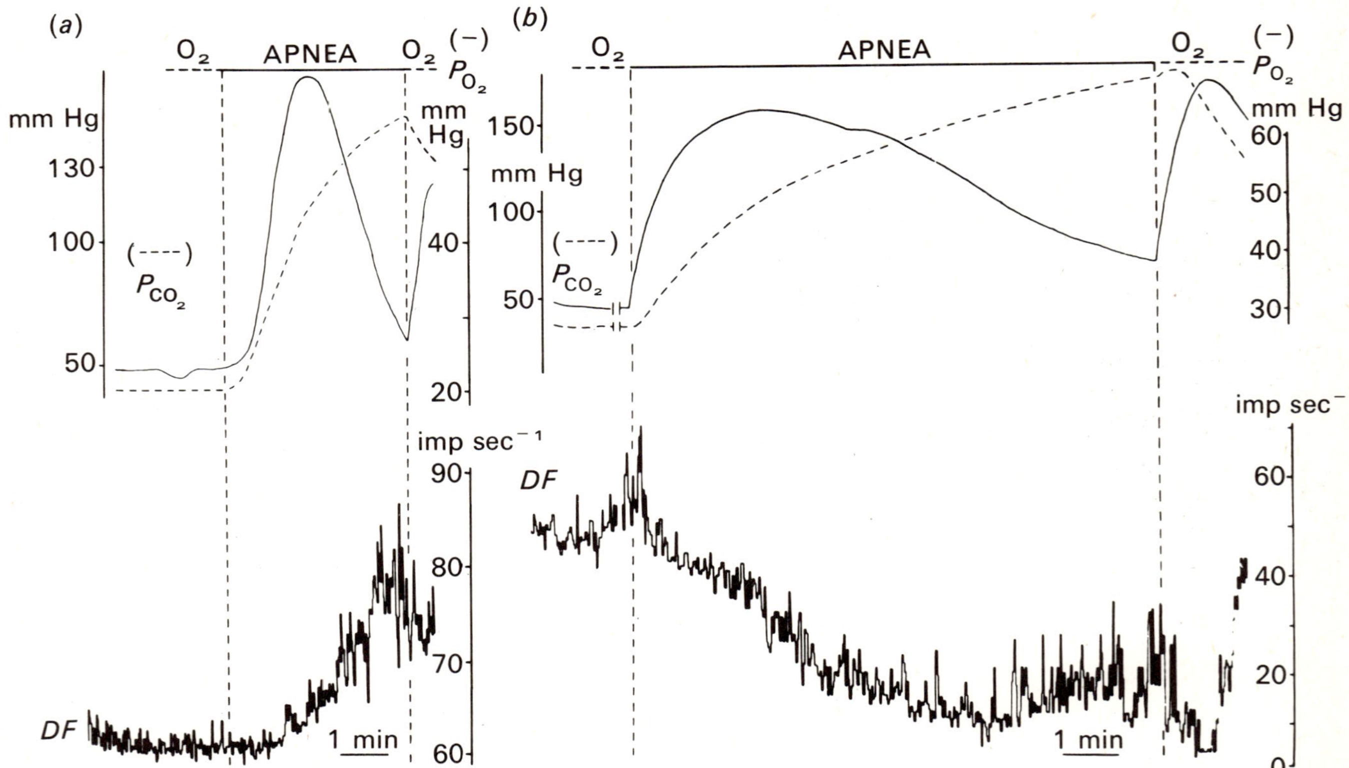

Fig. 2. Variations of discharge frequency (*DF*) of rat lumbar neurons, of local tissue P_{O_2} and of arterial P_{CO_2} during apnoeic periods (dashed vertical lines) after O_2 ventilation: (*a*) E-neuron; (*b*) I-neuron. (From Speckmann *et al.* (1970). *Pflügers. Arch. ges. Physiol.*, **319**, 122–38.)

charge rate was enhanced (Fig. 2*a*). These neurons were labelled as 'E-neurons' (for excited units; see Speckmann *et al.*, 1970). In the lumbar region of the rat spinal cord, they comprised 5–10 % of the impaled cells and all of them had the appearance of being interneurons. The vast majority of neurons, however, tested in all parts of the spinal cord and all units in the cerebral cortex, showed a different type of response to a rise in carbon dioxide. As shown in part *b* of Fig. 2, a rise in P_{CO_2} was associated with a decrease in discharge frequency often down to zero level. Because of this inhibition, such neurons have been called 'I-neurons' (Speckmann *et al.*, 1970). Since the majority of neurons in the mammalian CNS proved to be I-neurons, the reactions of this group only will be described in further detail. The responses of invertebrate neurons to changes in gas pressures have been described by Chalazonitis (1968) and by Chalazonitis & Arvanitaki (1970).

In I-neurons, a rise of P_{CO_2} was always accompanied by a *hyperpolarization* of the membrane (see also Caspers & Speckmann, 1971, 1972; Krnjevic, Randic & Siesjö, 1965; Speckmann & Caspers, 1969; Speckmann *et al.*, 1970). A typical tracing is shown in parts *c* and *d* of Fig. 3. In order to get further information on the mechanisms underlying the increase in membrane potential, the behaviour of post-synaptic potentials as well as of some membrane properties has been studied in more detail (Caspers & Speckmann, 1971, 1972; Papajewski, Klee & Wagner, 1969; Speckmann & Caspers, 1969; Speckmann *et al.*, 1970). At first, oligo- and poly-synaptic e.p.s.p. of lumbar moto-neurons evoked by peripheral and cortical stimulation were tested (Fig. 3*a*). As shown in Fig. 3*b*, averaged e.p.s.p. elicited by cortical stimulation decreased progressively in amplitude, duration and slope of rise during apnoea, and recovered with re-ventilation and fall in P_{CO_2}. Generally, the depressive effect of carbon dioxide became more pronounced with increasing number of synaptic relays interposed between stimulation and recording sites. The analysis of carbon dioxide actions on inhibitory post-synaptic potential (i.p.s.p.) was more difficult since the membrane potential was shifted towards the equilibrium potential of the i.p.s.p. which by itself would result in a reduction of these p.s.ps. However, simulating the carbon dioxide-induced increase in membrane potential by hyperpolarizing current injections through an impaled microelectrode, the changes of i.p.s.p. were found to be less pronounced. This finding as well as the observation that the membrane input resistance rose up to double the normal level during the apnoeic period suggest that the i.p.s.ps were reduced as well.

Thus far, the effects of carbon dioxide upon complex synaptic activities have been considered but clearly there could be *direct actions* of carbon

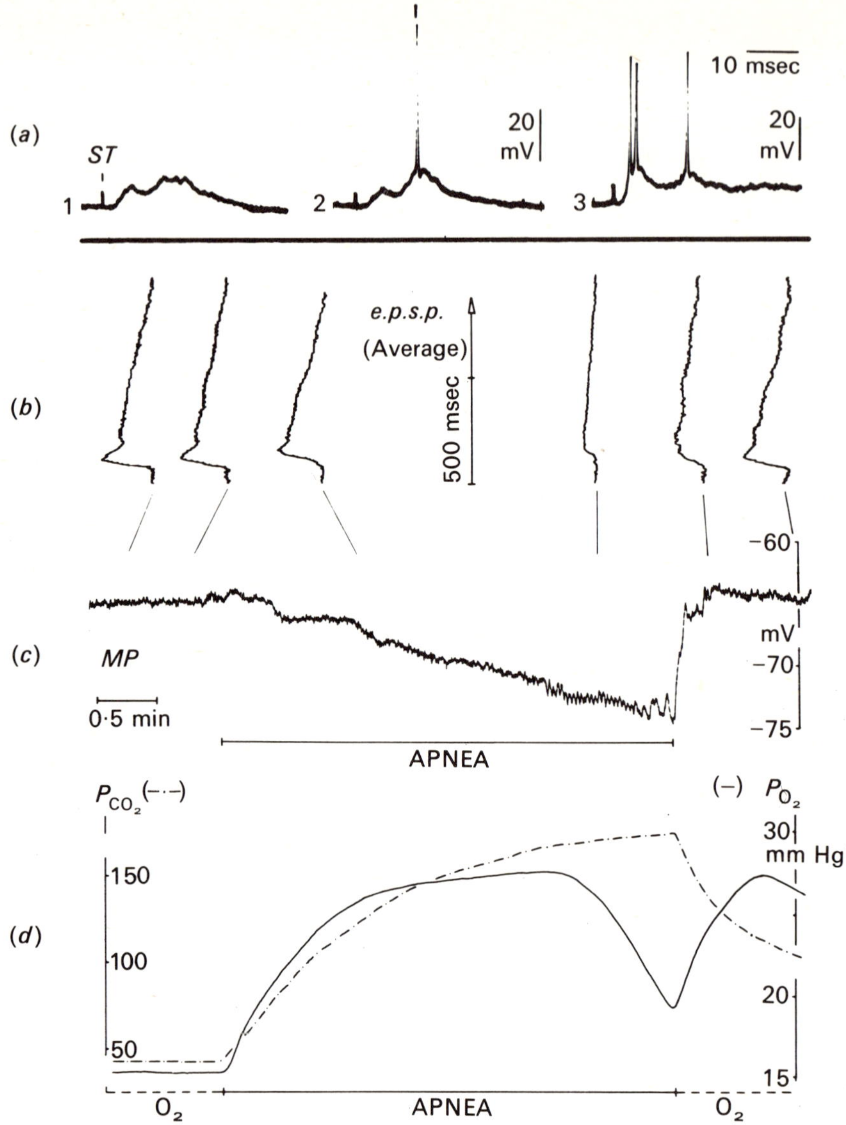

Fig. 3. Effects of an increased P_{CO_2} on membrane potential (*MP*) and on excitatory post-synaptic potentials (e.p.s.p.) of a rat lumbar I-neuron. (*a*) Sequences of e.p.s.p. elicited by single stimuli (*ST*.) applied to the sensorimotor cortex. (*b*) Averaged e.p.s.p. (20 sweeps) evoked by epicortical stimulation. The relation to the *MP* tracing is indicated by black lines. (*c*) Tracing of the *MP* performed by an inkwriter. (*d*) Simultaneous changes in arterial P_{CO_2} and tissue P_{O_2}. (From Caspers & Speckmann (1972). *Epilepsia*, **13**, 699–725.)

dioxide on the properties of neural membranes: and this question was considered in some further experiments. First, the influence of an isolated hypercapnia on antidromic action potentials evoked by ventral root stimulation was studied. As shown in Fig. 4*a* and *b*, the antidromic action potentials were blocked if an arterial P_{CO_2} of 100–150 mm Hg had been reached. In the course of this process the transient A–B blockade which was frequently observed during hypoxia (Fig. 1*e*) was never seen. Furthermore,

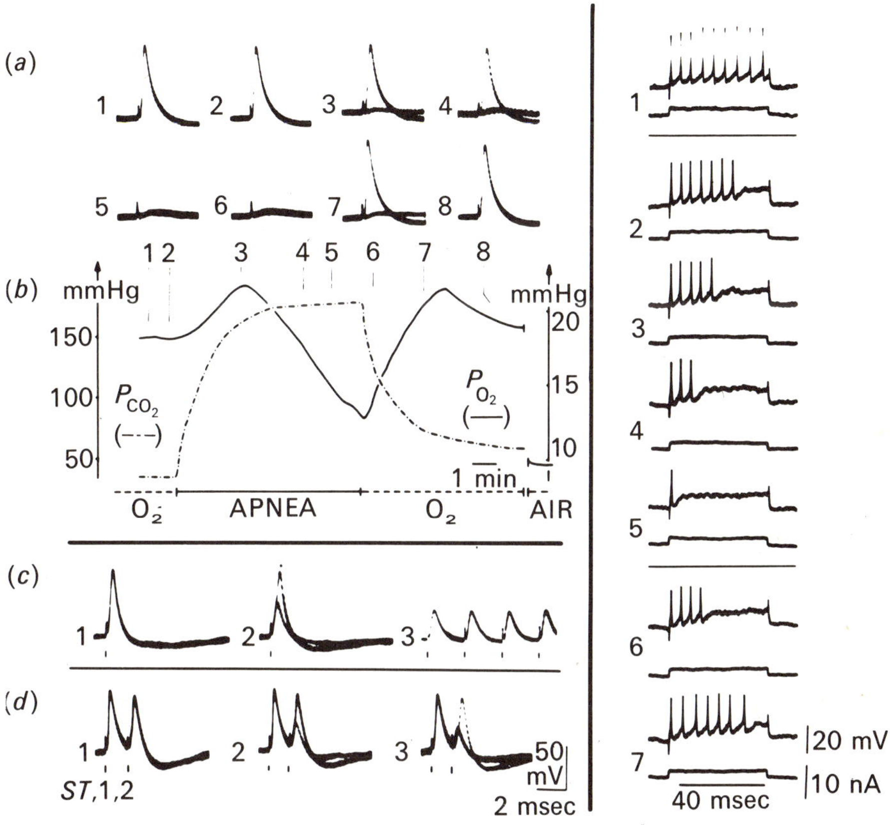

Fig. 4. (*a*) and (*b*) Blockade of antidromic action potentials by an increase in P_{CO_2} during an apnoea. Action potentials were evoked in a rat lumbar motoneuron by ventral root stimulation (*a*). The relation between antidromic action potentials and changes in arterial P_{CO_2} and tissue P_{O_2} (*b*) is indicated by corresponding numbers.

(*c*) and (*d*) Separation of the A and B components of antidromic action potentials by high-frequency stimulation (*c*) and by paired impulses (*ST*: 1,2) with decreasing intervals (*d*); same neuron as in (*a*).

(*e*) Depression of repetitive discharges in a rat lumbar motoneuron by an increase in P_{CO_2} during an apnoeic period. Repetitive activity was elicited by injecting rectangular current pulses through the microelectrode. The current is monitored in each second row. 1: Control recording at 40 mm Hg P_{CO_2}; 2–5: increase in P_{CO_2} up to 120 mm Hg; 6 and 7: decreasing P_{CO_2} after re-ventilation. (From Caspers & Speckmann (1972). *Epilepsia*, **13**, 699–725.)

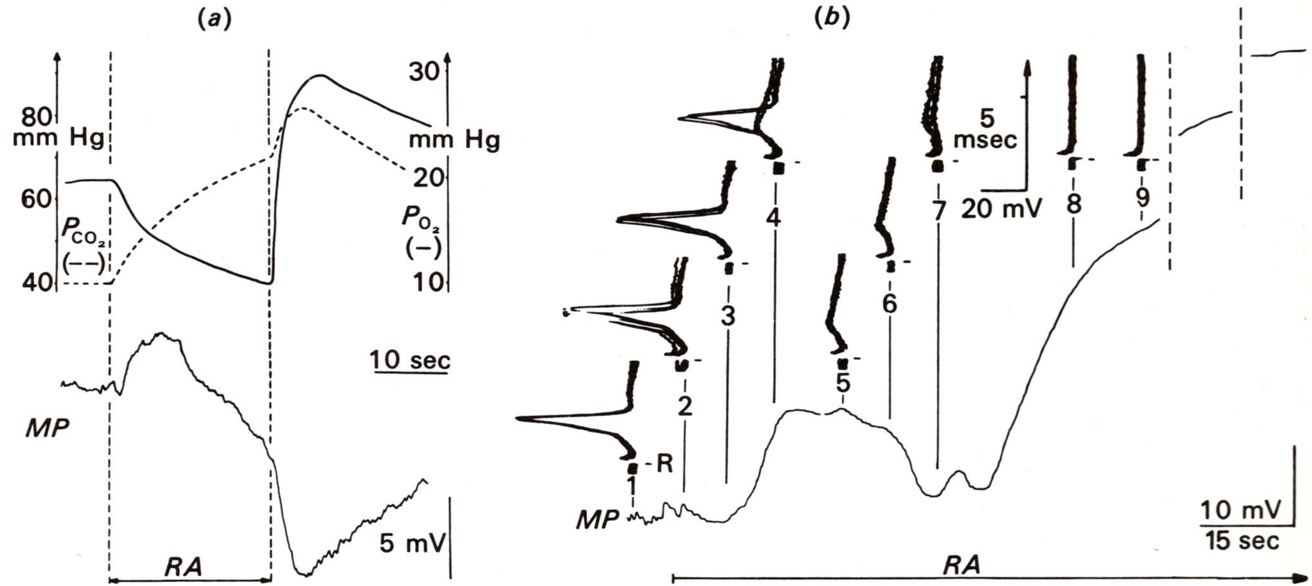

Fig. 5. Changes in membrane potential (*MP*) of rat lumbar I-neurons during a transient (*a*) and a persistent (*b*) respiratory arrest (*RA*) after air ventilation. The *MP* was traced by an inkwriter. (*a*) The upper part displays the concomitant variations in arterial P_{CO_2} and in local P_{O_2}. (*b*) *e.p.s.p.* and action potentials (5 responses superimposed) evoked by dorsal root stimulation (*R*) are related to the *MP* curve by vertical lines. Broken lines indicate interruptions of *MP* tracing for 30 sec each. (From Speckmann, Caspers & Sokolov (1970), *Pflügers. Arch. ges. Physiol.*, **319,** 122–38; and Caspers & Speckmann (1971), *Ärztliche Forschung*, **25,** 241–55.)

the repetition rate of action potentials elicited by rectangular current pulses was markedly reduced at high P_{CO_2} levels (Fig. 4*e*, 2–5). This depressive action is attributable mainly to a direct damping effect on the membrane. From comparative experiments with sustained artificial hyperpolarization, one can conclude that the carbon dioxide-dependent increase in membrane potential could be responsible only in part for this reaction and for the effect upon antidromic action potentials. These results therefore show that the depressive action of carbon dioxide on I-neurons can be mediated via synaptic processes and by direct actions on the cell membrane.

INTERACTIONS OF HYPOXIA AND HYPERCAPNIA

During primary asphyxia caused by turning off the respiratory pump after preliminary ventilation on air instead of 100% oxygen, some interactions between the effects due to hypoxia and hypercapnia described above were observed (Caspers & Speckmann, 1971, 1972; Speckmann & Caspers, 1967; Speckmann *et al.*, 1970). At first, a decrease in membrane potential occurred which was due to the early deficiency in oxygen (Fig. 5). In I-neurons, this depolarization was followed by an intermediate re- and sometimes hyper-polarization attributable to the rise in P_{CO_2} and to a simultaneous dis-facilitation. When the local P_{O_2} had reached a critical level, the intermediate repolarization progressed to a final breakdown of the membrane potential (Fig. 5*b*). It is remarkable that the action potentials had already disappeared at a time when the membrane was not yet critically depolarized. A successful re-ventilation was always accompanied by an increase in membrane potential, frequently beyond the initial level (Fig. 5*a*). This reactive hyperpolarization was usually associated with a rapid rise of P_{O_2} and an additional increase of P_{CO_2}. In E-neurons, on the contrary, a considerably enhanced activity was found in the post-asphyxial period (Speckmann *et al.*, 1970).

The rapid neural depolarization which terminates in a complete breakdown of the membrane potential is obviously attributable to a progressive lowering of tissue P_{O_2}. However, up to now the mechanisms underlying this effect have been poorly understood. Referring to findings in the literature (Kerkut & York, 1971; Ling, Miller & Ochsenfeld, 1973), a blockade of known active membrane processes, such as ion pumps, may not exclusively be responsible for this phenomenon.

SUMMARY

The described experimental findings show:

(1) A lowering of P_{O_2} in tissue exerts a biphasic effect on neural structures which consists of an initial increase in excitability and excitation and a later blockade of activity.

(2) An increase in local P_{CO_2} causes an excitation in a minority of neurons but in the majority, it leads to a depression of activity.

(3) During primary asphyxia, there are interactions of the two latter effects which predominate alternately.

REFERENCES

Caspers, H. & Speckmann, E.-J. (1971). Gleichspannungsverschiebungen an der Hirnrinde bei Asphyxie. *Ärztl. Forsch.*, **25**, 241–55.

Caspers, H. & Speckmann, E.-J. (1972). Cerebral P_{O_2}, P_{CO_2} and pH: changes during convulsive activity and their significance for spontaneous arrest of seizures. *Epilepsia*, **13**, 699–725.

Chalazonitis, N. (1968). Intracellular P_{O_2} control on excitability and synaptic activity in aplysia and helix identifiable giant neurons. *Ann. N. Y. Acad. Sci.*, **147**, 419–59.

Chalazonitis, N. & Arvanitaki, A. (1970). Neuromembrane Electrogenesis During Changes in P_{O_2}, P_{CO_2} and pH. *Adv. Biochem. Psychopharmacol.*, **2**, 245–84.

Collewijn, H. & van Harreveld, A. (1966). Intracellular recording from cat spinal motoneurons during acute asphyxia. *J. Physiol. Lond.*, **185**, 1–14.

Creutzfeldt, O., Kasamatsu, A. & Vaz-Ferreira, A. (1957). Aktivitätsänderungen einzelner corticaler Neurone im akuten Sauerstoffmangel und ihre Beziehungen zum EEG bei Katzen. *Pflügers Arch. ges. Physiol.*, **263**, 647–67.

Eccles, R. M., Løyning, Y. & Oshima, T. (1966). Effects of hypoxia on the monosynaptic reflex pathway in the cat spinal cord. *J. Neurophysiol.*, **29**, 315–32.

Glötzner, F. (1967). Intracelluläre Potentiale, EEG und corticale Gleichspannung an der sensomotorischen Rinde der Katze bei akuter Hypoxie. *Arch. Psychiat. Nervkrankh.*, **210**, 274–96.

Kerkut, G. A. & York, B. (1971) *The Electrogenic Sodium Pump*, Scientechnica: Bristol.

Kolmodin, G. M. & Skoglund, C. R. (1959). Influence of asphyxia on membrane potential level and action potentials of spinal moto- and interneurons. *Acta physiol. Scand.*, **45**, 1–18.

Krnjevic, K., Randic, M. & Siesjö, B. K. (1965). Cortical CO_2 tension and neuronal excitability. *J. Physiol., Lond.*, **176**, 105–22.

Ling, G. N., Miller, C. & Ochsenfeld, M. M. (1973). The physical state of solutes and water in living cells according to the association-induction hypothesis. *Ann. N. Y. Acad. Sci.*, **204**, 6–47.

Niechaj, A. & van Harreveld, A. (1968). Intracellular recording from cats spinal interneurons during acute asphyxiation. *Brain Res.*, **8**, 54–64.

Papajewski, W., Klee, M. R. & Wagner, A. (1969). The action of raised CO_2 pressure on the excitability of spinal motoneurons. *Electroenceph. clin. Neurophysiol.*, **27**, 618.

Speckmann, E.-J. & Caspers, H. (1967). Les modifications du potentiel continu cortical pendant l'arrêt respiratoire. *Rev. neurol.*, **117**, 5–19.

Speckmann, E.-J. & Caspers, H. (1969). Verschiebungen des corticalen Bestandpotentials bei Veränderungen der Ventilationsgröße. *Pflügers Arch.*, **310**, 235–50.

Speckmann, E.-J., Caspers, H. & Sokolov, W. (1970). Aktivitätsänderungen spinaler Neurone während und nach einer Asphyxie. *Pflügers Arch.*, **319**, 122–38.

DISCUSSION

MITCHELL I would like to extend this to other cells – respiratory cells in the medulla in the cat. These respond in the same way as your motor cells. They are hyperpolarized by carbon dioxide and depolarized as you lower carbon dioxide. There appear to be two populations of cells: high threshold cells which become silent when you lower carbon dioxide and low threshold cells which fire continuously if you lower carbon dioxide. The second effect of carbon dioxide upon these cells is on synaptic input, which is not random but highly synchronized, causing a spindle-like burst over the first two-thirds of depolarization with each inspiration. The frequency and amplitude of these synchronized inputs is increased with carbon dioxide so that this accounts for the increased firing of these cells with carbon dioxide. The direct effect is therefore one of increased membrane resistance and hyperpolarization, similar to that which you have described for motor cells in the spinal cord.

TORRANCE How do you relate your results with those of Chalazonitis (1963) who claimed that in simple nervous systems, he can get both raised P_{CO_2} and lowered P_{O_2} to depolarize the cell.

SPECKMANN It is very hard to correlate with the complex mammalian system. I am not sure that it can be done.

TORRANCE Is he in a more advantageous position to study this particular question, dealing as he does with a simple nervous system?

SPECKMANN Theoretically, he might seem to have the advantage dealing with fewer inputs and so on. But there is a difficulty because the nerve cells in the systems have neural membranes which are often highly specialized – that is they act like chemoreceptors, for example.

TORRANCE But maybe that is an advantage for studying chemoreceptors?

SPECKMANN Yes, but they will be less properly controlled.

FILLENZ How far do you think the results you have obtained are due to the direct action of carbon dioxide or hypoxia upon the cells you are recording from and how far do you think they may be due to transmitter release or something like that?

SPECKMANN It is hard to say. The direct effect is represented in the damping effect and in the blockade of the spike conduction of the axon hillock region by antidromic stimulation.

FILLENZ But that could be due to the action of a transmitter on the cell membrane.

SPECKMANN Under this condition, the membrane resistance would have to decrease by a large amount. But it was not the case.

FILLENZ But there are examples in the central nervous system of transmitters which cause a rise in membrane resistance: for example, Krnjevic, Pumain & Renaud (1971) have shown this in the cortex.

SPECKMANN But I think that is another story.

BISCOE What happened to blood pressure when you made the animals apnoeic?

SPECKMANN There was a rise in blood pressure at first by approximately 20–30 mm Hg: but by later in the apnoeic period, pressure had fallen to normal levels. We think this is one of the advantages of this method that blood pressure remains at or above normal levels throughout the testing period.

REFERENCES

CHALAZONITIS, N. (1963). Effects of changes in P_{CO_2} and P_{O_2} on rhythmic potentials from giant neurons. *Ann. N. Y. Acad. Sci.*, **109**, 451–79.

KRNJEVIC, K., PUMAIN, R. & RENAUD, L. (1971). The mechanism of excitation by acetylcholine in the cerebral cortex. *J. Physiol.*, **215**, 247–68.

Recent studies on the generation of chemoreceptor impulses

By C. EYZAGUIRRE, S. FIDONE and K. NISHI*
Department of Physiology, University of Utah College of Medicine, Salt Lake City, Utah 84132, USA

One of the main problems in carotid body function is to understand the mechanisms of chemosensory impulse initiation. In this respect, we have had a multitude of physiological and pharmacological studies utilizing reflexes and electrical recording from single or multiple sensory units during receptor stimulation with 'natural' (low P_{O_2}, high P_{CO_2} or acidity) or chemical (NaCN, ACh, KCl, etc.) stimuli. These studies have led to some understanding of the function of chemoreceptors; but this 'black box' approach has already yielded most of the information that these techniques can give. The information has been ample, but contradictory at times, and the essence of the problem, namely, a clear knowledge of the events leading to the generation of the sensory impulses, is still not available. It is true that there are a number of hypotheses trying to explain these phenomena; but unfortunately, these ideas have remained largely untested due to the complexity of these receptors. It is our feeling that the time has come to put to experimental tests the different hypotheses on chemoreceptor function. To do this, it is necessary to open the 'box' and analyse events occurring at the receptor level. This can be accomplished by utilizing chemical, histochemical, morphological and electrophysiological techniques. This presentation will deal exclusively with the latter, since a thorough review of other possible approaches would be too long to be included here.

To analyse electrical events occurring at the receptor level it is necessary to know what happens to the nerve endings and/or glomus cells during activity of the receptors. As shown elsewhere in this Symposium, the structure of the nerve endings is complex and the terminals have a great variety of shapes (see also Plate 1). The endings have an intimate relationship with the glomus Type I cells; furthermore, the glomus Type I cell–nerve ending complex is enveloped by another cell, the glomus Type II cell. Ideally,

* Present address: Department of Pharmacology, Kumamoto University, Kumamoto, Japan.

therefore, one must have direct experimental evidence on the behaviour of these three elements. Until recently this has been lacking. Consequently, the aim of this laboratory has been to explore the function and anatomical relationship of these structures to gain some further insight into the chemoreceptor mechanisms.

To study the electrical events occurring at the nerve endings, it is necessary to record the generator (receptor) potential which eventually leads to the all-or-none discharge of the nerve fibres. In other receptor organs, such as the Pacinian corpuscle (Gray & Sato, 1953; Alvarez-Buylla & Ramirez de Arellano, 1953), the frog muscle spindle (Katz, 1950), the crayfish stretch receptor (Eyzaguirre & Kuffler, 1955), photoreceptors (see Hartline, Wagner & McNichol, 1952), etc., this has been accomplished successfully and we have a fairly good understanding of the events leading to the development of the generator potential and the ensuing spike discharge. Unfortunately, this is not the case with chemical receptors, since generator (receptor) potentials have not been recorded until recently. Furthermore, their significance in terms of impulse origin is still less clear than the situation occurring in photo- or mechanoreceptors where a wealth of information about receptor mechanisms is available (for references, see Loewenstein, 1971). For instance, stimulation of olfactory receptors produces a slow negative wave which precedes or accompanies the increased discharge of the olfactory nerve (Ottoson, 1956; Gesteland, Lettvin, Pitts & Rojas, 1963). Similarly, activation of taste receptors induces an increased nerve discharge together with depolarization of the taste bud cells (Kimura & Beidler, 1961; Tateda & Beidler, 1964). However, in all these instances, there is still some uncertainty about the significance of these waves, since neighbouring epithelial cells which are not involved with either olfaction or taste also become depolarized by the stimuli, which appear to be specific for the receptors mentioned above (Eyzaguirre, Fidone & Zapata, 1972). In the case of the arterial chemoreceptors, receptor or generator potentials have not been recorded until recently (Eyzaguirre, Leitner, Nishi & Fidone, 1970), and these electrical events are still poorly understood, again due to the complexity of the chemoreceptor structure in the carotid body. This situation is even more difficult in the aortic chemoreceptors, due to the complexities of the anatomy of the region.

The recording of receptor potentials in carotid body chemoreceptors has been accomplished by using a variety of methods, such as placing a low resistance microelectrode on the nerve fibres as close as possible to their termination in the carotid body. Another method has been to place non-polarizable electrodes (calomel or Ag–AgCl) on the nerve across air, sucrose or oil gaps located between the neural pole of the carotid body and the end of the nerve. These techniques have been used for experiments *in vivo* and *in vitro* (Fig. 1). The recording leads are connected to high-input

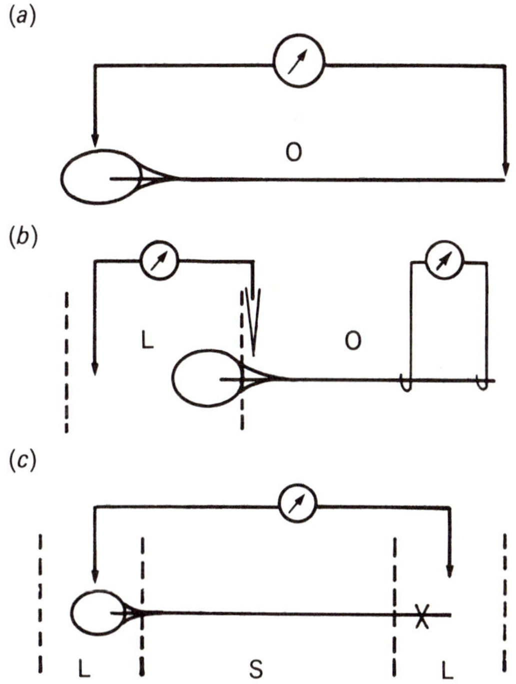

Fig. 1. Different methods used for recording the receptor potential of carotid body chemosensors. (*a*) Oil gap (O) between recording leads for *in vivo* recordings; (*b*) low resistance microelectrode placed extracellularly on nerve at its entrance into glomus. This portion of nerve is in oil, while indifferent electrode is in flowing Locke's solution (L) bathing the carotid body *in vitro*. Second set of electrodes placed near cut end of nerve to record sensory discharges; (*c*) sucrose gap (S) between containers having Locke's solution bathing carotid body and crushed end of nerve respectively. (Modified from Eyzaguirre *et al.*, 1970.)

impedance amplifiers used in conjunction with DC amplifiers of an oscilloscope or an ink-writing oscillograph. The 'active' electrodes have recorded slow baseline displacements which occur concomitantly with, or temporally very close to, changes in discharge frequency evoked by natural or chemical stimuli. In this communication, these baseline displacements will be referred to as slow negative waves, receptor depolarizations or mass receptor potentials since the baseline is always displaced in a negative direction during receptor stimulation.

Activation of the receptors by asphyxia (*in vivo*) or by chemical agents such as NaCN, ACh, KCl or acid (*in vivo* and *in vitro*) elicits receptor depolarization which appears concomitantly with an increase in sensory discharge frequency. An example of this is presented in Fig. 2.

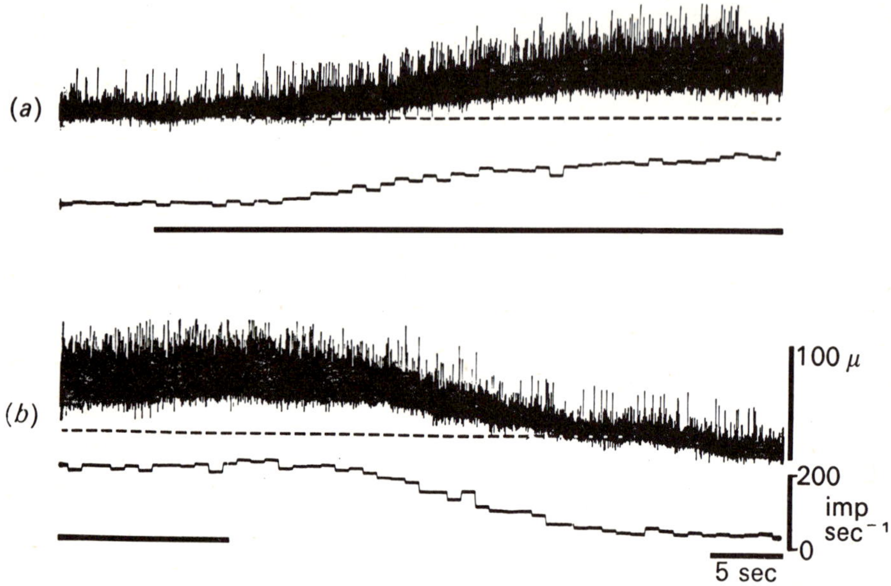

Fig. 2. Effect of asphyxia on receptor potential and sensory discharge frequency. *In vivo* experiment. Upper traces in (*a*) and (*b*), recordings from nerve; lower traces, discharge frequency. Proximal recording electrode on nerve at its entrance into glomus; distal electrode at cut end of nerve [see Fig. 1 method (*a*)]. Animal paralysed and under artificial ventilation. Horizontal bars in (*a*) and (*b*) indicate period of asphyxia when respirator was interrupted. (From Eyzaguirre & Nishi, 1974.)

Once these results were obtained it was important to determine the origin of this wave. It could have arisen by summation of negative afterpotentials when the nerve was discharging at relatively high frequency. This possibility was eliminated by applying tetrodotoxin (TTX) to the in-vitro preparation in concentrations of 4×10^{-8} g ml^{-1}. This substance eliminated the impulse discharge, but left the slow depolarization practically intact (Fig. 3); this shows that if in the preparation untreated with TTX negative after-potentials do sum, they do not contribute significantly to the slow negative wave. Similar results were obtained with local anaesthetics such as procaine or dibucaine; that is, maintenance of receptor depolarization but elimination of the sensory discharges. We were left then with two possibilities: namely, nerve ending depolarization and/or membrane potential changes in the glomus cells which are closely associated with the endings.

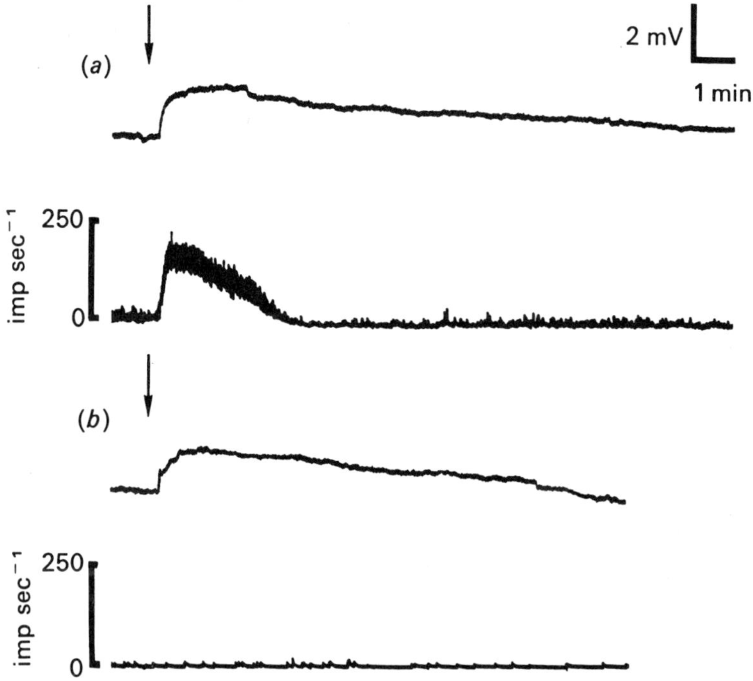

Fig. 3. Carotid body *in vitro*. Upper traces, slow depolarizing potentials recorded with Fig. 1 method (*b*). Lower traces, integrated spike discharge frequency. (*a*) Superfusion with normal Locke's solution. Arrow indicates injection of NaCN 15 μg. (*b*) Superfusion with Locke's solution containing TTX (4×10^{-8} g ml^{-1}). NaCN 15 μg at arrow. (From Eyzaguirre *et al.*, 1970.)

The latter possibility seemed unlikely, but as detailed below, it has not been left unexplored.

Another interesting characteristic of the receptor depolarization evoked by different stimuli is that it is a local potential. Thus, this wave can be recorded only if the proximal and active electrode is placed very close to the origin of the nerve; that is, very close to the carotid body. If this electrode is moved in steps away from the neural pole of the organ, the potential amplitude decays in an exponential fashion (Eyzaguirre & Nishi, 1974) and it disappears when this electrode is 0.7–1.5 mm from the carotid body. At the same time, the amplitude of the sensory discharges does not change significantly, thus eliminating the possibility that differences in inter-electrode distance are producing this effect (Fig. 4). The exponential decay in amplitude of receptor depolarization was accompanied by a slowing of its rise time. Furthermore, the evoked local depolarization of the receptors is graded, since its amplitude is dependent on the intensity of stimulation.

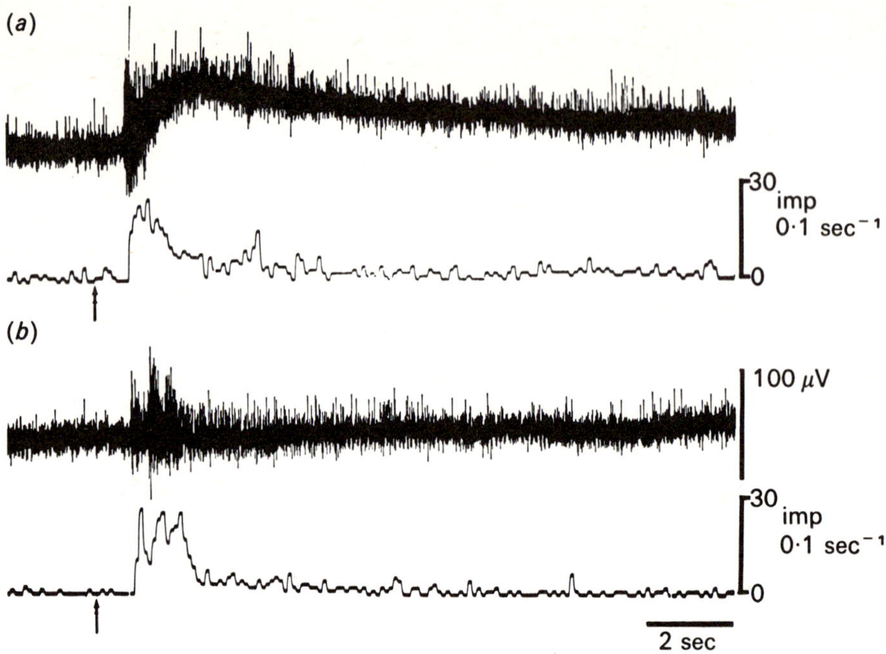

Fig. 4. Local nature of slow negative potential induced by intra-arterial injections of ACh 5 μg at arrows. (*a*) Record obtained with proximal recording lead at entrance of nerve into glomus. (*b*) Proximal electrode at 0.7 mm from glomus. In (*a*) and (*b*) distal electrode remained stationary near cut end of nerve. Upper traces, recordings from nerve; lower traces, impulse discharge frequency. (From Eyzaguirre *et al.*, 1972.)

There is a linear relationship between the stimulus intensity and the amplitude of the evoked response only within a narrow range. If stimulation is intense, linearity is lost (Fig. 5). Finally, if two stimuli are delivered at short intervals, the relationship between the amplitude of this potential and the sensory frequency changes is lost; the increased discharge elicited by the second stimulus is less marked when it is delivered very close to the first one. In contrast, the slow wave itself becomes larger. Thus, the slow potentials appear to have no refractory period (Eyzaguirre & Nishi, 1974).

All the characteristics mentioned above clearly indicate that the evoked receptor depolarization is a local (non-propagated) phenomenon. It was of interest then, to analyse further this potential in terms of ionic mechanisms involved in its production. To explore this, an air gap technique has been employed with the receptors *in vivo*.

Polarization of the receptor was profoundly influenced by the K^+ ion concentration in the bathing medium. Thus, an excess of K^+ ions (11.2–16.9 mM) induced receptor depolarization. When the carotid body was bathed in K^+-free Locke, marked hyperpolarization was observed. The

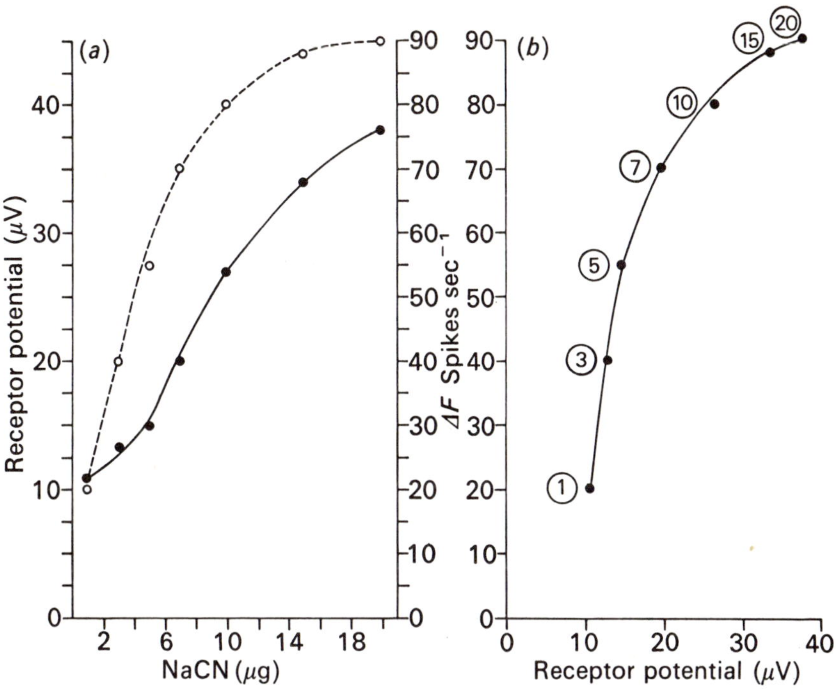

Fig. 5. (*a*) Effect of different doses of intra-arterial NaCN on receptor potential and chemosensory frequency; filled circles and solid line, amplitude of receptor potential in microvolts; open circles and broken line, increase in discharge frequency. (*b*) Relationship between receptor depolarization in microvolts and increased discharge frequency obtained after different intra-arterial injections of NaCN. The different doses of NaCN (in micrograms) are shown as circled numbers on the curve. (From Eyzaguirre & Nishi, 1974.)

amplitude of the responses evoked by ACh or NaCN followed predictably the different K^+ concentrations in the bathing solution: the evoked responses became of larger amplitude in a low K^+ medium and smaller in high K^+. These effects of K^+ may have been due to changes in polarization of the nerve endings and/or fibres.

In other experiments, either half or all of the NaCl was replaced with sucrose. At first, the receptor was hyperpolarized in low NaCl Locke and the responses evoked by ACh or NaCN became larger. Receptor polarization induced by low Na^+ may be due to an effect on the membrane potential of the nerve fibres (see Shanes, 1958) and/or an action on the nerve endings. However, as time proceeded, the depolarizing responses evoked by either ACh or NaCN became very small and this was accompanied by a reduction in spike discharge amplitude and frequency.

These experiments are promising, although rather difficult to interpret at present. For instance, it is possible that the effects of NaCl replacement are not due solely to a lack of Na^+ ions. The absence of Cl^- ions may also be important. It is known that Cl^- is important for the chemosensitivity of *Aplysia* neurones (Chalazonitis, 1961; Walker & Brown, 1970; Brown, 1972).* Further experiments are required to delineate clearly the role of Na^+ or Cl^- in chemoreceptor function. This may be accomplished by substituting glutamate for Cl^- and another cation for Na^+. The use of choline chloride does not seem to be feasible, since there is evidence that a cholinergic mechanism is important in the generation of the sensory discharges (Eyzaguirre & Zapata, 1968*b*; Eyzaguirre, Nishi & Fidone, 1972; Eyzaguirre & Nishi, 1974). In fact, large concentrations of choline have curarizing properties and tend to depress the chemoreceptor response (Eyzaguirre & Zapata, 1968*a*).

Elimination of Ca^{++} ions from the Locke's solution induced receptor depolarization and a reduction in the amplitude of the responses evoked by NaCN and ACh. Prolonged superfusion with Ca^{++}-free Locke induced irreversible receptor block. The depolarization induced by low Ca^{++} may be due to an effect on the nerve fibres as shown in other preparations by Stampfli & Nishie (1956). The reduction of the response evoked by NaCN or ACh in low Ca^{++} may be due to depolarization block and/or to interference with a cholinergic mechanism involved in the generation of this response (see also Eyzaguirre & Zapata, 1968*a*). An excess of Mg^{++} ions (2.16–4.32 mM) did not change the polarization of the receptor but it depressed the responses evoked by NaCN and ACh. At times, it was possible to depress only the NaCN response without altering the ACh-evoked potential. These effects of Mg^{++} fit with the idea that a cholinergic mechanism is involved in the generation of the sensory impulses. At the same time, it appears that the sensory endings are very sensitive to Mg^{++} ions, judging from the depressant action on the baseline discharge and on the response to ACh. It seems that Mg^{++} increases the threshold of the receptors to different forms of stimulation (C. Eyzaguirre & K. Nishi, unpublished observations).

The role of the glomus cells in the generation of sensory impulses is being investigated by using (1) intracellular recordings made by 'blindly penetrating' the whole carotid body, while recording the sensory discharges from the nerve and (2) impalements of glomus Type I cells under visual observation utilizing tissue slices and Nomarski optics.

* An artefactual component may have been present in these NaCl substitution experiments because of the use of sucrose. Thus, a small and undetected leak of sucrose into the air gap (which is usually humid) would tend to dry the gap, thus increasing the amplitude of the evoked response and inducing base-line shifts. This possible artefact would enhance the early effects of low Na^+ (membrane polarization, increased evoked depolarization and spike amplitude) but not the late effects of low Na^+, such as depression of the evoked response and elimination of the sensory discharges.

With respect to the first method of approach, impalements of glomus tissue with microelectrodes have yielded resting potentials of from 10–55 mV (Fig. 6) but these potentials have remained unchanged after applica-

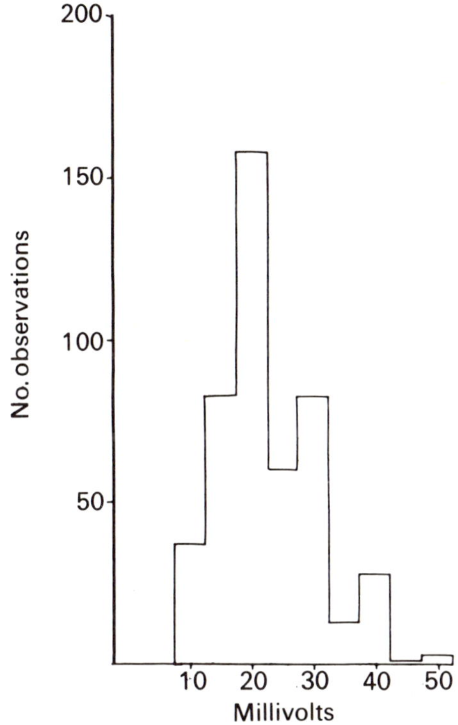

Fig. 6. Amplitude distribution of membrane potentials recorded by blind impalements of carotid body cells. Notice that the mean frequency of distribution is about 20 mV. (From C. Eyzaguirre & J. Lewin, unpublished record.)

tion of a number of physiological or pharmacological stimuli (Eyzaguirre & Koyano, 1965; Eyzaguirre *et al.*, 1970; Goodman & McCloskey, 1972). In studying this problem further we have obtained the following results: (1) There is depolarization of the penetrated elements when the preparation is bathed in high K^+, thus confirming Goodman & McCloskey (1972). Low K^+ has the opposite effect. (2) The transmembrane resistance is high (40–100 MΩ), which is similar to values found by Flock & Russell (1973) in hair cells of the acoustico-lateralis system of the fish. (3) High temperature (37–40 °C) polarizes the impaled cells while cooling (down to 29 °C) induced membrane depolarization. The latter phenomenon is accompanied by a reduction in membrane resistance (Fig. 7). (4) Ouabain ($2–5 \times 10^{-5}$ M) induces cell depolarization. The effects of temperature and ouabain

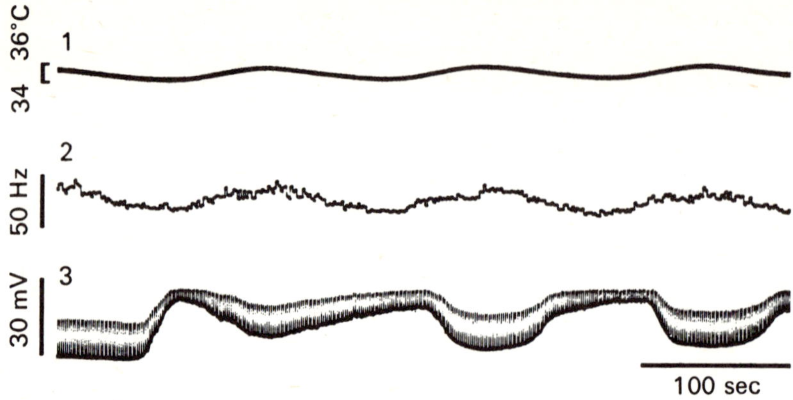

Fig. 7. Effect of temperature changes (1) on sensory discharge frequency (2) and on the membrane potential and resistance of glomus cell impaled by blind penetration of carotid body with high resistance microelectrode (3). Notice a reduction in membrane potential and resistance during cooling. With time there is a general shift of the membrane potential in a depolarizing direction, presumably due to cell damage. Membrane resistance measured by passing depolarizing current pulses through recording micropipette. (From M. Baron & C. Eyzaguirre, unpublished observations.)

strongly implicate a Na-pump. Whether this pump is electrogenic or neutral awaits further investigation; (5) Acidity and carbon dioxide induce cell depolarization while alkalinity polarizes the cells; (6) ACh (100–300 μg) applied to the bath does not change appreciably the resting potential of the impaled cells; (7) NaCN (50–100 μg) induces variable effects such as no changes in resting potential, depolarization or hyperpolarization. The nature of this variability is under investigation. Variations may be due, at least in part, to the alkaline pH of the NaCN solution. As shown above, alkalinity tends to increase the membrane potential of the impaled cells. Unfortunately, it is not possible to neutralize or buffer a NaCN solution to a neutral or slightly alkaline pH, since that would release HCN and make the solution ineffective; (8) Interruption of flow of the Locke's solution bathing the preparation induces intense chemoreceptor stimulation but does not change the membrane potential of the impaled cells. (M. Baron & C. Eyzaguirre, unpublished observations.)

The nature of the elements impaled in the carotid body with this method can be studied by using the technique employed by Flock & Russell (1973); that is, by ejecting procion navy blue from the recording micropipette. This will allow the identification of the cells after the experiment.

The second method of approach utilizes a special technique developed in this laboratory. Slices of carotid body tissue are prepared utilizing the sliding microtome described by Fidone, O'Leary & Eyzaguirre (1971). When the slices are sufficiently thin, they are mounted in a special experimental chamber through which flows Locke's solution at constant temperature (35–37 °C) at pH 7.43. The chamber is mounted on the stage of a Reichert inverted microscope equipped with Nomarski optics. The thin slice is then observed under oil immersion and at magnifications of × 1000–1600. This arrangement allows good visualization in the glomus Type I cells and nerve fibres (Plate 2). Other elements in the carotid body tissue slice can also be seen, although at present our experience with this technique is limited and proper identification of other elements will require further work. Once the desired elements are identified, one can advance a microelectrode (with a bent tip) under the condenser which has enough working distance (about 6 mm) to allow this. By closing the lamp diaphragm, one can detect a shadow made by the microelectrode which can then be directed to the desired target. Then, the cells are impaled and resting potentials of the same order of magnitude as seen in the whole carotid body are obtained. The diaphragm of the lamp is then opened and proper Nomarski effects are obtained again.

When the slices prepared with the sliding microtome are rather thick, the Nomarski effect is poor. It is desirable then, to tease the tissue (under a dissecting microscope) with the help of very fine needles to isolate groups of cells and nerve fibres before mounting the preparation on the stage of the microscope. Optical scanning of the tissue reveals the presence of relatively isolated Type I cells often connected to axons; occasionally one can discern the presence of structures which appear to be large nerve endings (Plate 3). As indicated above, one can penetrate these cells with the microelectrode and the uncertainties of cell identification no longer exist.

The membrane potential of the isolated glomus Type I cells is dependent on the K^+ concentration of the medium; they become readily depolarized when an excess of K^+ bathes the preparation. Removal of K^+ induces cell hyperpolarization. This is not surprising, since similar observations have been reported by a number of investigators working on different cells. At present, we do not know whether the membrane potential of the glomus Type I cells follows a Nernst slope. ACh, even in large concentrations, does not have marked effects on the resting potential of these cells although this point has to be elucidated further since we only have a few experiments investigating this point. NaCN in relatively large concentrations tends to hyperpolarize the glomus Type I cells, although as expressed above, this may be due entirely to shift in pH toward the alkaline side (Fig. 8).

It is hoped that this approach will give us some insight into the functioning of the glomus Type I cells. We still do not know if these cells are the receptor cells, since a causal relation between glomus Type I cells and nerve discharges has not yet been found. However, this type of preparation

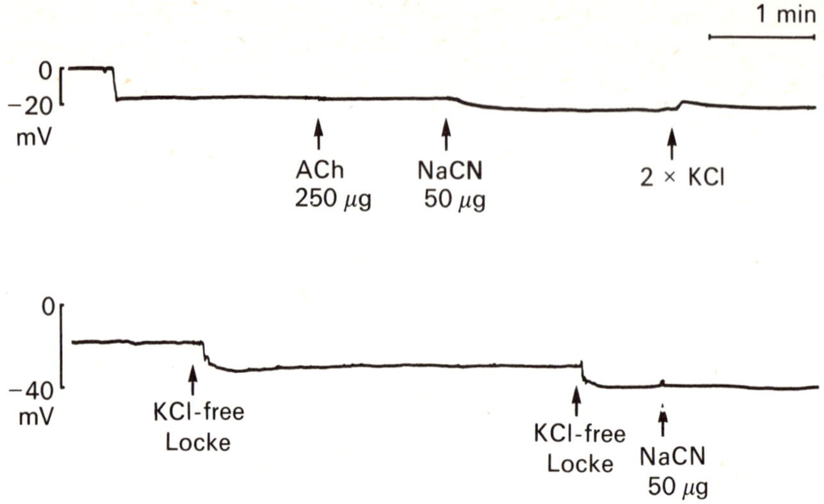

Fig. 8. Impalement of cell shown in Plate 3 with high resistance microelectrode gives resting potential of 20 mV. Continuous records. In the upper record, ACh 250 μg does not change the membrane potential of this cell; NaCN 50 μg induces hyperpolarization and high KCl produces depolarization. In the lower records, normal Locke is temporarily replaced with K^+-free Locke inducing membrane polarization. The same amount of NaCN does not induce membrane polarization when this substance is injected at a high resting potential induced by the low K^+ Locke. (From C. Eyzaguirre & S. Fidone, unpublished.)

lends itself to a direct study of this problem. For instance, one may be able to pass current through the micropipette inserted into the glomus cells and observe whether or not there are sensory discharge frequency changes in the nerve fibres connected to that cell. In this respect, it is interesting to mention that some indirect evidence is available. Utilizing the whole glomus and its nerve, it has been possible to inject current into a structure giving a stable membrane potential and a discharge increase has been observed from the carotid nerve (M. Baron & C. Eyzaguirre, unpublished observations). However, as mentioned above, the elements penetrated with the microelectrode still wait identification. Preferably, one may use tissue slices, penetrate the cells under high magnification and Nomarski optics and record extracellularly or intracellularly from the larger (calix-type) endings or the nerve fibres to see if there is a direct relationship between the injected current changing the membrane potential of the cells and the evoked nerve discharge. It appears possible to penetrate the larger endings (see Plate 1) with microelectrodes. In fact, blind penetrations into the carotid body have yielded occasionally large (40–60 mV) resting potentials in cells that show rhythmic activity in the form of action potentials, sometimes preceded by a slow depolarization or a complex waveform. The question is whether one is inside a nerve fibre or an ending. This problem has not been solved satisfactorily yet. However, there are some indications that in these cases the microelectrode has been inside a nerve ending: (1) antidromic stimulation of the nerve induced an evoked potential, picked up by the recording microelectrode, with a conduction velocity of less than 1 m sec^{-1} (Fig. 9). This indicates that one is inside either an

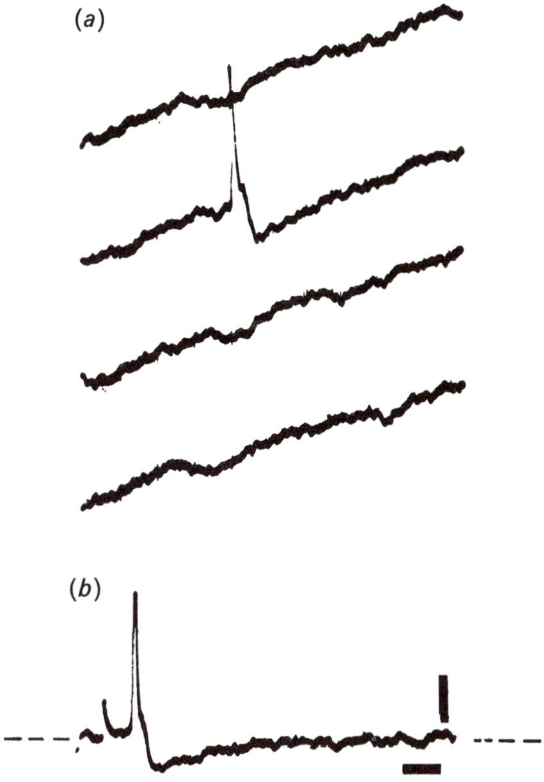

Fig. 9. Blind impalement of carotid body with high resistance microelectrode. (*a*) shows spontaneous activity recorded by the electrode. In (*b*) a single shock was delivered to the nerve trunk eliciting an evoked action potential in same cell.

Cal., 20 mV; time, 2 msec. (From H. Koyano & C. Eyzaguirre, unpublished observation.)

unmyelinated nerve fibre which innervates the carotid body (Fidone & Sato, 1969) or in a nerve ending. The first possibility may be discarded, since it is virtually impossible to hold an intracellular penetration of a C-fibre for very long (P. R. Burgess, personal communication) and we kept the electrode inside the rhythmically discharging structure for more than 40 min. The other, and more likely alternative, is that we were inside a large nerve ending, the conduction velocity being so slow because of extensive branching and thinning of the fibres before reaching the nerve terminal (see Eyzaguirre & Gallego, this Symposium); (2) ejection of methyl blue from the micropipette indicated that the electrode was inside (or extremely close to) a nerve ending. Unfortunately, these dye ejections have not given very clear histological pictures up to now (H. Koyano & C. Eyzaguirre, unpublished observations). Also, one may eject stimulating agents electrophoretically, such as ACh^+ and CN^- ions on different structures such as glomus cells, nerve endings or fibres and observe the effect on the nerve discharges. One problem with this technique is that if the nerve fibres are cut too close to a glomus cell, damage to the fibre and excessive depolarization may occur thus preventing the recording of healthy spikes. However, it is hoped that this shortcoming will be overcome in the future.

It is felt that these approaches are hopeful and that further investigation on carotid body function may eventually reveal the role of glomus cells in the generation of impulses in carotid nerve fibres.

The help of Mr P. Lenoir with the illustrations is gratefully acknowledged. This work was supported by Grants NS 05666 and NS 07938 from the US Public Health Service.

REFERENCES

Alvarez-Buylla, R. & Ramirez de Arellano, J. (1953). Local responses in Pacinian corpuscles. *Amer. J. Physiol.*, **172**, 237–50.

Brown, A. M. (1972). Effects of CO_2 and pH on neuronal membranes. *Fedn Proc.*, **31**, 1399–403.

Chalazonitis, N. (1961). Chemopotentials in giant nerve cells (*Aplysia fasciata*). In *Nervous Inhibition*, ed. Florey, E. New York: Pergamon.

Eyzaguirre, C., Fidone, S. & Zapata, P. (1972). Membrane potentials recorded from the mucosa of the toad's tongue during chemical stimulation. *J. Physiol.*, **221**, 515–32.

Eyzaguirre, C. & Koyano, H. (1965). Effects of electrical stimulation on the frequency of chemoreceptor discharges. *J. Physiol.*, **178**, 438–62.

Eyzaguirre, C. & Kuffler, S. W. (1955). Processes of excitation in the dendrites and in the soma of single isolated nerve cells of the lobster and crayfish. *J. gen. Physiol.*, **39**, 87–119.

Eyzaguirre, C., Leitner, L. M., Nishi, K. & Fidone, S. (1970). Depolarization of chemosensory nerve endings in carotid body of the cat. *J. Neurophysiol.*, **33**, 685–96.

Eyzaguirre, C. & Nishi, K. (1974). Further study on mass receptor potential of carotid body chemosensors. *J. Neurophysiol.*, **37**, 156–169.

Eyzaguirre, C., Nishi, K. & Fidone, S. (1972). Chemoreceptor synapses in the carotid body. *Fedn Proc.*, **31**, 1385–93.

Eyzaguirre, C. & Zapata, P. (1968*a*). Pharmacology of pH effects on carotid body chemoreceptors *in vitro*. *J. Physiol.*, **195**, 557–88.

Eyzaguirre, C. & Zapata, P. (1968*b*). The release of acetylcholine from carotid body tissues. Further study on the effects of acetylcholine and cholinergic blocking agents on the chemosensory discharge. *J. Physiol.*, **195**, 589–607.

Fidone, S., O'Leary, T. & Eyzaguirre, C. (1971). A simple method for the preparation of and recording from thin slices of living tissue. *Brain Res.*, **30**, 401–6.

Fidone, S. J. & Sato, A. (1969). A study of chemoreceptor and baroreceptor A and C fibers in the cat carotid nerve. *J. Physiol.*, **205**, 527–48.

FLOCK, Å. & RUSSELL, I. (1973). Efferent nerve fibres: postsynaptic action on hair cells. *Nature New Biology*, **243**, 89–91.

GESTELAND, R. C., LETTVIN, J. Y., PITTS, W. H. & ROJAS, A. (1963). Odor specificities of the frog's olfactory receptors. In *Olfaction and Taste*, ed. ZOTTERMAN, Y., Oxford: Pergamon.

GOODMAN, N. W. & MCCLOSKEY, D. I. (1972). Intracellular potentials in the carotid body. *Brain Res.*, **39**, 501–4.

GRAY, J. A. B. & SATO, M. (1953). Properties of the receptor potential in Pacinian corpuscles. *J. Physiol.*, **122**, 610–36.

HARTLINE, H. K., WAGNER, H. G. & MCNICHOL, E. F. (1952). The peripheral origin of nervous activity in the visual system. *Cold Spring Harbor Symp. Quant. Biol.*, **17**, 125–41.

KATZ, B. (1950). Depolarization of sensory terminals and the initiation of impulses in the muscle spindle. *J. Physiol.*, **111**, 261–82.

KIMURA, K. & BEIDLER, L. M. (1961). Microelectrode study of taste receptors of rat and hamster. *J. cell. comp. Physiol.*, **58**, 131–40.

LOEWENSTEIN, W. R. (1971). Principles of Receptor Physiology. In *Handbook of Sensory Physiology*, New York: Springer.

OTTOSON, D. (1956). Analysis of the electrical activity of the olfactory epithelium. *Acta physiol. Scand.*, **35**, 7–82.

SHANES, A. (1958). Electrochemical aspects of physiological and pharmacological action in excitable cells. *Pharmacol. Rev.*, **10**, 59–273.

STAMPFLI, R. & NISHIE, K. (1956). Effects of calcium-free solutions on membrane potential of myelinated nerve fibers of the Brazilian frog *Leptodactylus ocellatus*. *Helv. physiol. acta*, **14**, 93–104.

TATEDA, H. & BEIDLER, L. M. (1964). The receptor potential of the taste cells of the rat. *J. gen. Physiol.*, **47**, 479–86.

WALKER, J. L. & BROWN, A. M. (1970). Unified account of the variable effects of carbon dioxide on nerve cells. *Science*, **167**, 1502–4.

EXPLANATION OF PLATES

PLATE 1

Different types of nerve endings terminating on glomus Type 1 cells. Models made after serial electron microscope reconstructions. (Courtesy of K. Nishi & L. J. Stensaas, unpublished observations.)

PLATE 2

Carotid body tissue slice viewed with Nomarski optics. Notice cluster of glomus type 1 cells in the middle of picture. Some of the filaments in the picture probably correspond to nerve fibres. (From C. Eyzaguirre & S. Fidone, unpublished observations.)

PLATE 3

Isolated glomus Type 1 cell with its nerve. The arrow indicates possible presence of a large calix-type ending. Nomarski optics. (From C. Eyzaguirre & S. Fidone, unpublished.)

PLATE I

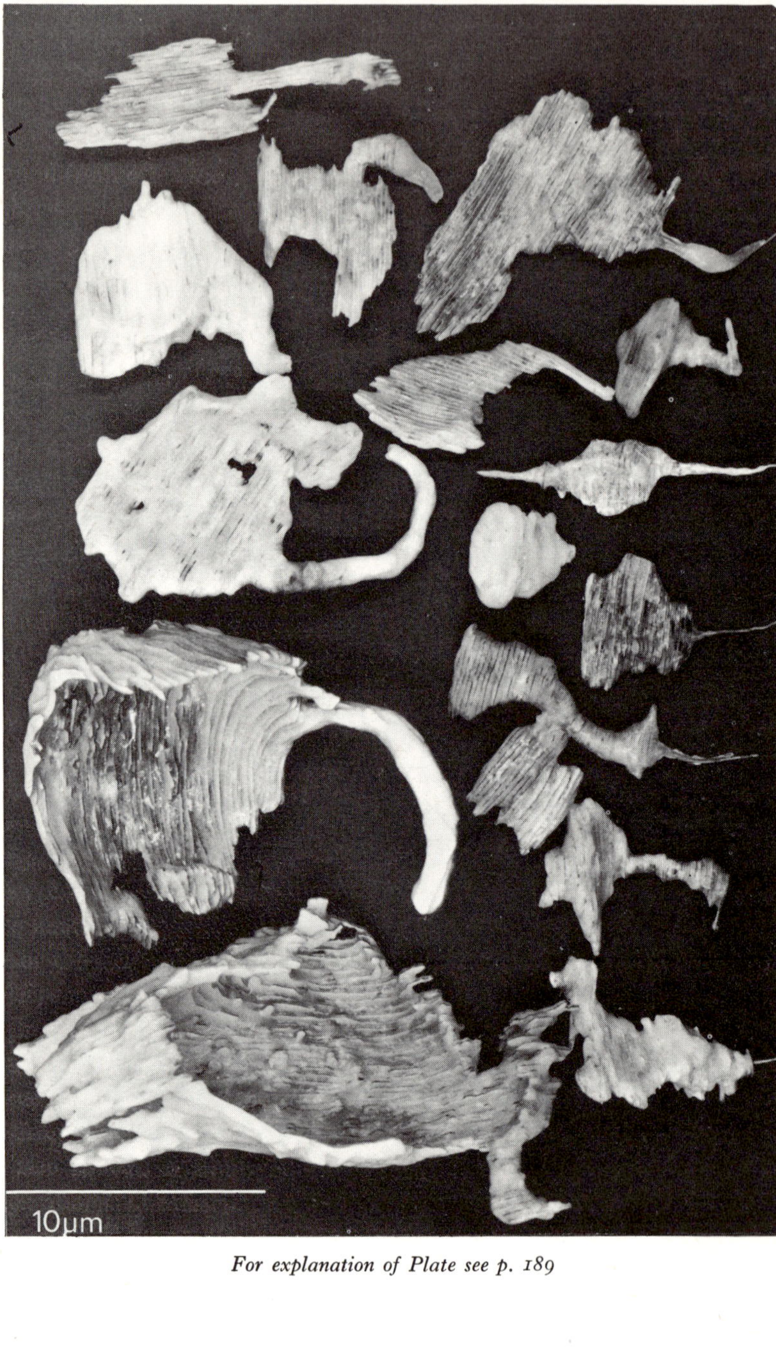

For explanation of Plate see p. 189

PLATE 2

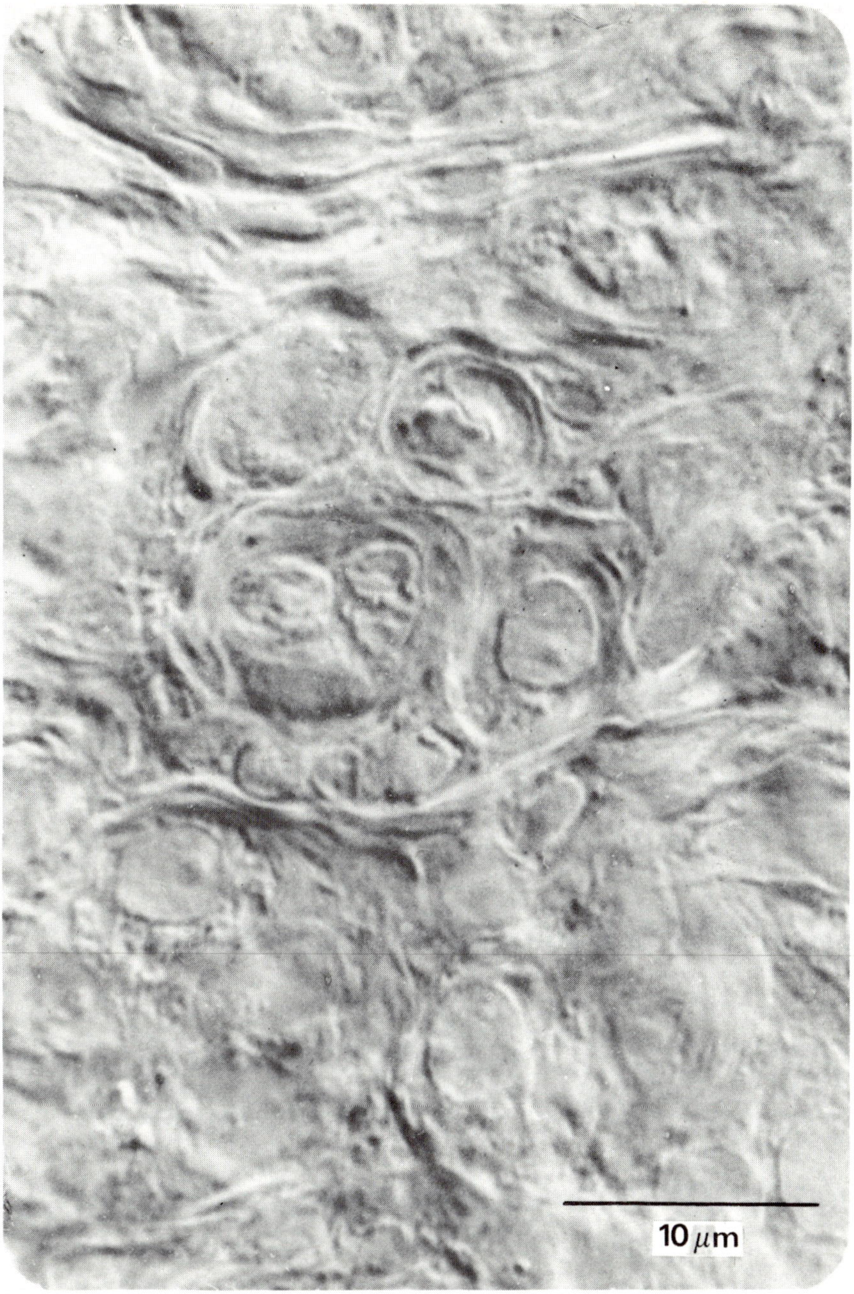

For explanation of Plate see p. 189

PLATE 3

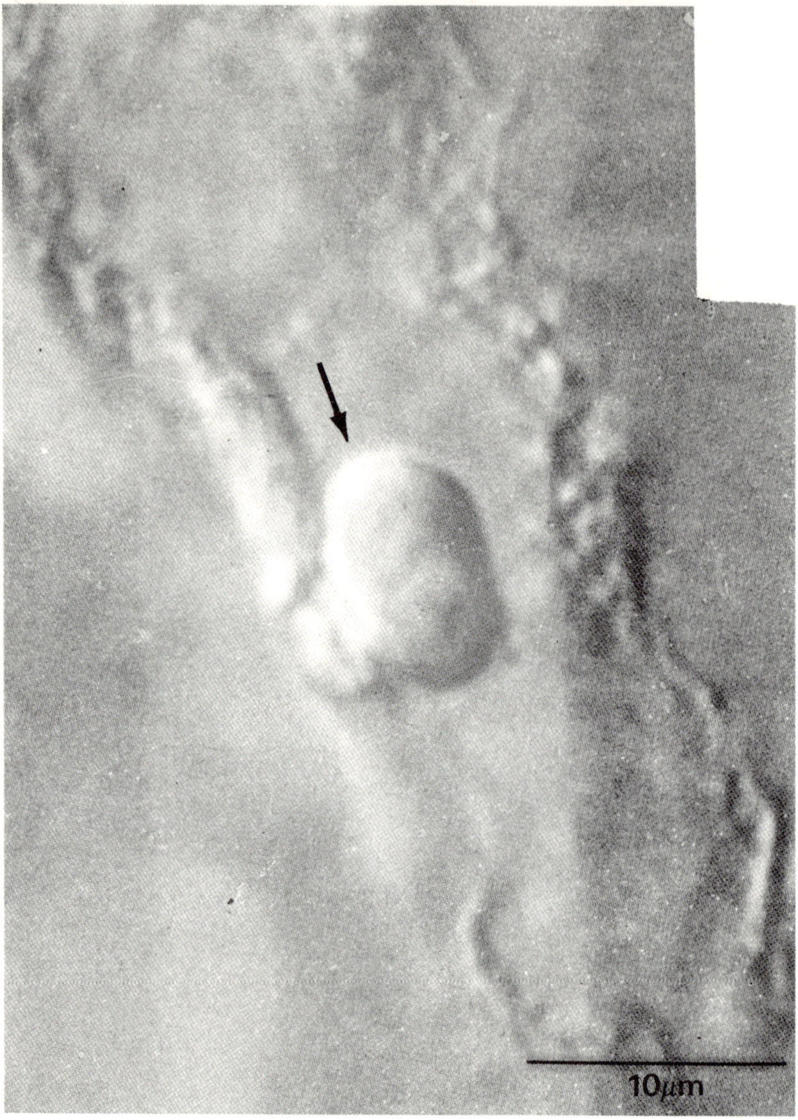

For explanation of Plate see p. 189

DISCUSSION

SAMPSON What solution do you use for bathing the tissue slices? Do you add protein, for example?

EYZAGUIRRE No: it is basically Locke solution and we haven't really developed tissue culture methods yet. But even with this rather primitive method, we have found that the electrode stays in for a long time and that the properties of the impaled cell remain constant. But I have presented only preliminary steps and I am certain that the whole method can and will be improved.

BISCOE How thick are your sections?

EYZAGUIRRE I am not sure.

BISCOE Did you try any of the commercial slicers? What about the non-vibrating slicers, e.g. Sorval Tissue Chopper, or is vibration important?

EYZAGUIRRE I have tried one or two of the commercial devices. They work satisfactorily for the superior cervical ganglion but not the carotid body unless it is fixed. I don't know: my carving knife may turn out to be the most useful.

CASPERS Could you speculate a bit about the origin of the slow potential changes which you saw.

EYZAGUIRRE I think they arise mainly in the nerve endings.

CASPERS And not in the so-called receptor cells?

EYZAGUIRRE No: I think not.

MITCHELL Have you tried the effect of dopamine?

EYZAGUIRRE No.

BINGMANN Another question about the slow potentials. If you record the potential with one electrode at the end of the nerve and with the other on the surface of the carotid body, you have shown the presence of these potentials. What happens if you leave the surface electrode but impale the carotid body with the other: are the potentials enhanced?

EYZAGUIRRE We haven't done that. But we have tried to record from the nerve and to probe the carotid body: but this is like looking for a needle in a haystack as you can imagine and I don't think that this type of random penetration is likely to be very helpful.

BINGMANN Is it possible that these DC potentials are generated in the walls of blood vessels?

EYZAGUIRRE It is certainly an idea: and I don't think I can exclude it. However, I don't think that blood vessel walls can be an important source because of the very large changes that are seen with ACh in the eserinized superfused preparations. Here, I still think the most likely candidates are the nerve endings.

SPECKMANN What experimental evidence do you have that the electrode placed just above the carotid body is really active for the potential recording?

EYZAGUIRRE Well, we see a depolarization and if you move the electrode by very small distances, there is a marked change in potential.

TORRANCE I must congratulate you on recording from within the nerve ending. When there was no spike occurring, there was some irregularity in the potential you recorded. Is this just noise or do you think that something is actually happening – for example, a transmitter acting?

EYZAGUIRRE I don't know. I can't exclude for example that it hasn't been due to damage because there has been a good deal of poking about. I think we have to develop the method of exposure of the ending more satisfactorily so that we can visualize more clearly the placing of the electrode tip.

SHEPHERD Have you tried the effect of a continuous perfusion of ACh rather than a sudden slug?

EYZAGUIRRE No: that is an interesting question. I suspect that we are getting into the situation uncovered by Matthews and the muscle spindle people. I suspect that we are not seeing the whole potential but only part of it. That is, the small change in potential may correspond to a dynamic phase and it may be very much easier to reproduce than if the stimulus is a slow one – say asphyxia, change of temperature, long-term perfusion with ACh or other drugs. So there may be some form of adaptation.

MILLS Do you have any idea of how that adaptation comes about – that is when the potential keeps rising but the frequency doesn't?

EYZAGUIRRE No.

MILLS Because it struck me that NaCN is likely to have many sites of action. Thus it could be at the node or where the impulses discharge but it could leave the generator alone. And this could also explain, in your 'exhaustion' experiments, that the drug is affecting the discharge but not the transmitter process.

WILLSHAW The use of hemicolineum appeared to slow down the speed of response to both ACh and NaCN: is this a regular finding?

EYZAGUIRRE I don't put very much weight on this type of pharmacological finding. There are too many variables. For example, with hemicolineum, blood pressure falls and this could clearly affect the response.

Penetration of horseradish peroxidase between all elements of the carotid body

By R. I. WOODS
Physiological Laboratory, Cambridge

These experiments were carried out on young rats weighing about 100 g. They were anaesthetized with pentobarbitone and horseradish peroxidase was injected into a femoral vein. The animals, while still under anaesthesia, were then killed from 30 seconds to 30 minutes after injection by sudden excision of the heart. At the moment of excision of the heart the bifurcations of the carotid arteries, which had already been exposed, were flooded with fixative. The carotid bodies were then dissected out, cut into four pieces each, and processed by the method of Graham & Karnovsky (1966).

Ten minutes after the injection, the carotid body was found to be heavily stained with peroxidase, although in other tissues, such as brown fat from the same region, it had not yet left the capillaries. In an artery within the carotid body the peroxidase had not penetrated the endothelium, but in the glomus tissue all the intercellular spaces were filled. Plate 1*a*,*b* show the staining between Type I and Type II cells. There are no spaces unfilled. The spaces around the nerve fibres are equally accessible as shown in Plate 2*a*–*c*. In Plate 2*a*, the axons in the Schwann cell cytoplasm are surrounded; in Plate 2*b* the axons wrapped in Type II cell cytoplasm are surrounded and in Plate 2*c* there is peroxidase between the nerve ending and the Type I cell. The synaptic thickenings are somewhat obscured by the staining for peroxidase which has penetrated all the way along the cleft, but the profile has the appearance of a nerve ending as it contains small, mainly clear, vesicles and no ribosomes.

In carotid bodies from animals killed 10 min after injection, there are what appear to be bristle coated pits on the Type I cell surfaces containing peroxidase staining. In untreated animals these are always empty. Plate 3*a*,*b*. show these on Type I cells and Plate 3*c*,*d* show them on nerve fibres as well. It is interesting to compare these with Plate 4 which is from the adrenal medulla of a 10-min animal. There is peroxidase in two pits extending into the medullary cell and also in two vesicles in the nerve ending. The next two Plates (5, 6*b*) show a possible succession of events.

Plate 5 shows two nerve endings synapsing with Type I cells, and contains a shallow pit (1), a deep pit (2), a just detached pit forming a vesicle (3), and a vesicle deep in the cytoplasm (4). Plate 6*b* from an animal killed 30 min after injection shows the possible fate of these vesicles fusing with a larger body, while Plate 6*a*, from an untreated animal, shows that this sequence is unlikely to have been induced by peroxidase.

This sequence of pit, free vesicle in the cytoplasm and fusion with a larger space, has been shown in skeletal muscle motor nerve endings, neurohypophyseal endings and adrenal medullary cells. These are all release sites and it is suggested that such a sequence shows vesicle recapture after release of the vesicle's contents. If this is so, then these pictures could well provide evidence that the Type I cells release some form of transmitter and the question at issue is whether this is as a result of activity in the nerve endings or in the cells themselves.

The one clear result of these experiments is that all regions and all intercellular spaces of the carotid body are accessible to a marker of mol wt 40 000 in the blood. However, it is not possible simply from these experiments to make calculations of the minimum time it might take some other substance, e.g. a bicarbonate ion, to pass from the blood to the receptor because it is not known whether the peroxidase is diffusing or whether it is being carried in a bulk flow of plasma being filtered through the fenestrated walls of the sinusoids. This latter idea would explain the paradox whereby peroxidase could be observed in the intercellular clefts after only 30 seconds while it had not left normal capillaries by 10 minutes.

I have examined an older rat to see whether the young rat's carotid body has any unusual features. In a 705 g rat the only visible difference was the presence of longer processes on the Type I cells. One of these formed a most interesting synapse with a nerve ending – Plate 7,*a*,*b*. It would be very difficult to distinguish the nerve ending from the process if the section had not been along the process. I wonder whether this type of synapse, which is reminiscent of an axodendritic spine one, has a different function from those on the cell body.

I should like to acknowledge Miss S. J. Hewitt's assistance during these experiments.

REFERENCE

GRAHAM, R. C. & KARNOVSKY, M. J. (1966). The early stages of absorption of injected horseradish peroxidase in the proximal tubules of mouse kidney: ultrastructural cytochemistry by a new technique. *J. Histochem. Cytochem.*, **14**, 291–301.

EXPLANATION OF PLATES

Abbreviations: I, Type I cell; II, Type II cell; SC, Schwann cell; A, Axon; NE, Nerve ending; AM, Adrenal medulla cell; IP, Type I cell process.
Plates 1 to 5. All from animals 10 min after injection of peroxidase.

PLATE 1

(*a*,*b*) Staining between Type I and Type II cells.

PLATE 2

(*a*) Staining between unmyelinated axons and their Schwann cell.
(*b*) Staining between axons and a Type II cell.
(*c*) Staining between a nerve ending and a Type I cell.

PLATE 3

(*a*,*b*) Coated pits in Type I cells.
(*c*,*d*) Coated pits in nerve fibres.

PLATE 4

Coated pits (lower arrows) in an adrenal medullary cell, and filled vesicles (upper arrows) in the nerve ending on it.

PLATE 5

A possible succession of events, 1, 2, 3 & 4 in the surface of Type I cells.

PLATE 6

(*a*) From an untreated animal. The final fate of recaptured vesicles?
(*b*) From an animal 30 min after injection. The final fate of recaptured vesicles?

PLATE 7

(*a*, *b*) From an untreated animal. A nerve ending on the end of a long process from a Type I cell. Low and high power.

PLATE I

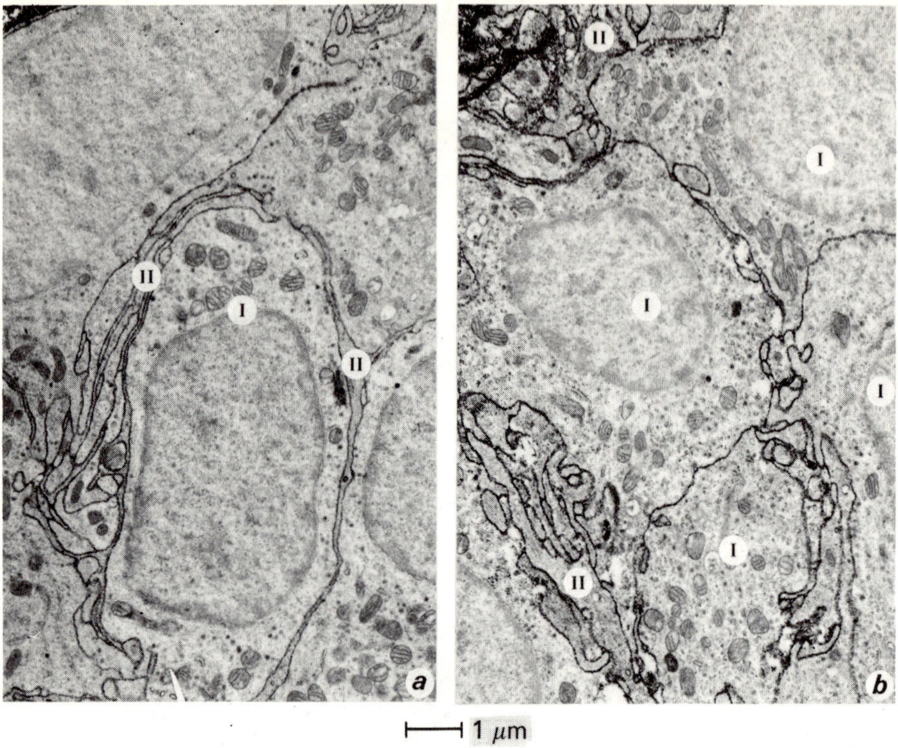

For explanation of Plate see p. 197

PLATE 2

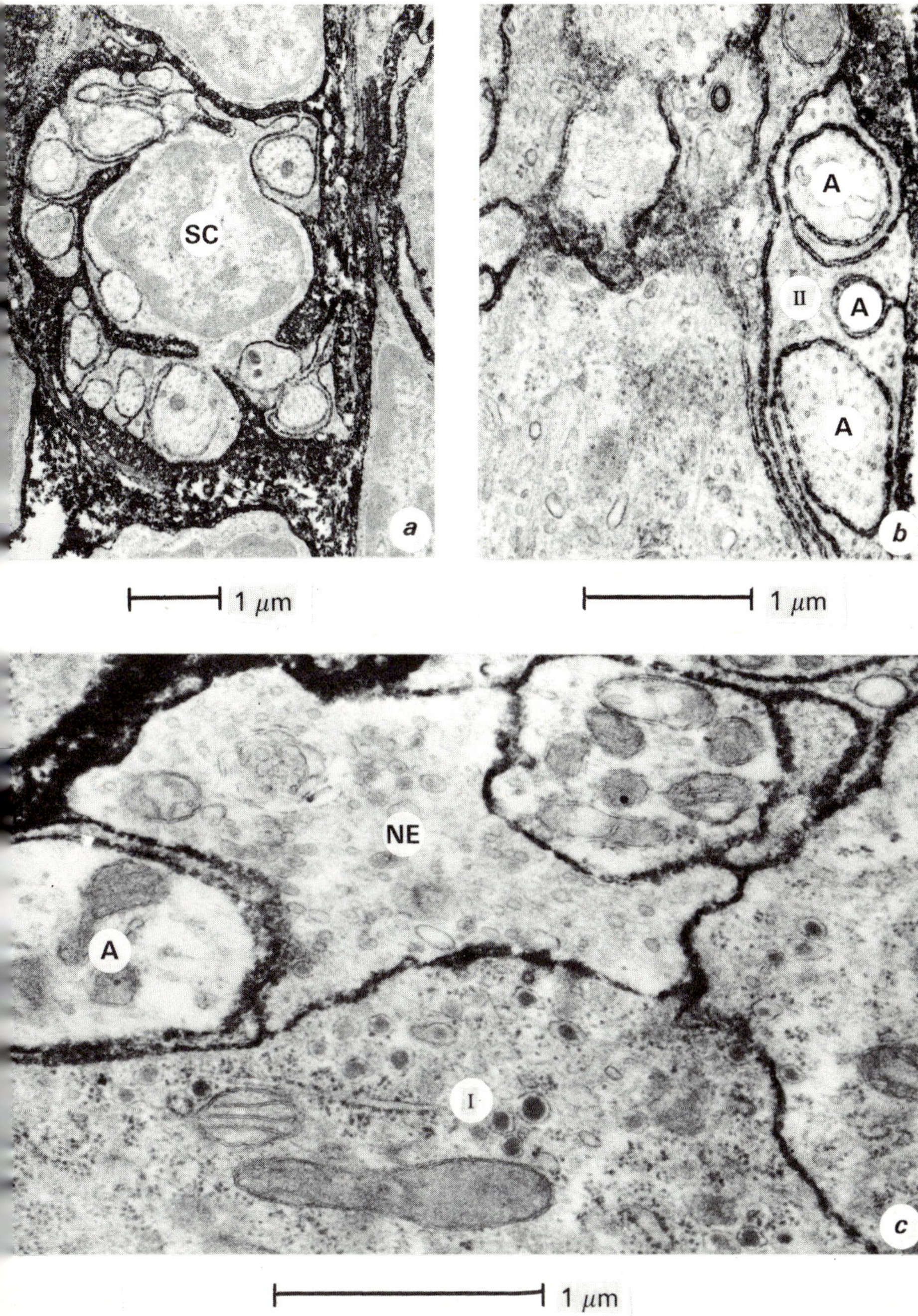

For explanation of Plate see p. 197

PLATE 3

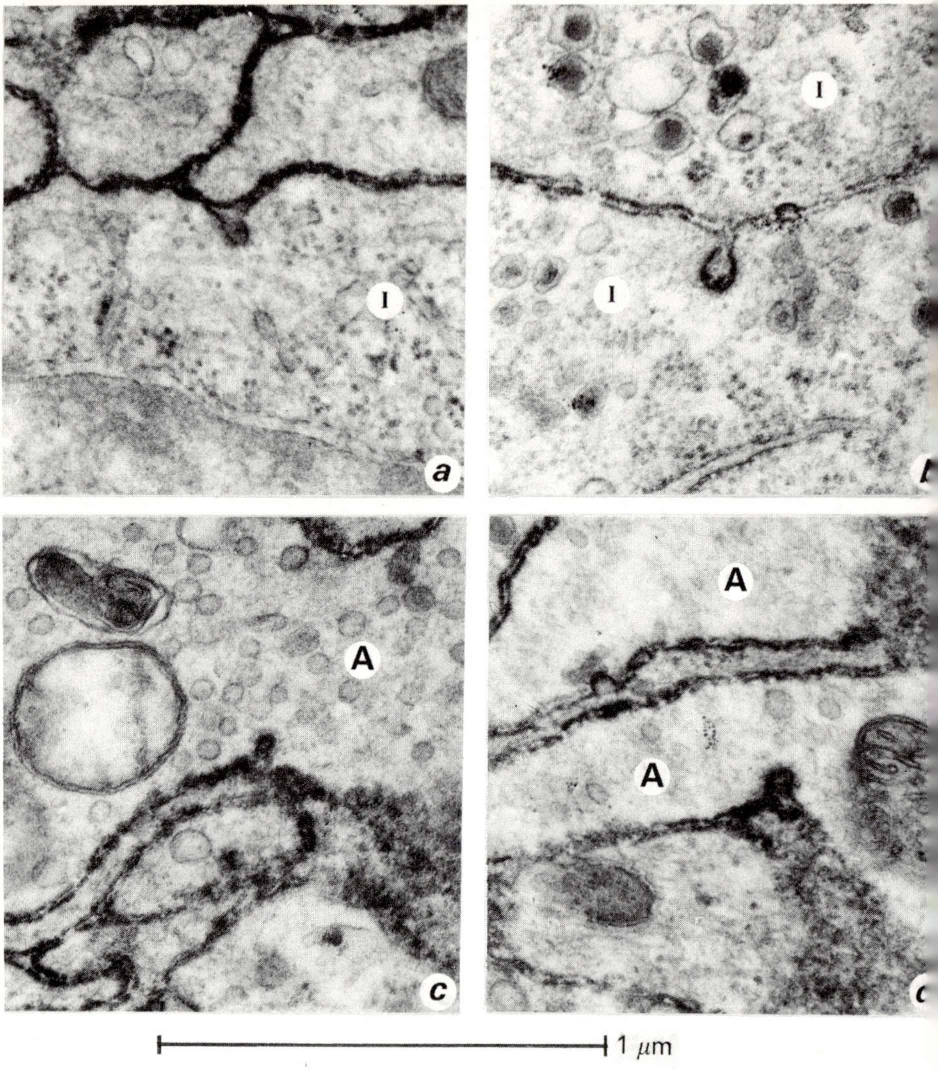

For explanation of Plate see p. 197

PLATE 4

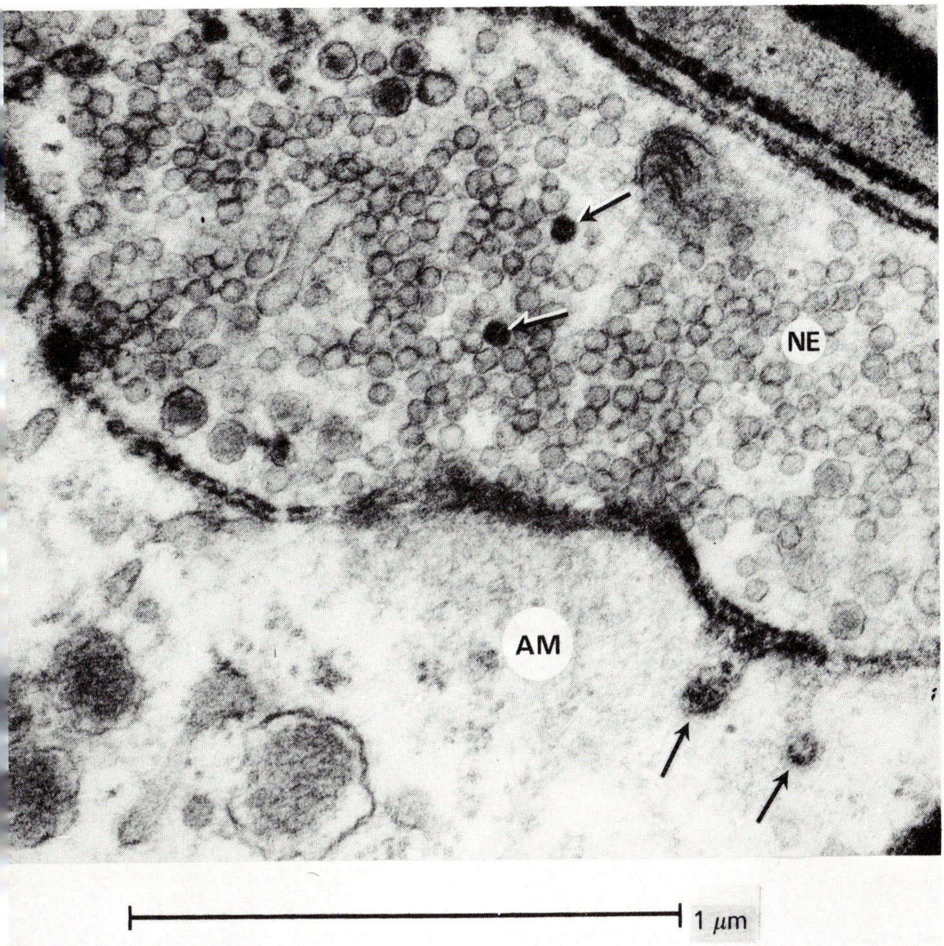

For explanation of Plate see p. 197

PLATE 5

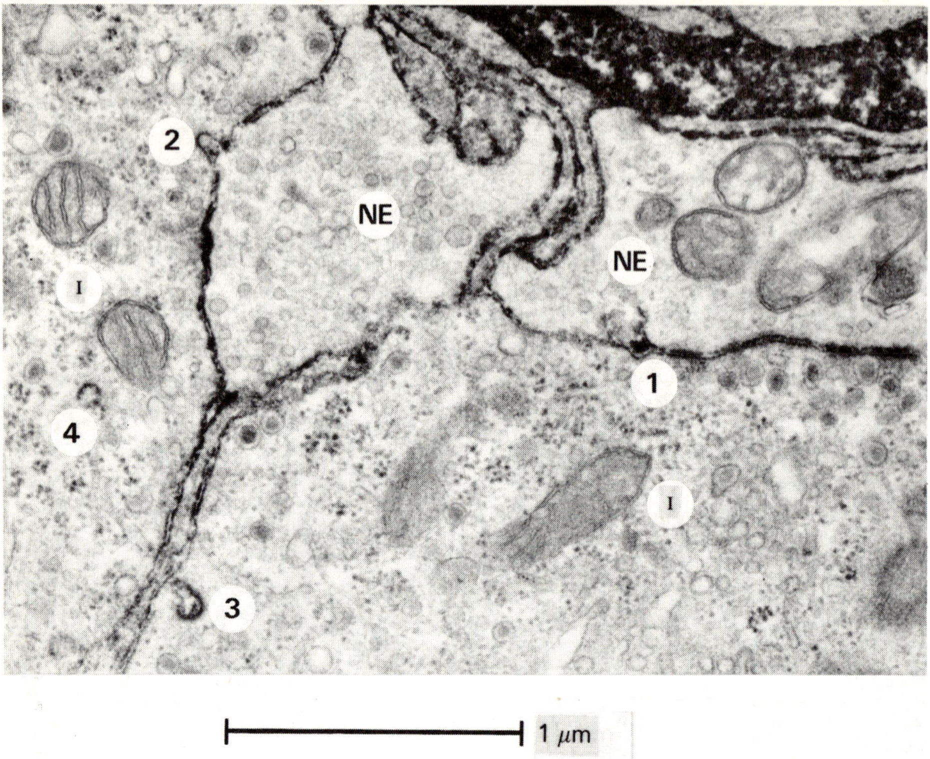

For explanation of Plate see p. 197

PLATE 6

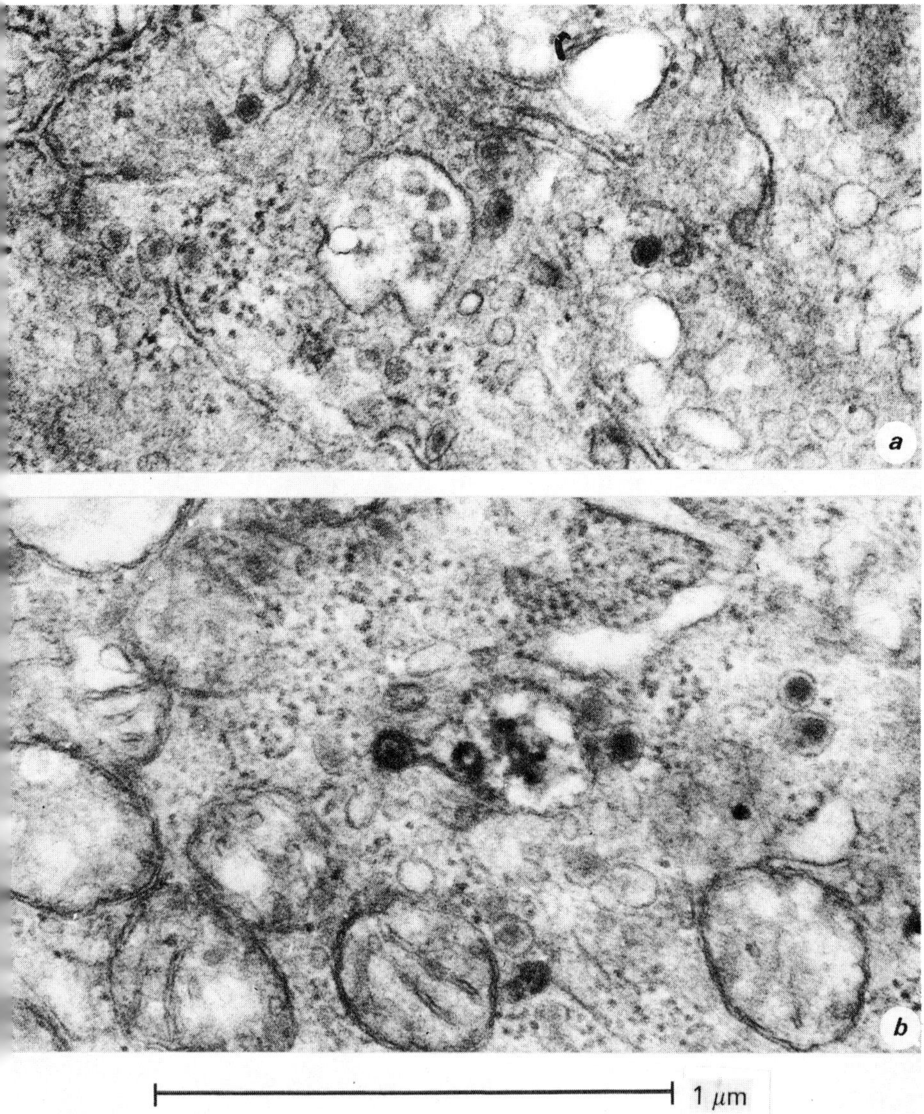

For explanation of Plate see p. 197

PLATE 7

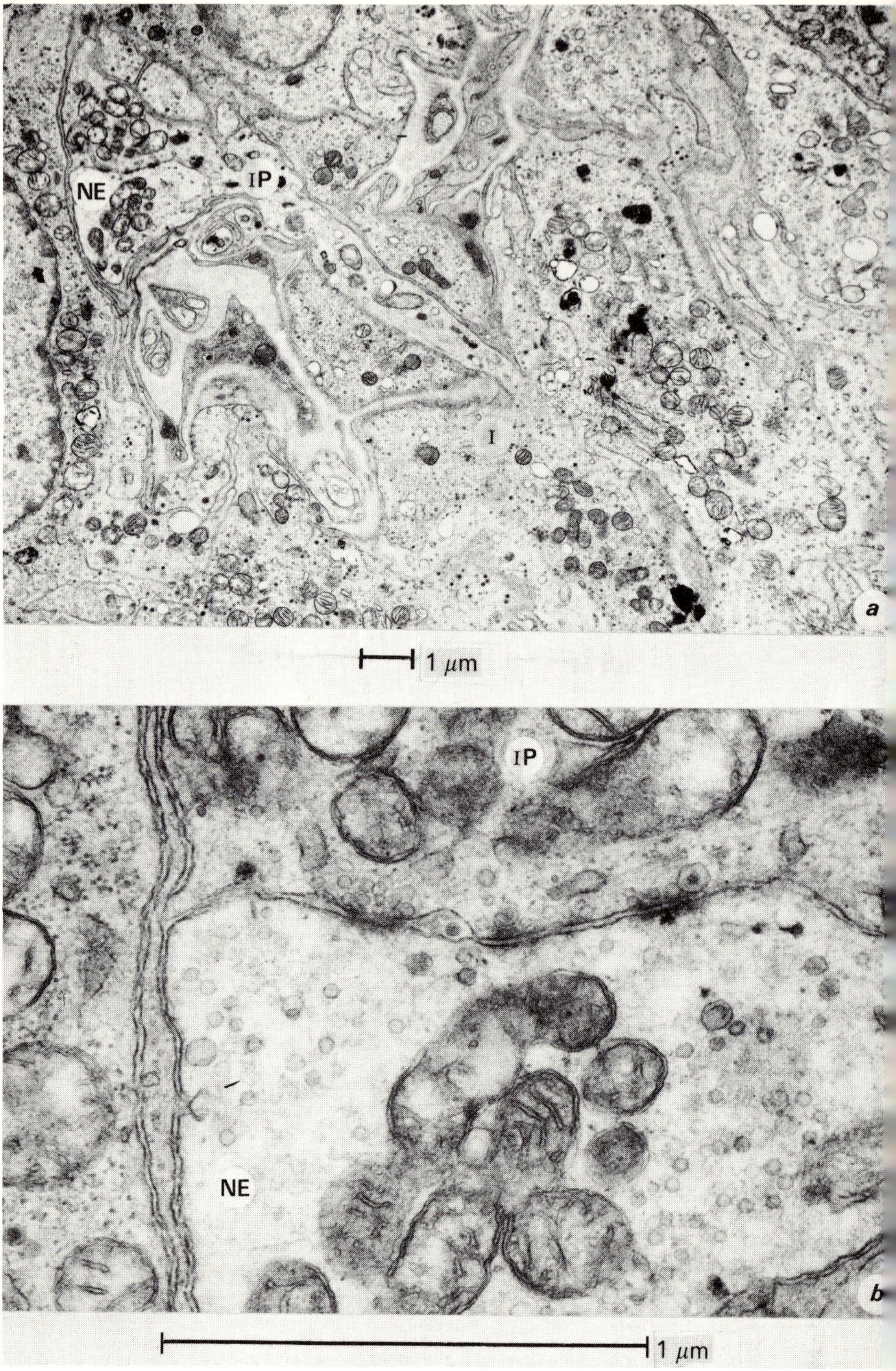

For explanation of Plate see p. 197

DISCUSSION

McDonald What are the pharmacological effects of peroxidase on the carotid body?

Woods The first people who used it didn't think that it had any effect and I've seen some quite heroic doses given. Certainly, in these doses, it can kill the animal and invariably causes a fall in blood pressure. I guess that it may let itself through as it were; but where it has been used in the kidney, for example, and has been checked with ferritin, it simply confirms the ferritin results. So this may be the case in the carotid body.

Torrance Hasn't peroxidase been given into the cerebral ventricles in animals in order to study the blood–brain barrier? As far as I can recollect it doesn't appear to affect the neural tissue so this is possibly some validation of your method.

On another point, you said that the movement of peroxidase gives no information about the diffusion of bicarbonate. I think it may. You obtained a time of a few minutes for diffusion of peroxidase throughout the carotid body. The diffusion constant is inversely proportional to the cube root of the molecular weight and the ratio of $\sqrt[3]{\text{mol wt}}$ for peroxidase and bicarbonate is about 10:1. Also if you have diffusion in a thin space, the problem arises of restricted diffusion if the width of the space is not much greater than the diameter of the molecule diffusing. I assume that the space between Types I and II cells in the carotid body is 200 Å as are most spaces in nervous tissue. The half time for penetration of peroxidase might then be one-tenth or one-twentieth of what you get for peroxidase. Gray (1971) found that bicarbonate takes 5–10 sec to have an effect on the discharge of the carotid body. This time relates well enough to the time you get for peroxidase penetration to suggest that HCO_3^- coming from the blood has its effect on nervous discharge at a point reached by your peroxidase.

Woods Well, that is alright if you think that HCO_3^- or any other substance passes from capillary to tissue simply by diffusion occasioned by a concentration gradient. But I have been impressed by the fenestration of the carotid body capillaries, and I think that there could equally well be a bulk fluid movement because of the high pressure and flow in these capillaries, possibly into some form of lymphatics. So if you had such a bulk type of flow, the movement of horseradish peroxidase and HCO_3^- would be the same.

REFERENCE

Gray, B. A. (1971). On the speed of the carotid chemoreceptor response in relation to the kinetics of CO_2 hydration. *Respir. Physiol.*, **11**, 235–46.

Pharmacology of feedback inhibition of carotid body chemoreceptors in the cat

By S. R. SAMPSON
Cardiovascular Research Institute and the Department of Physiology, University of California, San Francisco, California 94143, USA

We have heard a good deal about afferent and efferent synapses on Type I cells, but in order to conclude that true efferent synapses exist, we should consider whether or not there is any physiological basis for an efferent input to Type I cells in the carotid body. You might say that my contribution might be worded 'a cell in search of a function' or, more specifically, 'catecholamines in the carotid body in search of a function', because, as others have mentioned, it really is an assumption that the Type I cell is the receptor. At this stage people are very prone to select the evidence which fits with their prejudices. However, the fact is that in recent years a number of studies have shown that afferent activity from the carotid and aortic bodies can be inhibited by centrifugal activity in the nerves to these organs (Fidone & Sato, 1970; Neil & O'Regan, 1969*a*,*b*, 1971*a*,*b*; Sampson & Biscoe, 1970; Goodman, 1973). How this inhibition comes about is not known. Before discussing some of the possible mechanisms, I would like to review, very briefly, two techniques which are used to demonstrate what we call 'feedback inhibition'.

The procedure which I use and which has been subjected to a certain amount of criticism is the following. The carotid sinus nerve is exposed and dissected from surrounding tissue. I split off a small chemoreceptor strand and record from that. At the same time I lay the remainder of the nerve across a pair of stimulating electrodes in order to test the effects of centrifugal stimulation upon the discharge of the particular afferent fibre. Examples of the type of inhibition which can be obtained are shown in Fig. 2.

This shows not only inhibition of afferent discharges with centrifugal

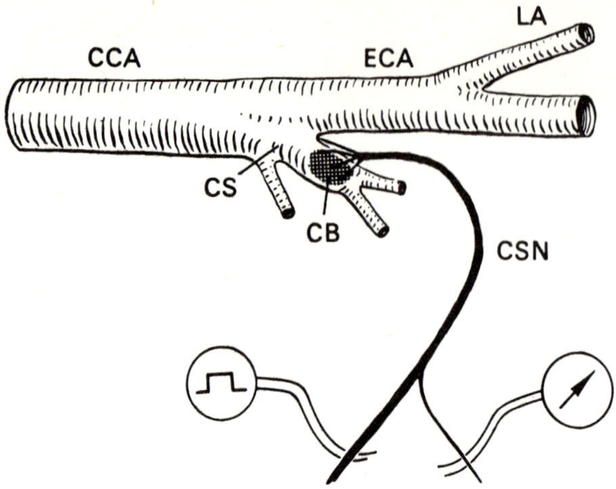

Fig. 1. Diagrammatic representation of experimental arrangement for eliciting feedback inhibition. The region of the carotid body and carotid sinus on the right side are presented from a medial view. Abbreviations: CCA, common carotid artery; ECA, external carotid artery; LA, lingual artery; CS, carotid sinus; CB, carotid body; CSN, carotid sinus nerve.

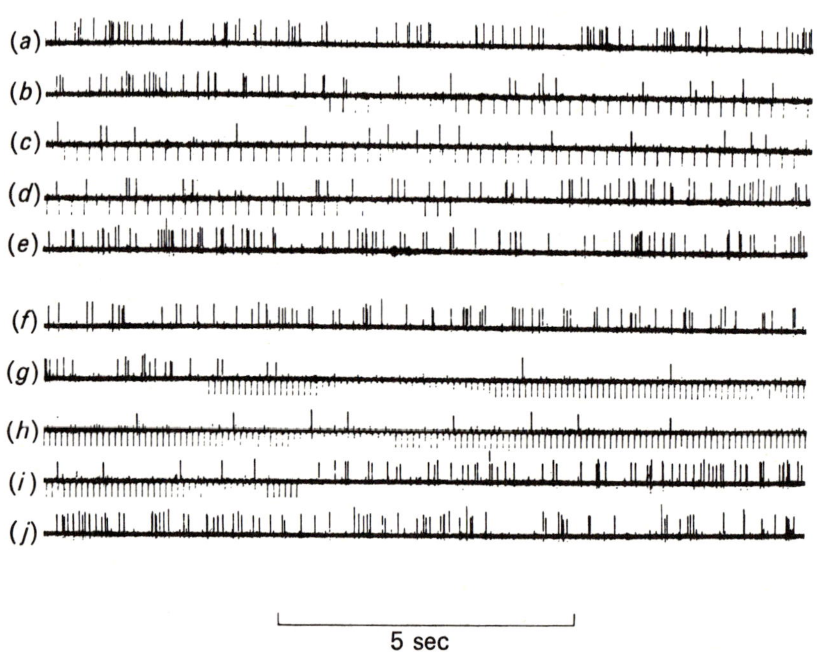

Fig. 2. Continuous filmed records of inhibition of chemosensory activity produced by electrical stimulation of remainder of CSN. Stimulation began in (*b*) and (*g*), and ended in (*d*) and (*i*), as indicated by the stimulus artifacts (vertical lines below baseline). The frequencies of stimulation were 5 per second in (*a*)–(*e*) and 10 per second in (*f*)–(*j*).

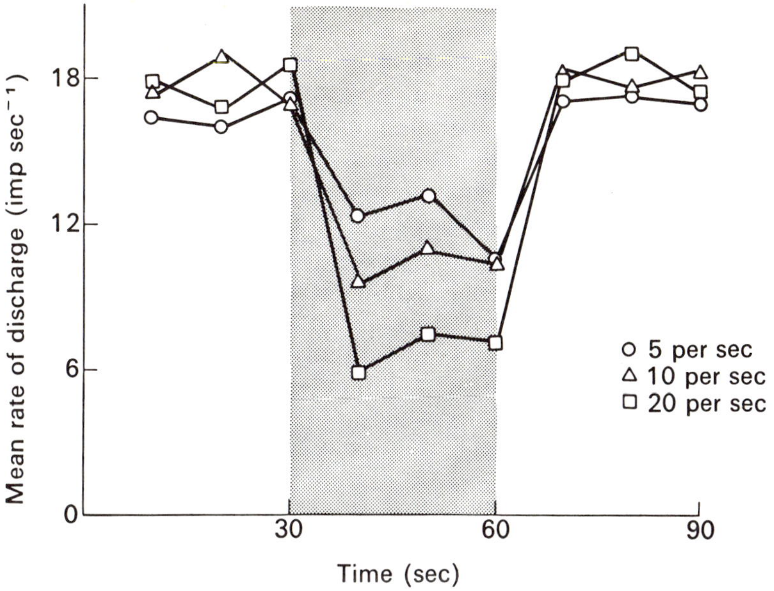

Fig. 3. Graphical representation of feedback inhibition. The symbols represented the mean rate of discharge (imp. sec^{-1}) occurring in successive 10-second periods before, during and after 30 seconds of stimulation, indicated by the hatched area from 30 to 60 seconds.

stimulation of the sinus nerve, but also an increase in the degree of inhibition as the frequency of stimulation is increased.

An alternative way of presenting this type of response is shown in Fig. 3. The mean rate of discharge is determined in 10-second periods, and the mean rate in successive periods is plotted as a function of elapsed time over an arbitrary period of 90 seconds; that is, 30 seconds of control, 30 seconds of stimulation and 30 seconds of recovery. This figure confirms that the degree to which afferent discharge is reduced is related to the frequency of stimulation.

The second technique which has been used to demonstrate feedback inhibition involves recording activity from a small chemoreceptor strand with the remainder of the carotid sinus nerve intact. The effects of subsequent section of the remainder of CSN, or of other procedures known to increase centrifugal discharges in CSN are then examined on chemoreceptor activity. Whereas none of the experiments I shall discuss were done this way, I believe that this approach is of critical importance for the establishment of feedback inhibition as a *bona fide* physiological phenomenon.

There are three possible explanations for this observed reduction in afferent activity with centrifugal stimulation. The first is that depression occurs as a result of an electrical artifact due to the spread of stimulating current from the portion of the nerve which is being stimulated. The second possibility is that stimulation of the remainder of the nerve causes the release within the carotid body of a substance which might cause an increase in blood flow and thereby depress chemoreceptor afferent discharge. The third possibility is that there is release within the carotid body of a substance or substances which hyperpolarize chemosensory fibres or in some other way depress their excitability.

Certainly the rate of discharge can be depressed by the spread of electrical current. This seems to be an established fact. And it should be recognized as a potential hazard in any of these experiments. It follows that whenever these kinds of experiments are done there are a number of precautions which must be taken. In my studies I have attempted to exclude this possibility by stimulating with a known electrical artifact and, as should become obvious later, I do not think that this has influenced the results obtained. Now, it is also true that there is a second type of inhibition which can be shown to be a very powerful depressant of chemosensory discharge, and I believe that a hypothesis regarding the mechanisms of this type of inhibition may be developed as follows. There is now good evidence that the Type I cell in the carotid body may be involved in the process of inhibition for the following reasons. First, there are the reports of efferent-type synapses on Type I cells (Biscoe & Stehbens, 1966; Kobayashi, 1971*b*; Biscoe, Lall & Sampson, 1970). Certainly it was the description of efferent-type synapses which provided the impetus to look for some form of modulation of chemosensory discharges by stimulating the sinus nerve towards the carotid body, but it is probable now that the origin of these synapses is a matter of dispute. Secondly, dopamine has been reported to inhibit respiration in the cat through an action on the carotid body and to depress chemosensory discharges (Black, Comroe & Jacobs, 1972; Sampson, 1972). Thirdly, dopamine is known to exist in Type I cells (Chiocchio, Biscardi & Tramezzani, 1966; Dearnaley, Fillenz & Woods, 1968; Chiocchio, King & Angelakos, 1971; Chiocchio, King, Carballo & Angelakos, 1971; Kobayashi, 1971*a*,*b*). A plausible hypothesis, therefore, is that dopamine is released from Type I cells of the carotid body under conditions of stimulation of the remainder of the nerve and, upon release, it does something to affect the rate of chemoreceptor discharge. My approach has therefore been to study the effects of adrenergic and cholinergic blocking agents on both the depression of chemoreceptor firing by catecholamines and the inhibition elicited by stimulation of the sinus nerve. In addition to the arrangement shown in Fig. 1, a catheter is inserted retrogradely into the lingual artery and advanced so that the tip is at about the level of the carotid sinus. The drugs which I shall be talking about – dopamine, noradrenaline, adrenaline, acetylcholine and sodium cyanide – are injected through this catheter; and the other drugs – the adrenergic and cholinergic blocking agents – are injected intravenously.

The effects of catecholamines are illustrated in Fig. 4.

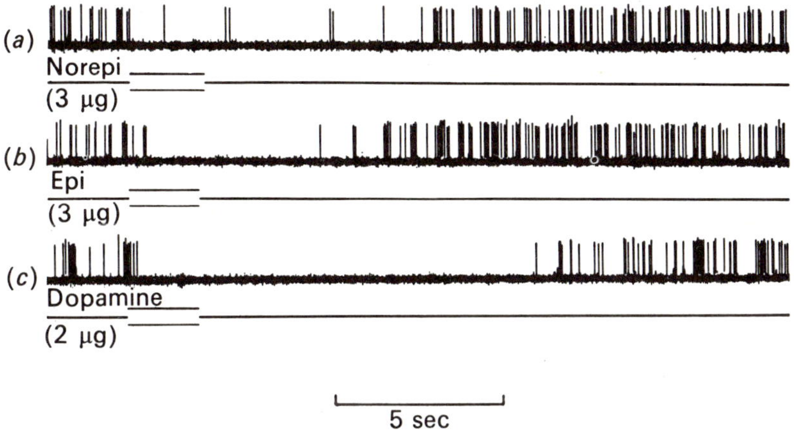

Fig. 4. Filmed records of effects of catecholamines on chemosensory activity. (*a*) Norepi; (*b*) Epi; (*c*) Dopamine. The dose of each drug, injected intra-arterially at the break in the lower baseline in each record, is indicated in parentheses.

From this it can be seen that on initial injection there is a depression of firing from the chemoreceptors and in the case of dopamine this depression is quite long; with adrenaline and noradrenaline there is a secondary rise in discharge after the initial depression. Because these drugs were made up and injected in a buffered Locke's solution, there is the obvious possibility of an artifactual depression of chemoreceptor discharge from the action of the diluent alone. I have tested this possibility by injecting buffered Locke's solution; this by itself does not cause depression of afferent discharge.

In the first series of experiments the effects of alpha-adrenergic blocking agents were studied, and the results are shown in Figs. 5–8.

Fig. 5 shows the effects of dibenziline upon the depressant action of catecholamines. After the injection of dibenziline (20 mg kg^{-1}) some 60 min previously, two things can be observed. First, there is an increase in the baseline discharge and secondly, the effects of catecholamines are abolished. Furthermore, if the animal was made to breathe an enriched oxygen mixture in order to bring the baseline level of discharge down to control levels, the effects of catecholamines were still abolished.

In Fig. 6 the effects of dibenziline upon feedback inhibition are shown – in this case inhibition was elicited by stimulation of CSN at 50 Hz.

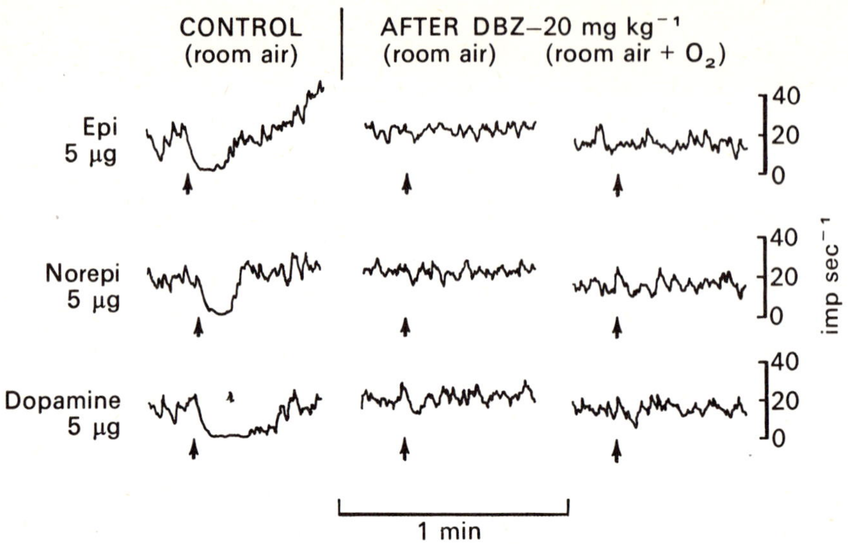

Fig. 5. Ratemeter recordings of effects of DBZ on depression of chemosensory discharges by catecholamines. Drugs were injected intra-arterially at the arrow in each record.

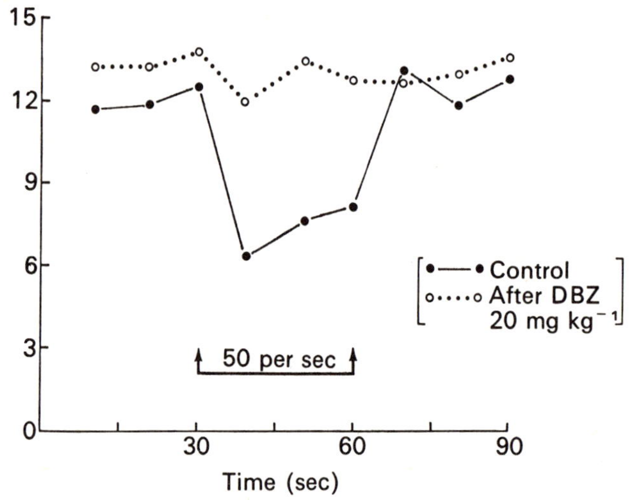

Fig. 6. Effects of DBZ on feedback inhibition of chemosensory discharges; same fibre as Fig. 5. Points as in Fig. 3.

In the control trace, this form of stimulation caused a reduction of the mean rate of firing from approximately 11 to 6 imp sec^{-1}; following treatment with dibenziline the effect of this stimulation was abolished.

There was the possibility in these experiments that, because dibenziline can be considered to be a toxic drug, and because its action only becomes effective after comparatively long periods following administration, the effects of CSN stimulation might have deteriorated with time. Therefore, these tests were repeated with another alpha-adrenergic blocker, dihydro-ergotamine, which is reported to be a competitive antagonist on alpha receptors and is also reversible. The results obtained with the use of this drug are summarized in Figs. 7 and 8.

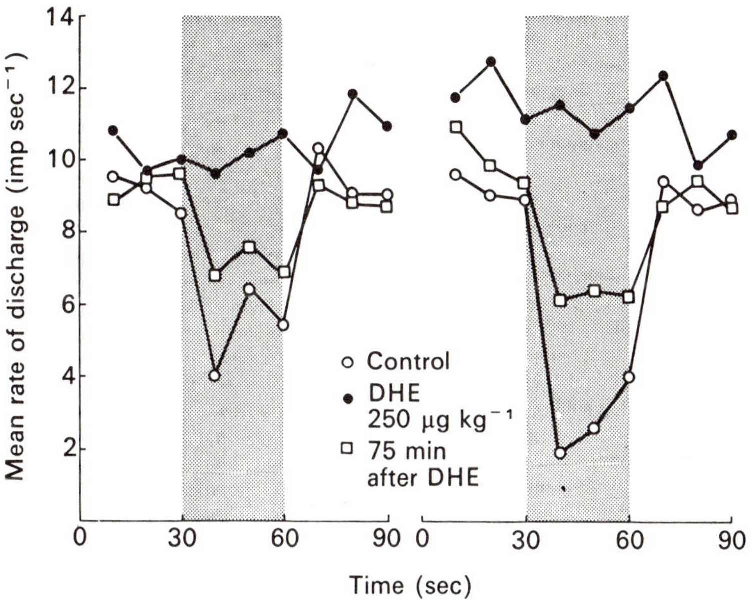

Fig. 7. Effects of dihydroergotamine on feedback inhibition of chemosensory discharges The frequency of stimulation was 10 per second in (*a*), 20 per second in (*b*).

Fig. 7 shows first, further examples of feedback inhibition caused by stimulation of the sinus nerve at 10 and 20 Hz. After dihydroergotamine, 250 μg kg^{-1}, had been given intravenously, the inhibitory effect was essentially abolished, and some 75 minutes later it had partially recovered. In Fig. 8 are shown the effects of dihydroergotamine on the depression of afferent discharge by close-arterial injection of dopamine, noradrenaline and adrenaline. There is a rise in the baseline discharge and virtually complete abolition of the depressant actions of the catecholamines.

In other experiments it was shown that dihydroergotamine had no discernible effects on the excitation of chemoreceptors by close-arterial injections of acetylcholine or sodium cyanide.

These results demonstrate that catecholamines, which are present in

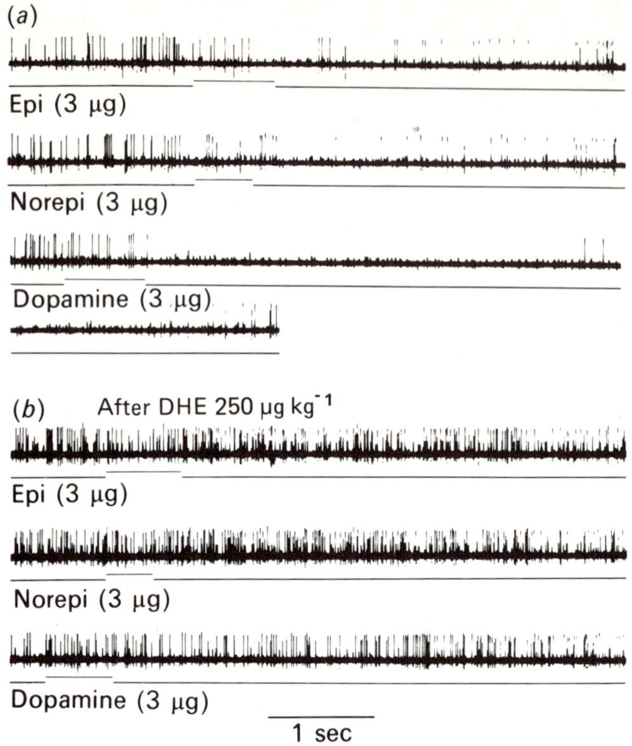

Fig. 8. Filmed records of effects of DHE (250 $\mu g\ kg^{-1}$) on catecholamine depression. Each drug (3 μg) was injected intra-arterially at the deflection of the lower baseline in each record. Same fibre as in Fig. 7.

carotid bodies of cats, can depress chemoreceptor afferent discharges and, thus, mimic feedback inhibition caused by stimulating the sinus nerve. Furthermore, both the electrically- and pharmacologically-induced depression of chemoreceptor afferent discharges are abolished by alpha-adrenergic blocking agents. These factors, taken in combination, may suggest that catecholamines, in particular dopamine, are involved in the phenomenon of inhibition. It is possible of course that the amines could be released from catecholaminergic nerve terminals, or they could be released from the Type I cells. I have assumed as a working hypothesis that the catecholamines are most likely to come from the Type I cells and that they may be released by the action of acetylcholine at the efferent synapse, because we know that there is acetylcholine in the carotid body. A further study was therefore made using cholinergic blocking agents to see whether these drugs affected feedback inhibition. The nicotinic blockers which were used, TEA and mecamylamine, failed to reduce feedback inhibition induced by electrical stimulation. However, atropine did block inhibition, and this effect of atropine given at the dose of 1 mg kg^{-1} intravenously is shown in Fig. 9.

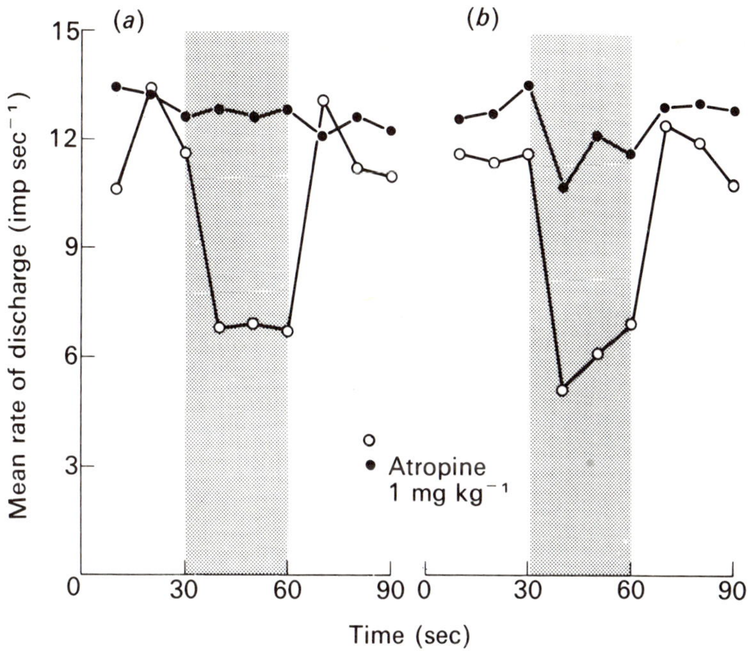

Fig. 9. Effects of atropine (1 mg kg^{-1}) in feedback inhibition. The frequencies of stimulation were 20 per second in (*a*) and 50 per second in (*b*) and occurred during the time indicated by the hatched area.

Feedback inhibition was induced with stimulation of the sinus nerve at 10 and 20 Hz. After administration of atropine, inhibition at the lower stimulating frequency was completely abolished, whereas that at the higher stimulating frequency was considerably attenuated. Other studies showed that atropine had little or no effect upon the effects of catecholamines, acetylcholine or sodium cyanide.

Blockade of feedback inhibition with atropine has been reported by Goodman (1973), and he has concluded that the most likely explanation of this is that atropine may block cholinergic vasodilatation. I believe that there are certain arguments against this interpretation. The first is that it does not take into account the inhibitory effects of catecholamines, especially dopamine. Secondly, it seems unlikely that changes in total carotid body flow can completely explain the inhibitory actions observed in the experiments reported here (Neil & O'Regan, 1971*a*; Biscoe, Bradley & Purves, 1970). It is, therefore, proposed that the phenomenon of feedback inhibition is best explained by the release of acetylcholine, or some other substance with muscarinic-like properties, at the synapse between nerve and Type 1 cell; that this substance acts on the Type 1 cell to release catecholamines, and that these mediate the inhibitory effect. How this inhibition comes about, even if it is allowed that it is mediated by the catecholamines, is poorly understood. I think it is unlikely that this is primarily a blood flow effect because dopamine is not a vasodilator except in a few special cases and then the vasodilatation occurs after alpha blockade (see Goldberg, 1972). In the present study it has been shown that the depressant action of dopamine is abolished by blockade of alpha receptors. It is therefore concluded that the action of dopamine upon the receptor is a direct one.

This work was supported by grant R HL-14201 from the National Heart and Lung Institute.

REFERENCES

BISCOE, T. J., BRADLEY, G. W. & PURVES, M. J. (1970). The relation between carotid body chemoreceptor discharge, carotid sinus pressure and carotid body venous flow. *J. Physiol.*, **208**, 99–120.

BISCOE, T. J., LALL, A. & SAMPSON, S. R. (1970). Electron microscopic and electrophysiological studies on the carotid body following intra-cranial section of the glossopharyngeal nerve. *J. Physiol.*, **208**, 133–52.

BISCOE, T. J. & STEHBENS, W. E. (1966). Ultrastructure of the carotid body. *J. Cell Biol.*, **30**, 563–78.

BLACK, A. M. S., COMROE, J. H., JR. & JACOBS, L. (1972). Species difference in carotid body response of cat and dog to dopamine and serotonin. *Am. J. Physiol.* **223**, 1097–102.

CHIOCCHIO, S. R., BISCARDI, A. & TRAMEZZANI, J. H. (1966). Catecholamines in the carotid body of the cat. *Nature, Lond.*, **212**, 834–5.

CHIOCCHIO, S. R., KING, M. P. & ANGELAKOS, E. T. (1971). Carotid body catecholamines. Histochemical studies on the effects of drug treatments. *Histochemie*, **25**, 52–9.

CHIOCCHIO, S. R., KING, M. P., CARBALLO, L. & ANGELAKOS, E. T. (1971). Monoamines in the carotid body cells of the cat. *J. Histochem. Cytochem.*, **19**, 621–6.

DEARNALEY, D. P., FILLENZ, M. & WOODS, R. I. (1968). The identification of dopamine in the rabbit's carotid body. *Proc. R. Soc. B*, **170**, 195–203.

FIDONE, S. J. & SATO, A. (1970). Efferent inhibition and antidromic depression of chemoreceptor A-fibers from the cat carotid body. *Brain Research*, **22**, 181–93.

GOLDBERG, L. I. (1972). Cardiovascular and renal actions of dopamine: potential clinical applications. *Pharmacol. Rev.*, **24**, 1–29.

GOODMAN, N. W. (1973). Efferent control of arterial chemoreceptors mediated by glossopharyngeal fibres and artifacts introduced by stimulation techniques. *J. Physiol.*, **230**, 295–311.

KOBAYASHI, S. (1971*a*). Comparative cytological studies of the carotid body. 1. Demonstration of monoamine-storing cells by correlated chromaffin reaction and fluorescence histochemistry. *Arch. Histol. Jap.*, **33**, 319–39.

KOBAYASHI, S. (1971*b*). Comparative cytological studies of the carotid body. 2. Ultrastructure of the synapses on the chief cell. *Arch. Histol. Jap.*, **33**, 397–420.

NEIL, E. & O'REGAN, R. G. (1969*a*). Effects of sinus and aortic nerve efferents on arterial chemoreceptor function. *J. Physiol.*, **200**, 69–71P.

NEIL, E. & O'REGAN, R. G. (1969*b*). Efferent and afferent impulse activity in the 'intact' sinus nerve. *J. Physiol.*, **205**, 20–21P.

NEIL, E. & O'REGAN, R. G. (1971*a*). The effects of electrical stimulation of the distal end of the cut sinus and aortic nerves on peripheral arterial chemoreceptor activity in the cat. *J. Physiol.*, **215**, 15–32.

NEIL, E. & O'REGAN, R. G. (1971*b*). Efferent and afferent impulse activity recorded from few-fibre preparations of otherwise intact sinus and aortic nerves. *J. Physiol.*, **215**, 33–47.

SAMPSON, S. R. (1972). Mechanism of efferent inhibition of carotid body chemoreceptors in the cat. *Brain Research*, **45**, 266–70.

SAMPSON, S. R. & BISCOE, T. J. (1970). Efferent control of the carotid body chemoreceptor. *Experientia*, **26**, 261–2.

DISCUSSION

ABBOUD I am concerned about the specificity of the blocking effect of phenoxybenzamine and dihydroergotamine. You did block the inhibitory effect of the catecholamines. Did you also block the excitatory effect of these drugs?

SAMPSON Yes. I would imagine that both of them are mediated by an action on alpha receptors and I have therefore assumed that these agents block the alpha receptor mediated vasoconstriction.

ABBOUD So you think that the inhibition of the afferent discharge is mediated through a direct effect on 'receptors' and that the excitation is mediated through a vascular effect?

SAMPSON Correct.

ABBOUD I would have been happier if you had used another substance or stimulus which is non-adrenergic and you still demonstrated an effect after the block such as, for example, hyperoxia. Was the effect of hyperoxia sustained after blockade?

SAMPSON I cannot answer that directly beyond commenting that hyperoxia will depress the rise in afferent discharge brought about by the blocking agents. But that does not completely answer your question.

ABBOUD The other question is that if you are postulating a muscarinic type of action at the synapse, as a result of a neurogenic stimulus, why don't you get the same response as with the injection of acetylcholine?

SAMPSON Well, acetylcholine acts directly upon chemosensory fibres to excite them, so there is a possibility of a dual effect. Perhaps the question is why is it that acetylcholine, after the administration of TEA or mecamylamine, doesn't cause depression of afferent discharge? I'll take that criticism one stage further and ask why it is that no muscarinic agent causes a depression of discharge. I don't know the answer to this. I have looked at three such agents and they all are without effect. There may be something in the technique of administration. I have done it by rapid close arterial injection and it may be that under those conditions the drugs are not there long enough to work.

GOODMAN One of the striking inconsistencies is the differences between the depth of inhibition which you obtained with electrical stimulation and that which Fidone & Sato and I obtained using the same frequencies of stimulation. You showed, I believe, a reduction of about 50 or 60 % in your Fig. 3, while Fidone & Sato (1970) had only just started getting any inhibition at all at this frequency. It seems to me to be important to look for the silent period after each stimulation especially if you are using multifibre strands because the adventitious can vary from fibre to fibre and also from moment to moment.

SAMPSON I'd be forced to agree with what you say and the only defence I'd offer is that the more I did these experiments, the better I got, the more easily I dissected the strands, the more impressive was the degree of inhibition.

GOODMAN Yes, but did you look for the silent period?

SAMPSON No, but I have carried out another type of control which is to stimulate the whole nerve and to record from a small bundle and there was of course inhibition of discharge but this was unaffected by adrenergic or cholinergic blocking agents.

WILSHAW Do you think it is possible that this dual effect of small amounts of adrenaline and noradrenaline could both be vasomotor? Thus, if you have a very fast transit time for flow through the arterio–venous anastomoses then a large proportion of noradrenaline would enter the arterio–venous anastomoses before it enters the sinusoidal circulation. And if there is vasoconstriction in the arterio–venous anastomoses there may be a redistribution of flow in favour of the sinusoids causing a transitory increase in sinusoidal flow and a depression of afferent discharge. The late apparent excitation would then be due to the correspondingly late vasoconstriction in the sinusoidal circulation.

SAMPSON Well, I would be prepared to accept that as an explanation for the action for noradrenaline but not for the action of dopamine which is really the substance I prefer to concentrate on.

MILLS The doses of the alpha-adrenergic blocking agents which you used, are they sufficient to produce a generalized alpha receptor blockade?

SAMPSON Certainly, if you give the phenoxybenzamine in the doses I gave, you get generalized blockade. I was surprised at first that I had to give such a high dose, but when you go back to the literature you find that if you wish to get blockade of the majority of alpha systems you may have to go to very high doses indeed.

MILLS Is there therefore some possibility that the increase in activity which you see after giving catecholamines is due to a reduction in flow through the carotid body consequent possibly on a reduction in systemic arterial pressure?

SAMPSON The pressure doesn't actually drop by very much.

GOODMAN Is dopamine found in the venous effluent from the carotid body when you stimulate the nerve?

SAMPSON I tried to look for that by labelling the carotid body but I never got sufficient labelling to detect dopamine in the effluent.

GOODMAN Did you see the inhibition on stimulation with every fibre that you used, or did you find some strands that seemed to show absolutely no effect?

SAMPSON Yes, I found both single and multi-fibres in which there was no obvious effect. But of course this may equally well have been due to some form of damage during dissection.

ACKER The critical closing pressure for carotid body flow of the cat normally lies in the region of 50–60 mm Hg. This holds only for physiological conditions. I think that it has been shown in other organs that adrenaline or noradrenaline raises the critical pressure to 80 mm Hg or more. Do you think that this is possible in your experiment?

SAMPSON I don't know.

ABBOUD I agree. The fall in pressure necessary to cause closure would be less after vasoconstriction, i.e. the critical closing pressure will rise in the presence of vasoconstricting agents. One last point, have you given dopamine after a beta receptor blocking agent?

SAMPSON No, I haven't done that.

REFERENCE

FIDONE, S. J. & SATO, A. (1970). Efferent inhibition and antidromic depression of chemoreceptor A-fibres from the cat carotid body. *Brain Res.*, **22**, 181–93.

The influences exerted by the centrifugal innervation of the carotid sinus nerve

By RONAN G. O'REGAN
Department of Physiology University College, Dublin

The existence of efferent nerve fibres in the sinus nerves of a variety of animals, including the cat, was reported by De Castro in 1926. Further evidence for the presence of centrifugal pathways stems from electroneurographic studies involving the recording of activity from efferent slips of both cut (Biscoe & Sampson, 1968; O'Regan, 1970) and otherwise intact sinus nerves (O'Regan, 1970; Neil & O'Regan, 1969*b*, 1971*b*). Hebb (1968) from estimations of choline acetyltransferase activity of sinus nerves, considered that efferent cholinergic fibres accounted for approximately 30 % of the total innervation.

Thus, efferents do exist in the sinus nerve but it is only recently that any serious attempts have been made to determine the influences they exert upon the carotid body. The majority of these studies have employed electrical stimulation of the cut sinus nerve in order to invoke propagated efferent discharge. The effects observed during such procedures include (1) a reduction of chemoreceptor afferent traffic (Neil & O'Regan, 1969*a*, 1971*a*; Fidone & Sato, 1970; Sampson & Biscoe, 1970; Sampson, 1972; Goodman, 1973), (2) an elevation of carotid body blood flow (Neil & O'Regan, 1969*a*, 1971*a*; Biscoe, Bradley & Purves, 1969), (3) a stimulation of the oxidative metabolism of the carotid body (Biscoe *et al.*, 1969; Mills & Jobsis, 1970; Mills, 1972), and (4) a decrease of the electron density of the Type I cell granules (Yates, Chen & Duncan, 1970). In my discussion here, I intend to limit myself to a consideration of the effects upon chemoreceptor discharge and carotid body blood flow, effects which were first demonstrated by Neil & O'Regan (1969*a*, 1971*a*).

The method used to investigate the influence upon the chemoreceptor discharge is open to criticism because of the likelihood of stimulus spread from the stimulating electrodes to the afferent slip causing, as a consequence, an antidromic depression of the chemoreceptor discharge and giving, therefore, an erroneous impression of true efferent influences (Fidone & Sato, 1970; Goodman, 1972, 1973). However, studies using afferent preparations of otherwise intact sinus nerves and in which efferent activity was engendered by procedures other than by the use of electrical stimulation have verified that centrifugal pathways do indeed depress chemoreceptor afferent discharge (Neil & O'Regan, 1969*b*; Sampson & Biscoe, 1970; Majcherczyk & Willshaw, 1973).

While there seems little doubt that efferents do have inhibitory influences, there exists considerable controversy as to the mechanism of this inhibition. Because excitatory effects upon chemoreceptor afferent discharge occur during increased activity of the sympathetic supply to the carotid body (Floyd & Neil, 1952; Eyzaguirre & Lewin, 1961; Biscoe & Purves, 1967), and because such excitation is associated with a reduction of carotid body blood flow (Daly *et al.*, 1954; Purves, 1970*a*), it seems reasonable to assume that the reduction of chemoreceptor afferent traffic which is seen during sinus nerve stimulation could have arisen from the accompanying carotid body hyperaemia. We considered such a mechanism to be very unlikely because the depression of afferent discharge during electrical stimulation of the sinus nerve (*a*) still occurred in circumstances of glomeral ischaemia, (*b*) was unaffected by doses of atropine which abolished the associated blood flow changes and (*c*) was, in its temporal pattern of change, unrelated to the accompanying alteration of blood flow (Neil & O'Regan, 1969*a*, 1971*a*). Recently, Goodman (1973) has criticized our interpretation considering that we were actually studying the effects of antidromic depression rather than true efferent influences and these effects would not be affected by atropine and would still be elicited during carotid body ischaemia. He repeated the findings of Fidone & Sato (1970) that antidromic depression can occur and that although true efferent inhibition of discharge is present during electrical stimulation of the nerve, its magnitude of effect is slight and, furthermore, such effects are blocked by atropine.

While we feel reasonably certain that we were studying true efferent effects, nonetheless, as we did not monitor stimulus-spread to the afferent slips in these experiments which were used to eliminate blood flow changes as the mechanism for the alteration of chemoreceptor discharge, there must exist doubt as to whether our interpretation is correct.

I decided, therefore, to reinvestigate this problem using an ischaemic carotid body but now studying the effects upon chemoreceptor discharge during an enhanced activity of sinus nerve efferent fibres, this activity being elicited by procedures which did not involve electrical stimulation. Obviously, under these circumstances any changes of chemoreceptor discharge could not be construed as having been the result of adventitious excitation of the afferent slip. Furthermore, any changes that occur in an ischaemic preparation could hardly have arisen from alterations of blood flow.

METHODS

Seven cats (1.1–3.75 kg) were anaesthetized with thiopentone sodium ('Intraval' Sodium, May & Baker Ltd) 50 mg kg^{-1}, i.p. Anaesthesia was maintained at a desired level by the i.v. injections of appropriate doses of either thiopentone sodium or chloralose. The femoral artery and its accompanying vein were cannulated, the cannulae being pushed cephalad until their tips lay within the abdominal aorta and inferior vena cava respectively. The trachea was also cannulated. The portions of the larynx, pharynx, oesophagus and trachea which lay between the lower border of the hyoid bone and the insertion of the tracheal cannula were removed. All the arteries arising from the common and external carotid arteries in the neighbourhood of the left carotid bifurcation were systematically identified and cut between ligatures, great care being taken not to damage either the arterial inflow or the venous outflow of the carotid body. The common and external carotid arteries remained patent throughout the dissection. The sympathetic nerves to the glomus were sacrificed but the sinus nerve was left intact. At the end of the dissection all that connected the area of the carotid bifurcation to the rest of the animal were the common and external carotid arteries, the carotid body vein or veins and the sinus nerve (Fig. 1). Cannulae were introduced into both lingual arteries and pushed into the external carotid artery until their tips lay at approximately the levels of the carotid bifurcations.

Production of carotid body ischaemia

When required, total abolition of carotid body blood flow could be readily achieved by occlusion of both the common and external carotid arteries using cross-action haemostats. At the end of the experiment, section of the artery within the occluded segment confirmed, as only to be expected, that total vascular isolation had been achieved. In order to prevent clotting during the periods of glomeral ischaemia, heparin (Pularin, Evans Medical Ltd) was administered every two hours in doses of 1000 IU kg^{-1}, i.v.

Preparation of afferent slips from otherwise intact nerves

The technique used here resembles that previously employed (Neil & O'Regan, 1969*b*, 1971*b*) and involved careful removal of the sheath of the sinus nerve followed by the dissection of slender afferent slips. The remainder of the nerve was left intact with the central neuraxis and its reflex

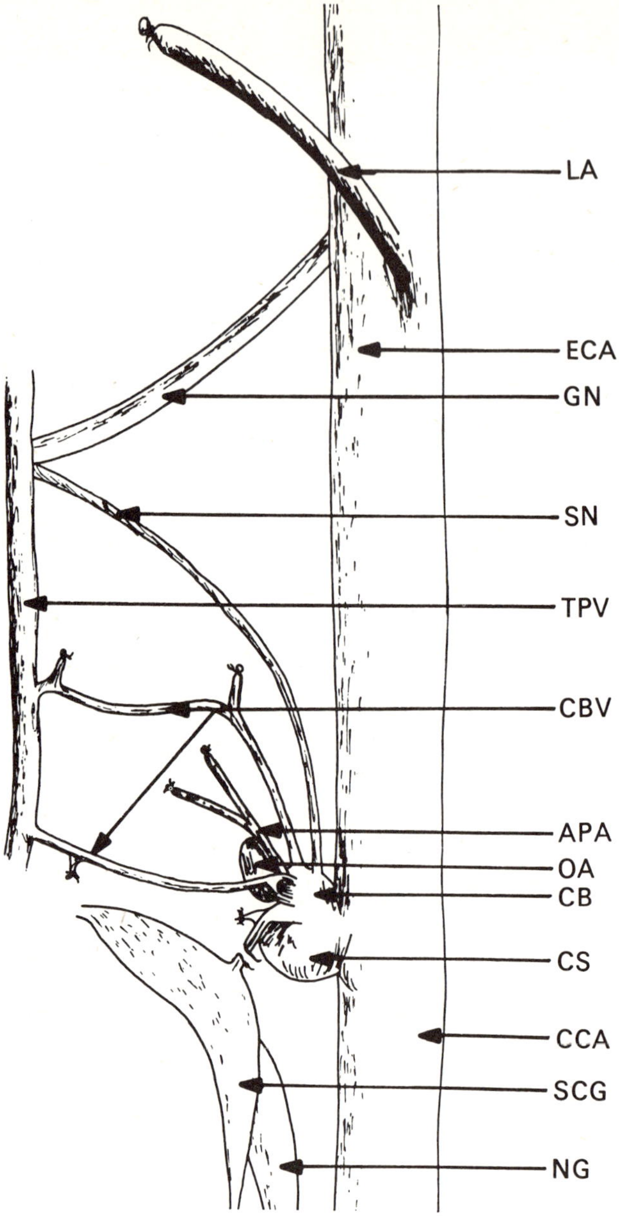

Fig. 1. Diagram of the medial aspect of the dissected carotid bifurcation (for details see text). APA = ascending pharyngeal artery; CB = carotid body; CBV = carotid body veins; CCA = common carotid artery; CS = carotid sinus; ECA = external carotid artery; GN = glossopharyngeal nerve; LA = lingual artery; NG = nodose ganglion; OA = occipital artery; SCG = superior cervical ganglion; SN = sinus nerve; TPV = transverse pharyngeal vein. Total abolition of carotid body blood flow was produced by occluding the common and external carotid arteries.

continuity confirmed either by the observation of a respiratory stimulation to close arterial NaCN (lingual artery) or, if this was not feasible, the demonstration of a pressor response to occlusion of the ipsilateral common carotid artery.

Recording procedures

Using saline-wick electrodes, and after suitable preamplification (Grass, Model P.16), the chemoreceptor afferent activity of few-fibre preparations was led to an audiomonitor and an oscilloscope (Tektronix 502A) for preliminary examination. The output of the preamplifier was also led to one channel of a direct-writing ink jet recorder (Mingograf 34, Sierex Ltd) from which permanent records were obtained. Systemic arterial blood pressure was measured using a pressure transducer (EMT 34, Sierex Ltd), together with an electromanometer (EMT 311, Sierex Ltd), displayed on the oscilloscope and recorded on another channel of the ink jet recorder. Impulse activity was considered to be of chemoreceptor origin if its frequency increased following either the close-arterial injections of NaCN (10–20 μg) or subsequent to the animal breathing an hypoxic gas mixture.

Engagement of sinus nerve efferent activity

Adrenaline, known to powerfully enhance the activity of slips peeled off from the central end of the cut sinus nerve (Biscoe & Sampson, 1968; O'Regan, 1970; Neil & O'Regan, 1971*b*), was used to engage efferent activity in the intact portion of the sinus nerve in all the experiments conducted here. The effects, during glomeral ischaemia, upon the chemoreceptor activity of the afferent slips were then studied. In two cats 10% oxygen:90% nitrogen was also administered under the same conditions; hypoxic hypoxia is also known to increase efferent discharge (Biscoe & Sampson, 1968; O'Regan, 1970).

Drugs

Atropine sulphate (Kochlight), adrenaline hydrochloride (Parke Davis) and NaCN were used. Gallamine triethiodide (Flaxedil, May & Baker Ltd) was used in three cats to prevent disturbing movements and consequently, it was necessary to mechanically ventilate these animals using a Starling pump (0.5–1 l min^{-1}).

RESULTS

The activities of twenty-seven chemoreceptor afferent slips of otherwise intact nerves were studied in these experiments.

Changes during glomeral ischaemia

Excitation of sinus nerve efferent discharge occurs during elevated chemoreceptor afferent activity in the same nerve (O'Regan, 1970; Neil & O'Regan 1969*b*, 1971*b*). The increased centrifugal discharge has the effect of suppressing chemoreceptor activity (Sampson & Biscoe, 1970). Consequently, it is to be expected that the responses of chemoreceptor afferent preparations during glomeral ischaemia should differ when these preparations are obtained from otherwise intact, rather than cut, sinus nerves. It was noted in these experiments that during glomeral ischaemia there was a slow build-up of chemoreceptor discharge to maximum values (5–15 minutes) when afferents of the otherwise intact nerve were used and that occasionally such build-up occurred in a series of increasing steps rather than in a gradual manner. However, adequate comparisons with the responses when the nerve was cut (central to the afferent slip) were not carried out and no significance can, as yet, be attached to these findings. Nevertheless, the chemoreceptor afferent traffic of slips of the otherwise intact nerve did eventually reach high levels of impulse frequencies, following abolition of carotid body blood flow, and these high levels were sustained throughout the further duration of the ischaemia. Occasionally, waxing and waning of the chemoreceptor discharge was observed during ischaemia but these preparations were not used in the present study.

Effects of adrenaline

During total abolition of carotid body blood flow the discharges of seven of twenty-four chemoreceptor preparations of otherwise intact sinus nerves were unequivocally reduced following i.v. injections of adrenaline (10–30 μg). Smaller doses (2–10 μg) injected into the contralateral lingual artery, so as to enter the stream of blood flowing in the external carotid artery, caused similar effects. The remaining seventeen preparations studied were unaffected by either procedure. Fig. 2 demonstrates the response of one of the preparations whose activity was altered by the drug. Track *a* (*AFF*) shows the level of activity during glomeral ischaemia. Following an i.v. injection of adrenaline (20 μg) the discharge was quickly depressed (Track *b*), remained depressed for a further thirty seconds or so and then recovered to levels close to those which prevailed before administration of the drug (Track *c*). The responses of systemic arterial blood pressure (*BP*) are also shown. A similar type of response was observed in five other preparations. The discharge of another chemoreceptor afferent slip behaved differently (Fig. 3). Here the reduction of discharge appeared quickly (Track *b*) after the injection of the drug but this reduction of impulse frequency remained for several minutes becoming more pronounced with

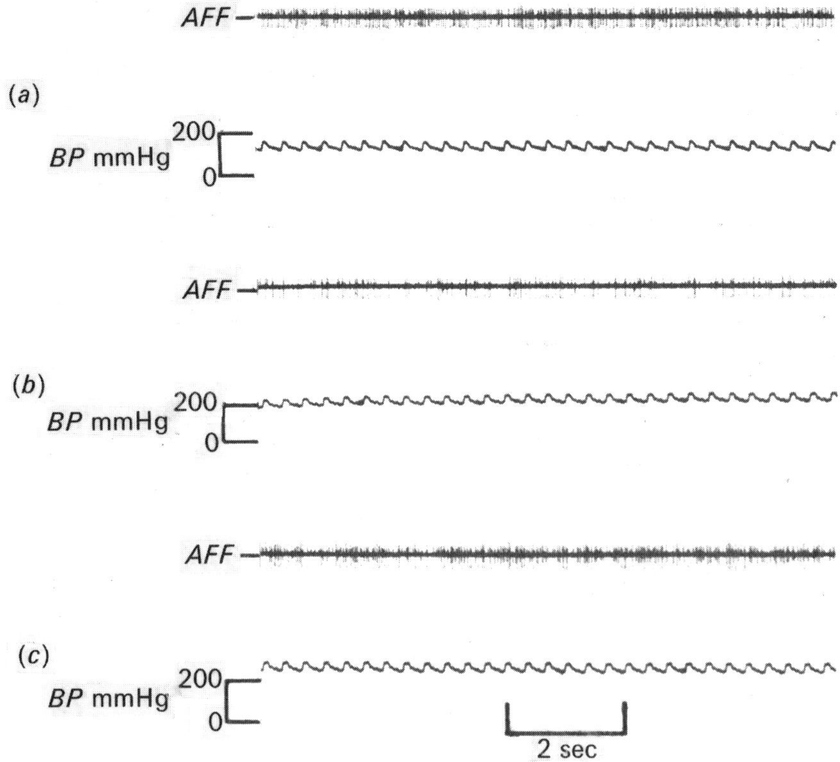

Fig. 2. Chemoreceptor activity (*AFF*) recorded from an afferent slip of the otherwise intact left sinus nerve; carotid body ischaemic. Track a = recording taken before adrenaline was injected; Track b = 20th–29th second after adrenaline (20 μg) i.v.; Track c = 61st–70th second after adrenaline. *BP* = blood pressure recorded from the descending aorta.

time (Track *c*); gradual recovery to control values then occurred (Track *d*).

Effects of central section of the sinus nerve on the responses to adrenaline

In three cats, section of the sinus nerve central to the chemoreceptor afferent slip was carried out. In all cases the depressive effects of adrenaline upon the chemoreceptor preparations were abolished.

Responses of efferent preparations to adrenaline

Following injections of adrenaline the responses of the activity of efferent slips of the cut sinus nerve vary considerably in temporal patterns of behaviour. Fig. 4 demonstrates the extremes noted in these cats and also in other experiments previously performed (O'Regan, 1970).

In the example shown in the upper part of this figure, the accentuation of discharge following, in this instance, an intralingual injection of the

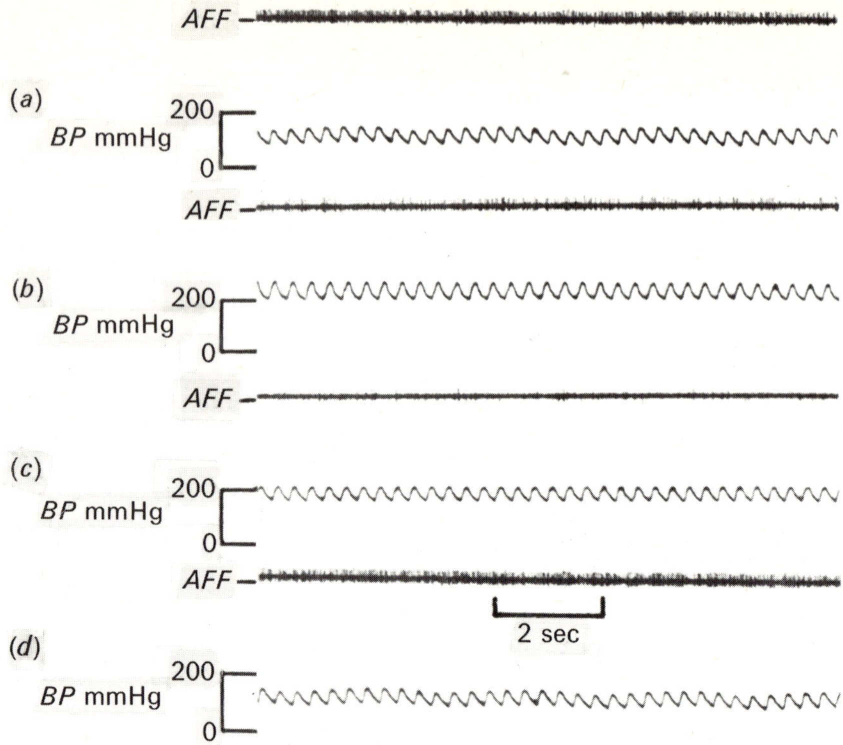

Fig. 3. Chemoreceptor activity recorded from an afferent slip of the otherwise intact left sinus nerve; carotid body ischaemic. Track a = before adrenaline was injected; Track b = 18th–28th second after adrenaline (30 μg) i.v.; Track c = 250th–260th second after adrenaline; Track d = 451st–461st second after adrenaline. *BP* and *AFF* as in Fig. 2.

drug, appeared quickly but the effect was short-lived while in the other example (lower part of Fig. 4) the build-up of activity was more gradual but sustained at high levels of discharge for several minutes. Because of these differing responses of efferent discharge to adrenaline it is to be expected that, following injections of this drug, the changes of chemoreceptor afferent activity of slips of the otherwise intact nerve should also vary widely. This was so in these experiments (Fig. 2, 3).

Effects of atropine

In one cat atropine (1 mg) was injected close-arterially via the cannula in the ipsilateral lingual artery. The increase of the activity of a chemoreceptor afferent preparation of the otherwise intact sinus nerve, to a bout of glomeral ischaemia subsequently produced, was much less than that noted before administration of atropine (Fig. 5*a*) but following the i.v. injection of adrenaline the activity was reduced, albeit not very markedly

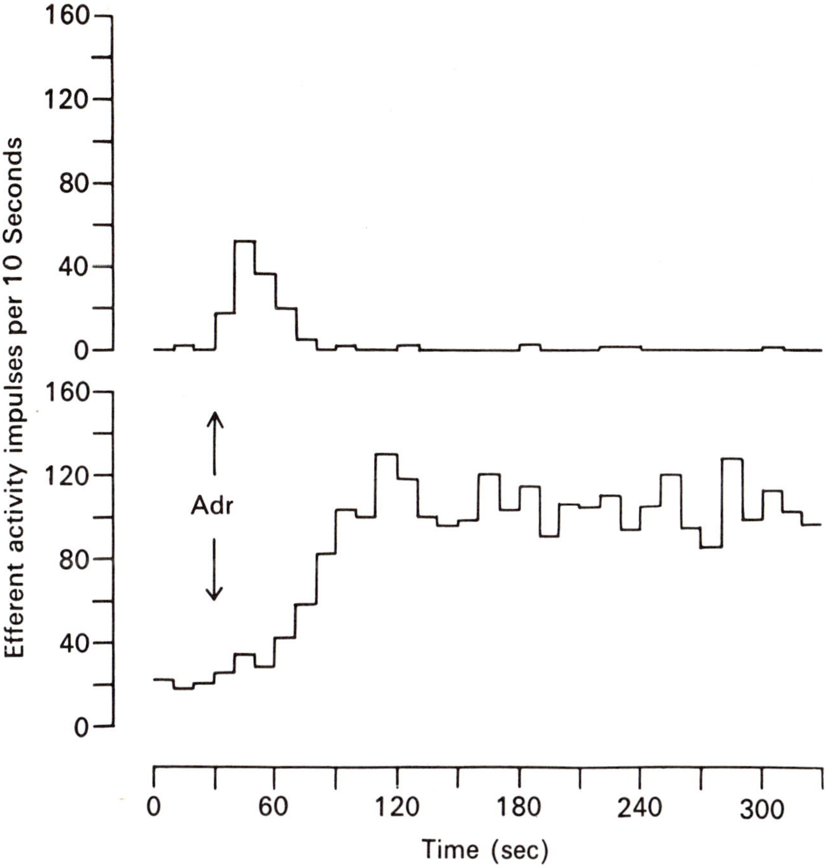

Fig. 4. Effects of adrenaline upon the activity of efferent slips of the cut sinus nerve. Upper = intralingual arterial injection of adrenaline (Adr) (5 μg); lower = i.v. adrenaline (30 μg).

(Fig. 5*b*). The effects of glomeral ischaemia and of adrenaline on the activity of the same preparation before atropine was injected and under the same conditions of experimentation can be seen in Fig. 2.

Effects of hypoxia

The effects of breathing a hypoxic gas mixture (10 % O_2 : 90 % N_2) upon the discharge of a chemoreceptor afferent slip of the otherwise intact nerve during conditions of glomeral ischaemia was tried on four occasions. In one preparation activity was depressed while breathing the gas mixture (Fig. 6*b*). The same preparation was also affected by adrenaline injections under the same experimental situation. The other afferent preparations were unaffected by either hypoxia or adrenaline injections.

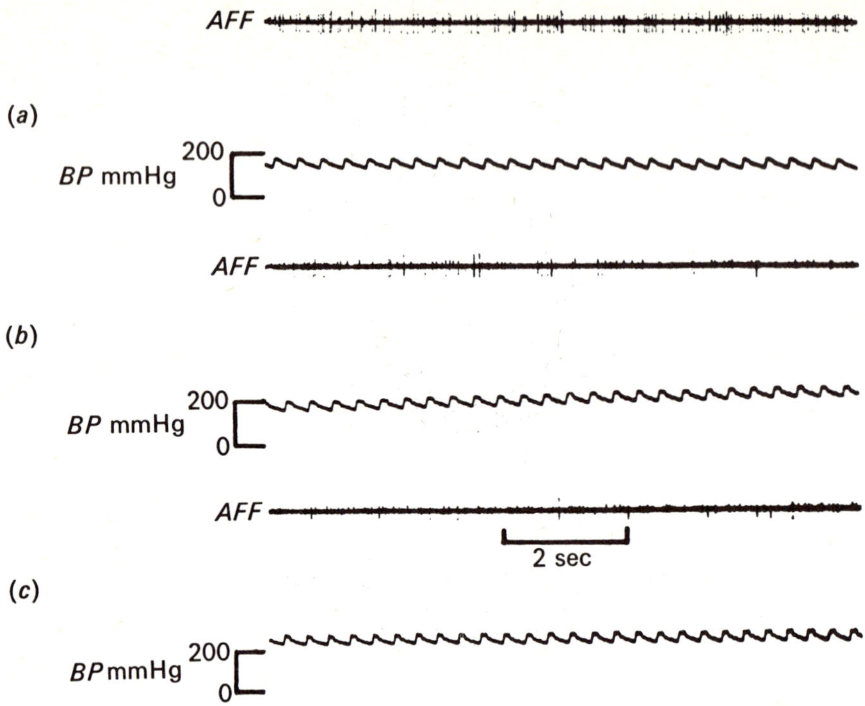

Fig. 5. Chemoreceptor activity recorded from an afferent slip of the otherwise intact left sinus nerve; carotid body ischaemic. Before production of ischaemia, atropine (1 mg) was injected into the area of the left carotid bifurcation via an intralingual cannula. Track a = before adrenaline was injected; Track b = 15th–24th second after adrenaline (20 μg) i.v. Track c = 60th–69th second after adrenaline. AFF and BP as in Fig. 2.

DISCUSSION

This investigation showed that, during total abolition of carotid body blood flow, both adrenaline and hypoxic hypoxia depressed the discharge of chemoreceptor afferent slips peeled off from otherwise intact sinus nerves. Under the experimental conditions, these effects could only have resulted from an engagement of efferent activity in the intact portions of these nerves which in turn inhibited the activity of the carotid body. It could be argued that the passage of the drug in a retrograde manner, via the carotid body veins, could have occurred under the experimental conditions and caused as a consequence an inhibition of discharge (Sampson, 1972). However, if this were so I would have expected the majority of chemoreceptor preparations to be affected by adrenaline which was not the case in these experiments. Furthermore, the effect of the drug upon the chemoreceptor activity was abolished when the sinus nerve was cut central to the afferent slip, a procedure which removed centrifugal influences. Addi-

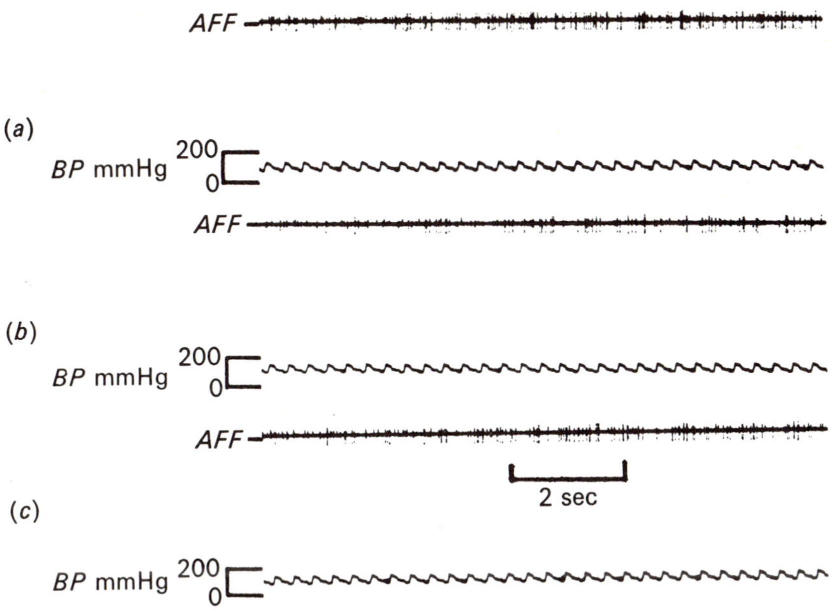

Fig. 6. Chemoreceptor activity recorded from an afferent slip of the otherwise intact left sinus nerve; carotid body ischaemic. Tracks *a* and *c* = animala rtificially ventilated with room air; Track *b* = animal ventilated with a 10 % O_2 : 90 % N_2 gas mixture. *BP* and *AFF* as in Fig. 2.

tionally, hypoxia reduced the afferent traffic during glomeral ischaemia and I would have expected the reverse if access to the chemoreceptor organ was present.

Thus, while there seems little doubt that increased centrifugal activity of the sinus nerve was responsible for the effects, it is puzzling that so few chemoreceptor preparations were affected; either there is an efferent innervation to small numbers of receptor complexes or the efferent fibres which would have been effective were destroyed during the dissection of the nerve, possibly by the inadvertent peeling off of these fibres with the afferent preparation. Similar findings occurred when efferent activity was invoked by electrical stimulation of the nerve, the majority of preparations being unaffected by this procedure (O'Regan, 1970).

Mechanisms of inhibition: vascular or non-vascular

It is possible that blood flow through the carotid body could continue for some time after occlusion of common and external carotid arteries because the pressure in the occluded segment would still be higher than that of the veins receiving the carotid body venous outflow. However, Purves (1970*b*) showed that, under similar conditions of experimentation, blood flow, as measured by the collection of the venous drainage of the carotid body, ceased within 20–30 seconds of occlusion of these carotid blood vessels. In the present study, no chemoreceptor preparations during glomeral ischaemia were tested until maximum discharge rates had built-up, which took between 5–15 minutes to occur. Flow would certainly have stopped after these intervals. It is also possible, if unlikely, that efferent innervation could have caused a relaxation or a constriction of the blood vessels of the carotid body, thereby effecting intraglomerular shunting of blood; such shunting could have altered the discharge. However, a constrictor effect would have involved an increase of sympathetic activity which is the reverse of what usually occurs after adrenaline injections. Also, the musculature of the carotid body blood vessels would be hypotonic from the stagnant asphyxia and any further relaxation would be of a small order if at all; hypotonia from stagnant asphyxia is suggested by reactive hyperaemia which ensues following a period of glomeral ischaemia (unpublished observations). Therefore, it is improbable that the effects observed in these experiments could have arisen from vascular changes and some other mechanism for the suppression of discharge must be envisaged.

Further evidence against a vascular causation stems from the work of Willshaw who will show in a subsequent paper at this Symposium that no alteration of carotid body blood flow occurs during superfusion of the brainstem with alkaline solutions, a procedure which increases the activity of efferent preparations of the sinus nerve and also suppresses the activity of afferent slips of otherwise intact sinus nerves (Majcherczyk & Willshaw, 1973). Furthermore, Sampson (1972) has shown that the effects upon chemoreceptor discharge during electrical stimulation of the cut sinus nerve are abolished by α-adrenergic blocking agents. The effect of sinus nerve efferents are dilatory (Biscoe, *et al.*, 1969), presumably cholinergic (Neil & O'Regan, 1969*a*, 1971*a*) and such effects would be unaltered by such agents.

Goodman (1973) suggested that blood flow changes were responsible for the effects upon discharge but did not monitor blood flow while he electrically stimulated the cut sinus nerve. The intensities of stimulation (with stimulus isolation) used by him, and which were shown to alter discharge, would not have had any effect whatsoever upon carotid body blood flow under the conditions of our experiments (Neil & O'Regan, 1969*a*, 1971*a*).

Taking all this evidence into account it would appear that our original interpretation was correct, in that the effects upon chemoreceptor activity which we observed, when we electrically stimulated the sinus nerve, could not have been the consequence of the associated blood flow changes. However, it is reasonable to assume that vascular changes can affect chemoreceptor activity but at present the functional significance of such changes are obscure.

Possible non-vascular mechanism of inhibition

A likely mechanism of inhibition is that proposed by Sampson (1972) that the efferent endings upon the Type I cell, when activated, cause the release of catecholamines from these cells; these agents in turn depress the chemoreceptor afferent terminals. His evidence is strong and is supported by the findings of Yates *et al.* (1970) that the electron densities of the catecholamine-containing vesicles (Lever, Lewis & Boyd, 1959; Chiocchio, Biscardi & Tremezzani, 1966; Dearnaley, Fillenz & Woods, 1968) of the Type I cells are reduced following a bout of electrical stimulation of the cut sinus nerve. While we found atropine to be ineffective in abolishing nerve-evoked inhibition, Goodman (1973) and Sampson (this Symposium) did not agree. On the one occasion I used atropine in the present series of experiments, the changes of chemoreceptor discharge induced by adrenaline were still present after atropine administration. However, the chemoreceptor response to glomeral ischaemia was reduced following the atropine injection either because of a local anaesthetic action (Heymans, 1955; Sheu, George & Jenden, 1969) or because of a true anticholinergic effect (Nishi & Eyzaguirre, 1970). Furthermore, the inhibition was not as marked as before the injection of the drug.

In any discussion of the mechanism of inhibition the axodendritic junctions described by Verna (this Symposium) must be given serious consideration and it may be that there are more than one means of suppressing chemoreceptor activity.

Anatomical possibilities in relation to physiological findings

Finally, I would like to discuss some anatomical possibilities that could account for those physiological findings which have been gleaned from experiments used in the study of the centrifugal innervation of the sinus nerve. Fig. 7 which is taken from my Ph.D. thesis (1970) illustrates the most likely possibilities which I considered at that time could best explain the available experimental data. The possibility shown in *a* of this figure considers that the efferent cell bodies reside in the brainstem and their axons proceed without interruption to end directly upon the Type I cell; chemoreceptor afferents can synaptically excite the efferent neurones. This possibility resembles other situations such as the olivocochlear innervation to the organ of Corti (Whitfield, 1967) or the efferent nerves to the lateral line organ (Russell, 1968). The findings of Biscoe *et al.* (1970) that the majority of the endings upon the Type I cells degenerate following supra-

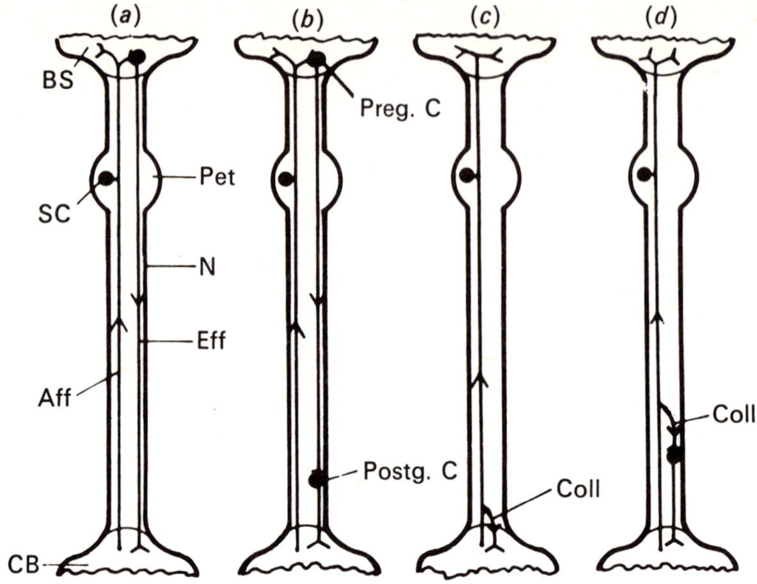

Fig. 7. Some anatomical possibilities (*a*, *b*, *c* and *d*) that could account for the physiological findings in relation to sinus nerve centrifugal pathways (see text). Aff = sensory axon; BS = brainstem; CB = carotid body; Coll = collateral of sensory axon; N = nervous pathways; Pet = petrosal ganglion; Preg. C = preganglionic cell; Postg. C. = postganglionic cell; SC = sensory cell. (From O'Regan, 1970.)

petrosal section of the glossopharyngeal nerve and, that despite such degeneration, chemoreceptor afferent activity could still be recorded from the sinus nerve innervating the affected organ suggested then that this possibility was by far the most likely. However, these consequences of suprapetrosal section were contrary to the results obtained by De Castro (1928) and De Castro & Rubio (1968) and recent work (Hess & Zapata, 1972) has failed to confirm the findings of Biscoe *et al.* Consequently, I would now consider this possibility to be unlikely. Possibility *b* states that the efferent cells reside within the brainstem but their axons relay with ganglion cells located at the periphery of the carotid body, within this organ and also within the glossopharyngeal and sinus nerves. This was the view of De Castro (1926, 1951) but he considered that the postganglionic axons innervated the blood vessels of the carotid body. However, following sinus nerve section, of many months' standing, some endings upon the Type I cells survive degeneration (Hess & Zapata, 1972; Biscoe & Stehbens, 1967) and may, as suggested by Hess & Zapata, represent the terminals of the axons of the postganglionic cell. While they considered these pathways to be of minor significance, such pathways could explain all the results which were noted both during electrical stimulation of the sinus nerve and when activity was recorded from slips of otherwise intact nerves. The third possibility (Fig. 7*c*) considers axon reflex mechanisms. These were suggested by Torrance (1968, 1969) as possibilities and could readily explain the effects during electrical stimulation if the efferent limbs were stimulated while recordings were made from their afferent counterparts. However, it is difficult to envisage how such a mechanism could account for the findings when chemoreceptor afferent slips of the otherwise intact nerve were used; adrenaline would be unlikely to excite antidromic impulses in afferent nerves. Nevertheless, chemoreceptor 'afferent' activity can be readily recorded from 'efferent' slips of otherwise intact nerves, this activity being still present following central section of the nerve but being no longer obtainable after distal section. The last possibility (Fig. 7*d*) envisages a collateral from a chemoreceptor afferent effecting a synapse with a ganglion cell, whose axon in turn terminates within the receptor complex from which the afferent arose. During chemoreceptor excitation, a negative feedback is exerted as in the manner of the Renshaw cell (Eccles, Fatt & Kotetsu, 1954), but involving here a sensory receptor rather than a motor neurone. Such a possibility cannot explain all the physiological findings but could exist together with the anatomical possibility depicted in Fig. 7*b*.

While other possibilities can exist, I would consider those in Fig. 7 to be the most likely to account for all of the physiological findings in relation to centrifugal pathways. However, none of these possibilities can fully explain all the morphological and physiological data which are at present available.

I wish to express my gratitude to the Wellcome Trust and the Medical Research Council of Ireland for help with the purchase of equipment used in the work presented here.

REFERENCES

Biscoe, T. J., Bradley, G. W. & Purves, M. J. (1969). The relation between carotid chemoreceptor activity and carotid sinus pressure in the cat. *J. Physiol.*, **203**, 40P.

Biscoe, T. J., Lall, A. & Sampson, S. R. (1970). Electron microscopic and electrophysiological studies on the carotid body following intra-cranial section of the glossopharyngeal nerve. *J. Physiol.*, **208**, 133–52.

Biscoe, T. J. & Purves, M. J. (1967). Factors affecting the cat carotid chemoreceptor and cervical sympathetic activity with special reference to passive hindlimb movements. *J. Physiol.*, **190**, 425–41.

Biscoe, T. J. & Sampson, S. R. (1968). Rhythmical and non-rhythmical spontaneous activity recorded from the central cut end of the sinus nerve. *J. Physiol.*, **196**, 327–38.

Biscoe, T. J. & Stehbens, W. E. (1967). Ultrastructure of the denervated carotid body. *Quart. J. exp. Physiol.*, **52**, 31–6.

Chiocchio, S. R., Biscardi, A. M. & Tramezzani, J. H. (1966). Catecholamines in the carotid body of the cat. *Nature, Lond.*, **212**, 834–5.

Daly, M. de B., Lambertsen, C. J. & Schweitzer, A. (1954). Observations on the volume of blood flow and oxygen utilization of the carotid body in the cat. *J. Physiol.*, **125**, 67–89.

DEARNALEY, D. P., FILLENZ, M. & WOODS, R. I. (1968). The identification of dopamine in the carotid body of the rabbit. In *Arterial Chemoreceptors*, ed. TORRANCE, R. W., pp. 189–92, Oxford: Blackwell.

DE CASTRO, F. (1926). Sur la structure et l'innervation de la glands intercarotidienne (Glomus caroticum) de l'homme et des mammiferes, et sur un nouveau systeme d'innervation autonome du nerf glossopharyngien. Etudes anatomiques et experimentales. *Trab. Lab. Invest. Biol. Univ. Madrid.*, **24**, 365–432.

DE CASTRO, F. (1928). Sur la structure et l'innervation du sinus carotidien de l'homme et des mammiferes. Nouveaux faits sur l'innervation et la fonction du glomus caroticum. Etudes anatomiques et physiologiques. *Trab. Lab. Invest. Biol. Univ. Madrid.*, **25**, 331–80.

DE CASTRO, F. (1951). Sur la structure de la synapse dans les chemorecepteurs: leur macanisme d'excitation et role dans la circulation sanguine locale. *Acta physiol., Scand.*, **22**, 14–43.

DE CASTRO, F. & RUBIO, M. (1968). The anatomy and innervation of the blood vessels of the carotid body and the role of chemoreceptive reaction in the autoregulation of the blood flow. In *Arterial Chemoreceptors*, ed. TORRANCE, R. W., pp. 267–70, Oxford: Blackwell.

ECCLES, J. C., FATT, P. & KOTETSU, K. (1954). Cholinergic and inhibitory synapses in a pathway from motor axon collaterals to motoneurones. *J. Physiol.*, **126**, 524–6.

EYZAGUIRRE, C. & LEWIN, J. (1961). The effect of sympathetic stimulation on carotid nerve activity. *J. Physiol.*, **159**, 251–67.

FIDONE, S. J. & SATO, A. (1970). Efferent inhibition and antidromic depression of chemoreceptor A-fibres from the cat carotid body. *Brain Res.*, **22**, 181–93.

FLOYD, W. F. & NEIL, E. (1952). The influence of the sympathetic innervation of the carotid bifurcation on chemoreceptor and baroreceptor activity in the cat. *Arch. Intern. Pharmacodyn.*, **91**, 230–9.

GOODMAN, N. W. (1972). Efferent control of arterial chemoreceptors. *J. Physiol.*, **225**, 31–32P.

GOODMAN, N. W. (1973). Efferent control of arterial chemoreceptors mediated by glosopharyngeal fibres and artifacts introduced by stimulation techniques. *J. Physiol.*, **230**, 295–311.

HEBB, C. O. (1968). In *Arterial Chemoreceptors*, ed. TORRANCE, R. W., pp. 138–9, Oxford: Blackwell.

HESS, A. & ZAPATA, P. (1972). Innervation of the cat carotid body: normal and experimental studies. *Fedn Proc.*, **31**, 1365–4.

HEYMANS, C. (1955). Action of drugs on carotid body and sinus. *Pharmacol. Rev.*, **7**, 119–42.

LEVER, J. D., LEWIS, P. R. & BOYD, J. D. (1959). Observations on the fine structure and histochemistry of the carotid body in the cat and rabbit. *J. Anat.*, **93**, 478–90.

MAJCHERCZYK, S. & WILLSHAW, P. (1973). Inhibition of peripheral chemoreceptor activity during superfusion with alkaline c.s.f. of the ventral brainstem surface of the cat. *J. Physiol.*, **231**, 26–27P.

MILLS, E. (1972). Spectrophotometric and fluorometric studies on the mechanism of chemoreception in the carotid body. *Fedn Proc.*, **31**, 1394–8.

MILLS, E. & JOBSIS, F. F. (1970). Simultaneous measurement of cytochrome a_3 reduction and chemoreceptor afferent activity in the carotid body. *Nature, Lond.*, **225**, 1147–9.

NEIL, E. & O'REGAN, R. G. (1969*a*). Effects of sinus and aortic nerve efferents on arterial chemoreceptor function. *J. Physiol.*, **200**, 69–71P.

NEIL, E. & O'REGAN, R. G. (1969*b*). Efferent and afferent impulse activity in the 'intact' sinus nerve. *J. Physiol.*, **205**, 20–21P.

NEIL, E. & O'REGAN, R. G. (1971*a*). The effects of electrical stimulation of the distal end of the cut sinus and aortic nerves. *J. Physiol.*, **215**, 15–32.

NEIL, E. & O'REGAN, R. G. (1971*b*). Efferent and afferent impulse activity recorded from few-fibre preparations of otherwise intact sinus and aortic nerves. *J. Physiol.*, **215**, 33–47.

NISHI, K. & EYZAGUIRRE, C. (1970). Effects of atropine on chemoreceptor discharges in the carotid body of the cat. *Brain Res.*, **23**, 292–7.

O'REGAN, R. G. (1970). Studies on oxygen usage of the carotid body and the effects of efferent nerves on arterial chemoreceptor function in the cat. Ph.D. Thesis, University of London.

PURVES, M. J. (1970*a*). The effect of hypoxia, hypercapnia and hypotension upon carotid body blood flow and oxygen consumption in the cat. *J. Physiol.*, **209,** 395–416.

PURVES, M. J. (1970*b*). The role of the cervical sympathetic nerve in the regulation of oxygen consumption of the carotid body of the cat. *J. Physiol.*, **209,** 417–32.

RUSSELL, I. J. (1968). Influence of efferent fibres on a receptor. *Nature, Lond.*, **219,** 177–8.

SAMPSON, S. R. (1972). Mechanism of efferent inhibition of carotid body chemoreceptors in the cat. *Brain. Res.*, **45,** 266–70.

SAMPSON, S. R. & BISCOE, T. J. (1970). Efferent control of the carotid body chemoreceptor. *Experientia, Basel*, **26,** 261–2.

SHEU, Y. S., GEORGE, R. & JENDEN, D. J. (1969). Comparison of the effects of atropine sulphate and procaine hydrochloride on nerve conduction. *Experientia, Basel*, **25,** 1053–4.

TORRANCE, R. W. (1968). Prolegomena. In *Arterial Chemoreceptors*, ed. TORRANCE, R. W., pp. 1–40, Oxford: Blackwell.

TORRANCE, R. W. (1969). The idea of a chemoreceptor. In *The Pulmonary Circulation and Interstitial Space*, ed. FISHMAN, A. P. & HECHT, H. H., p. 223, Chicago University Press.

WHITFIELD, I. C. (1967). *The Auditory Pathway*, pp. 124–41, London: Arnold.

YATES, R. D., CHEN, I-LI & DUNCAN, D. (1970). Effects of sinus nerve stimulation on the carotid body glomus cell. *J. Cell. Biol.*, **46,** 544–52.

DISCUSSION

GOODMAN How many of the receptors responded to the close injection of adrenaline?

O'REGAN I tested 24, of which seven responded, and this fits in very well with the results I obtained during electrical stimulation of the sinus nerve.

GOODMAN So in other words you are going to have to envisage some form of interaction in the carotid body between an efferent and an afferent.

O'REGAN No, I think there are other explanations. It may be, for example, that very few receptors are innervated; but it could also be that during dissection some of the efferents are destroyed.

BINGMANN I'd like to ask whether the carotid body chemoreceptors responded to the blowing of oxygen or nitrogen over the surface of the carotid body? The procedure of clamping all arteries might have changed the barrier properties of the outer zone described by Acker & Lübbers (1971).

O'REGAN I didn't try that.

BINGMANN A further question is: are you quite certain that all inflowing arteries were clamped?

O'REGAN It was a completely ischaemic preparation. All the arterial inflow had been clamped off.

TORRANCE Was it one final clamping off?

O'REGAN No, it was repeated clamps and I used plenty of heparin and there was no doubt that if one of these clamps was removed, the discharge disappeared very rapidly.

PURVES Further to Dr Bingmann's question. What did you use to cover the surface of the carotid body?

O'REGAN We used warmed saline equilibrated with air. I didn't use paraffin.

TORRANCE The question of critical closing pressure has been raised before. You said that when you clamped off the arteries, the venous outflow stopped within 20 seconds. Did you measure the pressure in the carotid segment? Is it conceivable that the pressure there falls to a critical closing pressure but that if you then superimpose upon the situation a nervous vasodilator effect, the critical closing pressure falls and there is a further run-off of blood through the carotid body?

O'REGAN That is possible, but as I have said, it does show pronounced reactive hyperaemia and I think it is very unlikely.

TORRANCE Does the carotid body show pronounced reactive hyperaemia?

O'REGAN Yes.

TORRANCE But let me repeat my question: if you have a steady state and you introduce a vasodilator effect, you might lower the critical closing pressure and so have further transient episode of blood flow.

O'REGAN That is true; but in that case I would have expected more than just a few chemoreceptors to respond.

TORRANCE But you would expect the afferent discharge to diminish during this period of blood flow?

O'REGAN Yes, but I must repeat that if it were a vascular mechanism, I would have expected the majority of the receptors to respond in this way. And that was not the case.

ACKER We have some experience with this problem. If you artificially perfuse the carotid body and then suddenly reduce perfusion pressure, you must make sure that the perfusion pressure in the carotid sinus is really zero. If you clamp off the artery the pressure only slowly decreases but you can by some suitable preparation make sure that the pressure and flow suddenly fall to zero – for example, by opening the external carotid artery to the atmosphere. If you do this you can also observe a virtually instantaneous stop of venous outflow. I would agree with Dr Torrance that if you do not take this precaution and perfusion pressure falls to, for example 10 mm Hg, it is still possible to get, under these experimental conditions, some flow through the carotid body.

O'REGAN You mean that flow could still be continuing?

ACKER Yes, if the pressure on the arterial side is not exactly zero.

O'REGAN I must confess that I have been largely dependent on the observation by Dr Purves in his paper of 1970 who showed that if the arterial inflow is clamped in very much the same way as in the present experiment, blood flow stopped within 20 seconds.

WILLSHAW I hope that we have got this question of critical closing pressure straight. I understand from previous discussion that the critical closing pressure will increase with vasoconstriction and from what Dr Torrance has been suggesting that with a dilator influence the critical closing pressure will decrease.

TORRANCE All I was saying was that if you carry out a standard Burton-type critical closing pressure procedure and allow pressure to fall until it falls no longer and if you have previously introduced a vasodilator, the final point at which pressure ceases to drop will be less than it would have been had you not introduced the vasodilator. So, if you have a vasodilator influence which you can introduce, I imagine – I repeat imagine – that if you have got the flow to stop under the condition of some vasomotor tone, the vasodilator would relax the vascular muscle and you would expect some further flow and the pressure would then drop to a new lower critical closing pressure. That is all I meant.

ACKER I would just reiterate that it is important to remember that under such experimental conditions, the critical closing pressure in the carotid body can be very low, that is about 10 or 20 mm Hg.

ABBOUD If you have all the arteries which supply the carotid body completely blocked, I am uncertain as to where the blood would come from which it is proposed still flows through the carotid body. Are you suggesting that there might be retrograde flow from the veins?

BISCOE As I understand it O'Regan is saying that, by 20 seconds flow has stopped. Therefore isn't this discussion largely irrelevant?

TORRANCE I think not. I am considering the possibility that the flow out into veins is zero but that the pressure in the arterial segment is still higher than in the vein because critical closure has taken place. That is, there is some potential for blood flow through the carotid body tissue from the distended carotid artery segment. Since the arteries are clamped there can clearly be no further flow into the carotid body except from the blood which is held in the carotid segment. If you now dilate the vessels in the carotid body, you might transiently get some flow which would account for the depression of afferent discharge observed by O'Regan and others.

REFERENCES

ACKER, H. & LÜBBERS, D. W. (1971). Eine Sauerstoffundurchlässige Zone in der Wand der Arteria Carotis der Katze. *Experientia*, **27**, 394–5.

PURVES, M. J. (1970). The effect of hypoxia, hypercapnia and hypotension upon carotid body flow and oxygen consumption in the cat. *J. Physiol.*, **209**, 395–416.

Efferent inhibition of arterial chemoreceptors and stimulation of the sinus nerve

By N. W. GOODMAN*
University Laboratory of Physiology, Oxford

The experiments involving stimulation of the carotid sinus nerve (Neil & O'Regan, 1971*a*) will form the main theme of this contribution. The procedure has been described by the previous speakers in this session and is illustrated by Sampson. The scale of the situation should be realized. Using the sinus nerve the two sets of electrodes are rarely more than 1 cm apart, and the fine slip from which one is recording may be only 0.1 mm long. I shall briefly discuss:

1. The hazards of such stimulation.
2. The inhibition of the discharge that is seen during electrical stimulation of the glossopharyngeal efferents.
3. Some features of antidromic depression of chemoreceptor endings.
4. Some other observations relevant to a discussion of efferent control.

Much of the experimental work covering 1–3 above has already been published (Goodman & McCloskey, 1972; Goodman, 1973), and only the conclusions will be repeated. Most of the work has been done on the carotid body, although the few experiments that were done on the aortic body do not suggest any major differences. All the experiments were in the cat.

THE HAZARDS OF CLOSE STIMULATION

Fidone & Sato (1970) pointed out, and I confirmed (Goodman, 1973), that there is a danger of stimulus escape causing adventitious excitation of the fine slip from which one is recording. The resulting nerve impulse, travelling antidromically, drastically reduces the mean rate of discharge of the fibre, over and above the reduction expected because of collision between the antidromic impulses set up by the stimulus and spontaneous, orthodromic, impulses set up by the receptor.

* MRC scholar during the course of this work.

For reasons described in the above papers, one must use an indirect method to monitor the occurrence of adventitious excitation in chemoreceptor fibres. Basically this is to trigger a storage oscilloscope from each pulse in the stimulus train and, if adventitious excitation has occurred, there is a silent period of about 10–20 msec visible on the screen between the stimuli and the stored spontaneous impulses that followed them.

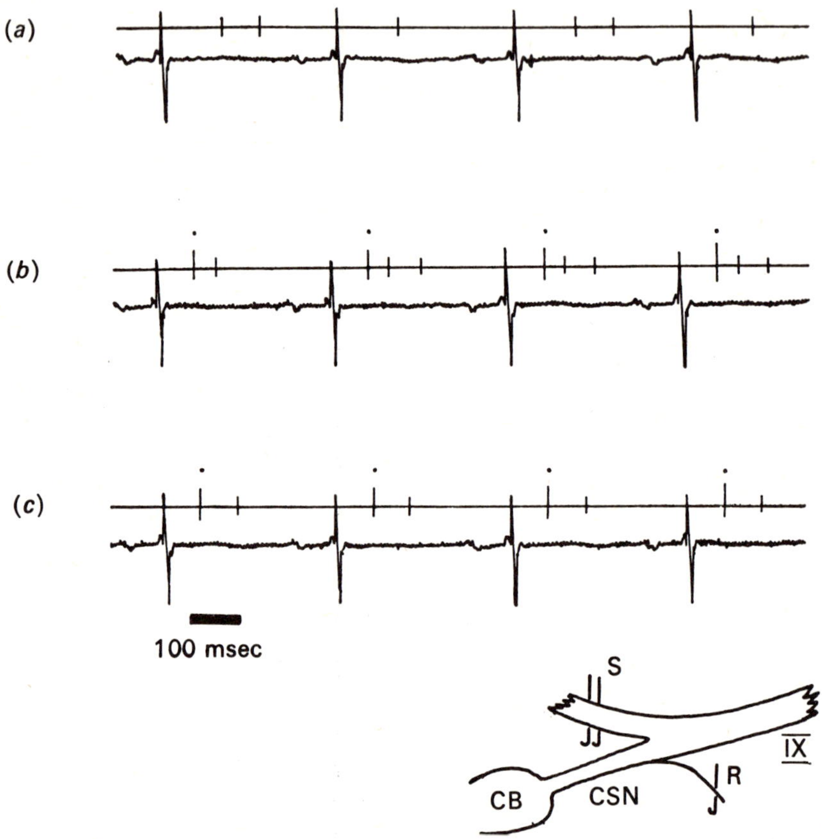

Fig. 1. Cat: nembutal. Recording from a single baroreceptor fibre (upper trace of each pair) as in the diagram. CB = carotid body and bifurcation; CSN = carotid sinus nerve; IX = glossopharyngeal nerve; R = recording electrode; S = stimulating electrodes. Note that IX is cut both centrally and peripherally and S is applied to the cut distal end so that the stimulus is applied to a portion of nerve with an electrical, but not a neural, connection with the baroreceptor recorded from.

The stimulus is triggered with a constant delay from the ECG (the lower trace of each pair).

(*a*) No stimulus. The first baroreceptor impulse occurs approximately 120 msec after the ECG. (*b*) Stimulus on, 5 V 0.25 msec. The stimulus artefact is denoted by the mark above the trace. The first impulse still occurs at 120 msec. (*c*) Stimulus now 10 V 0.25 msec. The first impulse is delayed and now occurs 145 msec after the ECG. There has been adventitious excitation of the recorded slip of nerve and the resulting antidromic impulse has 're-set' the baroreceptor nerve ending. (Records retouched.)

I add here an example of adventitious excitation affecting a receptor which fires more predictably than a chemoreceptor as a way of demonstrating the reality of the phenomenon we have been discussing. Fig. 1 shows re-setting of a baroreceptor, analogous to that shown for the muscle spindle by Matthews (1933), with the stimulating electrodes arranged so that there was unlikely to be any neural connection between them and the baroreceptor nerve ending. If the ECG-stimulus delay was increased, the baroreceptor impulse could be delayed more until, if the stimulus was delayed more than the normal ECG-impulse delay, the impulse would occur before the stimulus.

The conditions for adventitious excitation are difficult to define, and may even vary between two, apparently identical, stimuli only a few minutes apart. It is also important to use *only single fibres* because it is possible in a few-fibre slip that one fibre will be affected but another not. It is also possible to affect the amplitude of recorded nerve impulses in a reversible manner (Fig. 2).

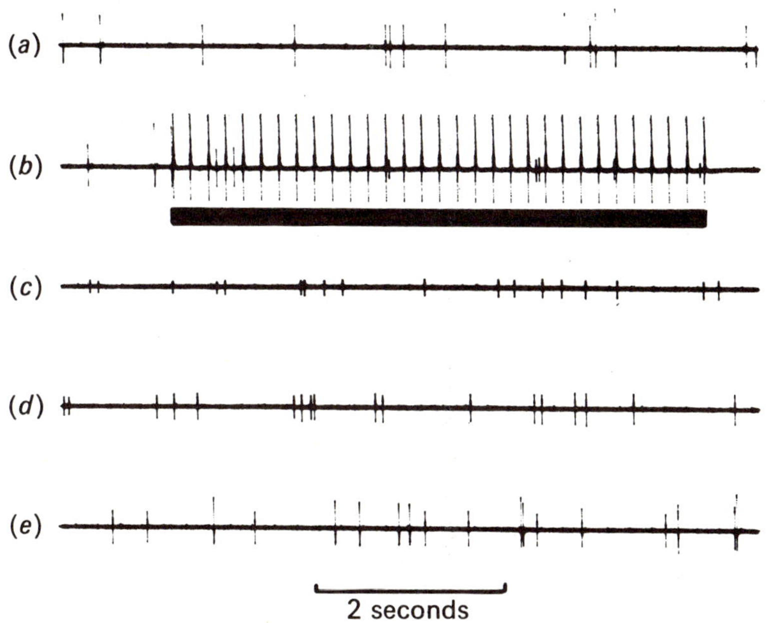

Fig. 2. Cat: nembutal. (*a*)–(*e*) Continuous recording from a few-fibre preparation of chemoreceptors from the sinus nerve, probably containing three fibres. Stimulus applied to the rest of the sinus nerve in the Neil & O'Regan (1971*a*) situation, 14 V, 0.06 msec, 5.2 Hz as indicated by the bar under (*b*).

Note the reduction in amplitude of the recorded nerve impulses caused by the stimulus, and the gradual recovery over (*c*)–(*e*). The apparent frequency of discharge is reduced over the period of stimulation, but it is not possible to say whether the real frequency has been reduced as conduction may be completely blocked in one of the fibres in the preparation.

THE 'TRUE' INHIBITION

Keeping a careful watch, *all the time*, for the appearance of the silent period, as Fidone & Sato (1970) found, there is still evidence of an inhibition when one stimulates the sinus nerve supramaximally for the 'C' fibres within it.

In my experience (Goodman, 1973) the three main features of it are: its variability, both between fibres and between responses of the same fibre at different times and to different stimulus intensities; it is only a mild inhibition, to about 70% of control values; and it is blocked by atropine in the same dose that Neil & O'Regan (1971*a*) found to block the hyperaemia that occurs when the sinus nerve is stimulated.

The possibility of some synaptic connections in the carotid body had been raised by Neil & O'Regan (1971*a*,*b*) and Dr Ian McCloskey and I looked for some electrophysiological evidence of this both *in vitro*, using the superfused preparation, and also *in vivo*. Using microelectrodes we were able to impale cells but we could not affect the membrane potentials in any way when stimulating the sinus nerve (Goodman & McCloskey, 1972): no junctional potentials, no i.p.s.ps, not even any slow drifts with prolonged bouts of stimulation.

This is not of course evidence of lack of inhibition, but it might suggest that there is no synaptic inhibition as we tend to think of it. In this connection it is interesting that Flock & Russell (1973) have recently demonstrated hyperpolarizing potentials of 2–7 mV arising in hair cells during analogous stimulation of efferent fibres in the lateral line nerves which causes inhibition of the afferents at the receptor level. If there had been this order of change in the carotid body we would have seen it in our experiments. (We have also heard from Dr Fillenz of the SIF cells, which resemble the Type I cells of the carotid body, and that electrophysiological evidence exists for a postulated inhibitory role that these cells could play.)

ANTIDROMIC DEPRESSION

In the microelectrode studies we were stimulating at an intensity above maximal for exciting the 'C' fibres and so would have been stimulating all the 'A' fibres antidromically. Another conclusion, therefore, is that there seems to be no change in the membrane potential of Type I cells on antidromic excitation of the chemoreceptor afferent fibres.

The chemoreceptors are far more susceptible to antidromic depression than are the baroreceptors. The comparison was made by stimulating the sinus nerve distal to the point at which the recorded slip left it, and thus the recorded afferent was stimulated directly. With the chemoreceptors, a train of stimuli of about 2 Hz per second could lead to a long-lasting and profound depression of rapid onset, often abolishing the discharge altogether. It could outlast the stimulus for a few seconds. Fidone & Sato (1970) reported that rebound was common on turning off the stimulus, but I saw this only rarely. I could not dissociate these profound inhibitions of discharge from the silent period mentioned above, and I assume that they occurred as a result of adventitious excitation.

Baroreceptors, on the other hand, although they could be re-set, could not be depressed beyond the first spontaneous impulse that followed a stimulus, unless the stimulus was a large, tetanic one, e.g. 90–270 Hz for half a second. The reason for this difference in the characteristics of antidromic depression is not clear. It may reflect the differences in size and ramification of the two types of fibre terminations.

SOME PROPERTIES OF THE EFFERENT DISCHARGE

Non-rhythmical efferent discharge was recorded from the cut central end

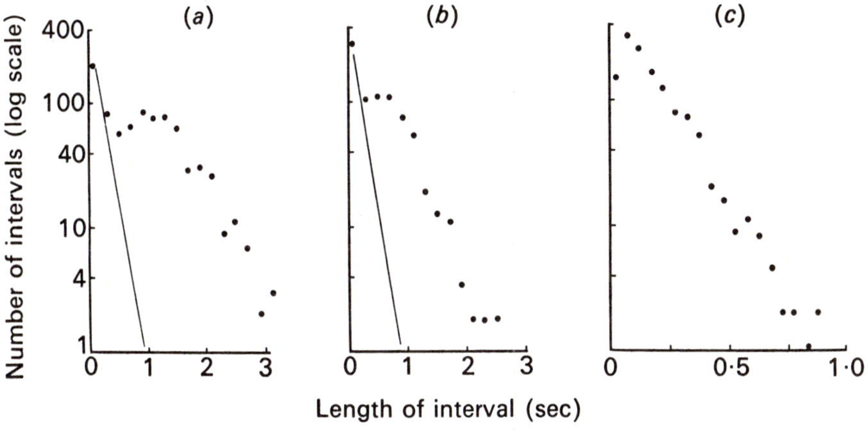

Fig. 3. Cat: nembutal. Interval distributions constructed from the discharge of single fibre preparations of the carotid sinus nerve. Ordinate: log. number of intervals. Abscissae: length of interval, note that the bin widths are longer in (*a*) and (*b*) (200 msec) than in (*c*) (50 msec).

(*a*) An efferent fibre. Animal breathing air. Total number of impulses (n) = 819, mean rate of discharge (f_{IMP}) = *c.* 0.1 per sec. (*b*) The same efferent fibre. Animal breathing 5 % CO_2+21 % O_2 in N_2. n = 780, f_{IMP} = *c.* 0.2 per sec. (*c*) A chemoreceptor afferent fibre from another cat. n = 1297, f_{IMP} = 5.1 per sec.

The distribution of intervals from the afferent can be described essentially as exponential (Biscoe & Taylor, 1963). The efferent shows exponential form for the first 300 msec only. The lines drawn in on (*a*) and (*b*) are the exponentials constructed from interval distributions using the same intervals but a smaller bin width (20 msec) to show that the exponential process is seemingly unaffected by the increase in the mean rate of discharge of the fibre, the half-time being about 100 msec in both (*a*) and (*b*).

of the sinus nerve, as has been done by other workers (Biscoe & Sampson, 1968; Neil & O'Regan, 1971*b*), with the object of getting out single fibres and looking at the discharge characteristics. All the known chemoreceptors, both carotid bodies and both aortic bodies, were denervated. Activity was sparse, and in five cats, only one unambiguously single-fibre preparation was obtained that gave a good-sized potential for computer analysis (Nail, 1972). Fig. 3 shows that the interval distribution declined exponentially for about 300 msec and then had a marked hump on it. The exponential was unaffected by an increase in the mean frequency of discharge, but the hump peaked at a shorter interval length. This may suggest that there are two processes generating efferent activity. Perhaps there is a basic, rather irregular, generative process, and, when an impulse arises, there is an increase in the probability, declining with time, that another impulse will occur.

The efferent discharge does not show a cyclic variation with respiration as the chemoreceptor afferents do. Neither do the efferents appear to alter the oscillations in afferent single-fibre activity (Goodman, Nail & Torrance, unpublished observations).

CONCLUSION

I must proceed from Fidone & Sato (1970). They found two types of inhibition when they stimulated the sinus nerve as did I: one accompanied by the silent period, the other not. That not accompanied by the silent period is blocked by atropine. It is very variable. No changes in membrane potentials could be demonstrated within the carotid body on stimulating the nerve. The simplest conclusion from this is that the inhibition that I saw was vasomotor. This does not rule out a more complex analysis based on further data, nor that another mechanism exists that I did not uncover in my experiments.

The general conclusion to draw is that electrical stimulation in this situation is a most unreliable method, and more physiological experiments, of the type described by some of the other contributors in this session, must be used for further investigation of the efferent control of the carotid body.

REFERENCES

BISCOE, T. J. & SAMPSON, S. R. (1968). Rhythmical and non-rhythmical spontaneous activity recorded from the central cut end of the sinus nerve. *J. Physiol.*, **196**, 327–38.

BISCOE, T. J. & TAYLOR, A. (1963). The discharge pattern recorded in chemoreceptor afferent fibres from the cat carotid body with normal circulation and during perfusion. *J. Physiol.*, **168**, 332–4.

FIDONE, S. J. & SATO, A. (1970). Efferent inhibition and antidromic depression of chemoreceptor A-fibers from the cat carotid body. *Brain Res.*, **22**, 181–93.

FLOCK, A. & RUSSELL, I. (1973). Postsynaptic action of efferent nerve fibres on hair cells. *Nature New Biol.*, **243**, 89–91.

GOODMAN, N. W. (1973). Efferent control of arterial chemoreceptors mediated by glosso-pharyngeal fibres and artifacts introduced by stimulation techniques. *J. Physiol.*, **230**, 295–311.

GOODMAN, N. W. & MCCLOSKEY, D. I. (1972). Intracellular potentials in the carotid body. *Brain Res.*, **39**, 501–4.

MATTHEWS, B. H. C. (1933). Nerve endings in mammalian muscle. *J. Physiol.*, **78**, 1–53.

NAIL, B. S. (1972). On-line biomedical data analysis. *J. Physiol.*, **225**, 5–7P.

NEIL, E. & O'REGAN, R. G. (1971*a*). The effects of electrical stimulation of the distal end of the cut sinus and aortic nerves on peripheral arterial chemoreceptor activity in the cat. *J. Physiol.*, **215**, 15–32.

NEIL, E. & O'REGAN, R. G. (1971*b*). Efferent and afferent impulse activity recorded from few-fibre preparations of otherwise intact sinus and aortic nerves. *J. Physiol.*, **215**, 33–47.

DISCUSSION

SPECKMANN What experimental evidence is given by your findings that the hyperpolarization that you show really represents i.p.s.ps?

GOODMAN I should point out that that particular slide was from Flock & Russell (1973). It is not my own work. The point is, there is definite efferent inhibition in this situation and some form of potential which is picked up with microelectrodes in what seems to be an analogous situation to that in the carotid body. When McCloskey and I prodded the carotid body with microelectrodes, we were looking for anything at all but we found absolutely nothing.

BISCOE I am not clear how you think that this antidromic depression occurs. You say that when you stimulate you are getting activation and an impulse in the strand from which you are recording?

GOODMAN That is right.

BISCOE If that is the case, then why is it that the impulse is not recorded?

GOODMAN I think that the adventitious excitation occurs at the paraffin-saline interface. The slip from which one is recording is not usually very long and the conduction time is so short that any impulse is lost in the stimulus artifact.

BISCOE Yes, but I would draw your attention to the slide that Sampson showed where he had the artifact quite clearly shown.

GOODMAN Yes, but one does not necessarily see the whole of the artifact, and, even if one does, it may not be possible to resolve the action potential.

BISCOE You mean that it is off the top of the record?

GOODMAN It could well be. But more usually the action potential is lost within the artifact. It is a hideous shape. If you look hard enough, it may be possible to find the action potential.

O'REGAN There is a considerable difference in the threshold levels between Fidone & Sato and yourself. I am prepared to admit that adventitious excitation may occur but I cannot see how you can say that it definitely occurred in our experiments. The only reason that you suggest that we were getting antidromic invasion was that we were getting a greater than 50% reduction in discharge. What also surprised me in view of the conclusions which you draw is that you didn't monitor carotid body blood flow. When you used the superfused preparation with McCloskey (1972) you got an inhibition of chemoreceptor afferent activity during electrical stimulation of the sinus nerve. Did you check that this inhibition was due to adventitious excitation?

GOODMAN McCloskey and I (1972) obtained what appeared to be an inhibition, but we were not certain that it wasn't adventitious excitation.

O'REGAN I am surprised you did not explore this further because as Eyzaguirre & Lewin (1961) have shown, the superfused preparation allows one to eliminate vascular changes as the cause for any modifications of discharge which occur as a consequence of a particular procedure.

GOODMAN I didn't fancy measuring blood flows very much.

SAMPSON Did you only use a stimulus duration of 1 msec?

GOODMAN No, it varied. For adventitious excitation the duration of the stimulus pulse made no difference. The shortest stimulus pulse was 60 μsec. Presumably the recording side of the apparatus is charged up by the stimulus and, as the time constant of the capacitance is less than 60 μsec, I could never reduce the effective voltage being put through the slip that was recorded from.

MILLS I am now rather confused. Could you tell me what structures depolarize as a consequence of that stimulus?

GOODMAN I don't know. There is certainly antidromic depression.

MILLS The thing that concerns me is that when we stimulate the nerve antidromically, we get a very large transient decrease in NADH – the sort of thing you see when nerve is depolarized.

BISCOE How long is the silent period?

GOODMAN Generally about 10 msec, which corresponds to the refractory period of the endings (Biscoe & Taylor, 1963) plus the conduction time.

SHEPHERD Did you try anything else besides atropine and phenoxybenzamine?

GOODMAN No.

BULLER Did you make any calculation as to what the probability of seeing a post-synaptic potential was? What strikes me is that you admit you saw very small depressions when recording from afferent fibres. This would imply that there are relatively few structures and you said, as I recall, that you made 160 penetrations with microelectrodes.

GOODMAN Fifty per cent of the fibres were affected by the stimulus I used, and we looked at 180 cells. (These were separate studies.) We stimulated the carotid sinus nerve whilst holding each cell. If one assumes that 50 % of those cells were not affected by this type of stimulation, it still leaves 90 and I would have thought that if something definite was going on we would have seen it if it was a transient response of more than 0.5 mV, or a slow drift of about 3–4 mV.

TORRANCE Flock & Russell (1973) have shown in recent work that i.p.s.ps in sensory cells correlate with efferent inhibition. This makes the failure of McCloskey & Goodman to find any change in carotid body cells more significant. By this I mean it could have been simply a failure to pick up something but if in a very similar situation i.p.s.ps were obtained by Flock & Russell – and obtained regularly and easily – then not getting them in the carotid body becomes more significant.

SAMPSON You did report that potassium depolarizes the cells which you had impaled.

GOODMAN Yes. This was the evidence we used for penetration of a cell.

SAMPSON Nothing else did that you tested?

GOODMAN Absolutely nothing. We used acetylcholine, sodium cyanide, carbon dioxide, stopped the flow; and we saw no changes at all.

O'REGAN Is it necessary to have a change in membrane potential accompanying transmitter release?

GOODMAN No, I don't suppose it is. Negative results with microelectrodes is not a damning piece of evidence against inhibition. It shows that if there is any inhibition it is not seemingly by the same mechanisms as have been described for other inhibitions.

EYZAGUIRRE I think that this situation is probably complex. We may have here more than one mechanism. We may have additional evidence of a vascular effect due to some work being done by Belmonte in my laboratory. We have been using a superfused preparation. It has also been perfused at the same time. When the carotid body is superfused inhibition caused by stimulating a nerve disappears; when the carotid body is perfused, the inhibition reappears. It is possible to go back and forth between these procedures, many times in one experiment. That evidence of course does not rule out the possibility of a direct form of inhibition where, for example, you may be washing out a transmitter but I think that it does indicate that there is an important vascular component. It is also worthwhile commenting that in the superfused preparation, dopamine works if applied from without but its effects are variable – that is, sometimes it inhibits and sometimes it excites. And it is also possible to show that the inhibitory effect is reversed after two or three injections and Dr Zapata has also found evidence of cross-tachyphylaxis with adrenaline and noradrenaline.

SAMPSON Let me add to that that if you add dopamine to the carotid body you have a terrible time washing it out. And you can get spurious depressant effects after dopamine.

BISCOE I take it that we are agreed that there *is* efferent inhibition and that there may be several mechanisms, possibly three, one being the artifact, one being a vasomotor mechanism and the third being something else.

MITCHELL I'm a bit concerned about the tendency for making deductions on negative results. My own experience when looking for efferent inhibition is such that I have a very good idea of what I am going to find after the initial dissection and indeed I don't bother looking for the inhibition when

I know I am not going to get it. That is to say, I know that if the dissection goes well and the nerve splits easily and I get a slip with one or two active fibres in it I can get efferent inhibition at around 2 V and 2 cycles per sec and the discharge will be 50–60 % down on control. On the other hand I know that if I have to start sifting through the nerve with more and more dead fibres, if I get a fibre out I do not get good efferent inhibition. I just have a feeling that one good positive or a series of good positives is worth a whole mass of negatives.

BISCOE But you are not arguing from this that there are several classes of efferent fibre.

MITCHELL We'll come to that later.

REFERENCES

BISCOE, T. J. & TAYLOR, A. (1963). The discharge pattern recorded in chemoreceptor afferent fibres from the cat carotid body with normal circulation and during perfusion. *J. Physiol.*, **168**, 332–44.

EYZAGUIRRE, C. & LEWIN, J. (1961). Effect of different oxygen tensions on the carotid body *in vitro*. *J. Physiol.*, **159**, 238–50.

FLOCK, A. & RUSSELL, I. (1973). Post synaptic action of efferent nerve fibres on hair cells. *Nature New Biol.*, **243**, 89–91.

GOODMAN, N. W. & MCCLOSKEY, D. I. (1972). Intracellular potentials in the carotid body. *Brain Res.*, **39**, 501–4.

Sinus nerve efferents as a link between central and peripheral chemoreceptors

By P. WILLSHAW*

Department of Physiology, Middlesex Hospital Medical School, London W.1

The earlier papers devoted to the topic of efferent inhibition have shown only too clearly that electrical stimulation of the sinus nerve can give rise to artifacts and difficulties of interpretation about the origin of the inhibition of afferent discharge. In this contribution I want to consider an alternative method of causing sinus nerve efferent fibres to discharge without there being any chance of stimulating the afferent fibre from which chemoreceptor activity is being recorded and which does not excite antidromically the afferent fibres in the main nerve trunk.

The earlier experiments were carried out in collaboration with Dr S. Majcherczyk of Warsaw and resulted in a communication to the Physiological Society given earlier this year (Majcherczyk & Willshaw, 1973). In summary, we showed that superfusion with an alkaline mock cerebrospinal fluid (CSF) of the ventral medullary surface in the cat resulted in enhanced efferent activity recorded from the central end of the cut carotid sinus nerve. When afferent chemoreceptor activity was recorded from a slip peeled off the otherwise intact sinus nerve, alkaline superfusion of the medulla resulted in a marked inhibition of afferent chemoreceptor activity. As previous workers (Biscoe & Sampson, 1968) have reported, single-fibre efferent activity is difficult to record from the central end of the cut sinus nerve and so most of the recordings of efferent activity have been made from multifibre preparations. In these preparations it was necessary to cut the ipsilateral cervical sympathetic to ensure that sympathetic activity running centrifugally in the sinus nerve was not recorded.

The majority of the cats were anaesthetized with a mixture of α-chloralose (50 mg kg^{-1}) and urethane (500 mg kg^{-1}) given intraperitoneally, although in some cases pentobarbitone sodium (50 mg kg^{-1}) was used. A number of cats were decerebrated at midcollicular level before paralysis with gallamine and all the cats were artificially ventilated during the experimental period. Fig. 1 shows a few-fibre recording made from the

* Present address: Sherrington School of Physiology, St Thomas's Hospital, London.

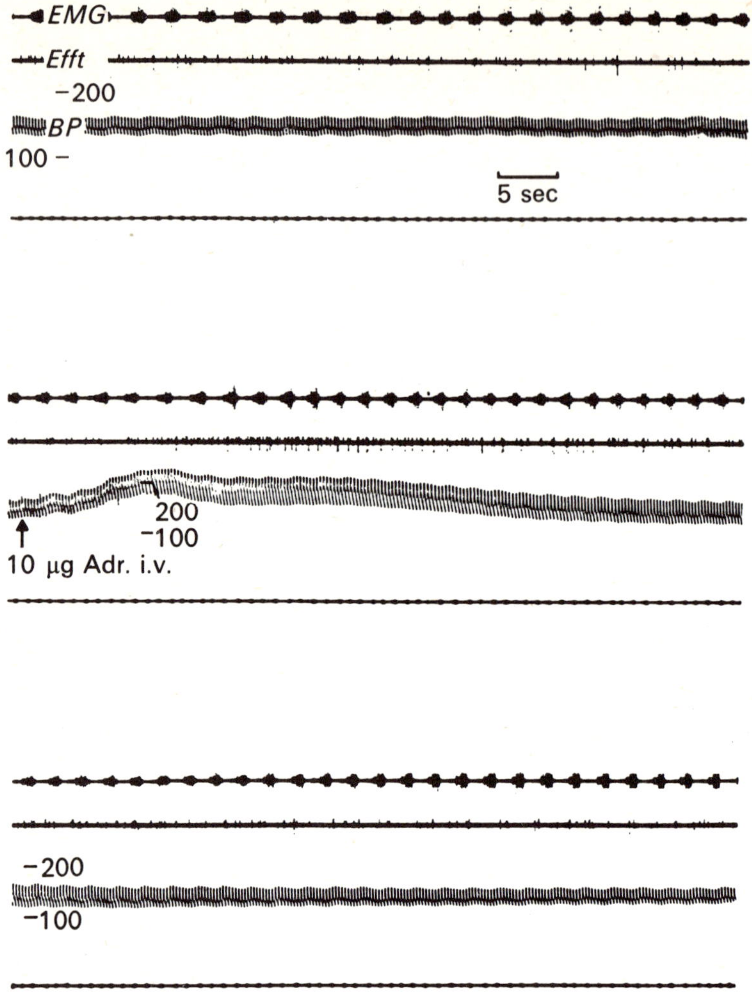

Fig. 1. Three recordings are presented, which were originally in continuity.
EMG, diaphragmatic electromyogram; *Efft*, sinus nerve efferent activity recorded from the central cut end of the left sinus nerve. Ipsilateral vagosympathetic trunk intact; *BP*, femoral arterial blood pressure: calibration in mm Hg. 10 μg adrenaline injected intravenously at the arrow.
Cat: 'nembutal' anaesthesia: artificial ventilation.

central end of the cut carotid sinus nerve in a cat anaesthetized with pentobarbitone. During the control period, it can be seen that spontaneous efferent activity is sparse and irregular. The injection of 10 μg adrenaline caused a marked excitation of the efferent activity, as shown initially by Biscoe & Sampson (1968) who also showed that the response persists after section of the other sinus nerve and both vago-sympathetic trunks. It is possible that the adrenaline acts beyond the blood–brain barrier and this possibility was tested by injecting the same dose (10 μg) of adrenaline into a pool of CSF overlying the area on the ventral surface of the medulla which was superfused with mock CSF in the main part of the experiment. The result of this adrenaline stimulus is shown in Fig. 2. This figure shows a continuous record of few-fibre efferent activity and it can be seen that the

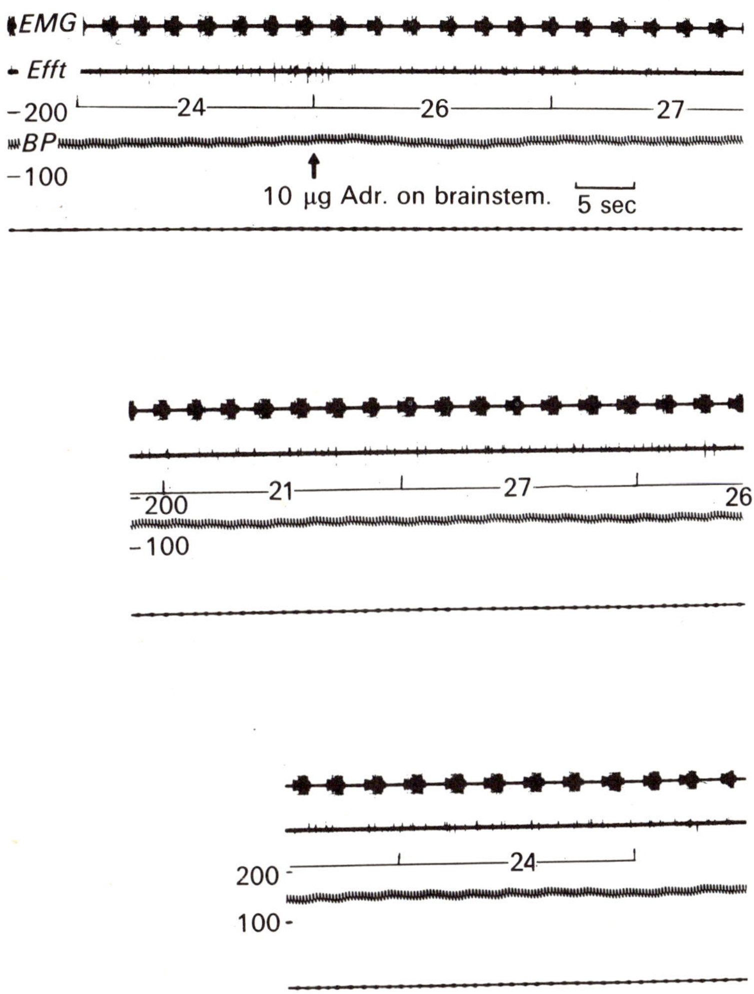

Fig. 2. Three recordings, originally in continuity, are presented. Traces identical with those in Fig. 1. The numbers below the middle trace indicate the total number of impulses counted in the 20-second periods between the vertical bars. 10 μg adrenaline were injected into a pool of artificial CSF lying on the brainstem. Moment of injection is indicated by the arrow.

Cat: 'nembutal' anaesthesia: artificial ventilation.

adrenaline stimulus had no effect at all when administered by this route. A fair conclusion to draw from this experiment is that the effect of adrenaline on efferent discharge is not mediated by its actions on superficial structures in the ventral surface of the medulla. In contrast to the above finding, Fig. 3 shows the effect of alkaline CSF superfused over the same

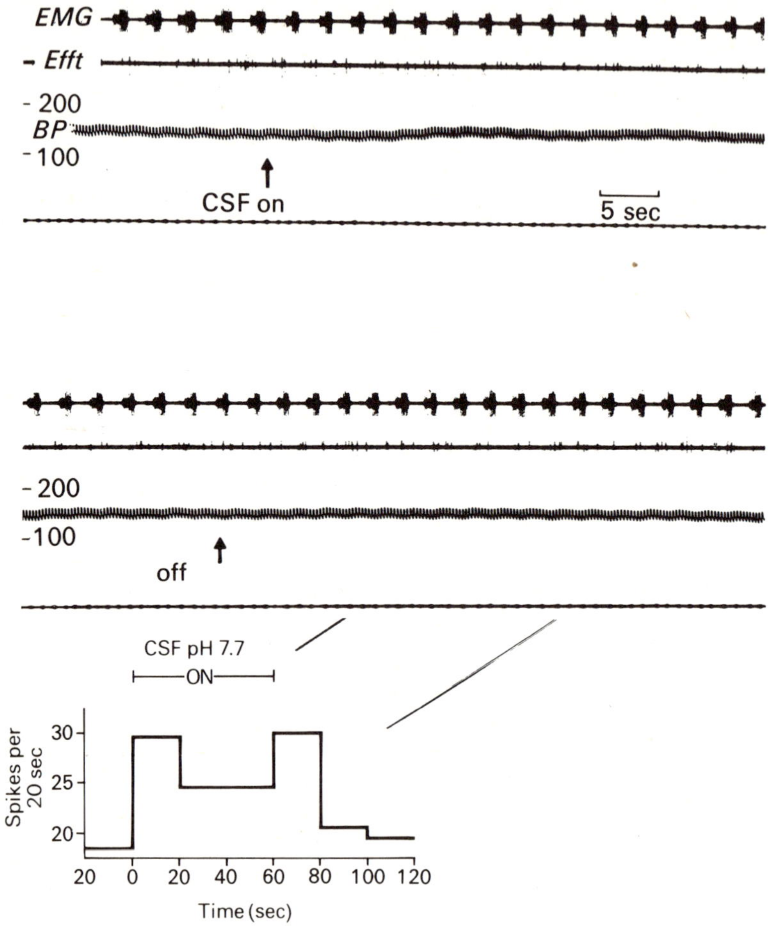

Fig. 3. Two recordings, originally in continuity, are presented together with a graph showing the changes in nerve impulse frequency counted over 20-second periods. Traces identical with those in Fig. 1. Between the arrows, the brainstem was superfused with an artificial CSF at pH 7.7.
Cat: 'nembutal' anaesthesia: artificial ventilation.

area on the ventral surface of the medulla. The pH of the mock CSF was altered by varying the bicarbonate concentration in the solutions at constant P_{CO_2} (5% in air). In the experiment shown in Fig. 3 the pH was 7.7, and its application resulted in a rapid increase of efferent activity from a control level of 20 impulses per 20-second period to 25 impulses per 20-second period, with transient increases to 30 impulses per 20-second period at the beginning and end of the superfusion. Fig. 4 shows the response of a

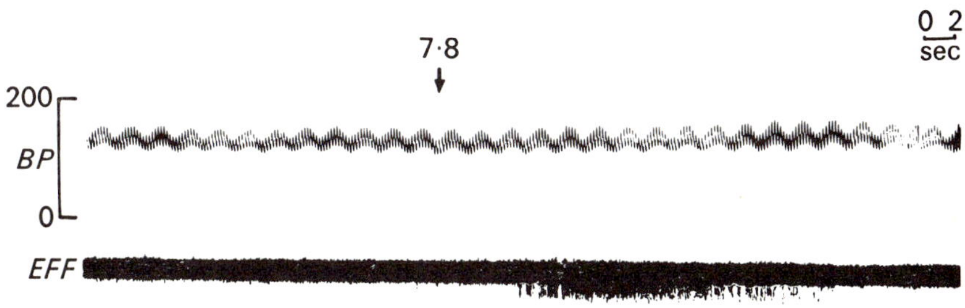

Fig. 4. Top trace: Femoral arterial pressure (*BP*), calibration in mm Hg. Lower trace: Recording of multifibre activity (*EFF*) from the central end of the cut left sinus nerve. Ipsilateral vagosympathetic trunk cut. At the arrow, 0.2 ml artificial CSF at pH 7.8 was applied to the brainstem (near area M).

Cat: chloralose/urethane anaesthesia: artificial ventilation.

multifibre preparation of the central end of the cut sinus nerve to application of 0.2 ml of mock CSF at pH 7.8 in a cat anaesthetized with chloralose/urethane mixture. A marked burst of efferent activity is seen, which gradually diminishes towards the end of the trace. I would like to draw your attention to the fact that in most experiments the increase in efferent discharge started within 20 seconds of the start of superfusion with alkaline CSF.

The next series of experiments involved the recording of afferent chemoreceptor activity from the sinus nerve by peeling a small slip away from the main nerve trunk. The activity recorded is usually described as arising from the 'otherwise intact' sinus nerve. Two difficulties were encountered. Firstly, the pH of the artificial CSF which produced effects in anaesthetized cats was very high, in the region of pH 7.8. It was found that a more physiological pH of 7.5 produced the effects in midcollicularly decerebrate cats. Secondly, in many cases hypertensive response was present during superfusion. It was possible that any decrease in chemoreceptor afferent activity seen during alkaline superfusion of the medulla may have been caused by increased carotid body blood flow consequent upon the hypertension. It was therefore necessary in many cases to isolate the carotid bifurcation vascularly and to perfuse it artificially with warm arterial blood at constant pressure.

Fig. 5 shows a typical result from a decerebrate preparation with an

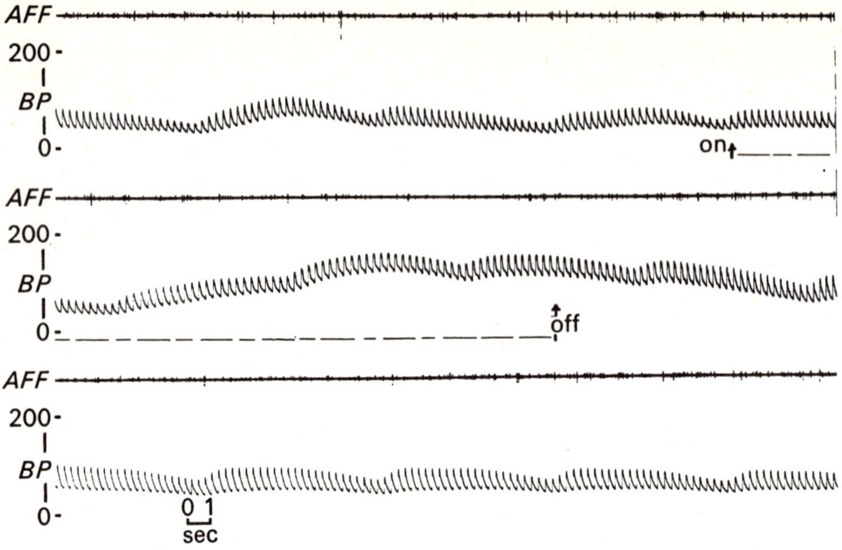

Fig. 5. Three recordings, originally in continuity, are presented. Upper trace: afferent chemoreceptor activity (*AFF*) recorded from the otherwise intact left sinus nerve. Ipsilateral vagosympathetic trunk cut; relevant carotid bifurcation arterially isolated and perfused at constant pressure with warm arterial blood. Lower trace: Femoral arterial pressure (*BP*). Calibration in mm Hg. Between 'on' and 'off', superfusion with artificial CSF at pH 7.52.

Cat: decerebrate: paralysed and artificially ventilated.

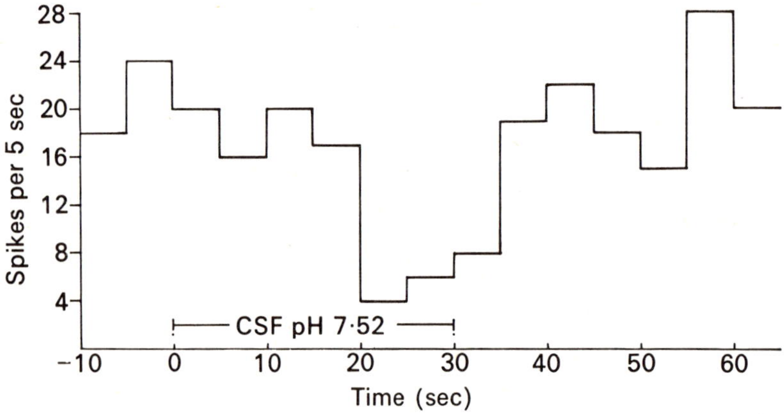

Fig. 6. Graph showing the time-course of changes in afferent chemoreceptor activity sampled from the otherwise intact sinus nerve, during superfusion with artificial CSF at pH 7.52. The ordinate shows the number of impulses counted in successive 5-second periods, from the few-fibre original recording.

Animal prepared as in Fig. 5.

arterially isolated blood-perfused carotid bifurcation. The upper trace shows afferent chemoreceptor activity sampled from the otherwise intact sinus nerve and the lower trace shows systemic arterial blood pressure. Artificial CSF at pH 7.52 was superfused over the medulla between the marks 'on' and 'off'. It is clearly seen that afferent chemoreceptor activity is inhibited in response to the superfusion. Notice that in this example, the peak of the hypertensive response coincides with the period of greatest chemoreceptor inhibition, underlying the necessity for arterial isolation in the preparations showing the hypertensive response. Fig. 6 shows in graphical form the changes in the chemoreceptor activity from a similar experiment, and the upper panel of Fig. 9 shows an unusually prompt inhibitory response to superfusion, the inhibition taking place some time before the hypertensive phase of the response (not shown). In all experiments of this type, sectioning the sinus nerve central to the afferent slip resulted in a loss of the inhibitory response to alkaline superfusion. The integrity of the vagosympathetic trunk did not affect the inhibition, and so it may be concluded that the inhibition resulted from the enhanced sinus nerve efferent activity during alkaline superfusion.

MEASUREMENT OF CAROTID BODY FLOW

In the next series of experiments, the mechanism of the inhibitory effect was investigated. Initial studies were made on carotid body blood flow before, during, and after a period of superfusion with alkaline CSF. The most successful experiments utilized the technique of measuring carotid body blood flow by monitoring the rate of inflow of blood into the arterially isolated carotid bifurcation at constant pressure, the blood flowing in from a pipette graduated in microlitres. The blood flow during the control period was first measured, the pipette recharged and flow then measured during the period of superfusion with alkaline CSF. The hypertensive response to superfusion had the advantage that it has never been seen in the absence of chemoreceptor inhibition, and could therefore be used as a reliable indicator that inhibition was taking place. In animals which did not show the hypertensive response, a series of flows was measured during several superfusion periods and then an afferent-going slip peeled from the otherwise intact sinus nerve so that inhibition could be recorded directly, the bifurcation being perfused naturally. The experiment of simultaneously recording blood flow and electroneurogram has not yet been achieved.

Two results will be presented. Fig. 7 shows a result from an animal displaying the hypertensive response, the systemic arterial pressure changes

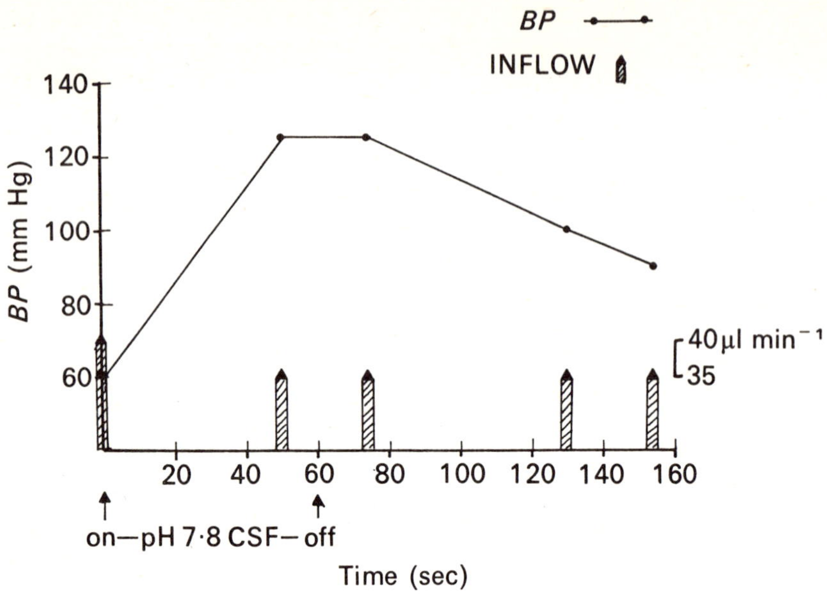

Fig. 7. Graph showing changes in mean systemic arterial pressure (left ordinate) and carotid body blood flow (right ordinate) (inflow method). Relevant carotid bifurcation arterially isolated and perfused at constant pressure with warm arterial blood. Sinus nerve intact: sympathetic trunk sectioned: superfusion with artificial CSF at pH 7.8 between arrows. Cat: chloralose/urethane anaesthesia: artificial ventilation.

being indicated by the continuous line. The vertical bars show carotid body blood inflow rate at constant pressure (usually 100–120 mm Hg). The flow rate during the control period was 42 $\mu l\ min^{-1}$ and remained constant at 36 $\mu l\ min^{-1}$ throughout the superfusion period which resulted in a hypertensive response, indicating that inhibition was present.

Fig. 8(*a*) shows the percentage inhibition of chemoreceptor afferent activity and (*b*) the percentage changes in carotid body blood flow in a cat which did not display a hypertensive response to superfusion. The change in blood flow was less than 10% compared with the control value of 65 $\mu l\ min^{-1}$ and varied less than 5% during the experimental run. These changes are insignificant with the method used. The percentage inhibition of afferent chemoreceptor activity was considered to be significant above the 10% level, and as Fig. 8*a* shows there was inhibition in response to alkaline superfusion.

To summarize, superfusion of the ventral medullary surface with alkaline CSF results in increased sinus nerve efferent activity, inhibition of chemoreceptor afferent activity, and no significant change in total carotid body blood flow to account for the inhibition.

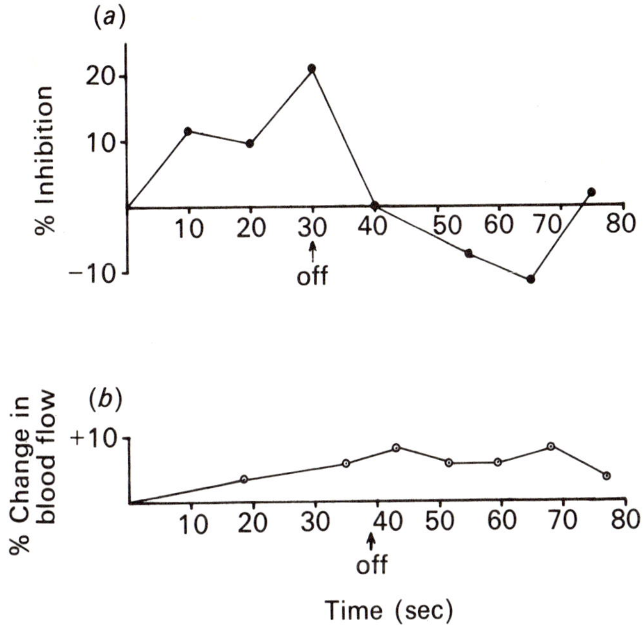

Fig. 8. Two graphs are presented. In each the abscissa shows time in seconds, with superfusion starting at time zero and ending at 'off'. The CSF pH was 7.8.

(*a*) Shows percentage inhibition of afferent chemoreceptor activity sampled from the otherwise intact sinus nerve. Ipsilateral sympathetic supply to carotid body cut.

(*b*) Shows percentage change in carotid body blood flow measured by the inflow method Sinus nerve intact: ipsilateral sympathetic supply to carotid body cut.

Cat: chloralose/urethane anaesthesia.

THE EFFECT OF BLOCKING AGENTS

Some further experiments were carried out similar to those reported by Sampson (1972) in which the effects of phenoxybenzamine on the inhibitory process were tested.

The results are shown in Fig. 9. Before the slow intravenous injection of 50 mg phenoxybenzamine, a very fast and profound inhibitory response to alkaline superfusion was obtained. After the phenoxybenzamine injection, superfusion no longer resulted in chemoreceptor inhibition. The effects of close-arterial injection of 600 μg atropine were also tested, and the results are shown in Fig. 10. Atropine injected close to the carotid body has been shown (Neil & O'Regan, 1971) to block the hyperaemia of electrical stimulation of the sinus nerve but to leave the inhibitory effects untouched. Fig. 10 shows that close-arterial injection of 600 μg atropine did in fact block the inhibitory effects of alkaline CSF superfusion.

Apart from the results of the pharmacological experiments, this study

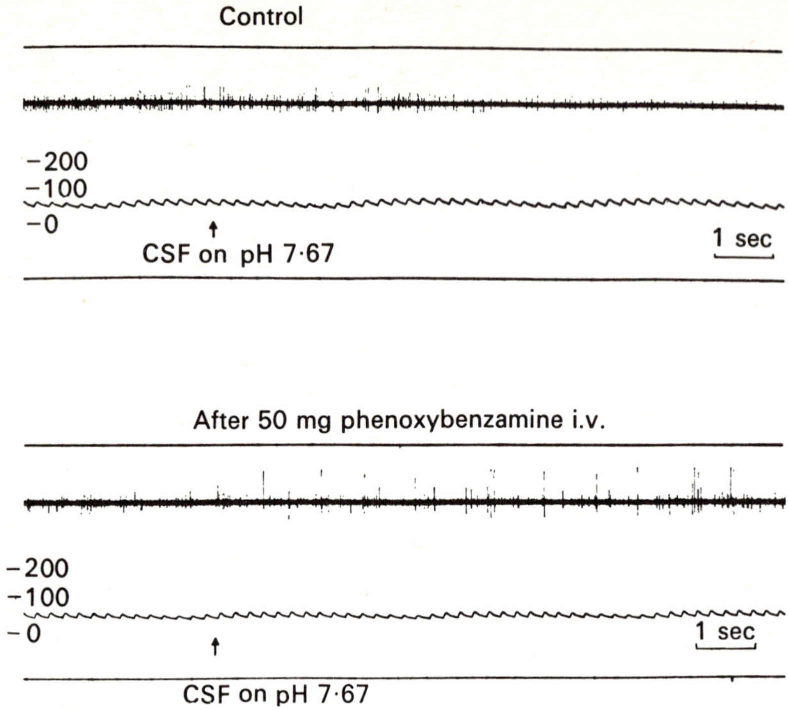

Fig. 9. Two recordings are presented. In each: Upper trace: afferent chemoreceptor activity sampled from otherwise intact left sinus nerve; sympathetic intact. Lower trace: femoral arterial pressure; calibration in mm Hg. The upper recording was made before, and the lower recording after, the slow intravenous injection of 50 mg phenoxybenzamine. Superfusion with artificial CSF at pH 7.67 began at the arrow on each recording.
Cat: decerebrate: artificial ventilation.

indicates that if the extracellular fluid of part of the medulla is made alkaline, the cell bodies of the efferent fibres in the sinus nerve are excited and that this causes inhibition of the chemoreceptors. We might ask what possible physiological significance this sequence has. I would like to suggest one possibility and it has to do with the respiratory response to acute hypoxia. Some 80 years ago, Haldane & Lorraine Smith (1893) showed that when inspired oxygen was reduced from 20 % to 14 % the subject showed no significant increase in ventilation. Later, von Euler, Liljestrand & Zotterman (1939) showed from sinus nerve recording that with this reduction of inspired oxygen, afferent chemoreceptor discharge progressively increased. This discrepancy could be reconciled if we propose that during hypoxia, hyperventilation causes an alkaline CSF which in turn through the efferent mechanism inhibits chemoreceptor afferent activity and this will have the effect of limiting any further rise in ventilation. This may be

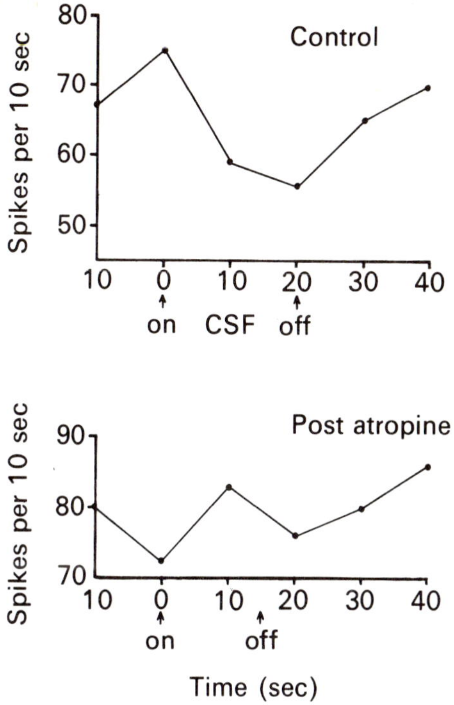

Fig. 10. Two graphs are presented. In each, the abscissa shows a period of superfusion with artificial CSF at pH 7.54 between 'on' and 'off'.

The ordinates show the number of chemoreceptor afferent impulses in 10-second periods, sampled from the otherwise intact left sinus nerve. Ipsilateral vagosympathetic supply intact. The upper graph shows the control response to superfusion, and the lower graph the response after close-arterial injection of 600 μg atropine.

Cat: chloralose/urethane anaesthesia: artificial ventilation.

of particular importance during the early stages of altitude acclimatization, that is until the transport mechanisms at the blood–brain barrier bring about the return of pH_{CSF} to normal levels.

I would like to thank Dr S. Majcherczyk for supplying the previously unpublished record shown in Fig. 4.

REFERENCES

Biscoe, T. J. & Sampson, S. R. (1968). Rhythmical and non-rhythmical spontaneous activity recorded from the central cut end of the sinus nerve. *J. Physiol.*, **196**, 327–38.

Euler, U. S. von, Liljestrand, G. & Zotterman, Y. (1939). The excitation mechanism of the chemoreceptors of the carotid body. *Skand. Arch. Physiol.*, **83**, 132–52.

Haldane, J. S. & Smith, J. L. (1893). The Physiological effects of air vitiated by respiration. *J. Path. Bact.*, **1**, 168–86.

Majcherczyk, S. & Willshaw, P. (1973). Inhibition of peripheral chemoreceptor activity during superfusion with an alkaline c.s.f. of the ventral brainstem surface of the cat. *J. Physiol.*, **231**, 26–27P.

Neil, E. & O'Regan, R. G. (1971). The effects of electrical stimulation of the distal end of the cut sinus and aortic nerves on peripheral arterial chemoreceptor activity in the cat. *J. Physiol.*, **215**, 15–32.

Sampson, S. R. (1972). Mechanism of efferent inhibition of carotid body chemoreceptors in the cat. *Brain Res.*, **45**, 266–70.

DISCUSSION

ACKER Have you tested to see whether the effect of perfusing the medulla affects the chemoreceptor response to hypoxia?

WILLSHAW I haven't tested that.

ACKER Then why do you call this inhibition?

WILLSHAW Simply because the chemoreceptor afferent discharge decreases at constant blood gas tensions.

McDONALD Could you tell us what part of the surface of the medulla is bathed by the artificial medium? Does the bath reach the roots of either the glossopharyngeal or vagus nerve?

WILLSHAW I'm not certain on this point. However, I have, at the end of the day, injected indian ink into the perfusion cannula and allowed it to flow and remain *in situ* for some time. I have then fixed the head in formalin and have been able to confirm that the area which is stained is fairly local just beneath the cannula and that it certainly does not affect cranial nerve roots.

TORRANCE Sometimes you did these experiments on decerebrate animals and sometimes on anaesthetized animals; did you observe any effect of anaesthesia upon the response?

WILLSHAW Yes. For example, in the decerebrate animals it is possible to obtain the effect when the pH of the perfusing fluid is about 7.5 units. In the anaesthetized animals it may have to be about 7.8.

MITCHELL Could you refresh my memory Dr Biscoe, about the experiments in which you measured efferent activity and its response to hypoxia before and after a section of the remaining sinus nerve. I seem to recall that the curve relating discharge and oxygen tension was only very slightly shifted.

BISCOE That is right.

WILLSHAW If you look at that paper carefully (Sampson & Biscoe, 1970) and draw lines on the graph and work the values out, you find that the inhibition is of the order of 28 % per 100 mm Hg oxygen tension.

GOODMAN I know little about the CSF but, as I understand it, here we have evidence of an efferent pathway which is stimulated by various things, adrenaline, systemic hypoxia, but also hypercapnia and an *alkaline* CSF.

WILLSHAW In the original Biscoe & Sampson paper (1968) they showed a response curve to hypercapnia of efferent activity recorded from the central cut end of the sinus nerve, represented by two graphs. In the first graph the efferent activity increased from low to very high carbon dioxide values and this was when all the other chemoreceptor nerves were intact. When they cut all the other chemoreceptor nerves and repeated the test they only increased the carbon dioxide up to 60 mm Hg and didn't go any further. I am speculating – but it is quite possible that during hypercapnia cerebral blood flow increases and that the increase in cerebral blood flow is inordinately large for the increase in P_{CO_2}. So it is just possible that over a certain range of P_{a,CO_2}, venous P_{CO_2} may actually be lowered and of course venous P_{CO_2} is supposed to be representative of interstitial or extracellular carbon dioxide.

GOODMAN But, in my experience, these efferent fibres are also stimulated by the intra-cranial injection of saline equilibrated with high carbon dioxide.

WILLSHAW How did you know that they were efferent fibres?

GOODMAN Because they had all the characteristics of efferent fibres previously described.

WILLSHAW Right, I have to agree that the response to carbon dioxide worries me and I cannot at present reconcile it with the evidence from the alkaline CSF perfusion studies.

O'REGAN In my experience if you put adrenaline on to the surface of the medulla it causes very little effect.

WILLSHAW I agree. The response is very small compared to the effect of giving it intravenously.

BINGMANN I wonder if you could tell me how the area on the ventral surface of the medulla which you perfused compares with the areas described by Mitchell *et al.* (1963) and by Loeschcke *et al.* (1970) which appear to be sensitive to the local application of hydrogen ion?

WILLSHAW I can get effects from stimulating both these areas.

BISCOE What about the surface medial to Mitchell's area?

WILLSHAW There was very little effect.

ACKER Have you tried the alternative experiment of perfusing the medulla with acid CSF and does this give rise to a decrease in efferent activity and an increase in afferent activity?

WILLSHAW If you superfuse with an acid CSF, you see very little change in efferent activity.

CASPERS But this is exactly the opposite result to that which Loeschcke *et al.* (1970) described. They found that when you apply acid CSF you have an increase in chemoreceptor activity.

WILLSHAW Do you mean from this that he recorded from the intact sinus nerve?

CASPERS No. I should not have used the word chemoreceptor but rather respiratory neurones in the medulla which are clearly activated. Yet, as I understand it, if the efferent activity which you are describing in the sinus nerve does anything in response to superfusion of the medullary surface with acid CSF, it would tend to diminish. Is that right?

WILLSHAW Yes.

PURVES I don't understand how you localize the areas which are perfused by fluid on the surface of the medulla and therefore how you can be so certain that you are only stimulating these areas which have been discussed.

WILLSHAW I can't be sure. The fluid is in fact covering the whole surface of the medulla. What I normally do is to squirt the fluid through a fine metal cannula over the area I am interested in but of course then the fluid immediately diffuses.

PURVES So in fact you can't really be certain about this localization, unless for example, you use a small localizing device such as Schläfke (1967) has used.

WILLSHAW Yes, I suppose that is correct except that if you do leave a fold of dura around the area you are interested in it may help to limit the spread of the fluid.

MITCHELL Have you looked at this response after the sympathetics have been cut?

WILLSHAW Yes, there is no change.

MITCHELL And what about the output of other parasympathetic nerves?

WILLSHAW No, I haven't looked into that.

MITCHELL Have you perfused the surface of the medulla with fluids of varying pH to see whether there is some kind of threshold for the response?

WILLSHAW No, I have no information on that and I agree that it would be an experiment worth doing.

REFERENCES

BISCOE, T. J. & SAMPSON, S. R. (1968). Rhythmical and non-rhythmical spontaneous activity recorded from the central cut end of the sinus nerve. *J. Physiol.*, **196**, 327–38.

LOESCHCKE, H. H., DE LATTRE, J., SCHLÄFKE, M. E. & TROUTH, C. O. (1970). Effects on respiration and circulation of electrically stimulating the ventral surface of the medulla oblongata. *Respir. Physiol.*, **10**, 184–97.

MITCHELL, R. A., LOESCHCKE, H. H., SEVERINGHAUS, J. W., RICHARDSON, B. W. & MASSION, W. H. (1963). Regions of respiratory chemo-sensitivity on the surface of the medulla. *Ann. N.Y. Acad. Sci.*, **109**, 661–81.

SAMPSON, S. R. & BISCOE, T. J. (1970). Efferent control of the carotid body chemoreceptor. *Experientia*, **26**, 261–2.

SCHLÄFKE, M. E. & LOESCHCKE, H. H. (1967). Lokalisation eines an der Regulation von Atmung und Kreislauf beteiligten Gebietes an der ventralen Oberfläche der Medulla oblongata durch Kälteblockade. *Pflügers Arch. ges. Physiol.* **297**, 201–20.

Adjustment of chemoreceptor sensitivity in the cat carotid body by reciprocal synapses

By ROBERT A. MITCHELL and DONALD M. McDONALD

Departments of Physiology, Anatomy, Anesthesia, and the Cardiovascular Research Institute, University of California, San Francisco, California 94143, USA

The function of the carotid body, although examined extensively for several decades, still eludes precise definition. Several hypotheses have been proposed to explain the process of chemoreception, but none successfully integrates all of the existing physiological and anatomical data. The first hypothesis, which developed from the anatomical studies of De Castro (1928) and the physiological studies of Heymans, Bouckaert & Dautrebande (1930), held that (Type I) glomus cells are receptors sensitive to hypoxia. When stimulated, they release a transmitter which initiates action potentials in sensory nerves. In recent years, Eyzaguirre, Koyano & Taylor (1965) have supported this hypothesis and the concept that the transmitter released by glomus cells is acetylcholine. However, this hypothesis is unacceptable to many because the excitation produced by acetylcholine can be blocked pharmacologically without blocking the responsiveness to hypoxia (Sampson, 1971).

In the late 1960s reports appeared which suggested that glomus cells were innervated by efferent – but not afferent – nerves. Biscoe & Stehbens (1966) reported that many nerve endings in the carotid body contained synaptic vesicles and appeared to be presynaptic to glomus cells. Coincident with these studies, it was demonstrated that antidromic electrical stimulation of most of the carotid sinus nerve depressed chemoreceptor activity recorded from a small strand split from the nerve (Sampson & Biscoe, 1970; Neil & O'Regan, 1971; Fidone & Sato, 1970; Goodman, 1973). This electrically-evoked inhibition was called 'efferent inhibition'.

In 1967, Biscoe & Stehbens reported that nerve endings on glomus cells degenerate when the glossopharyngeal nerve is cut peripheral to the petrosal ganglion, and Biscoe, Lall & Sampson (1970) reported that the endings also degenerate (but very slowly) when the glossopharyngeal nerve is cut central to the petrosal ganglion. From these results, Biscoe and his associates concluded that large nerve endings on glomus cells belonged not to afferent axons whose cell bodies were in the petrosal ganglion, but to efferent axons whose cell bodies were in the brain stem. They also concluded that chemoreceptive nerves did not end on glomus cells.

The demonstration of 'efferent inhibition' together with the assumption that glomus cells have only an efferent innervation, led Biscoe (1971) to postulate that the chemoreceptive property of the carotid body resides not in glomus cells, but in nerves ending nearby. Furthermore, he postulated that efferent nerves and glomus cells form an inhibitory mechanism which regulates the sensitivity of chemoreceptive nerves. Sampson studied the putative transmitters of this inhibitory circuit (1972, and this Symposium). He found that dopamine, which exists in high concentration in glomus cells, depressed chemoreceptor activity and that the depression was blocked by drugs (dihydroergotamine, phenoxybenzamine) that abolish the inhibitory action of dopamine. He postulated that efferent nerves caused glomus cells to release dopamine which in turn decreased the sensitivity of the chemoreceptive nerves to hypoxia.

Hess & Zapata (1972) repeated some of the morphological experiments of Biscoe *et al.* and found that nerves ending on glomus cells degenerate when the glossopharyngeal nerve is cut peripheral to the petrosal ganglion, but not when cut central to the ganglion. They concluded that glomus cells are innervated only by afferent nerves. This conclusion supported the original observations of De Castro, but obviously was at variance with the findings of Biscoe and his associates.

Our hypothesis differs from both those of De Castro and Biscoe. We postulate that afferent nerve endings are receptors sensitive to hypoxia and that glomus cells are interneurons which adjust the sensitivity of the afferent nerve endings. However, unlike Biscoe's hypothesis, we postulate that glomus cells have principally an afferent innervation and that the feedback mechanism which regulates afferent nerve sensitivity is based on reciprocal synapses between afferent nerves and glomus cells. In an accompanying paper (McDonald & Mitchell, this Symposium), we give ultrastructural evidence for such reciprocal synapses. In the present paper, we present four lines of evidence suggesting that afferent nerve endings are chemoreceptors and that glomus cells are dopaminergic interneurons – not chemoreceptive transducers.

CHEMORECEPTIVE PROPERTIES OF REGENERATED AFFERENT AXONS IN THE CAROTID SINUS NERVE

To test the hypothesis that afferent nerve endings are chemoreceptive transducers, we studied the chemoreceptive properties of axons regenerated from the cut end of the carotid sinus nerve (Mitchell, Sinha & McDonald, 1972). In these experiments, we exposed the right carotid sinus nerve in four cats anaesthetized with sodium pentobarbital and cut it close to the carotid body. In three cats the end distal to the cut was ligated, and the proximal end was inserted into a 1 cm segment of silastic tubing which in turn was sutured to connective tissue 0.5–1.0 cm from the carotid body (2 cats) or buried in muscle (1 cat). In the fourth cat the distal segment of the carotid sinus nerve and the carotid body were removed, and the proximal end of the nerve was buried in connective tissue.

Twelve to 18 months later, we anesthetized the cats, cut the right carotid sinus nerve near its junction with the glossopharyngeal nerve, separated the carotid sinus nerve into fine strands, and recorded action potentials from afferent axons. We found many spontaneously active fibres in each cat. Other fibres could be stimulated by pressing on the neuroma but did not have spontaneous activity and did not respond to chemical stimuli. Rhythmically firing baroreceptors, which dominate the activity in the normal carotid sinus nerve, were absent. The activity of afferent nerves was increased by hypoxia and hypercapnia in all cats (Fig. 1). In three cats we measured the afferent nerve activity at four different levels of P_{a,O_2}. The P_{a,O_2}/response curve described by these points was hyperbolic. Increased P_{a,CO_2} potentiated the effect of hypoxia. Sodium cyanide, acetylcholine, and nicotine injected intravenously or into the right common carotid artery also increased activity in the carotid sinus nerve. Dopamine decreased nerve activity in both cats receiving it by intracarotid injection.

At the end of the experiment, the carotid sinus nerve and its neuroma were removed for histological studies. Serial sections of neuromas from three cats were prepared for fluorescence microscopy according to the histochemical technique of Falck (1962) for demonstrating catecholamines. We found numerous fluorescent nerves in the neuroma both around blood vessles and in isolated groups. However, none of the specimens contained cells having the intense yellow–green fluorescence characteristic of glomus cells. Serial sections of the neuroma from the fourth cat, stained with hematoxylin and eosin, showed large bundles of nerves with an organisation characteristic of a neuroma but no normal carotid body tissue. Consequently, we do not believe the chemosensitivity of axons in the neuroma resulted from innervation of aberrant or newly-formed glomus cells. It is similarly unlikely that the response to hypoxia resulted from re-innervation of the carotid body, because the carotid body was removed from one cat during the initial surgery, and in the other three cats we cut tissue between the carotid body and the neuroma before doing neurophysiological studies.

Furthermore, chemosensitivity does not appear to be a non-specific property of sensory nerve endings in neuromas for two reasons. First, we did not find chemosensitive nerves in neuromas formed from the superior laryngeal nerve; and second, endings of fibres in the carotid sinus nerve

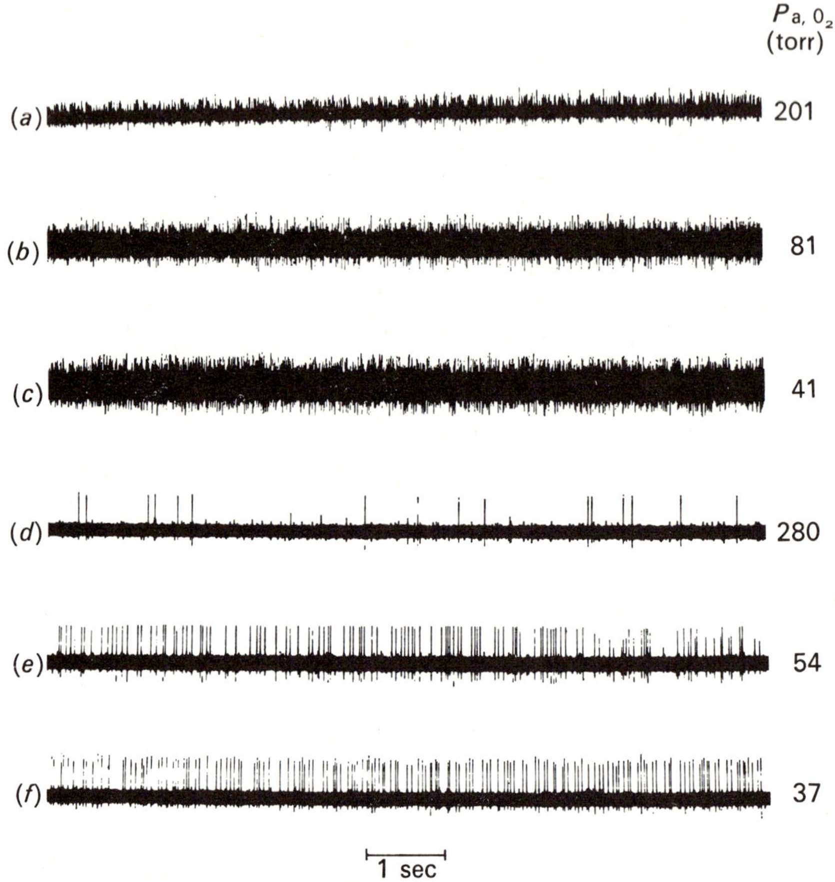

Fig. 1. The effect of graded hypoxia (at constant P_{a,CO_2} and arterial pH) on the rate of action potentials recorded from the whole carotid sinus nerve (*a*, *b*, and *c*) and from a small strand of the nerve (*d*, *e*, and *f*). (From Mitchell, Sinha & McDonald, 1972.)

which could be stimulated by pressing on the neuroma (probably baroreceptors) were not chemosensitive. Therefore, we believe our results can be explained best by the hypothesis that the nerve endings – not glomus cells – are chemoreceptive transducers.

INHIBITORY MODULATION OF CHEMORECEPTORS PRODUCED BY AFFERENT NERVE-GLOMUS CELL RECIPROCAL SYNAPSES

Results from our experiments in the rat (McDonald & Mitchell, this Symposium) support the concept that most nerves ending on glomus cells have their cell bodies in the petrosal ganglion and are afferent. These data support those of De Castro (1928) and Hess & Zapata (1972) but oppose those of Biscoe *et al.* (1970). To examine the discrepancy between the findings of De Castro, Hess & Zapata on the one hand and Biscoe *et al.* on the other hand (all of whom studied cats), we compared the ultrastructure of the cat and rat carotid bodies. To make this comparison we studied six cats. The first two cats were normal controls. In the second two cats, we cut the right glossopharyngeal nerve below (peripheral to) the petrosal ganglion, and in the remaining two cats we cut the right glossopharyngeal and vagus nerves above (central to) their sensory ganglia. Ten days after the operation, we anesthetized the cats with an intraperitoneal injection of chloralose (40 mg kg^{-1}) and urethane (200 mg kg^{-1}) and cannulated the trachea. We exposed the right carotid sinus nerve, cut it near its junction with the glossopharyngeal nerve, removed the sheath and separated the carotid sinus nerve into strands containing a few active chemoreceptor fibres, and then did our neurophysiologic studies (described below). After these studies, we perfused the cats with fixative via a cannula inserted into the ascending aorta through an incision in the left ventricle. The fixatives, method of preparing tissue for electron microscopy, and the method of identifying and quantitating nerve endings and glomus cells are described in an accompanying paper (McDonald & Mitchell, this Symposium). The effect on glomus cell innervation of cutting the glossopharyngeal nerve above or below the petrosal ganglion is shown in Table 1. Cutting the nerve below the ganglion eliminated nearly all nerve endings on glomus cells, whereas cutting it above the ganglion did not reduce the number of endings. These findings are identical to those of our experiments done on rats (McDonald & Mitchell, this Symposium). We conclude that most nerves ending on glomus cells in the cat and rat carotid body are afferent.

In our neurophysiologic studies on these cats, we recorded chemoreceptor activity from the carotid sinus nerve with platinum electrodes and displayed the amplified signal on an oscilloscope for photographic recording. The amplified signal also was fed to an audio monitor. To quantitate chemoreceptor activity, the signal was fed to a pulse-height selector and a pulse counter. The output of the pulse counter, arterial blood pressure, and tracheal P_{CO_2} were recorded on a Grass polygraph. After identifying a

Table 1. *Effect of supra- and infra-petrosal section of the glossopharyngeal nerve on the number of nerve endings on glomus cells 10 days after surgery*

	Glomus cells counted	Nerve endings/ glomus cells Mean	Range
Control	200	1.7	(1.4–1.9)
Supra-petrosal section (10 days)	204	1.7	(1.5–1.8)
Infra-petrosal section (10 days)	269	0.05	(0.0–0.1)

chemoreceptor fibre, we ventilated the cat mechanically, paralysed it with 20 mg gallamine triethiodide intravenously, and maintained paralysis with 10 to 20 mg gallamine per hour. To test for 'efferent inhibition' we stimulated all of the carotid sinus nerve, except for the strand from which we were recording, at a frequency of 5–50 Hz, a duration of 0.25 msec, and a voltage from 1–7 V.

We found no difference in the response of chemoreceptors to hypoxia between control cats and cats whose glossopharyngeal nerve was cut above the petrosal ganglion. Electrical stimulation of the carotid sinus nerves caused a 10–98% reduction in the frequency of chemoreceptor impulses in both groups of cats (Figs. 2, 3, 4). The magnitude of the response depended on the frequency and intensity of the stimulus (Fig. 5). Inhibition was greatest when the nerve was not injured during the dissection. In such

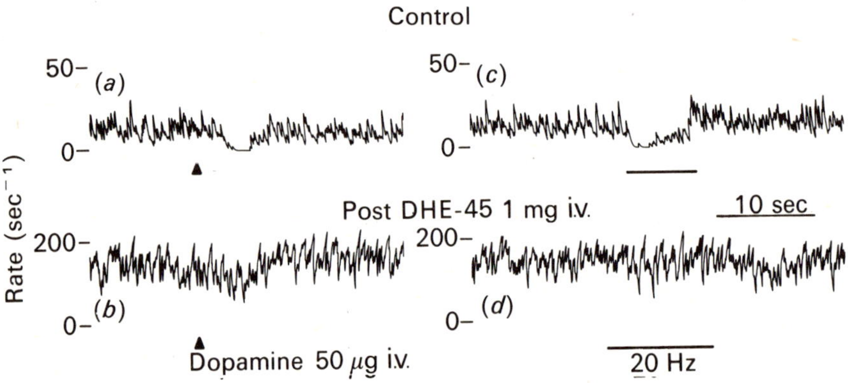

Fig. 2. Rate meter recordings of carotid sinus nerve chemoreceptive activity from one cat showing the effects of dopamine (*a* and *b*) and electrically-evoked suppression of activity (*c* and *d*) before and after i.v. injection of DHE (1 mg). Duration of stimulus indicated by solid bar in *c* and *d*. Stimulus parameters: 5 volts, 0.25 msec, 20 Hz. Note change in scale of firing rate in *b* and *d*.

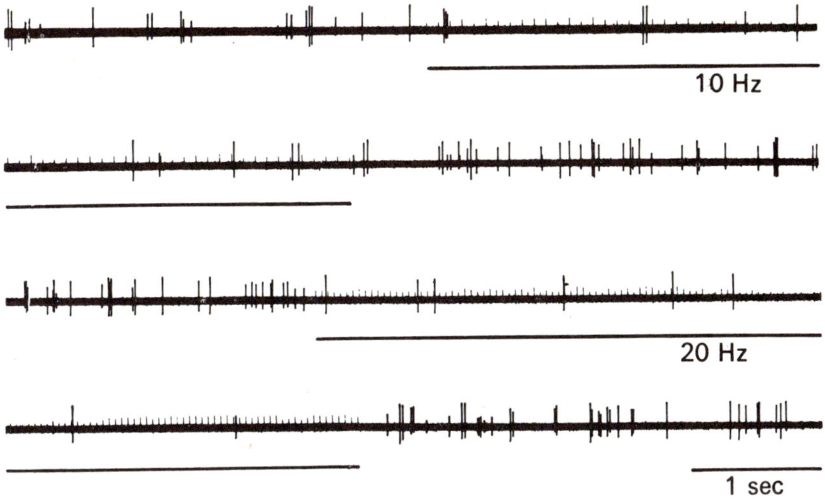

Fig. 3. Filmed record of carotid chemoreceptive activity in a strand from the carotid sinus nerve (10 days after suprapetrosal section of the glossopharyngeal nerve) showing electrically-evoked suppression. Stimulus duration indicated by solid bars. Stimulus parameters: 5 volts, 0.25 msec, 10 Hz and 20 Hz.

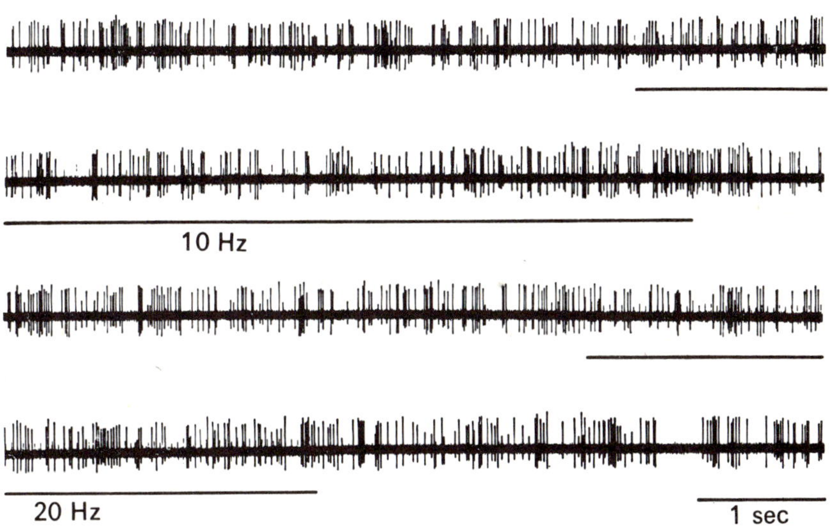

Fig. 4. Filmed record of the same strand as shown in Fig. 3 after injection of DHE (2 mg (i.v.). Description as in Fig. 3.

preparations we obtained 50–60% depression of chemoreceptor activity with a stimulus as small as 1 volt at 5 Hz. Dopamine (50 μg i.v.) mimicked the effect of electrical stimulation, depressing chemoreceptor activity for

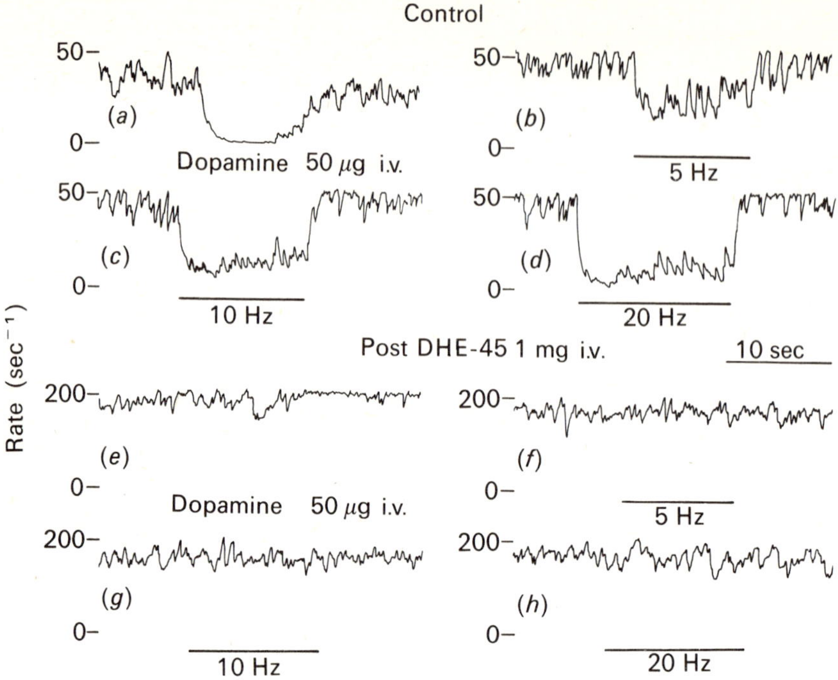

Fig. 5. Rate meter recording of carotid chemoreceptive activity showing the effects of dopamine (*a* and *e*) and electrically-evoked suppression of activity (*b*, *c*, *d*, *f*, *g*, and *h*) before and after DHE (1 mg) (10 days after suprapetrosal section of the glossopharyngeal nerve). Duration of stimulus indicated by solid bars. Stimulus parameters: 5 volts, 0.25 msec, 5–20 Hz.

10–20 seconds (Figs. 3, 5). Dihydroergotamine (DHE) (0.25–0.30 mg kg^{-1} i.v.) not only blocked (reversibly) the electrically- and dopamine-evoked depression of chemoreceptor activity, but also increased the sensitivity of chemoreceptors to hypoxia both in control cats and in cats whose glossopharyngeal nerve was cut above the ganglion (Figs. 2, 4, 5). (We assume that drugs which block the inhibitory action of dopamine do not stimulate chemoreceptors directly.)

Suppression of chemoreceptor activity produced by electrical stimulation of the carotid sinus nerve has been postulated to be due to dopamine released by glomus cells when stimulated by efferent axons from the sinus nerve (Biscoe *et al.*, 1970), increased carotid body blood flow produced by efferent nerves ending on blood vessels (Goodman, 1973), or spread of current from the stimulating electrode to afferent nerves (Fidone & Sato, 1970; Goodman, 1973). We do not believe our data are explained by these hypotheses. Most of the nerves in synaptic contact with glomus cells are afferent. We found no efferent axons from the carotid sinus nerve that end on glomus cells. Therefore, we doubt that electrically-evoked inhibition results from activation of efferent nerves ending on glomus cells. Furthermore, we do not believe that electrically-evoked inhibition results from spread of current from the stimulus to afferent nerves because DHE blocks the inhibition and cannot eliminate current spread.

We conclude that 'efferent inhibition' (a better term for this phenomenon is inhibitory modulation) is produced by antidromic stimulation of some afferent nerves, which initiates the release of dopamine from glomus cells; this release in turn suppresses chemosensitivity of other afferent nerves. We postulate that reciprocal synapses between afferent nerve endings and glomus cells form the basis of this local feedback loop.

BLOCKING INHIBITORY MODULATION CHANGES THE RESPONSE OF CHEMORECEPTORS TO HYPOXIA

Using the methods described in the previous sections, we measured chemoreceptor activity in carotid sinus nerves of four normal cats at several levels of P_{a,O_2}, then determined the maximum amount of depression of

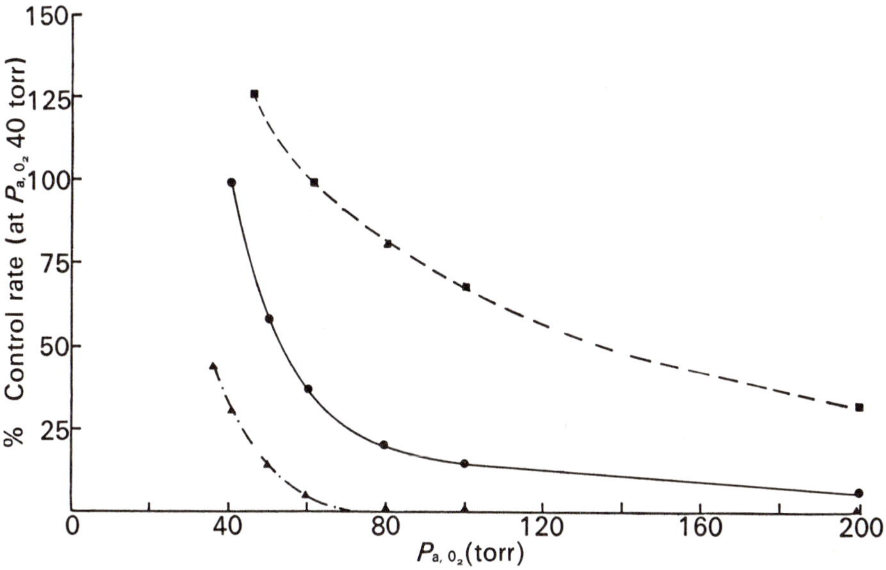

Fig. 6. Plot of the average chemoreceptive activity in 4 cats as a function of P_{a,O_2}. Chemoreceptive activity is plotted as per cent of the firing rate of chemoreceptive axons in control cats (●) having a P_{a,O_2} of 40 torr. The post PBZ curve (■) was plotted from data obtained after injecting PBZ (10 mg/kg i.v.). The CSN stim. curve (▲) was plotted from data obtained while chemoreceptor activity was suppressed by electrically-evoked inhibition. Stimulus parameters: 5 volts, 0.25 msec, 20 Hz.

chemoreceptor activity that electrical stimulation of the carotid sinus nerve (5–7 V, 0.25 sec and 20 Hz) produced at each P_{a,O_2}. After making these measurements, we infused phenoxybenzamine (PBZ) (20 mg kg^{-1} i.v.) over 30–60 minutes; then we repeated the studies. The results of these experiments are summarized in Fig. 6. Electrically-evoked suppression of chemoreceptor activity could be elicited at all levels of P_{a,O_2}. At P_{a,O_2} levels above 80 torr, the suppression was almost complete; at lower levels, activity persisted during stimulation. PBZ injected intravenously not only blocked the suppression of chemoreceptor activity produced by dopamine and electrical stimulation, but increased chemoreceptor activity five- to ten-fold in cats breathing either air (P_{a,O_2} 80–90 torr) or hyperoxic gas mixtures ($P_{a,O_2} > 190$ torr). However, at P_{a,O_2} levels less than 80 torr, PBZ increased chemoreceptor activity relatively less. The nonuniform disinhibition produced by PBZ yielded a P_{a,O_2}/response curve that was more linear than that obtained in controls.

In three cats we tested the effect of atropine (1–4 mg kg^{-1}) on the P_{a,O_2}/response curve (Fig. 7). Atropine reduced electrically-evoked inhibition, abolishing it in one cat, but did not reduce the inhibition produced by injected dopamine. Atropine, like PBZ also increased chemoreceptor

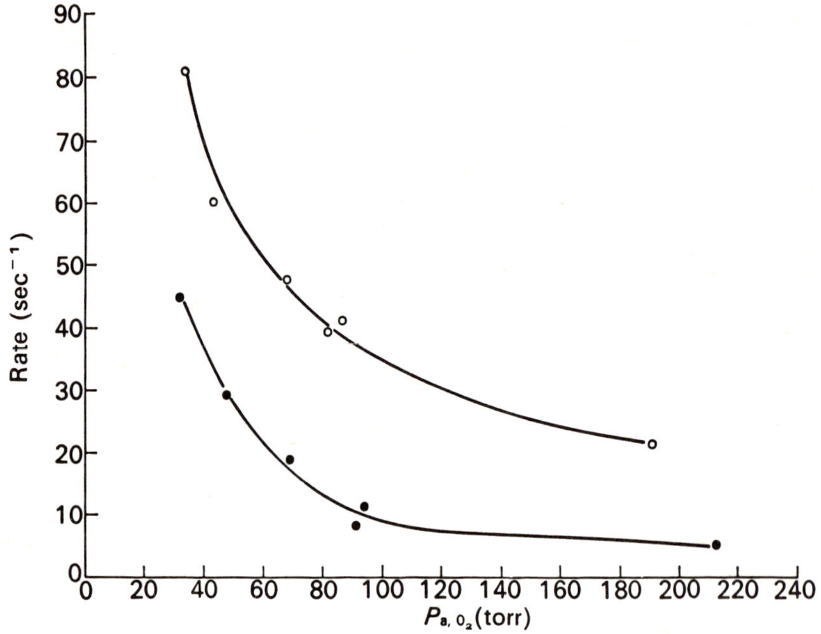

Fig. 7. Plot of chemoreceptive activity as a function of P_{a,O_2} before (●) and after injecting atropine (1 mg kg^{-1}) (○).

activity over control levels at all values of P_{a,O_2}, but unlike PBZ, it did not make the P_{a,O_2}/response curve more linear. Also, the amount of increase in chemoreceptor activity produced by atropine in the three cats was more variable than that produced by PBZ. The reason for the difference in the effects of these two blocking agents is not evident from our studies. Perhaps the blocking action of atropine was incomplete or too brief to produce a reproducible degree of inhibition. Variations in circulating catecholamines occurring while data for the P_{a,O_2}/response curve was being obtained may have influenced our results, since their effects on the chemoreceptor activity presumably are not blocked by atropine.

We conclude from this study that opening the glomus cell–afferent nerve inhibitory feedback loop, which normally suppresses chemoreceptor activity, increases the sensitivity of chemoreceptors to hypoxia and converts the normally hyperbolic P_{a,O_2}/response curve to a more linear one. Consequently, although inhibitory modulation exists at all levels of P_{a,O_2}, the inhibition appears to be greater at higher levels of P_{a,O_2}. While we consider this interpretation reasonable, it probably is an oversimplification, because the P_{a,O_2}/response curve of some chemoreceptive nerves in neuromas (in the absence of glomus cells) tended to be hyperbolic.

SIMILARITY OF GLOMUS CELLS AND INTERNEURONS IN SYMPATHETIC GANGLIA

Small cells, resembling glomus cells in structure, exist in sympathetic ganglia. These cells, like glomus cells, have a high content of catecholamine and an intense fluorescence when treated according to the histochemical method of Falck (1962). Consequently, they have become known as small intensely fluorescent (SIF) cells. These SIF cells are post-synaptic to incoming pre-ganglionic axons and pre-synaptic to sympathetic ganglion cells (Matthews & Raisman, 1969). Glomus cells and SIF cells have striking structural similarities and may have functional similarities as well. According to Libet (1970), SIF cells are dopamine-releasing, inhibitory interneurons. When SIF cells are stimulated by their pre-ganglionic synaptic input, they release dopamine, which in turn hyperpolarizes sympathetic ganglion cells. Dopamine also inhibits chemoreceptors, mimicking electrically-evoked inhibition. DHE and phenoxybenzamine not only block the dopamine-induced, long-lasting hyperpolarization of sympathetic ganglion cells (Libet, 1970), but also block electrically-evoked inhibition of chemoreceptors. These drugs increase chemoreceptor activity when given in doses sufficient to block electrically-evoked inhibition in normal cats (Sampson, this Symposium) and in cats whose glossopharyngeal nerve has been cut above the petrosal ganglion (Figs. 2–6). Consequently, endogenous dopamine from glomus cells probably suppresses chemoreceptive nerves at normal arterial oxygen tensions.

A further pharmacologic similarity of glomus cells and SIF cells is seen in the effect of atropine. This muscarinic blocking agent not only blocks the pre-ganglionic sympathetic input to SIF cells, but also blocks (or diminishes) electrically-evoked inhibition of chemoreceptors. Atropine does not diminish the effect of dopamine, but it does increase chemoreceptor activity (Sampson, this Symposium). Our data support those of Sampson which suggest that atropine causes these effects by blocking the excitation of glomus cells produced by the transmitter released by sensory nerve endings. Like the drugs that block the action of dopamine, atropine opens the feedback loop between glomus cells and afferent nerve endings. Consequently, the two different types of drugs have some similar effects, in spite of having different sites of action.

SUMMARY

We have presented four lines of evidence in support of our hypothesis that afferent nerve endings, connected by reciprocal synapses to glomus cells, are chemoreceptors sensitive to hypoxia and that glomus cells are dopaminergic interneurons which modulate the sensitivity of afferent nerve endings.

First, glomus cells are not essential for the chemoreceptive properties of endings of regenerated axons from the carotid sinus nerve.

Second, most nerves which end on glomus cells are afferent, and afferent nerve endings are connected to glomus cells by reciprocal synapses. Antidromic electrical stimulation of some afferent axons in the carotid sinus nerve depresses the activity of other chemoreceptive afferent axons in the nerve (inhibitory modulation or 'efferent inhibition'). Cutting the glossopharyngeal nerve central to its sensory ganglia (to eliminate efferent – but not afferent – axons from the nerve) does not eliminate electrically-evoked inhibition of chemoreceptors. Dopamine injected intravenously mimics electrically evoked inhibition, and drugs that block the inhibitory action of dopamine abolish the phenomenon. We postulate that reciprocal synapses between afferent nerves and glomus cells form the basis of the feedback loop responsible for electrically-evoked inhibition of chemoreceptors.

Third, drugs which abolish electrically-evoked inhibition of chemoreceptors change the response of chemoreceptors to hypoxia. These drugs make the normally hyperbolic P_{a,O_2}/response curve more linear. We postulate that dopamine released from glomus cells depresses the sensitivity of chemoreceptors to hypoxia more at high P_{a,O_2} levels than at low levels. Drugs which block the action of dopamine open the inhibitory feedback loop between glomus cells and afferent nerves, thereby eliminating inhibitory modulation and producing a more linear P_{a,O_2}/response curve.

Fourth, glomus cells and SIF cells have structural and functional similarities. SIF cells are considered dopaminergic interneurons which inhibit ganglion cells in sympathetic ganglia. We postulate that glomus cells are dopaminergic interneurons which inhibit chemoreceptive nerve endings. When afferent nerves are stimulated by hypoxia, not only are action potentials generated, but also an excitatory transmitter is released from their terminals. This transmitter causes glomus cells to release dopamine, which in turn decreases the sensitivity of chemosensitive nerves.

We wish to thank Miss D. A. Herbert and Mr R. W. Blewett for their excellent technical help. This work was supported in part by USPHS, NIH Grants GM 15571, HL-06285, 5-K3-1319411 and I-FO-3-HD-45487.

REFERENCES

Biscoe, T. J. (1971). Carotid body: Structure and function. *Physiol. Rev.*, **51**, 437–95.

Biscoe, T. J., Lall, A. & Sampson, S. R. (1970). Electron microscopic and electrophysiological studies of the carotid body following intracranial section of the glossopharyngeal nerve. *J. Physiol.*, **208**, 133–52.

Biscoe, T. J. & Stehbens, W. E. (1966). Ultrastructure of the carotid body. *J. Cell. Biol.*, **30**, 563–78.

Biscoe, T. J. & Stehbens, W. E. (1967). Ultrastructure of the denervated carotid body. *Quart. J. Exptl. Physiol.*, **52**, 31–6.

De Castro, F. (1928). Sur la structure et l'innervation du sinus carotidien de l'homme et des mammifères. Nouveaux faits sur l'innervation et la fonction due glomus caroticum. Etudes anatomiques et physiologiques. *Trav. Lab. Rech. Biol. Univ. Madrid*, **25**, 331–80.

Eyzaguirre, C., Koyano, H. & Taylor, J. R. (1965). Presence of ACh and transmitter release from carotid body chemoreceptors. *J. Physiol.*, **178**, 463–76.

Falck, B. (1962). Observations on the possibilities of cellular localizations of monoamines by fluorescence methods. *Acta. physiol. Scand.*, **56**, *Suppl.* **197**, 1–25.

Fidone, S. J. & Sato, A. (1970). Efferent inhibition and antidromic depression of chemoreceptor A-fibers from the cat carotid body. *Brain Research*, **22**, 181–93.

Goodman, N. W. (1973). Efferent control of arterial chemoreceptors mediated by glossopharyngeal fibers and artifacts introduced by stimulation techniques. *J. Physiol.*, **230**, 295–311.

Hess, A. & Zapata, P. (1972). Innervation of the cat carotid body: Normal and experimental studies. *Fedn Proc.*, **31**, 1365–83.

Heymans, C., Bouckaert, J. & Dautrebande, L. (1930). Sinus carotidien et réflexes respiratoires. II. Influences respiratoires réflexes de l'acidose, de l'alcalose, de l'anhydride carbonique, de l'ion hydrogène et de l'anoxémie. Sinus carotidiens et échanges respiratoires dan les poumons et de la des poumons. *Arch. Intern. Pharmacodyn.*, **39**, 400–48.

Libet, B. (1970). Generation of slow inhibitory and excitatory post-synaptic potentials. *Fedn Proc.*, **29**, 1945–56.

Matthews, M. R. & Raisman, G. (1969). The ultrastructure and somatic efferent synapses of small granule-containing cells in the superior cervical ganglion. *J. Anat.*, **105**, 255–82.

MITCHELL, R. A., SINHA, A. K. & MCDONALD, D. M. (1972). Chemoreceptive properties of regenerated endings of the carotid sinus nerve. *Brain Research*, **43**, 681–5.

NEIL, E. & O'REGAN, R. G. (1971). The effects of electrical stimulation of the distal end of the cut sinus and aortic nerves on peripheral arterial chemoreceptor activity in cat. *J. Physiol.*, **215**, 15–32.

SAMPSON, S. R. (1971). Effects of mecamylamine on responses of carotid body chemoreceptors *in vivo* to physiological and pharmacological stimuli. *J. Physiol.*, **212**, 655–66.

SAMPSON, S. R. (1972). Mechanism of efferent inhibition of carotid body chemoreceptors in the cat. *Brain Res.*, **45**, 266–70.

SAMPSON, S. R. & BISCOE, T. J. (1970). Efferent control of the carotid body chemoreceptor. *Experientia*, **26**, 261–2.

DISCUSSION

FILLENZ Just on your last point I can't see why this horizontal connection should be necessary since presumably the hypoxia is affecting all the sensory terminals?

MITCHELL What I am saying is that it is essential for the electrically induced inhibition.

HESS With regard to those neuroma experiments I wouldn't be surprised if nerve fibres severed from their receptors, could still transmit the appropriate sensations; and this could be true for example, of pain fibres still connected to their ganglion cells. So I don't think it is surprising that the fibres from the neuroma retain their original properties.

MITCHELL This was the point I was trying to make. Furthermore, the Type I cell does not appear to be an essential part of the receptor apparatus when viewed simply from the ability of the receptor to respond to hypoxia.

HESS But I wonder if, for example, you separated taste fibres from taste buds whether the fibres would still respond to taste.

MITCHELL Well, this may be complex because it has been suggested that there are two forms of triggering of taste, one directly on the nerve, the other through the specialized ending. However, it is interesting that in many other secondary nerve endings such as taste or lateral line in various fish if you cut the nerve, the specialized receptor degenerates, whereas in the carotid body this is not so; and if you look at the glomus cells a year or more after cutting the sinus nerve, they did not degenerate. In fact, we got the impression that the glomus cells had developed rather more processes than normal. This may be analogous to processes extended by adrenal medullary cells, denervated and transplanted into the eye.

HESS It was also possible to demonstrate inhibition of afferent discharge when the sinus nerve was stimulated?

MITCHELL No, this is one of the things we didn't do and should.

HESS But you tried the effects of dopamine and other catecholamines?

MITCHELL Yes.

HESS And what about the synapses?

MITCHELL We have only studied this with light and fluorescent microscopy. We do, however, have a second group of chronically denervated cats and these are at present being prepared for electron microscopic study of the neuromas.

HESS Dopamine stimulates the dog carotid body, isn't that right?

MITCHELL Yes, that is so, and we therefore feel obliged to repeat a number of these studies in the dog to see whether it is a genuine exception. It is worth commenting that occasionally in the cat after inhibition with dihydroergotamine, dopamine will cause excitation, and we may be dealing here with different populations of receptors, some of which are excited and others inhibited by dopamine. The response in the dog has some peculiarities. If you are recording from a single fibre in the dog, dopamine gives rise to a sharp burst of activity followed by inhibition. Both of these responses are blocked by dihydroergotamine. So it may be that the dual response which is seen in the dog can be elicited in the cat, but in the cat the inhibitory part of the response predominates.

HESS How many reciprocal synapses are there in the cat and do you find a piling up of the granulated vesicles as you see in the rat?

McDONALD They are less frequent in cats than in the rat.

HESS If dopamine is the inhibitor and if carotid body glomus cells fluoresce, presumably the dopamine is to be found throughout the cell in the granulated vesicle and, as we have heard in previous papers, the granulated vesicles close to the sensory terminal are different from those in other parts of the cell. Certainly in size and possibly also in kind. So my question is do you think that the dopamine which is involved in inhibition is that which is in the vesicles piled up close to the terminal?

McDONALD I think so, but I do not have definitive evidence. Nonetheless, I don't agree that the dense-cored vesicles close to the synaptic junction are different from the population of dense-cored vesicles as a whole. I think that there are two types of vesicles: one is found throughout the glomus cell, is large and has a dense core; the other is small and does not have a dense core. The small vesicles acquire dense cores in rats receiving 5-hydroxydopamine. This suggests that they may have the capacity to accumulate catecholamines. There is no direct evidence that either type of vesicle contains dopamine, but I suspect both do.

WILLSHAW As I understand the feedback system, you should be able to repeat Biscoe & Sampson's experiments with the oxygen response curves.

MITCHELL I think this is difficult to predict because we have found, and other people who have commented on this today appear to agree with us, that measurable efferent activity in the sinus nerve is definite but sparse. When we cut the glossopharyngeal nerve rostral to the petrosal ganglia, one of the factors that makes us feel secure about true degeneration is the fact that the axons which synapse on the ganglion cells degenerate. We consider these parasympathetic. There are also sympathetic nerve endings, as McDonald mentioned, which are pre-ganglionic sympathetic and which synapse directly upon the Type I cells. If you stimulate the sympathetic you get rather a mixed response, usually starting with an inhibitory response and followed by excitation of the chemoreceptors. And if you give such a blocking agent as phenoxybenzamine, the preliminary inhibition is abolished. This suggests that there is yet another inhibitory pathway acting through the Type I cell, this one pre-ganglionic sympathetic.

WILLSHAW How in your system would you account for the changes which you get with a step change with, for example carbon dioxide?

MITCHELL I can't really answer that question because I do not know the time constants of individual pathways that may be involved. You are thinking of the problem of overshoots and damping I take it?

WILLSHAW Yes. That might be predicted from your system.

MITCHELL Well, perhaps. But it is worth remarking that with these inhibitory pathways we are not talking about something that is necessarily producing action potentials. The Type I cell may be an interneurone that produces graded responses. Graded responses have been shown to occur in the retina and olfactory bulb at reciprocal synapses (Reese & Shepherd, 1972). These produce a graded type of response, not an all-or-none type of response.

GOODMAN In one of your slides (Fig. 3) you showed a stimulation in which the impulses stopped very abruptly at the start of the stimulation and there are no impulses during the stimulation. How, then, can you say that there was no silent period? Also, you admit that it is a multifibre preparation, but with these it is possible to have adventitious excitation in some of the fibres of the strand whilst the remainder are unaffected. Then there is an apparent true inhibition of discharge, but no silent period is seen because of the unaffected fibres.

MITCHELL Let me make this clear. We have all these records on tape at stimulation frequencies of between 2 and 50 Hz. Where we have blocking at a low stimulation level, we have re-evaluated the inhibition in multifibre preparations. Although we did not have long periods of recording, these fibres clearly fired early after the stimulus artifact and there was no evidence of a silent period. We then purposely increased the stimulus level to try and develop the silent period you have described, and we were not able to do this. When we had finished, as I mentioned earlier, we picked apart the nerve and although we could get active fibres, the majority of the fibres were dead. If you now stimulated the nerve at high intensity, you start getting evidence of current spread and so you see inhibition. In fact, our most successful experiments were those in which we obtained active chemoreceptor fibres after only two or three splits and it was because of this that we tended to use multifibre preparations. Further or extensive dissection to get single active fibres reduced the chance of getting inhibition.

EYZAGUIRRE Did you mention that when you used the superior laryngeal for the neuroma type of experiments you didn't get very clear responses?

MITCHELL That is right.

EYZAGUIRRE Well, I can turn your argument around because when Zapata, Hess and I did this with the carotid body we got very clear responses when we innervated the carotid body with the superior laryngeal nerve.

MITCHELL What I am saying is that my responses were similar to yours using, as far as I can recall, asphyxia as the stimulus.

EYZAGUIRRE I am not saying that our experiments provided evidence for the notion that the glomus cells are the receptor cells. But I think that one has to watch for the logic in the argument. The other thing that worries me is that people have found inhibition with a stimulus intensity necessary to activate C-fibre or small myelinated fibre pathways, while we are talking here about reciprocal inhibition being a general phenomenon occurring with stimulation of every A-fibre.

PALLOT Could you just remind me what you know about the histology of the sinus nerve neuroma?

MITCHELL We have done principally fluorescent microscopic studies, and so I have no idea of what the sensory nerve terminals look like. While I do not know what we will find when we study the neuroma by electron microscopy, I anticipate considerable difficulty in interpreting these findings. For example, sympathetic axons form part of the neuroma. This was obvious from the fluorescence microscope studies. Also, there are baroreceptor fibres with their endings, and possibly true efferents. We may be dependent on the quantitative analysis that Dr McDonald was talking about to help us identify the various types of terminals in the neuroma. Another problem was that in three of the animals the neuroma was discrete, but in the fourth animal it fanned out in the scar tissue. The latency of the response to hypoxia of nerve from the latter neuroma, which was not highly vascular, was much longer than the other three.

MILLS I cannot recall, did you test the hypoxic response of these neuromas and if so what did the curves look like?

MITCHELL Yes. I cannot recall the precise details but one of the things that bothers me is that the hypoxic response curves were very variable. We would expect that if the efferent modulation were effective, the hypoxic response curve would be rather flat. We certainly saw that but we also saw one that looked very normal. However, we have insufficient controls and perhaps I ought to wait until this problem has been looked at systematically.

BINGMANN Would you like to comment on the pattern of vessels supplying the neuroma and have you any idea about the internal oxygen tension?

MITCHELL No. We could certainly identify the vessels in two of the discrete neuromas since they were lying very close to the external carotid artery and it was interesting that if you put a clamp on the external carotid artery just above the bifurcation, that is, a clamp which would not affect the normal carotid body, this produced a burst of firing from the neuroma; so it was quite clear that the blood flow to the neuroma was a local one and had not developed from the site of the original carotid body.

NAIL Did you test the ventilatory responses of these cats with neuromas to hypoxia?

MITCHELL Unfortunately, we didn't do this except in the last one. Actually we tested it with cyanide and there was a ventilatory response. It was quite weak and was certainly less than the response to a similar stimulation of the normal carotid body on the other side.

ACKER I wonder if it is possible that, because the blood supply to the neuroma is so poor compared to the normal carotid body, the uptake of catecholamines by the chief cells is correspondingly reduced. The chief cells might be present in the neuroma but possibly did not stain sufficiently strongly for you to see them in the fluorescent study?

MITCHELL Perhaps, but I am not dependent on our fluorescence microscopical studies. I did take the precaution of sending sections of the first neuroma we studied to a number of people – some of whom are in this room – because it seemed to me on first examination that there were a number of Type I cells present. However, when these sections were studied systematically it was clear that these cells had quite different dimensions from the Type I cells; but this and the fluorescent studies do not positively exclude alternative possibilities. First, there might have been aberrant carotid body tissue which was responsible for the discharge and secondly, there may be cells which are modified Type I cells which do not contain catecholamines and do not fluoresce. Our planned ultrastructural studies will help us evaluate these possibilities.

MCDONALD The structure of the neuroma as seen by light microscopy does not resemble that of the carotid body. The neuroma is composed principally of groups of axons arranged in whorls surrounded by connective tissue. Compared to the carotid body, the neuroma contains few cells.

TORRANCE How do you propose that propagation through the cell takes place?

MITCHELL I do not know. I doubt that it is by the generation of an action potential as I mentioned earlier.

BISCOE Do you ever find efferent activity after supra-petrosal section? We never tested for that.

MITCHELL We never tested for it.

BISCOE And you said that after supra-petrosal section, there was no degeneration at 10 days whereas the infra-petrosal section did cause degeneration, and you say that there is no true motor response at 10 days?

MITCHELL We selected this period, 10 days, from the results of experiments in rats.

BISCOE But how is it that if you don't see degeneration you can say that there is no true motor response at 10 days? Was that on the assumption that 10 days is long enough for things to degenerate?

McDONALD Yes, we learned from studying the rat carotid body that afferent and efferent nerves ending on glomus cells, preganglionic nerves ending on ganglion cells, and sympathetic nerves ending on blood vessels degenerate within about 2 days of the severance of their axons from their cell bodies.

BISCOE So am I right in assuming that what you mean is that there is no change in structure or physiology at 10 days?

MITCHELL Yes.

HESS When you carried out a decentralization of the petrosal cells did you find degenerated synapses on the ganglion cells within the carotid body?

McDONALD Yes, at one day after cutting the nerve. But at 10 days the endings were absent.

HESS So if you have seen degenerated synapses or other evidence of degeneration on ganglion cells within the carotid body following supra-petrosal section then those must be parasympathetic cells.

McDONALD Yes, that was our conclusion.

BISCOE What I am a little confused about is the relationship between your results to the previous work on the efferent pathway. These workers have been claiming that there is an efferent pathway going into the carotid body from which they can record. I understand now there is a claim that the efferent pathway has an effect which is not artifactual and which is not vascular. And I further understand from your diagram that your efferent pathway is parasympathetic.

FILLENZ But isn't that efferent?

BISCOE But they are saying that their parasympathetic is a *vascular* pathway.

FILLENZ I understood that they are saying that there are preganglionic fibres going to Type I cells. They must be efferent.

TORRANCE As I understand it, Mitchell is agreeing entirely with Goodman in that he is accepting that there may be artifacts but he is saying that he has not got them; he is accepting that there is a vasomotor mechanism through the carotid sinus nerve; he is accepting that when you have an artifact from excitation by stimulus escape, there may be a complicated process of antidromic depression set up. So what Mitchell is talking about is what Goodman calls antidromic depression but Mitchell says that antidromic depression is an aspect of a more interesting phenomenon than Goodman supposes it to be.

BISCOE But how do you explain Willshaw's findings?

TORRANCE I am not trying to explain anybody else's findings, I am just saying that I don't think that Mitchell and Goodman are that far apart in that they are both discussing the effects of antidromic impulses in sensory fibres.

WILLSHAW The blood flow changes which I believe Goodman sees may be caused by antidromic stimulation of the low-threshold chemoreceptor afferent fibres because I do not see this when I stimulate the efferents in what I believe to be an orthodromic manner.

O'REGAN I don't think there is really any dispute about the fact that there are vascular efferents. I must reiterate, however, that you can miss a post-ganglionic parasympathetic innervation of glomus cells and if there is such an animal this might explain some of the results.

BISCOE But you are saying that there is an efferent pathway in the sinus nerve which produces these effects. Now where is the efferent pathway in your picture?

MITCHELL This is one efferent pathway. This other pathway is parasympathetic and is vascular.

BISCOE That I understand. What I am asking is: where is the non-artifactual and non-vascular efferent pathway? Willshaw's evidence certainly suggests that there must be such an efferent inhibitory pathway which presumably is in the sinus nerve.

ABBOUD Of course it is still possible that the changes that Willshaw sees are vascular, that is to say they could be due to a redistribution of blood flow within the carotid body.

WILLSHAW I agree but certainly the changes in overall blood flow were very small, of the order of 5 μl per minute but even with this I still observed inhibition of afferent discharge.

TORRANCE The ending in Mitchell's model is functionally the same as the classical axon reflex in that you have the point on it at which it responds to a stimulus and another point on it where it produces an effect upon another cell. Ordinarily of course, you think of an axon reflex having a sensory ending at which impulses are set up and then are conducted along a nerve fibre to affect a blood vessel; but I do think Mitchell's model works like an axon reflex. If you stimulate a dorsal root you can get a vasodilatation in the periphery: in Mitchell's model antidromic impulses cause inhibition in the carotid body.

BISCOE But this still doesn't explain the efferent non-vascular, non-artifactual depression.

TORRANCE It seems to me that O'Regan and Willshaw are on the right lines for studying whether that really does exist for they are setting up efferent impulses in the carotid sinus nerve by methods which do not involve stimulating it electrically and we have heard a lot from Goodman about the risks of electrical stimulation close to recording electrodes.

O'REGAN May I remind you, Dr Biscoe, of experiments which you did in which, after you sectioned the sinus nerve, 3 months later you found apparently normal endings. Dr Hess repeated these experiments and he suggested that these were post-ganglionic parasympathetic. This is in fact what I have assumed and would agree with. You haven't shown this but two other workers have and I think that it is possible that you did miss these endings, particularly because there is so much branching of afferent fibres.

MCDONALD I missed the logic of that last sentence.

O'REGAN Let us start from a supra-petrosal section. The afferent fibres do not degenerate but there are so many endings compared to the relatively few fibres that even if the efferent fibres did degenerate, you might not notice it.

HESS They could most logically be post-ganglionic parasympathetic fibres because with a supra-petrosal section, the axons of parasympathetic ganglion cells, even if the fibres run in the sinus nerve would still be intact because transneuronal degeneration does not occur.

MCDONALD One reason we took out the cervical sympathetic ganglion was to learn if efferent axons that persist on glomus cells after the sinus nerve is cut enter the carotid body with the sympathetics. I think that you have to remove the sympathetic ganglion and cut the sinus nerve simultaneously to show that nerves arise from parasympathetic ganglion cell bodies intrinsic to the carotid body.

HESS But of course these endings could be sympathetic and of course they would survive supra-petrosal section of the sinus nerve.

GOODMAN It seems to me from this answer that we are back to square one. Can I change the direction of the question. May I try and elicit an answer to the problem that although acetylcholine is proposed as a transmitter for this very powerful inhibition, it normally excites the carotid body chemoreceptors? We know from Wood's demonstration that peroxidase can go anywhere; I assume that acetylcholine can also go anywhere.

WOODS I certainly can't answer that one. I don't know if it helps but in looking at the rabbit, I have seen no evidence of post-ganglionic parasympathetic endings on glomus cells. There are some fibres like that around blood vessels but there are no such endings on the Type I cells and I say this with some confidence because I have looked at this type of fibre in many organs.

McDONALD Is there any evidence that a sensory axon under normal conditions can conduct action potentials in either direction? In other words, is it possible that sensory neurons conduct information from the carotid body to the brain *and* from the brain to the carotid body?

WILLSHAW If that was so how would the central end of the afferent fibre know that the periphery wasn't setting an action potential up?

TORRANCE I would refer you back to the axon reflex. A powerful vasodilation is brought about by stimulating electrically the peripheral end of the cut dorsal root. If, however, the dorsal roots are connected to the spinal cord and all the ventral roots are cut, it is not possible by afferent stimulation to bring the dorsal root vasodilator mechanism into play. That is a purely peripheral mechanism. But there are cells, for example in the retina, in which the discharge can go in any direction.

REFERENCES

REESE, T. S. & SHEPHERD, G. M. (1972). Dendro-dendritic synapses in the central nervous system. In *Structure and function of synapses*, pp. 121–36, eds PAPPAS, G. D. and PURPURA, D. P., Raven Press.

ZAPATA, P., HESS, A. & EYZAGUIRRE, C. (1969). Reinnervation of carotid body and sinus with superior laryngeal nerve fibers. *J. Neurophysiol.*, **32**, 215–88.

VASCULAR AND METABOLIC ASPECTS OF CHEMORECEPTORS

On the morphology of the vascular system of the carotid body of cat and rabbit and its relation to the glomus Type 1 cells

BY ELLEN SEIDL

Max-Planck Institut für Systemphysiologie, Dortmund, West Germany

It is a well-known saying that the function of an organ is closely connected to its morphology. But when examining the carotid body we recognize that between species, it can vary from being compact to being disseminated, and the essential question then arises as to how far the above general statement may be substantiated for the carotid body.

Plate 1 shows, as an example of the compact type, the vascular system of the carotid body of the cat. The vascular arrangement is shown by a scanning electron micrograph of a cast preparation (Keller, Schäfer & Lübbers, 1972). For this purpose the carotid body was prepared as described by Purves (1970) and perfused for up to ten minutes with physiological sodium chloride solution until it was blood-free. Then the perfusion medium was immediately switched to 'fluid Technovit 8001' a methacrylic methylester that has a polymerization time of about 60 minutes. After being macerated and cleaned, the model was mounted with silver dag on an object slide, and coated with carbon gold in the EPA100. The picture reveals most distinctly the convoluted and very complicated network of the capillary system. Furthermore, on the surface, venous branches draining into two collecting veins can be seen.

The carotid body of the cat, as shown in Plate 2, has a polar structure with an arterial and a venous side. The artery of supply runs through the well developed connective tissue which passes into the adventitia. After entering the carotid body it branches dichotomously at acute angles forming its capillary terminals, which travel through the specific tissue.

Here it should be mentioned that Goormaghtigh & Pannier (1939) did not regard the terminals as being real capillaries, but spoke of 'pre-venous segments of arterio–venous lobular anastomoses about which the cells of the paraganglion are clustered'. On the venous side, there are only irregular, thin layers of connective tissue merging with clearly shaped, broad layers of fat. A connective-tissue structure like a capsule, however, cannot be found. Besides, with this carotid body the enlargement of the venous sinus is striking, as compared with other preparations. Arterial capillaries and venous vessels are also filled with blood. A possible reason for this phenomenon is that this carotid body has been prepared during anoxia. In some cases it was observed that the carotid body of the cat was divided in sub-departments; and this type of arrangement is seen more distinctly below in the rabbit. Within the sub-departments we find, as functional units, the 'glomoids'. We have chosen this name because of the tortuous course of its vessel. Two pecularities observed with the vascular supply of the cat's carotid body should be emphasized. First, as shown on Plate 3, the lumen of the vessels can vary over very short distances. Narrow capillaries merge at acute angles into the sinus which is recognized by having a diameter ten or more times larger. Under such conditions, blood flow velocity must change considerably.

The second peculiarity was that arterio–venous anastomoses exist outside the carotid body on the arterial side. We were able to find two different kinds of A–V anastomoses: (*a*) bridge-anastomosis, and (*b*) spiral-anastomosis, as described by Staubesand (1968).

(*a*) With a cast preparation we observed a very short bridge anastomosis located immediately on the entry into the carotid body between the glomus artery and the veins situated laterally. This is illustrated in Plate 4 where the common initial part of the pharyngeal and occipital arteries is displayed. From here, two vessels proceed into the widely branched meshwork of vessels of the carotid body. At the left upper side of the figure, parts of the venous outflow are to be observed. Since the arterial inflow can be followed up to the venous outflow outside the carotid body, we conclude that there is a direct connection from the arterial supply to the venous drainage system. The picture obtained resembles strikingly a picture of an injection preparation published by De Castro (1951) (Plate 5).

(*b*) We were able to observe in about a quarter of the glomera studied a further vascular connection between the carotid body artery and the veins into which the carotid body veins are draining. These anastomoses departed immediately behind the origin of the carotid body artery and travelled by various windings through the connective tissue of the arterial pole. This vessel represents a characteristic arterio–venous anastomosis in so far as it shows the thickening of the media accomplished by epithelioid cells arranged in longitudinal vascular direction. Plate 6 shows the origin of such an arterio–venous anastomosis and Plate 7 several transverse sections made through the anastomosis with its characteristic signs.

We do not know why these anastomoses can be observed only in a part of the carotid bodies; also we do not know the conditions under which the bridge anastomosis and the tight spiral one, respectively, will be found. But from their positions, they could clearly shunt blood away from the carotid body as a whole.

The following part of this paper is concerned with the rabbit's carotid body, since the carotid body of this species is different from that of the cat. Here, large connective-tissue cords can be found separating the functional units, the glomoids, very clearly. They are called 'disseminated glomoids'.

In Plate 8 disseminated glomoids are to be seen with their specific cells as discrete vascular islands.

Since compact, lobular or disseminated carotid bodies can be found in different species, our first question has to be repeated: how far is the morphologic appearance of the total organ related to its function, always assuming that the functional role of the carotid body of cat and rabbit is similar. It could be that the glomoids are responsible for the function and that therefore they are composed in a similar way in all the different types.

Since the isolated glomoid of the rabbit is analysed more easily, we have attempted to reconstruct its vascular course and the location of its specific cells with the aid of uninterrupted serial slices, the thickness of which was 7 μm. With the aid of interference contrast according to Nomarski, the slice could be optically divided into 2–4 layers resulting in a thickness of 7/2 or 7/4 μm (Seidl, 1973).

In my experience, two types of glomoids can be distinguished:

(*a*) the multiple glomoid where a sequence of glomus cell aggregations relate to a strongly convoluted vessel,

(*b*) more rarely, the singular glomoid where only a single aggregation of glomus cell belongs to the network of vessels.

Plate 9 shows such a reconstructed system of vessels. There are bifurcations, hairpin bends, thick and thin vascular dissections, but there is no splitting into capillaries as described by Muratori (1943), where a large number of capillaries depart from one point and are recollected after having left the glomus cell aggregation.

In particular, we can see with this model – as was clear in the preparation of the cat's carotid body – that vessels having a large lumen suddenly narrow prior to merging into the sinus. A funnel-shaped vessel is to be seen merging laterally into another the diameter of which is considerably larger (Plate 10). The dark-coloured parts of the vessel demonstrate the localization of Type I cells. It seems that the specific cells are clustered on one side of the vessel. The contact areas run along the vessels as more or less large stripes (Plate 11).

The way in which specific glomus tissue relates to the blood vessel is clearly shown in Plate 12. Plate 13 shows that the glomoid of the carotid body of the cat might be constructed in a similar way. The reconstruction of a singular glomoid is given in Plate 14. Here, in addition, the specific tissue is reconstructed. Again we find that a single vessel is passing tortuously through the glomoid. Inflow and outflow vessels are running in parallel. The aggregation of specific cells meets the vessel on one side. There are four independent contact sites between vessel and cell aggregation (Plate 15). The diameter of the glomoids is about 70 to 100 μm. As our measurements have shown the area of the contact surface relates to the volume of the specific tissue as does the surface of a cube to its volume (1:6).

From our experiments we might deduce as a working hypothesis that the basic functional unit of the carotid body, the glomoid, consists of the convoluted vessel accompanied on one side by the aggregation of glomus cells Type I. Type II cells envelop the Type I cells and contact the vessels with their processes.

REFERENCES

De Castro, F. (1951). Sur la structure de la synapse dans les chemocepteurs: Leur méchanisme d'excitation et rôle dans la circulation sanguine locale. *Acta physiol. scand.*, **22**, 14–43.

Goormaghtigh, N. & Pannier, R. (1939). Les paraganglions du coeur et des zones vaso-sensibles carotidienne et cardio-aortique chez le chat adulte. *Archs Biol., Paris*, **50**, 455–533.

Keller, H.-P. Schäfer, D. & Lübbers, D. W. (1972). The structure of the network of the vessels of the carotid body of the cat and its evaluation by means of the scanning electron microscope. *Z. Naturf.* **27 b,** 1118

Muratori, A. (1943). Ricerche anatomiche sulla vascolarizzazione del glomo carotico. *Archivio dell'Istituto biochimico italiano*, **15**, 145–69.

PURVES, M. J. (1970). The effect of hypoxia, hypercapnia and hypotension upon carotid body blood flow and oxygen consumption in the cat. *J. Physiol.*, **209**, 395–416.

SCHÄFER, D., SEIDL, E., ACKER, H., KELLER, H.-P. & LÜBBERS, D. W. (1974). Arterio-venous anastomoses in the cat carotid body. *Z. Zellphysiol.* **142**, 515–24.

SEIDL, E. (1973). Die Anwendung des Interferenzkontrast-Verfahrens am gefärbten Schnitt als Hilfsmittle bei der Rekonstruktion von Organellen. *Leitz-Mitteilungen für Wissenschaft und Technik*, **69**, 57–8.

STAUBESAND, J. (1968). Zur Orthologie der arteriovenösen Anastomosen. In *Die arterio-venösen Anastomosen*, eds. HAMMERSEN, F. & GROSS, D., Berne & Stuttgart: Verlag Hans Huber.

DISCUSSION

McDonald I wonder if you could define your use of the term glomoid?
Seidl The glomoid is a functional unit and is characterized by a vessel with an aggregation of cells round it, or a longer vessel with a number of such clusters of specific cells. In the rabbit's carotid body, these clusters are clearly separated by connective tissue.
Abboud Do you see fluorescence of the A–V shunts?
Seidl No, we did not carry out histochemistry on the A–V shunt sections.
Abboud You mentioned seeing a number of vessels without red cells. I gather that they were in sections where the vessel was surrounded by Type I cells.
Seidl Yes.
Abboud Was there any structural arrangement of the vessels which accounts for this segregation of red cells, the skimming. How do you explain this?
Seidl I cannot explain this. I thought I would find a connection between arterial and venous sides and perhaps much blood in the venous side and little in the arterial. But there was no such distinction: they appeared to be independent.
Torrance Could I ask you about your measurements of the size of the vessels. What was the diameter of the A–V anastomoses and how long were they?
Seidl The 'bridge' anastomosis is smaller than the arteriole, say about 15 μm.
Torrance And the diameter of the capillary?
Seidl About 7 μm.
Torrance And what is the distance of the most remote Type I cell from the capillary?
Seidl It is variable but could be up to 20 μm. It is also worth commenting that there appears to be thickening of the capillary wall where it is in contact with Type I cells.
Pallot Have you looked at the cat carotid body in this sort of detail to define the arrangement of cells around the capillary? If so, is it similar to the rabbit?
Seidl Yes, it appears to be very similar when you take into account the known different arrangement of specific cells and capillaries in the two species.
Kobayashi What kind of resin did you use for the mould?
Seidl We used 'Technovit' methacrylic methylester 8001 (Kulzer, Bad Homburg). It sets in about one hour.

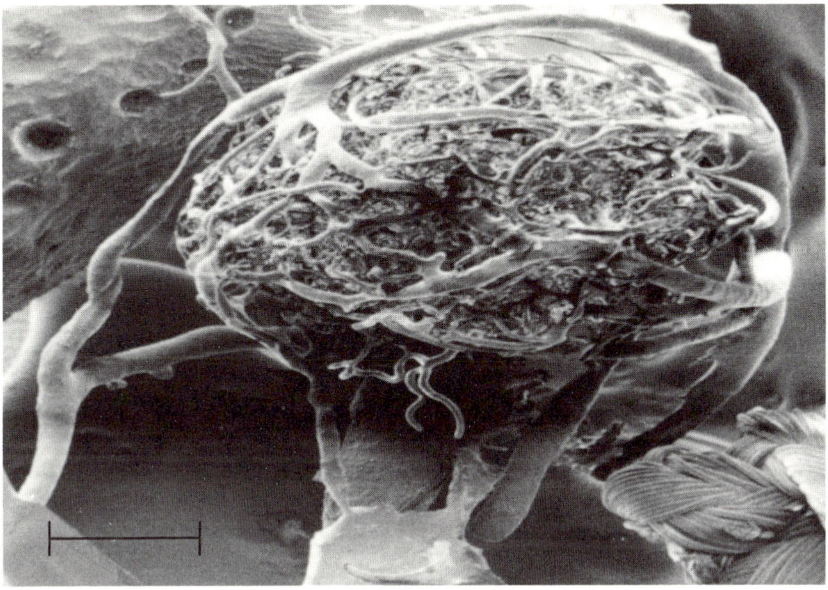

PLATE 1: Corrosion casting model of the vessels of the cat's carotid body as enlarged and demonstrated by scanning electron microscopy; bar: 500 μm.

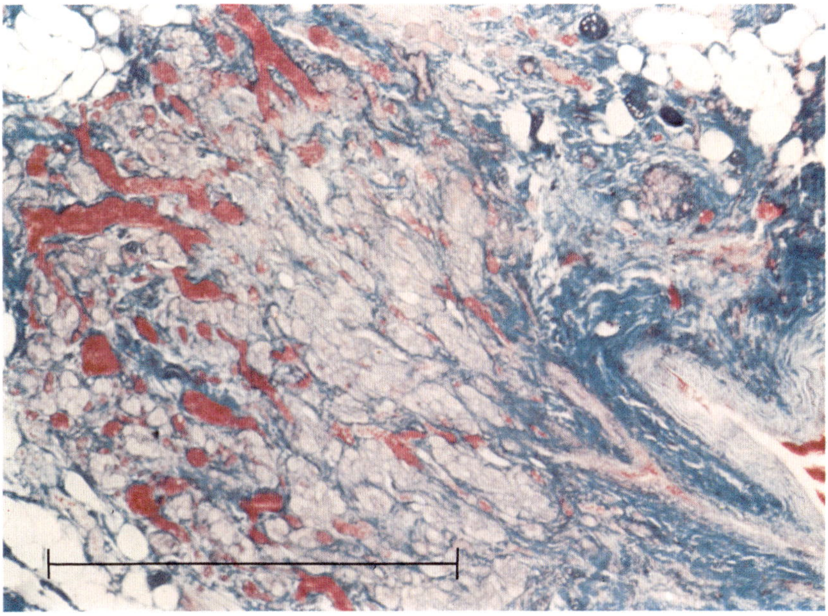

PLATE 2: Section through a cat's carotid body showing the polar structure with an arterial (right) and a venous pole (left). The dichotomous branching of the glomus artery can be seen departing from the right lower corner. Fixation: Bouin's fluid; Azan staining; bar: 500 μm.

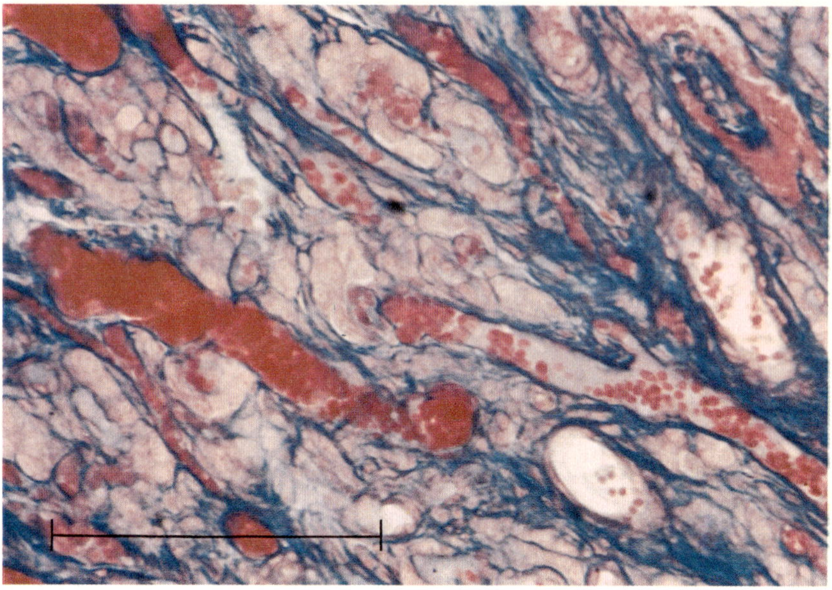

PLATE 3: Capillaries merging with sharp bendings into a blood-filled sinus. 7 μm section. Fixation: Bouin's fluid; bar: 100 μm.

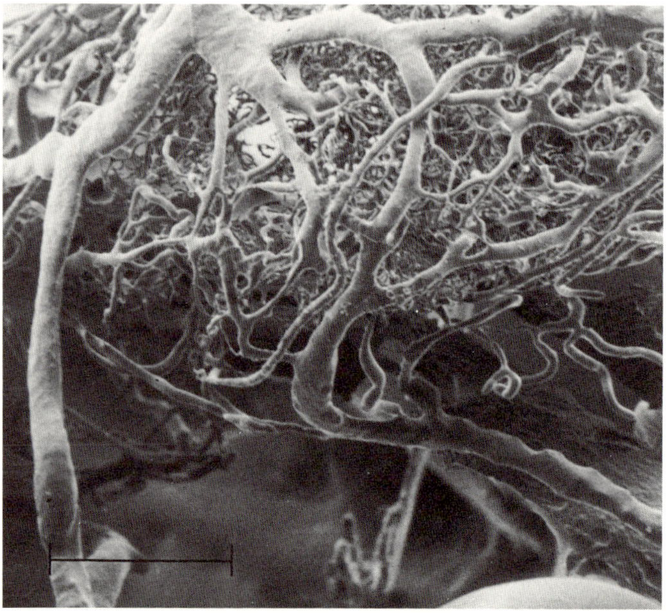

PLATE 4: Part of the corrosion casting model with arterial inflow coming from the right lower corner, and venous outflow on the left and upper side of the figure. Between both vessels a direct transition is to be seen; bar: 500 μm.

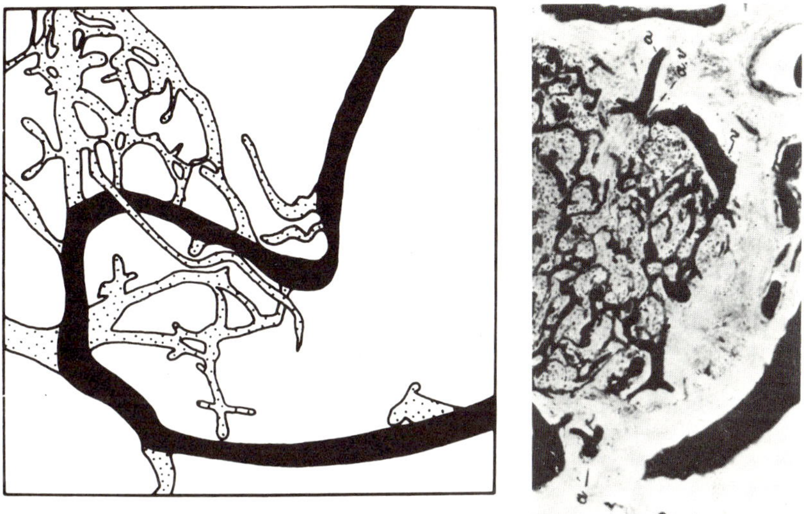

PLATE 5: Comparison of the Indian ink-injection slice of De Castro (1951) with a drawing of our corrosion casting model. The drawing shows (in black) the arterial inflow coming from the right and a short bridge anastomosis (top left) which connects artery with vein.

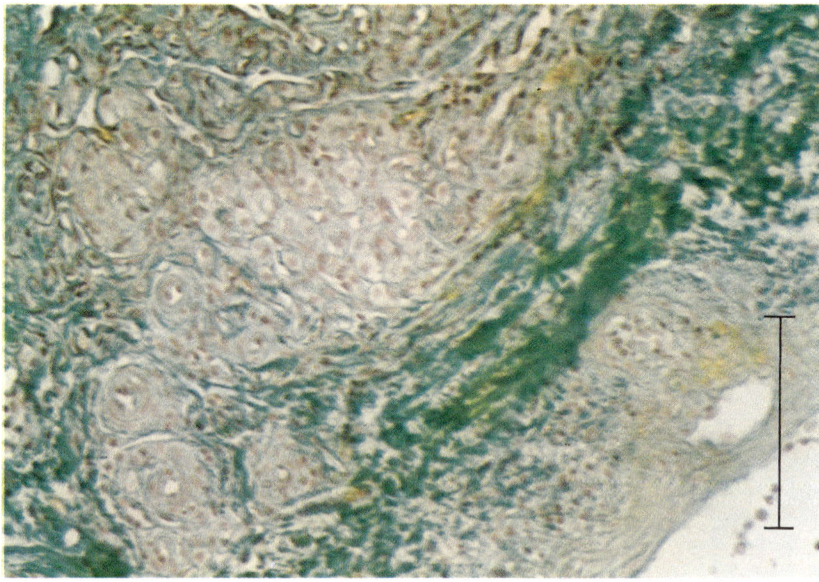

PLATE 6: Origin of the strongly wound anastomosis beside the origin of the glomoid artery. On the right and upper side: glomus tissue. Fixation: Regaud's fluid; Masson-Goldner staining; bar: 100 μm.

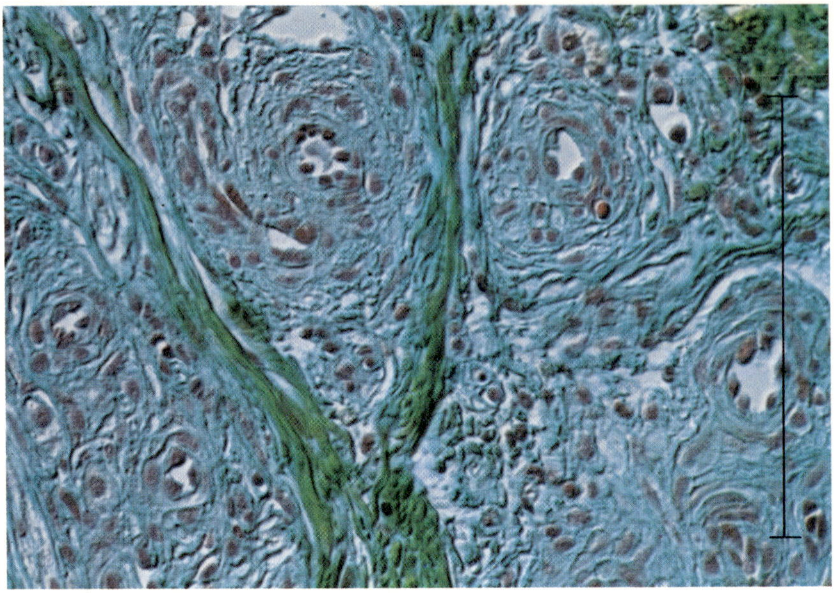

PLATE 7: Several transverse sections made through the anastomosis with characteristic endothelium and layers of epithelioid cells. Interference-contrast microscopy; Goldner staining; 7 μm section. Fixation: Regaud's fluid; Masson-Goldner staining; bar: 100 μm.

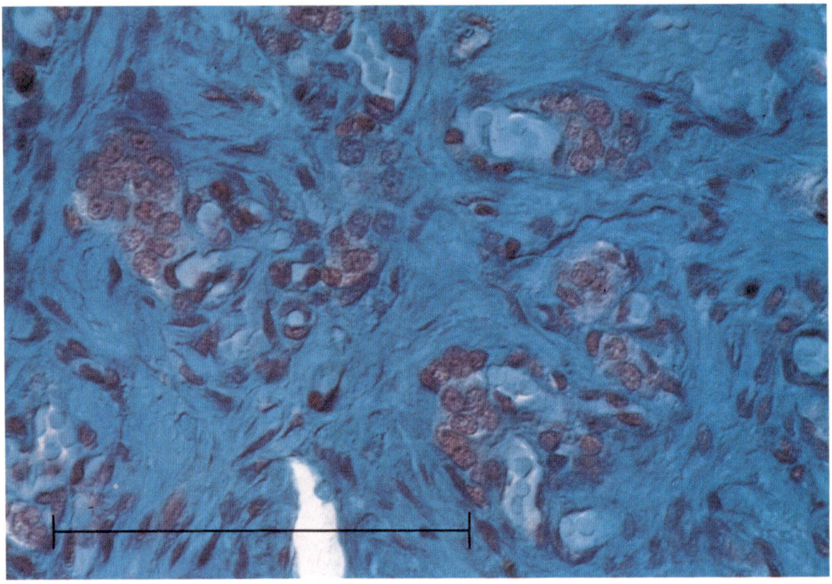

PLATE 8: Islands of glomoids (disseminated glomoids) in the carotid body of the rabbit. Interference-contrast microscopy. Fixation: Bouin's fluid; Azan staining, red: Type I cells, blue: connective tissue; bar: 100 μm.

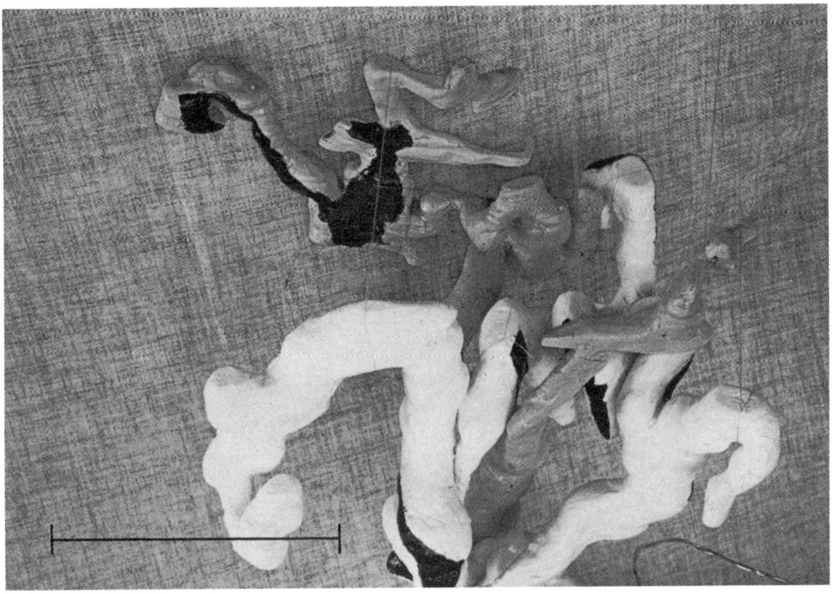

PLATE 9: Reconstruction of the complicated meshwork of vessels belonging to a multiple glomoid of the rabbit. Two independent, wound vessels (white and grey) are shown crossing each other in the same space. Black: contact areas of Type I cells. Diameter in horizontal direction: ≈ 180 μm.

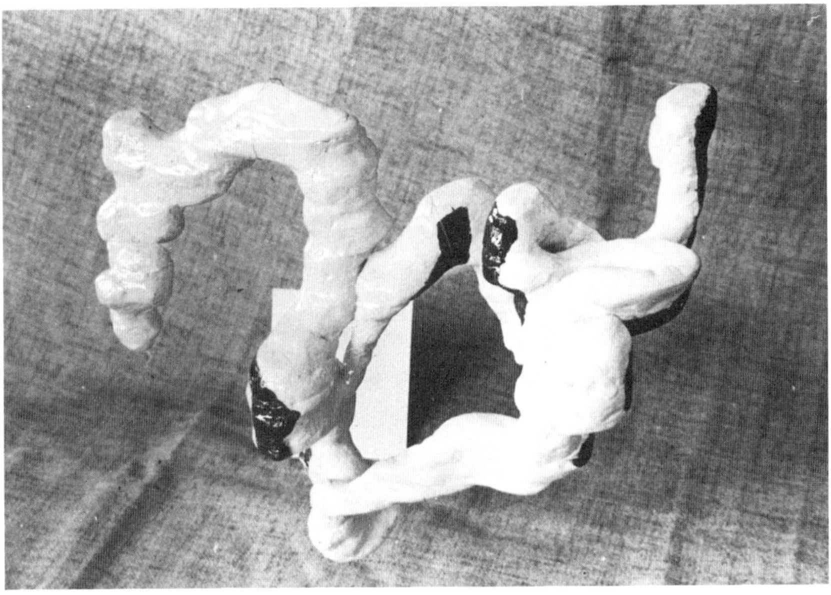

PLATE 10: Part of the reconstruction. In front of the square card, a funnel-shaped vessel enters a large vessel. Diameter in horizontal direction: ≈ 180 μm.

PLATE 11: Areas of contact (black) between glomoid vessel and the Type 1 cells.

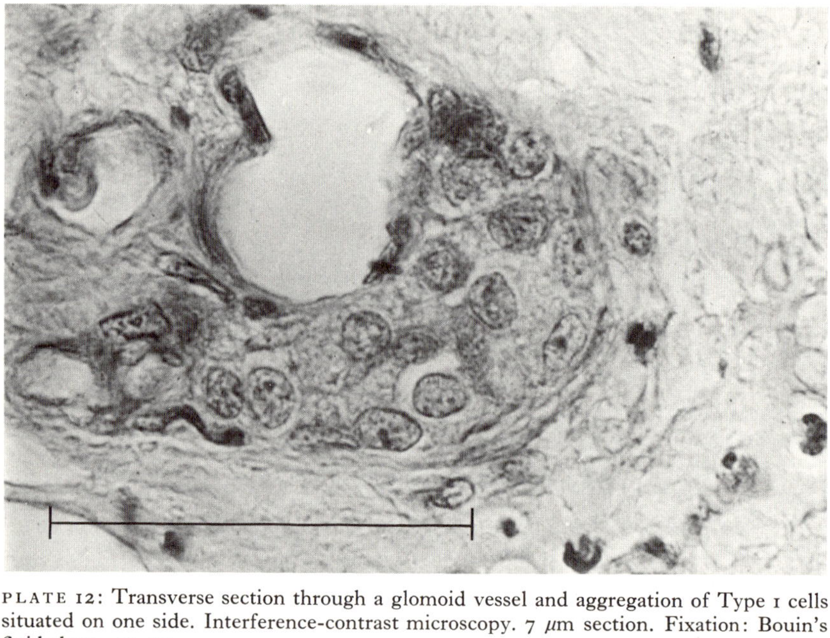

PLATE 12: Transverse section through a glomoid vessel and aggregation of Type 1 cells situated on one side. Interference-contrast microscopy. 7 μm section. Fixation: Bouin's fluid; bar: 100 μm.

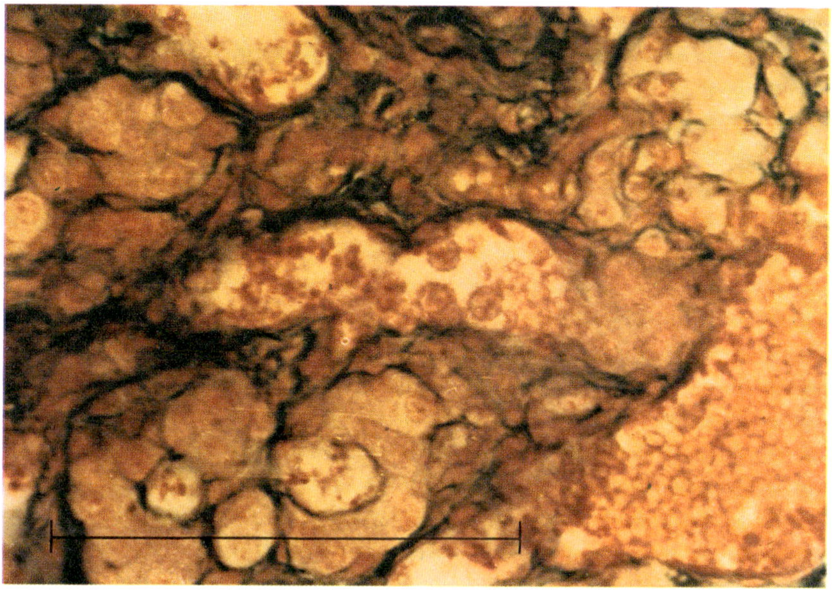

PLATE 13: Transverse section through the glomoids of a cat's carotid body. 7 μm section. Fixation: Bouin's fluid; Azan staining; bar: 100 μm.

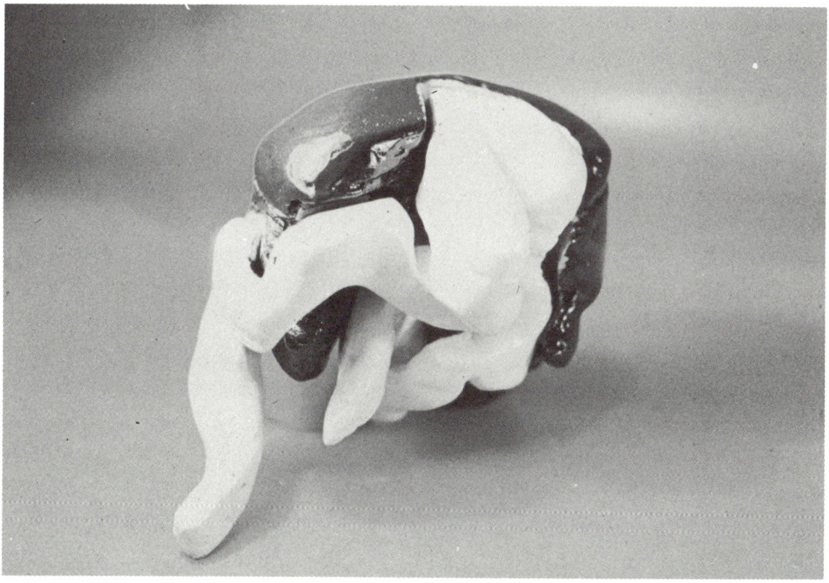

PLATE 14: Reconstruction of a singular glomoid. Diameter ~70 μm. The vessel is shown as white and the aggregation of Type 1 cells as a dark mass behind.

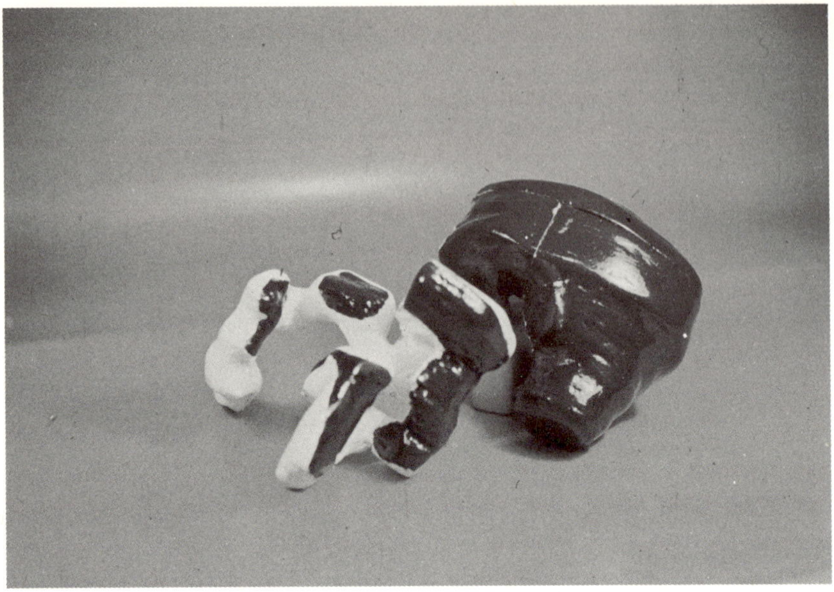

PLATE 15: Reconstruction of a glomoid showing a different pattern. On the right is an aggregation of Type 1 cells shown as a dark mass and on the left, the blood vessel (white) with black patches indicating areas of contact with further Type 1 cells.

PURVES Do I understand from your reconstruction that there are blind endings – backwaters as it were?

SEIDL It seems so but they do not have a connection with the Type I cells.

MILLS Does the post-capillary part or the venous drainage of one glomerulus pass close to the arterial inflow of another glomerulus?

SEIDL I can't answer that.

SHEPHERD It certainly raises the question of whether oxygen might diffuse directly from arterial to venous vessels.

SEIDL It may be a complex matter because, for example, I have found a number of instances where there is parallel arrangement of arterial and venous vessels.

ACKER But, as I recall, this parallel arrangement is not nearly so frequent or complete as that seen in the kidney?

SEIDL That is so.

Scanning electron microscopic observation of the fine three-dimensional distribution of the blood vessels in the frog carotid labyrinth

By S. KOBAYASHI*† and T. MURAKAMI‡

† Department of Human Morphology, University of Nottingham

‡ Department of Anatomy, Okayama University School of Medicine, Okayama, Japan

The amphibian carotid labyrinth is the characteristic swelling at the termination of each common carotid artery from which the internal and external carotid arteries arise. Whether this organ is homologous with the carotid body of mammals has long been in dispute (Adams, 1958). However, it has now been established that there occur cells which show a marked similarity in their ultrastructure to the chief cells of the mammalian carotid body (Rogers, 1963; Kobayashi, 1971*b*) and the presence of biogenic amines in the cells has been demonstrated by fluorescence histochemical method (Banister, Portig & Vogt, 1967; Kobayashi, 1971*a*). Furthermore, chemoreceptive activity similar to that of the mammalian carotid body has been demonstrated in the toad (Ishii, Ishii & Honda, 1966; Ishii & Oosaki, 1969). Thus, it is becoming increasingly evident that the amphibian carotid labyrinth may be the homologue of the mammalian carotid body. However, while in mammals only a part of the blood carried through the common carotid artery enters the carotid body, in most amphibian species all the blood in the common carotid artery passes into the carotid labyrinth (Adams, 1958). Therefore it may be assumed that this organ in the amphibian species is not solely a chemoreceptor but also plays an important role, as an effector, in the regulation of the volume of the blood which enters the brain via the internal carotid artery.

One of us has recently introduced the study of the resin-mould of blood vessels with the scanning electron microscope (Murakami, 1971). By this method, it is possible to show the three-dimensional distribution of the

* Dr Kobayashi is a Wellcome Japanese Research Fellow.

fine vasculature on a two-dimensional picture because of the deep focal length and high resolving power of the scanning electron microscope. In this study, the method of scanning electron microscopy of the resin mould was, for the first time, applied to the study of the fine internal structure of the amphibian carotid labyrinth.

MATERIALS AND METHODS

Twelve adult *Rana catesbeiana* were used. The animals were killed by pithing, pinned face up and the thorax widely opened. A thin polyethylene tube connected with a plastic syringe was inserted into the bulbus arteriosus and through it, the fixative – 2.5 % glutaraldehyde in a phosphate buffer of pH 7.4 – was injected. Some specimens were removed from the animal body immediately after the perfusion fixation, placed in a large volume of frog normal saline and then dehydrated in a graded series of acetone. They were allowed to dry in air, coated with a thin film of carbon followed by gold in a vacuum evaporator and observed and photographed in a Hitachi SSM-2 scanning electron microscope. Photographs shown in Plates 6 and 7 were prepared by this method.

For the preparation of the resin mould of the carotid labyrinth, methyl-methacrylate was supplemented with about 0.5 % of 2,4-dichlorobenzoyl peroxide as a catalyst, heated to 80 °C, held at nearly the same temperature for 5 minutes and then rapidly cooled until the viscosity of the mixture reached 70–80 centipoise (slightly less than glycerine) by the falling bowl method at 20 °C. Then benzoyl peroxide as a catalyst (1.5–2.0 %) and dimethylaniline as an accelerator (1.2–1.5 %) were added just before the injection. The half polymerized resin thus prepared was then injected into the perfusion fixed animal from the bulbus arteriosus through a polyethylene tube which was the same as that used for perfusion fixation of the animal. After the vessels were filled with the resin, the carotid labyrinth, together with surrounding tissues, was removed and dipped in a water bath in order to facilitate the polymerization of the resin at 60–70 °C for 1–3 hours or longer, placed in a concentrated solution of sodium hydroxide overnight to destroy the tissue and washed repeatedly in water to remove the corroded dusts. Then, the carotid labyrinth was dissected carefully and washing with water was further repeated to remove the dusts firmly attached to the mould. Then, the specimen was dried in air, mounted on a metal block with an electricity-conducting paint (Dotite), coated in a vacuum evaporator with a thin film of carbon followed by gold and studied in a scanning electron microscope.

One of the most important properties of the resin mould described here was its fragility. Thus we could break or cut some portion under the dissecting microscope to investigate the detailed internal structure. The specimens shown in Plates 4 and 5 were prepared by this method. Furthermore, the resin-mould became soft if it was treated with ethanol for about 3 hours at 35–40 °C so that it was possible to compress or to draw out some part of the resin-mould if required.

RESULTS AND COMMENTS

The bulbus arteriosus which originates from the ventricle of the heart divides into right and left trunci arteriosi at the anterior margin of the atrium to form a Y-junction. Each truncus arteriosus soon splits into three branches: the most caudal one is the pulmo-cutaneous artery, the middle one the aorta and the cranial one the common carotid artery. These three arteries run in a cranio-lateral direction for a considerable distance and then diverge from each other. The common carotid artery, which is the thinnest of the three, runs cranio-dorso-laterally and ends in a prominent, ovoid swelling in the axillary region. This swelling is the carotid labyrinth which in the adult *Rana catesbeiana* is approximately 2 mm long and 1.5 mm across. The external carotid artery arises recurrently from the proximal part of this organ and runs cranially. On the other hand, the internal carotid artery arises from the distal part of this organ and at first runs dorso-laterally and then turns medially.

Plates 1–5 are scanning electron micrographs of the resin-mould of the carotid labyrinth. The common carotid artery swells a little before entering the carotid labyrinth and then suddenly becomes thin to form a 'neck' as previously noted by Ishida (1954) in *Bufo vulgaris* and *Rana nigromaculata* (Plate 1). The neck is surrounded by the origin of the external carotid artery and ends in an enlarged space called the central chamber which corresponds to the central vessel of Ishida (1954) and the main chamber of Carman (1955), Plate 6.

The central chamber is about 0.5 mm long and 0.7 mm wide. Its wall is formed by a complex network of small blood vessels called the capillary plexus which communicates with its lumen by a number of small openings. These blood vessels have a relatively constant calibre of about 10 μm. Since they subsequently reunite to give off the internal carotid artery, the network formed by them is a true *rete mirabile.*

The tissue between the central chamber and the internal carotid artery (Plates 5 and 6) which contains the capillary complex, is relatively even in thickness and forms a kind of cap over the distal aspect of the central chamber.

As seen in Plates 1 and 3, in the distal portion of the capillary plexus, small blood vessels unite together to form thicker vessels. The latter join again and again to form several radicles of thicker calibre by the union of which the internal carotid artery is formed. Tributaries of the internal carotid artery remind us of the roots of a tree (Plates 1 and 3).

In the specimen shown in Plate 4, the internal carotid artery has been removed at its origin in order to show the structure covered by it. It is obvious that the internal carotid artery originates by the union of six radicles which are arranged in a horse-shoe. Thus, no radicle is seen on the central portion of the convex or distal aspect of the capillary plexus.

The external carotid artery arises by several radicles of diameter about 30 μm (Plates 1 and 5). Some of these radicles originate from the proximal portion of the capillary plexus while others, especially those in the area covered by the stem of the external carotid artery open directly in the central chamber. These radicles, after a very short course, anastomose with each other and end in trunks with thicker calibre, which form a complete ring around the neck of the common carotid artery (Plate 1). The trunk becomes thicker as it approaches the stem of the external carotid artery.

Another finding of interest is that, in the present study, small branches of blood vessels were sometimes observed to come out from various parts of the carotid labyrinth. In the example shown in Plate 1, such a branch arose from one of the radicles of the external artery. In another case, such a branch was given off from the capillary plexus.

Plates 6 and 7 are scanning electron micrographs showing the interior of the carotid labyrinth. These pictures were obtained from a carotid labyrinth perfusion-fixed with glutaraldehyde, cut into two pieces through the axis of the common carotid artery, dried and coated with carbon and gold. The termination of the common carotid artery invaginates the central chamber and is surrounded by small vessels which drain into the external carotid artery. The distal wall of the central chamber is perforated by a number of small orifices which lead to a network of small vessels i.e. a capillary plexus.

Previous studies on the fine three-dimensional distribution of blood vessels in the amphibian carotid labyrinth were carried out mainly by the use of serial sections and reconstruction. Among several papers published, the illustrations shown by Ishida (1954) and Carman (1955) are the best. These authors clearly showed that in most amphibian species, the common carotid artery ended in an enlarged space (the central chamber) which gave off a compact *rete mirabile* from which the internal carotid artery arose distally. They also pointed out that the origin of the external carotid artery was greatly different from species to species. The results of the present scanning electron microscopic study in *Rana catesbeiana* correspond essentially to those obtained by Ishida and Carman.

One of the advantages of the present study is that, by the method used here, the complicated distribution of the fine vasculature could be clearly and accurately shown on a photo-micrograph without losing the three-dimensional image. Furthermore, when the resin-mould is tilted at a suitable angle in the scanning electron microscope or cut at a suitable plane or partly broken mechanically, the originally hidden part of the specimen could be observed. This paper suggests that scanning electron microscopy of the resin-mould may provide a useful and easy method for the three-dimensional study of the fine vasculature, not only in the amphibian carotid labyrinth but also in the carotid body of mammals.

REFERENCES

ADAMS, W. E. (1958). *The comparative morphology of the carotid body and carotid sinus.* Springfield: Thomas.

BANISTER, R. J., PORTIG, P. J. & VOGT, M. (1967). The content and localization of catecholamines in the carotid labyrinth and aortic arches of *Rana temporaria. J. Physiol.*, **192**, 529–35.

CARMAN, J. B. (1955). The carotid labyrinth in *Hyla aurea*, with a note on that in *Leiopelma hochstetteri. J. Anat.*, **89**, 503–25.

ISHIDA, S. (1954). So-called carotid body of the amphibia. *Igaku-Kenkyuu, Fukuoka*, **24**, 1024–50.

ISHII, K., ISHII, K. & HONDA, K. (1966). Evidence for adrenergic transmission of the carotid chemoreceptor impulses in the toad. *Nature, Lond.*, **212**, 1499–1500.

ISHII, K. & OOSAKI, T. (1969). Fine structure of the chemoreceptor cell in the amphibian carotid labyrinth. *J. Anat.*, **104**, 263–80.

KOBAYASHI, S. (1971*a*). Comparative cytological studies of the carotid body. 1. Demonstration of monoamine-storing cells by correlated chromaffin reaction and fluorescence histochemistry. *Arch. Histol. Jap.*, **33**, 319–39.

KOBAYASHI, S. (1971*b*). Comparative cytological studies of the carotid body. 2. Ultrastructure of the synapses on the chief cell. *Arch. Histol. Jap.*, **33**, 397–420.

MURAKAMI, T. (1971). Application of the scanning electron microscope to the study of the fine distribution of the blood vessels. *Arch. Histol. Jap.*, **32**, 445–54.

ROGERS, D. C. (1963). Distinctive cell types in the amphibian carotid labyrinth. *Nature, Lond.*, **200**, 492–3.

EXPLANATION OF PLATES

PLATE 1

A scanning electron micrograph of a resin-mould of the left carotid labyrinth, seen from a nearly ventral aspect. Notice that a small branch is given off from a radicle of the external carotid artery and then turns in a cranial direction: common carotid artery (Cca), external carotid artery (Exa), internal carotid artery (Ica), capillary plexus (Cpl). × 130

PLATE 2

High-power scanning electron micrograph showing detail of the area marked by a box on Plate 1. Small, elongated dents on the surface of the resin-mould (arrows) are probably due to the presence of endothelial nuclei. × 800

PLATE 3

Ventral aspect of the left carotid labyrinth. Tributaries of the internal carotid artery are shown: common carotid artery (Cca), external carotid artery (Exa). × 160

PLATE 4

Distal aspect of the rete mirabile and the tributaries of the internal carotid artery. The main stem of the internal carotid artery has been removed in order to show the area covered by it. Notice that radicles of the internal carotid artery (R_1–R_6) are arranged in a horseshoe. × 160

PLATE 5

A scanning electron micrograph showing the cut surface of a resin-mould of the carotid labyrinth. The cut was made with a razor blade after cooling the specimen in a cryostat. Montage of three scanning electron micrographs: common carotid artery (Cca), external carotid artery (Exa), internal carotid artery (Ica). × 160

PLATE 6

A scanning electron micrograph showing the interior of the dorso-medial half of the left carotid labyrinth which was perfusion-fixed with glutaraldehyde and dried in air. Seen from the cranio-ventro-lateral aspect: common carotid artery (Cca), radicles of external carotid artery (Exa), internal carotid artery (Ica), central chamber (Cch), capillary plexus (Cpl). × 130

PLATE 7

High-power scanning electron micrograph showing the area marked by a box on Plate 6. × 700

PLATE I

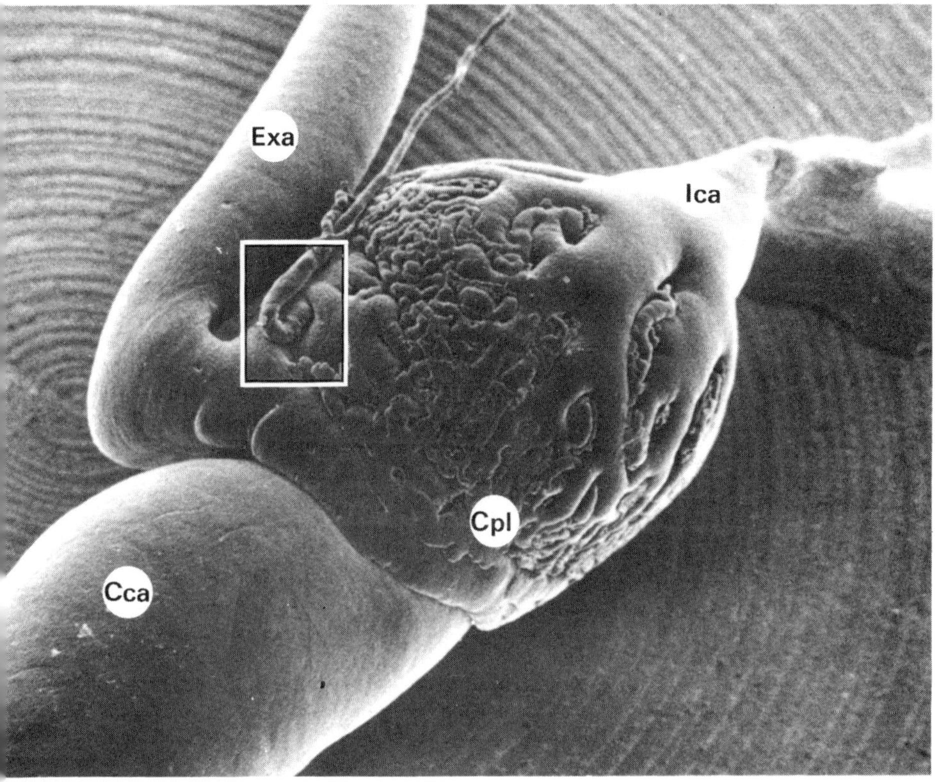

For explanation of Plate see p. 305

PLATE 2

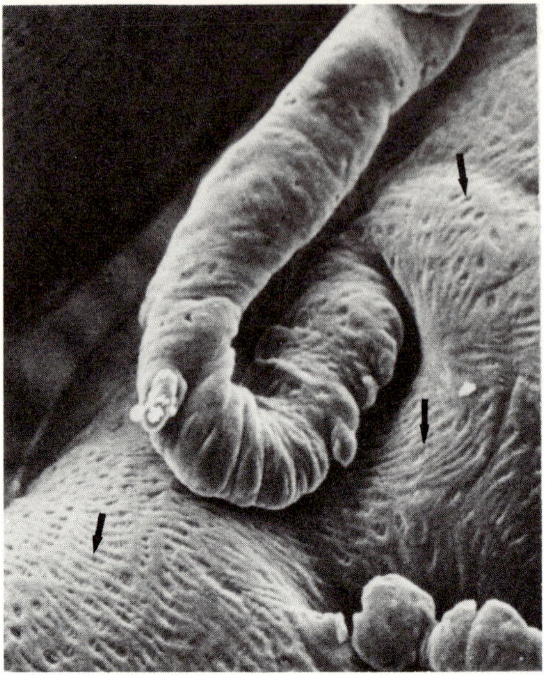

For explanation of Plate see p. 305

PLATE 3

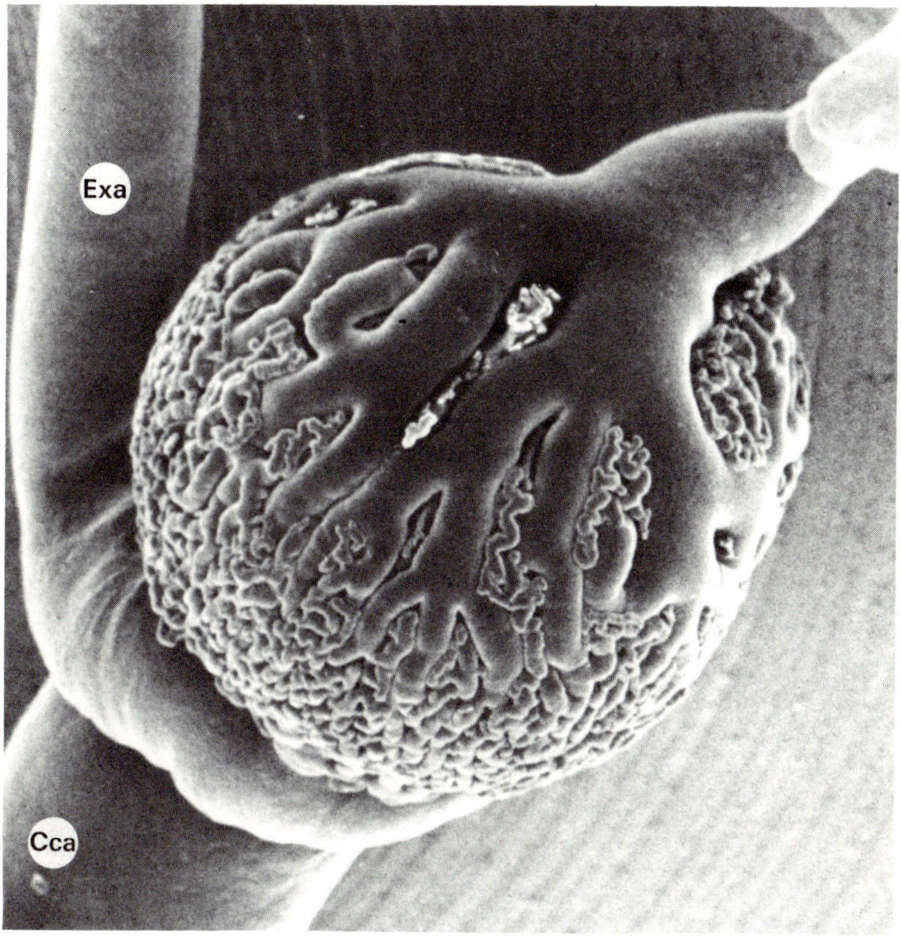

For explanation of Plate *see p. 306*

PLATE 4

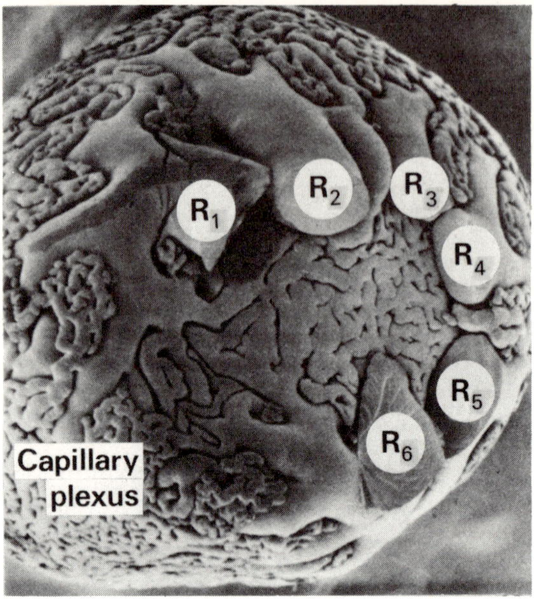

For explanation of Plate see p. 306

PLATE 5

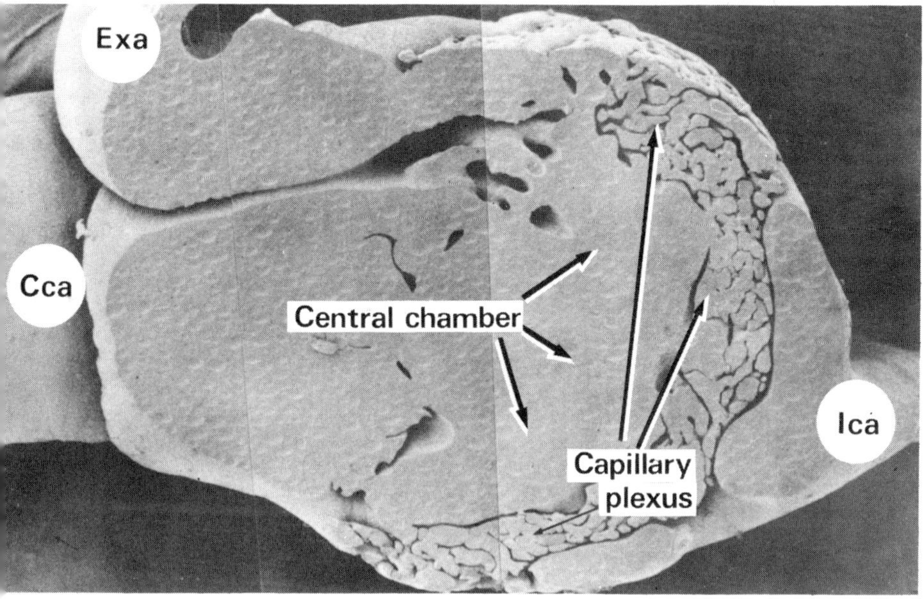

For explanation of Plate see p. 306

PLATE 6

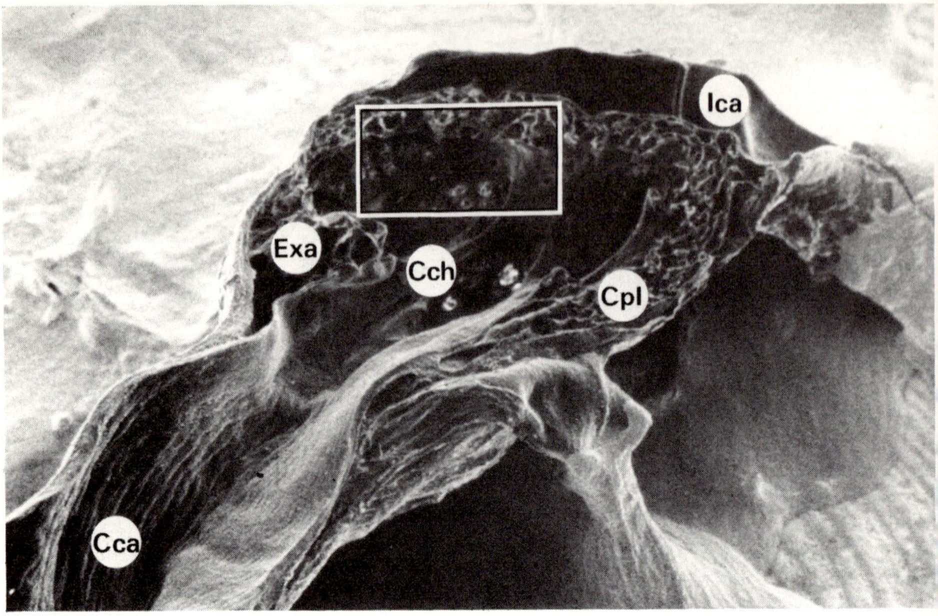

For explanation of Plate see p. 306

PLATE 7

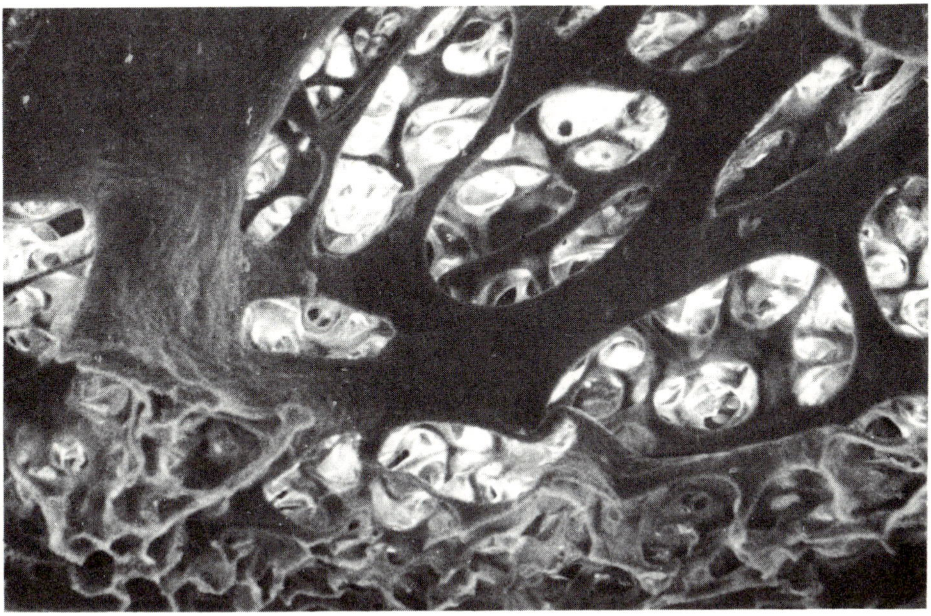

For explanation of Plate see p. 306

Vascular flows in the carotid labyrinth of *Rana*

By JEAN BANISTER, CLAIRE HERPOL* AND B. S. NAIL†

University Laboratory of Physiology, Oxford

INTRODUCTION

The functional significance of the complex vascular networks found in the carotid labyrinth of certain amphibia is unknown. Our interest in this organ was aroused by the discovery that it contained numerous elements which

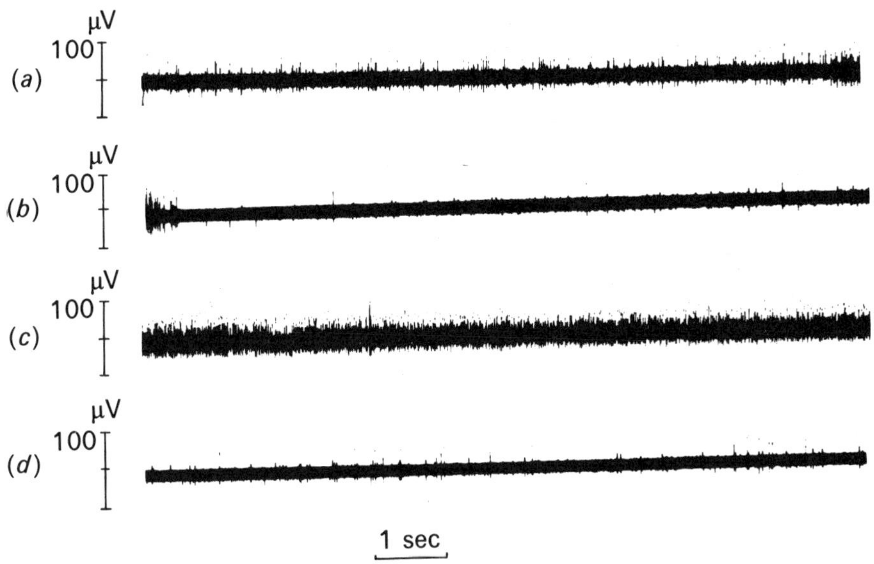

Fig. 1. Inhibitory effects of perfusion on the multi-fibre discharges recorded from the peripheral end of the nerve to the labyrinth. *Rana ridibunda* (11.8.71); pressure drop 37.5 cm perfusate.

(*a*) Discharge recorded 5 min after stopping inflow.

(*b*) Discharge when labyrinth was distended by rapid injection of 0.4 ml perfusate.

(*c*) Discharges induced by Dinitrophenol (100 μg in 0.5 ml perfusate) injected 5 minutes previously.

(*d*) Suppression of discharge from DNP within 10 seconds of the start of inflow.

* On leave from the Department of Zoophysiology, State University of Ghent, Belgium.

† Present address Department of Physiology, Bristol University, England.

fluoresced under ultra-violet light, under conditions which suggested that these elements contained rather high concentrations of catechol amines.

There is a wide distribution of fluorescence in the labyrinth. The elements were found to be cells, scattered in clusters throughout the intervascular stroma of the channels converging to form the two exit arteries (Banister, 1968). Ishii, Honda & Ishii (1966) succeeded in recording chemoreceptor discharges from the nerve to the labyrinth from Japanese toads and frogs. Since then records have been obtained of chemoreceptor activity from the carotid nerve in European frogs (Banister, Ishii & Ishii, unpublished observations).

Fig. 1 shows the remarkable sensitivity of discharges, recorded from the carotid nerve, to changes in the conditions of perfusion. Since so little information was available relating to the conditions of circulation through the carotid arterial tree and, since the relation between vascular flows and chemoreceptor activity from the labyrinth was not known in detail, we decided to undertake some studies of pressure–flow relations under very simple hydrodynamic conditions. It seems quite likely that the fluorescing cells might be responsible for controlling blood flow through the vascular networks instead of acting primarily as chemoreceptors.

METHODS

We isolate the carotid system of arteries of a pithed frog, having prepared the sympathetic trunk between ganglia 1 and 2, the jugular ganglion, and nerves IX and X before cannulation. We cannulate the common, external and internal carotid arteries respectively and perfuse the vascular beds at steady pressures with a modified Ringer's solution (Banister, Herpol, Johnston & Nail, 1973).

We have tested the reliability of counting the number of drops as an estimate of outflow under varying conditions of resistance to outflow in a model, a glass T-tube made from precision capillary tubing into which specimen cannulae are sealed. Fig. 2 shows that the relation between outflow and drop number remains consistent over the range of pressures tested. It also shows that outflow from one cannula (circles) changes by only 15% when outflow through the second channel is halved (triangles) by increasing the resistance using another cannula (open triangles).

We had some difficulty in guessing how the outflow resistance of the two vascular beds compare *in vivo*, since the internal carotid artery gives origin to branches which anastomose with arteries which stem from both the pulmonary-cutaneous and the aortic trees. The peripheral vascular bed supplied by the external carotid, however, is that of a normal systemic

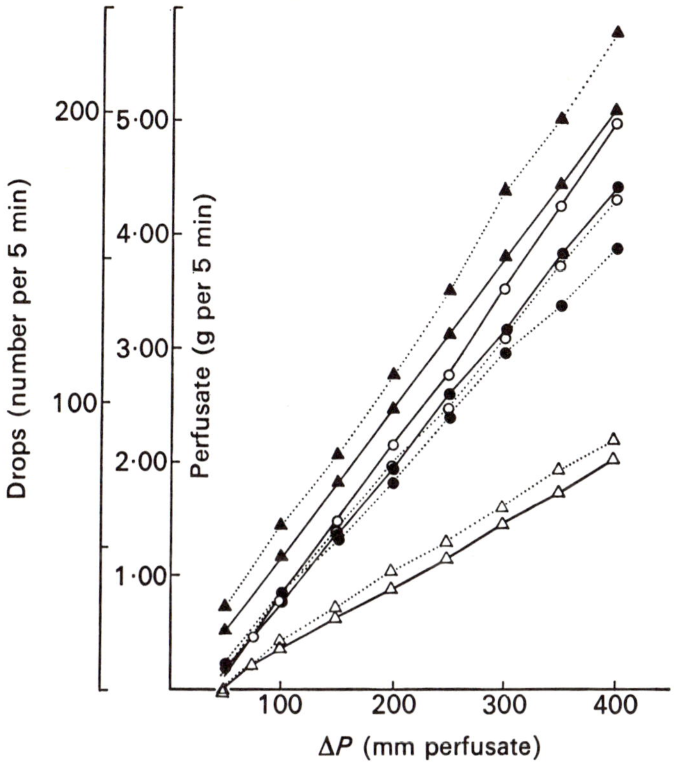

Fig. 2. The relation between pressure gradient, ΔP (mm perfusate) and flow plotted as weight (g) of perfusate or number of drops formed per 5 minute intervals obtained from a model experiment in which the carotid labyrinth was replaced by a glass T-tube to which the inflow and outflow cannulae were fixed. Curves from model 1, closed symbols; from model 2, open symbols. Triangles, external carotid flow; circles, internal carotid flow, g per 5 minutes (continuous line), drop per 5 minutes (dotted line).

artery. We decided to use the same two outlet cannulae on all the animals of one species and to match the resistances of the two cannulae as closely as possible.

RESULTS

Pressure–flow curves

We obtained curves on *Rana ridibunda* (five animals) and on *Rana catesbeiana* (five animals). The results from two animals are given in Fig. 3 and are typical of those obtained in both species.

Over the range of pressures tested (50 to 500 mm perfusate) the curves were linear up to about 300 mm pressure. Greater outflow was obtained from the internal carotid artery than the external in four out of five cases in each species. Above 300 mm perfusate pressure, outflow from the external carotid continued to increase linearly but flow from the internal carotid

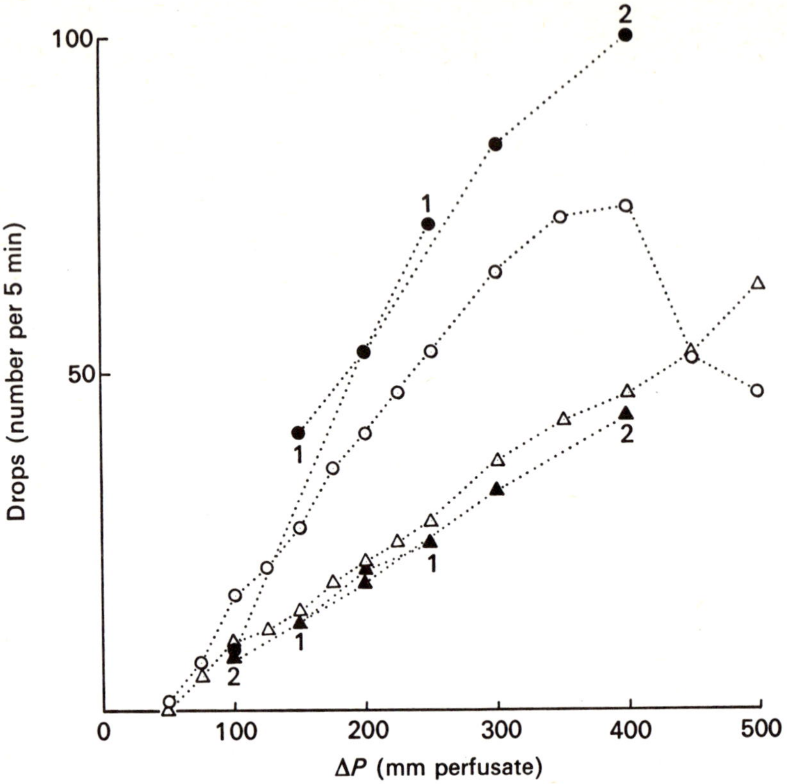

Fig. 3. The relation between ΔP (perfusate) and flow expressed as number of drops per 5 minutes, *Rana ridibunda*. Open and closed symbols denote experiments on separate animals: circles, outflow from internal carotid artery, triangles, outflow from external carotid artery. Curves labelled 1 and 2 were obtained respectively at start and finish, 2 hours later, of the experiment.

artery diminished progressively. We do not know whether the reduction in outflow is caused by the formation of interstitial oedema, auto-regulation or the blocking of channels by minute emboli. We have found little evidence of auto-regulation since we have not been able to restore the outflow to its original value by perfusing at low pressures for long periods, nor have we seen variations in drop intervals develop after sudden and large changes in perfusion pressure.

We were anxious to establish conditions under which outflow from both arteries could be stabilized for long periods before proceeding to investigations of the effects of nerve stimulation, anoxia or the infusion of drugs. Since we were able to establish stable conditions by perfusing at low pressures we studied the effects of nerve stimulation at pressures not exceeding 300 mm perfusate. The flow rates in the three experiments illustrated below varied from 7.3–12.5 litres 100 g^{-1} $minute^{-1}$. The approximate weight of a single labyrinth is 4.0 mg (*Rana catesbeiana*).

The effects of nerve stimulation

The carotid labyrinths of *Rana catesbeiana* were perfused at steady low pressures and the effects of nerve stimulation on outflow from the two arteries were compared with control periods before and after stimulation. It is claimed that the labyrinth is supplied by sympathetic fibres which join the jugular ganglion, travel on in the glossopharyngeal nerve (N. IX) finally reaching the labyrinth in a small nerve, the carotid nerve which is a branch

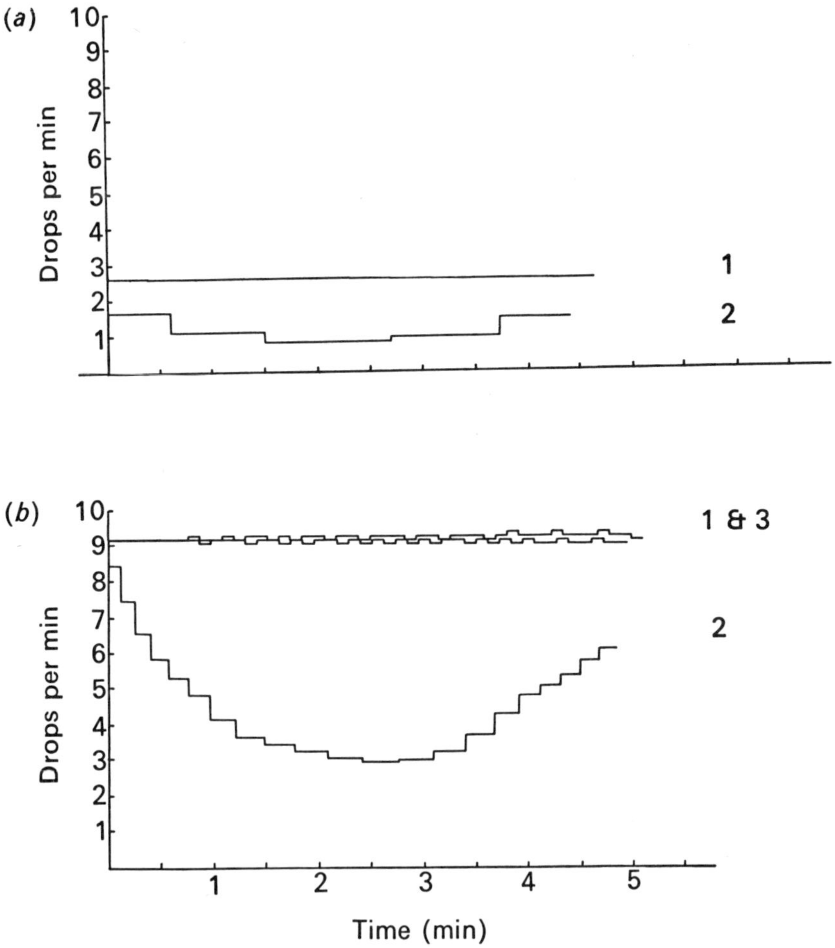

Fig. 4. The effect of stimulating the right sympathetic ganglion upon outflow from carotid labyrinth. (*a*) Flow from external carotid artery and (*b*) flow from internal carotid artery. In each panel, curves 1 or 3, control; curve 2 during and recovery from stimulation from $t = 0$ to $t = 2.5$ minutes, maximum strength at 5 Hz. *Rana catesbeiana*.

of N. IX. The carotid nerve carries fibres from N. IX as well. There could be some nervous contribution to the labyrinth from the vagus also (Ishii, *et al.*, 1966). We stimulated the sympathetic at the level of first or second ganglion in order to reduce the chance of stimulus spread to the cells and fibres of N. IX in the jugular ganglion.

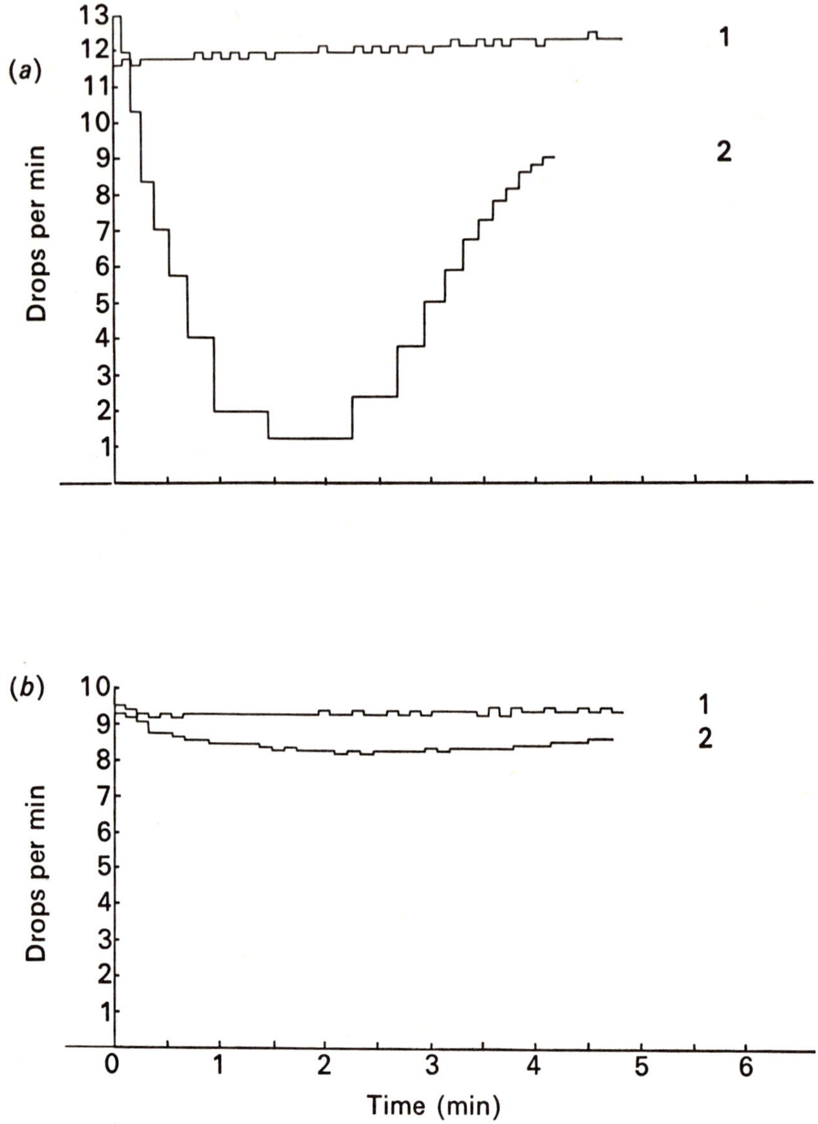

Fig. 5. The effect of stimulation of the peripheral trunk of ninth cranial nerve at maximum intensity, 10 Hz from $t = 0$ to $t = 2.5$ minutes. (*a*) Outflow from external carotid artery; (*b*) outflow from internal carotid artery. In each panel, curve 1 is control, curve 2 is during and recovery from stimulation. *Rana catesbeiana*.

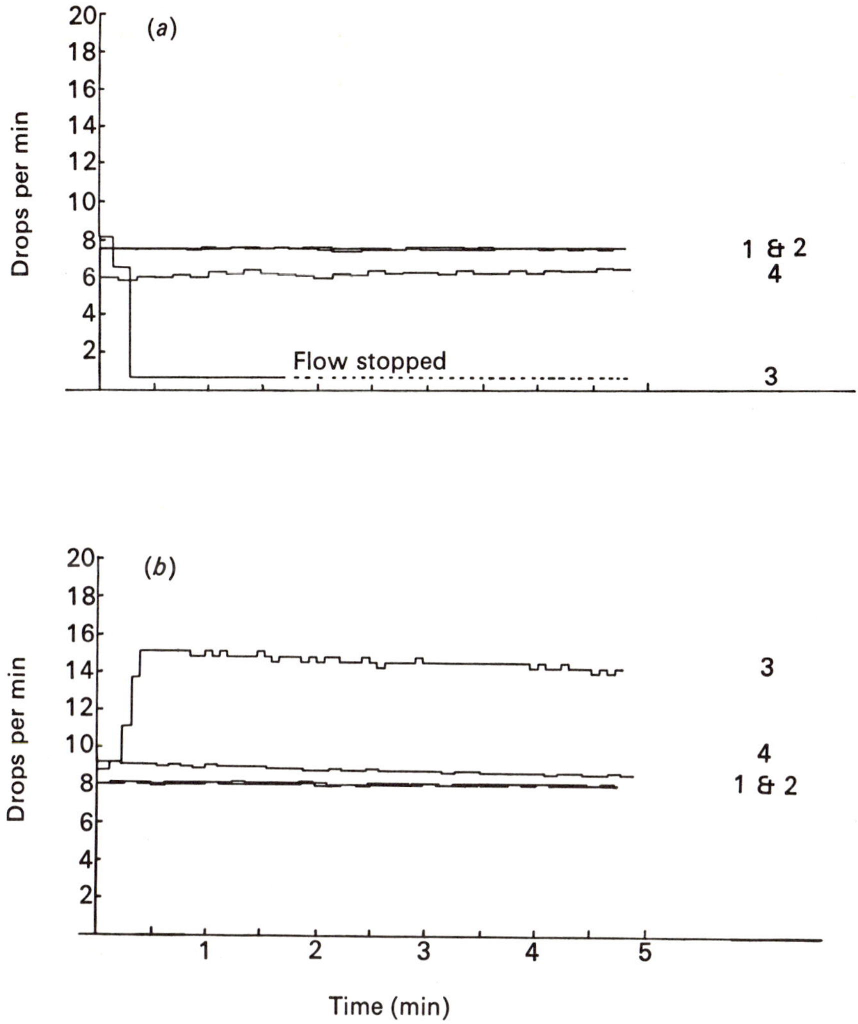

Fig. 6. Effects of stimulation of peripheral trunk of ninth cranial nerve at maximal intensity, 5 Hz from $t = 0$ to $t = 2.5$ minutes. (*a*) Outflow from external carotid artery; (*b*) outflow from internal carotid artery. For description of curves, see text. *Rana catesbeiana.*

Some results are illustrated in Figs 4–6. They were obtained from the taped records of the drop intervals which were measured and converted to instantaneous flow rates on a PDP 12A computer. Stimulation was started at time zero and was stopped after $2\frac{1}{2}$ minutes.

Stimulation of the sympathetic trunk

Fig. 4 shows an example of the marked reduction in outflow which occurred in both arteries (2) compared with control periods (1 and 3). The recovery began as soon as stimulation ceased. In some individuals we failed to obtain such marked responses in the two arteries; the outflow from external carotid artery was reduced by a greater amount than that from the internal carotid artery. Sympathetic stimulation, when N. IX was cut distal to the jugular ganglion, failed to give any response at all, which suggests that the sympathetic fibres reach the labyrinth via the carotid nerve (see Fig. 6).

Stimulation of the glossopharyngeal nerve (*N.* IX)

Stimulation of the distal end of N. IX gave results which, though consistent in any individual, varied from animal to animal. The responses were also greater than those seen on stimulating the sympathetic.

Fig. 5 shows marked reduction in outflow from the external carotid artery and a small effect on the internal: but, on some other occasions, the reverse effect was seen – marked reduction in outflow from the internal carotid artery and a small effect on the external.

Fig. 6 illustrates another type of response, which was obtained rarely. A marked increase in outflow from the internal carotid artery was obtained while outflow decreased and finally stopped from the external carotid. Curves 1 are the controls, curves 3 the stimulations and curves 4 the recoveries. Curves 2 show the effect of stimulating the sympathetic trunk when N. IX is cut distal to the jugular ganglion. In one experiment on *Rana ridibunda* (performed with Dr and Mrs Ishii), we also obtained a marked increase in outflow from the internal carotid artery and a reduction in the outflow from the external. The dilatation still occurred when the external carotid was tied off.

Stimulation of the distal end of the vagus

We frequently obtained small reductions in outflow from the internal carotid artery and sometimes small reductions in outflow from the external carotid artery also.

We shall need to do more experiments before the effect of nerve stimulation can be predicted. We have not yet been able to discover why in some animals the outflow is about half that in the others. It is in such animals that an increase in outflow on nerve stimulation may be obtained. The effects of nerve stimulation are certainly much less definite than those on the heart.

REFERENCES

BANISTER, JEAN (1968). Catechol amines in the carotid labyrinth of *Rana temporaria*. In *Arterial Chemoreceptors*, ed. TORRANCE, R. W., pp. 193–4, Oxford: Blackwell.

BANISTER, JEAN, HERPOL, CLAIRE, JOHNSTON, DAPHNE & NAIL, B. (1973). Perfusion of the vascular bed of the carotid labyrinth of *Rana catesbeiana*. *J. Physiol.*, **234**, 20–22P.

ISHII, K., HONDA, K. & ISHII, K. (1966). Function of the carotid labyrinth in the toad. *Tohoku J. exp. Med.*, **88**, 103–16.

DISCUSSION

WILLSHAW Could you tell me what is normal arterial pressure in the frog?
BANISTER About 10–20 cm blood, in pithed frogs.
O'REGAN Did you record any baroreceptor discharge?
BANISTER No: that is interesting, we never found any.
O'REGAN But even to find the sinus nerve to record from it is a very considerable achievement in the frog.
BANISTER Well the credit for that goes to the Ishiis.
ABBOUD You observed an increase in flow through the internal carotid artery when the ninth nerve was stimulated: but no responses to stimulation of the sympathetic nerve?
BANISTER Well, we did find this previously in some unpublished experiments with the Ishiis: but we also observed an increase in flow through the internal carotid artery with sympathetic stimulation: which is baffling. But, I have to confess that the nerves run very close to each other and there appears to be considerable admixture within the ganglion. Although we have tried different types and strengths of nerve stimulation, it has not been possible to get totally consistent results from IX and the sympathetic nerves.
ABBOUD Could you amplify the anatomy? The common carotid artery goes to the labyrinth and then there is branching off of the internal and external carotid arteries. It would seem that the changes in distribution of flow in each artery will depend on the resistance of the vascular beds distal to these two main branches.
BANISTER No, at least my hypothesis is that the regulation of flow is more local: that is, the proportion of flow in the internal and external carotid arteries is regulated by the nerves to the labyrinth. And I think that the fluorescent elements are quite central to this regulation.
ABBOUD If that is so, it would be unusual: I would have guessed that the flow would be determined by the behaviour of the vascular beds distal to the labyrinth.
DALY What is the weight of the labyrinth and the normal flow?
BANISTER The weight is about 4 mg and the flow about 7.5 litres 100 g^{-1} min^{-1}. But of course, this is perfused and I do not know what the actual flow is *in vivo*.
DALY But from these values, it is probable that the flow is in excess of the metabolic demands of the labyrinth.
BANISTER I would guess so.

The meaning of the tissue P_{O_2} of the carotid body for the chemoreceptive process

By H. ACKER and D. W. LÜBBERS
Max-Planck Institut für Systemphysiologie, Dortmund

It is known that the carotid body has the property of responding to a decrease in oxygen tension of arterial blood by increasing the chemoreceptor afferent discharge, see for example, Heymans & Neil (1958), Joels & Neil (1963), Torrance (1968), Biscoe (1971), Howe & Neil (1972). However, it is still unknown how this organ is able to transform a chemical stimulus into an electrical signal. If we assume that for this transducing process, tissue oxygen pressure (P_{t,O_2}) represents the essential parameter, then a study of its behaviour and reaction should permit us to obtain a better understanding of the chemoreceptive process.

THE RELATION BETWEEN ARTERIAL P_{O_2}, TISSUE P_{O_2} AND NEURAL DISCHARGE

In experiments carried out in collaboration with D. Bingmann, H. Schulze and H. Caspers of the University of Münster (Acker, *et al.* 1973), we measured arterial P_{O_2}, tissue P_{O_2} and neural discharge in parallel in order to determine whether the discharge was more closely related to P_{t,O_2} or P_{a,O_2}. It was necessary to define the area in which P_{t,O_2} was measured. It has been shown (Acker, Lübbers & Purves, 1971; Acker, Lübbers, Weigelt, Bingmann, Schäfer & Seidl, 1973) that the carotid body is surrounded by an oxygen-impermeable zone. The evidence for this is that if the tip of the oxygen micro-electrode is on the surface of the carotid body – that is a tissue which contains fat cells, some connective tissue, blood vessels and some nerves – oxygen tension is high – near or at arterial levels – and can be easily and rapidly affected by different concentrations of oxygen or nitrogen. If the electrode penetrates the tissue of the carotid body proper, the oxygen tension falls and the output of the electrode is unaffected by changes in external P_{O_2}. It appears that oxygen cannot diffuse from the outside and this has proved to be a valuable test to show in each experiment that we are actually in carotid body tissue.

All the measurements reported in this section were made in this zone of low P_{O_2} and it can be shown (Fig. 1) that if the cat is given varying con-

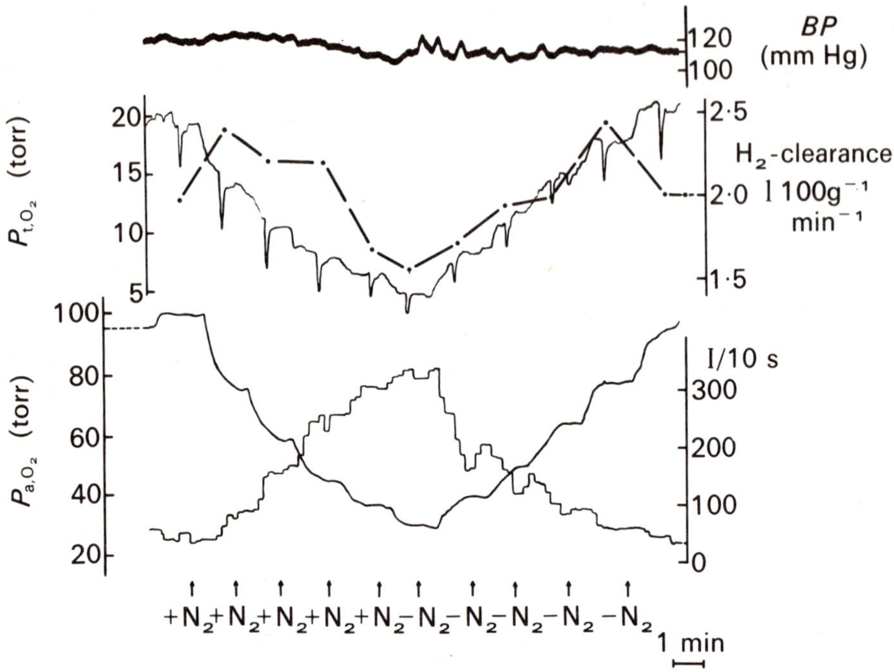

Fig. 1. Behaviour of tissue oxygen partial pressure (P_{t,O_2}) and chemoreceptor activity (I/10 sec = impulses per 10 sec) of the carotid body of the cat with changes in the inspiratory nitrogen content (arrows) that induce stepwise changes in arterial oxygen pressure (P_{a,O_2}). Furthermore, reaction of local blood flow of the carotid body (H_2-clearance), and arterial blood pressure (*BP*) are shown. Arterial P_{CO_2} (P_{a,CO_2}) is 37–8 Torr during this experiment. The dips in the P_{t,O_2} curve are caused by bolus injection of H_2-saturated Ringer solution into the common carotid artery.

centrations of nitrogen to inhale, there is a close relation between P_{a,O_2}, P_{t,O_2} and an inverse relation with chemoreceptor afferent discharge. As we have found in earlier experiments, the relation between P_{a,O_2} and P_{t,O_2} appears to be linear. Immediately following the period of hypoxia, however, the relation between P_{a,O_2} and P_{t,O_2} is altered, that is at the same P_{a,O_2}, P_{t,O_2} is increased although in most cases, P_{t,O_2} falls later to its initial value. As is also shown in Fig. 1, there is also a close inverse relationship between chemoreceptor discharge and P_{a,O_2} or P_{t,O_2}, and this relation is maintained after hypoxia. These measurements do not therefore allow us to decide whether P_{a,O_2}, local P_{t,O_2} or the total oxygen field is the important parameter for the neural output (Acker, Keller, Lübbers, Bingmann, Schulze & Caspers, 1973).

The oxygen field

Our measurements have shown (Acker *et al.* 1971) that the tissue P_{O_2} of the carotid body of the cat is distributed systematically. In the upper part of Fig. 2 tissue P_{O_2} is plotted over the depth of punctures. It can be seen that

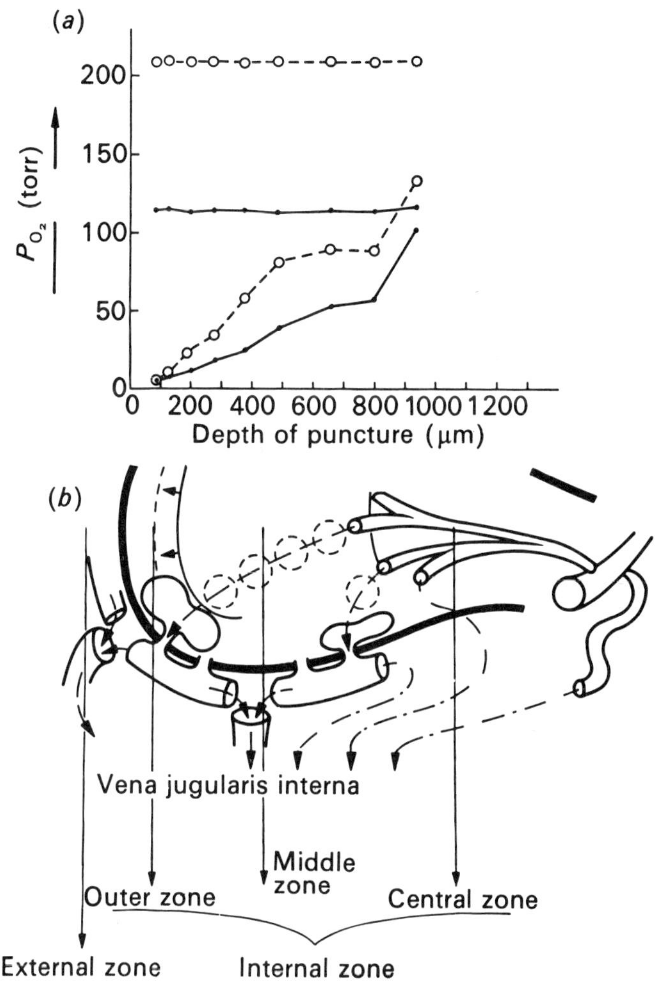

Fig. 2. Oxygen tension field in the carotid body of the cat (*a*) and schematic drawing of the corresponding tissue (*b*).

(*a*) Solid line with dots, air respiration; broken line with open circles, respiration of 40% O_2 in N_2. The upper two traces (*a*) are the record of arterial P_{O_2} throughout the experiment. Local tissue P_{O_2} increases progressively from the periphery to the centre.

(*b*) Model of the capillary network of the carotid body. We distinguish between three zones (outer, middle and central zone). The P_{O_2} values referring to these zones are given in (*a*). (From D. W. Lübbers, H. Acker, H.-P. Keller & E. Seidl (1973). *Neurohumoral and Metabolic Aspects of Injury*, ed. A. G. B. Kovach, H. B. Stones & J. J. Spitzer, Plenum: New York.)

with increasing depth of puncture tissue P_{O_2} rises from very low values up to nearly arterial ones. This systematic distribution is maintained also when arterial P_{O_2} increases, so that, in the central part the greatest increase is observed, whereas it is barely measurable in the periphery. This reaction is indicated by the dotted line. Thus, with changing arterial P_{O_2} the slope of the systematic P_{O_2} distribution is altered. It becomes steeper with increasing and flatter with decreasing arterial P_{O_2}. In the lower part of Fig. 2 (Lübbers, Acker, Keller & Seidl, 1973) we try to demonstrate in diagrammatic fashion the way in which this P_{O_2} distribution could be brought about. From right to left the inflowing artery is seen with its ramifying branches, it forms a central zone with anastomoses bypassing the tissue. We assume (Lübbers *et al.* 1973) that the capillaries have different lengths and contact different numbers of glomoids. The blood is collected in sinusoids and leaves the carotid body through the outside veins. The dash-and-dot lines show the direction of blood stream. This may explain why the venous collecting blood of the carotid body shows nearly arterial values, whereas in the carotid body tissue very low P_{O_2} values occur. Along the vascular bed aggregations of chemoreceptor cells, which we call glomoids (Seidl this Symposium), are formed. We think that the chemoreceptor cells situated in the low oxygen pressure field produce the nervous signal. The nervous output will depend on the number of cells situated in the low P_{O_2} zone. We call the low P_{O_2} zone, the outer zone; and the zone between central and outer zone, the middle zone. Outer, middle, and central zones form the internal zone which contains most of the chemoreceptor tissue. The outer zone is surrounded by the oxygen-impermeable zone. Around the oxygen impermeable zone an external zone is found which consists mainly of fat cells, connective tissue, and blood vessels and does not belong to the specific chemoreceptor tissue. To analyse and understand the systematic P_{O_2} distribution more thoroughly, we have studied in further experiments the influence of other parameters besides P_{a,O_2} which are known to influence tissue P_{O_2}, such as blood flow, oxygen transport capacity of the perfusion medium and oxygen consumption.

Local blood flow

Local blood flow is measured by the hydrogen clearance technique. Hydrogen pressure is recorded with palladinized platinum needle electrodes (tip diameter of 1 to 3 μm) (Lübbers & Baumgärtl, 1967). Hydrogen is electrolytically generated over the surface of the common carotid artery or injected (Keller & Lübbers, 1972; Acker & Lübbers, 1973). Fig. 3 shows that, with increasing depth of puncture, local flow rises from the periphery to the centre. Thus, a local flow profile exists which is analogous

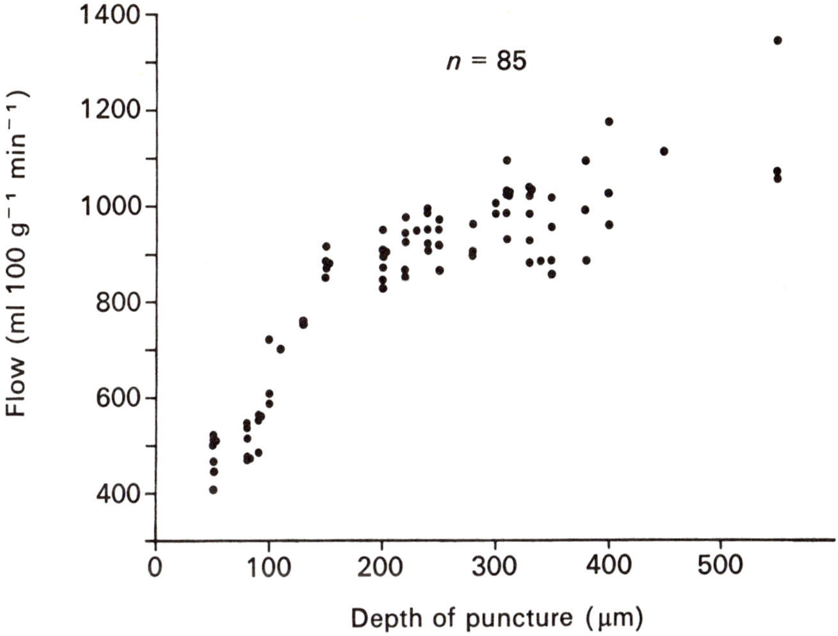

Fig. 3. Relationship between local flow through the carotid body of the cat and depth of puncture. With increasing depth of puncture local flow rose from the outer zone to the central zone: *n*, number of measurements. (From H. Acker, D. W. Lübbers (1973). *Nova Acta Leopoldina*, **211**, 277–86.)

to that observed with tissue P_{O_2}. In normoxia, local blood flow is in the range of 2000 ml 100 g^{-1} min^{-1}. In hypoxia (Fig. 1) local blood flow increases in most cases with the first nitrogen step, though blood pressure has remained nearly unchanged. At a lower oxygen level local blood flow decreases with decreasing P_{a,O_2}, though at that time arterial blood pressure exceeds its initial value. After adding oxygen to the gas mixture the rise of local blood flow is delayed with respect to the increase in P_{a,O_2} and P_{t,O_2}.

Oxygen transport capacity

Another parameter which determines tissue P_{O_2} is the oxygen transport capacity of local flow. The following method is used to obtain data concerning this question. With an appropriate preparation method (Purves, 1970) it is possible to perfuse the carotid body *in situ* alternately either with Krebs–Henseleit solution or naturally with blood. When the perfusion pressure is suddenly switched to zero the carotid body flow stops immediately. Under these conditions an oxygen pressure decrease is measured in tissue. We assume that this fall in P_{O_2} is caused by the oxygen consumption of tissue, since it can be poisoned by CO and KCN. From the slope of the relation between oxygen pressure and time measured during perfusion with Krebs–Henseleit solution, oxygen consumption (u) is calculated according to formula $u = \alpha . \frac{\Delta P_{O_2}}{\Delta t}$. (Leniger–Follert & Lübbers, 1973), where α is the solubility coefficient in ml O_2 ml^{-1} (solution), Atm^{-1}; ΔP_{O_2} is given in torr. Our experiments have shown that oxygen consumption is a function of P_{O_2}. Numerous determinations have substantiated that this function remains fairly constant. Thus, knowing tissue oxygen consumption, it is possible to measure the apparent coefficient of local solubility, α^* (Thews, 1968) under different experimental conditions. α^* is mainly influenced by hemoglobin of the perfusion medium. A quotient α^*/α of 1 indicates that no hemoglobin or a similar substance is near the tip of the electrode. In the naturally perfused carotid body α^* equals often α found for tissue perfused with Krebs–Henseleit solution. This can be seen in Fig. 4 where the quotient α^*/α is plotted against arterial oxygen pressure. It should be noted that (a) many measurements have values near 1, and (b) with decreasing P_{O_2}, more α^*/α values are found whose factor is far above 1. From these experiments we might conclude that, under our

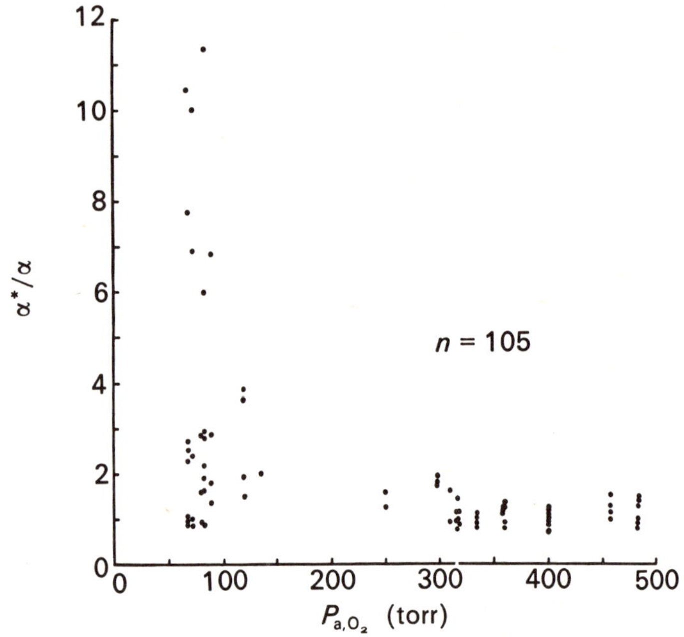

Fig. 4. Relationship between arterial P_{O_2} (P_{a,O_2}) and the quotient α^*/α. A quotient α^*/α of 1 indicates that the carotid body is mainly perfused by plasma. With decreasing P_{a,O_2} the quotient increases, i.e. the carotid body is perfused by a larger number of red cells: n, number of measurements.

experimental conditions, the number of red cells flowing through the specific carotid body tissue depends on oxygen pressure. Furthermore, it is possible to influence the quotient α^*/α by papaverine. After local application of papaverine to the surface of the carotid body, for example, the quotient rises to about 30.

Oxygen consumption

The next parameter which has been studied *in vitro* is local oxygen consumption. Fig. 5 shows the results of 47 local measurements made after

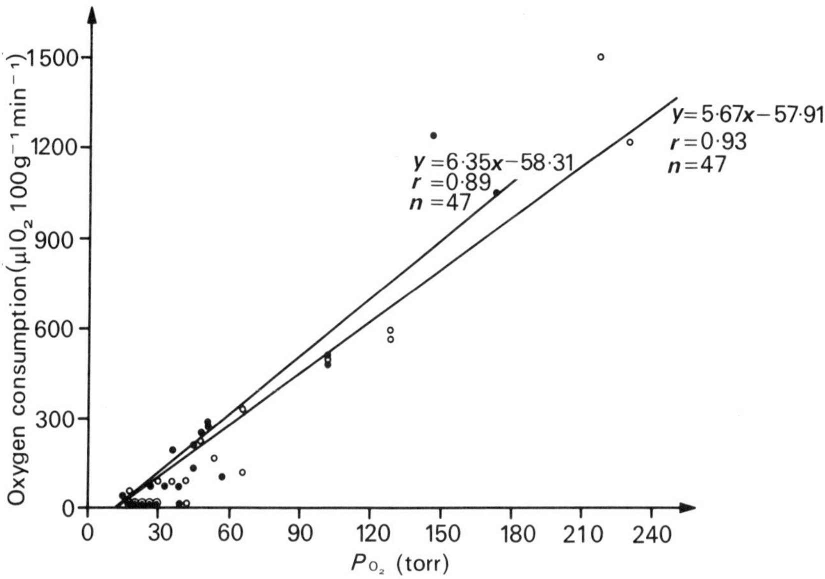

Fig. 5. Relationship between O_2 consumption (μl O_2 100 g^{-1} min^{-1}) and tissue P_{O_2} (P_{t,O_2}) in the carotid body of the cat. O_2 consumption increases with increasing P_{O_2}. Two curves are shown. One (open circles, $y = 5.67x - 57.91$) represents the mean value of measurements made in the first two seconds, the other (dots, $y = 6.35x - 58.31$) measurements made during the first 20 seconds after perfusion had been stopped. r = correlation coefficient; n = number of measurements.

perfusion had been stopped. We have found that, under our experimental conditions, oxygen consumption varies linearly with oxygen pressure. Oxygen consumption can be influenced in the following ways:

(1) It is possible to stop oxygen consumption with carbon monoxide as well as with 0.5 mM KCN;

(2) Standard poisons of the respiratory chain, such as rotenon (30 nM g^{-1} protein), amytal (2–5 mM), and antimycine (2–5 nM) do not

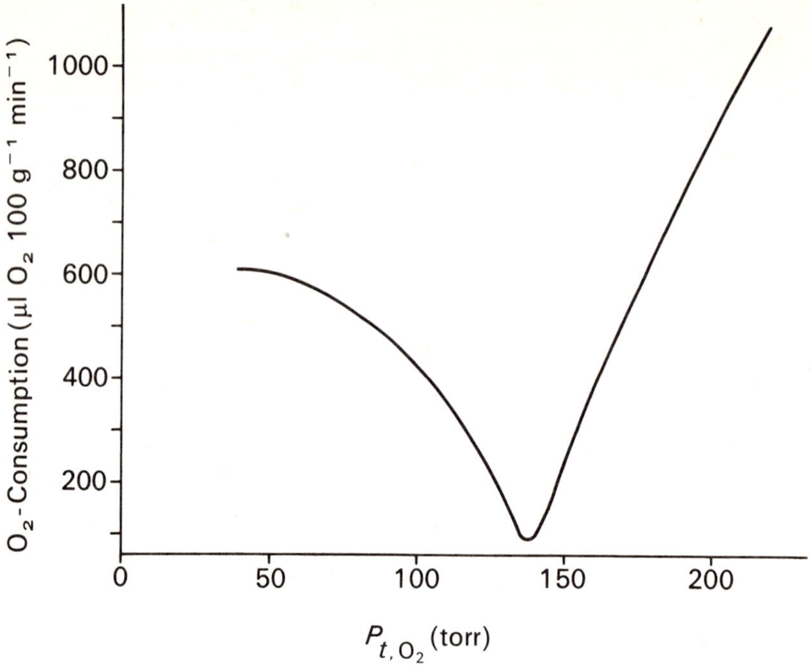

Fig. 6. O_2 consumption oscillates after poisoning with rotenon, amytal and antimycine. The linear dependence of O_2 consumption on oxygen pressure is disturbed and the relation between P_{t,O_2} and O_2 consumption now fluctuates.

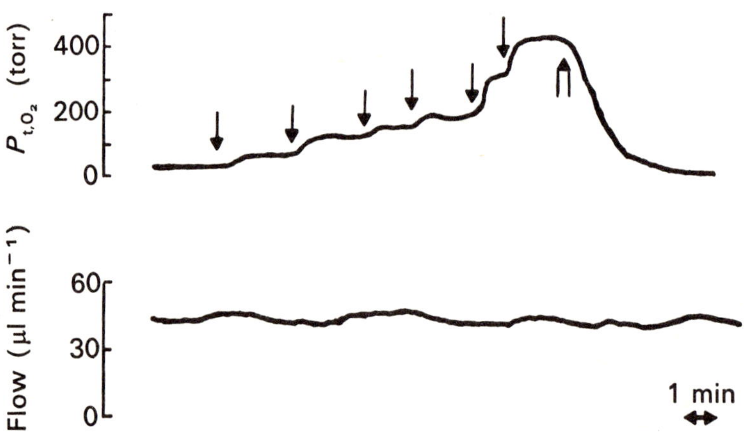

Fig. 7. Influence of Ca^{++} on tissue P_{O_2} (P_{t,O_2}) demonstrated with the hemoglobin-free perfused carotid body of the cat. With stepwise addition of a 'relaxation solution' (arrows), tissue P_{O_2} increases. After re-perfusion with Krebs–Henseleit solution (open arrow), the 'relaxation solution' is washed out and P_{t,O_2} returns to its initial value. Venous flow, measured with a micro balance, remains constant. During the experiment P_{O_2} of the perfusion medium (P_{a,O_2}) is 600 torr; CO_2 4.1 % and temperature (T) 38 °C.

abolish oxygen consumption. Fig. 6 shows that these substances can produce oscillations of oxygen consumption. This indicates that the linear dependence of oxygen consumption on oxygen pressure is disturbed, and points to more specialized oxidase reactions.

(3) Oxygen consumption and tissue P_{O_2} are found to be influenced also by calcium ions. By stepwise addition of a 'relaxing solution' it is possible to alter tissue P_{O_2} stepwise (Fig. 7). The 'relaxing solution' consists, according to Huxley (1963), of 0.1 M KCl, 10 mM ATP, 6 mM $MgCl_2$, 1 mM EGTA, 10 mM trismaleate buffer. This solution binds especially Ca^{++} ions.

These measurements are carried out on the isolated hemoglobin-free perfused carotid body of the cat (Acker, 1972). The changes of tissue P_{O_2} are accompanied by a decrease in oxygen consumption as measured by the microelectrode technique. The calcium effect can be abolished by washing out the relaxing solution and by re-perfusing with Krebs–Henseleit solution. In this case tissue P_{O_2} drops down to its initial values. Generally, our

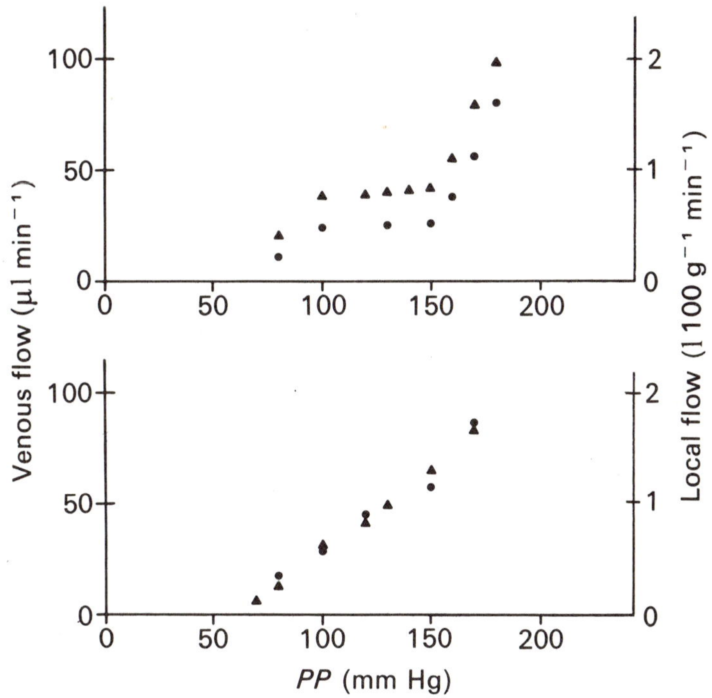

Fig. 8. Relationship between flow and perfusion pressure (*PP*) in the hemoglobin-free perfused carotid body of the cat. The points stand for local flow (ml 100 g^{-1} min^{-1}) measured with hydrogen clearance, the triangles indicate venous flow (μl min^{-1}) measured with a micro-balance. Flow autoregulation drawn on the top of the figure, is abolished by nembutal or papaverine (bottom).

experiments – *in vitro* and *in vivo* – show that tissue P_{O_2} of the carotid body of the cat decreases after adding calcium ions and increases after removing them.

Having experimentally shown the behaviour of the different parameters assumed to influence tissue oxygen tension we now discuss their actual influence on local oxygen tension under different conditions. Ideally, the signal of the carotid body should depend only on the arterial P_{O_2}. The question arises, how constant remains tissue oxygen pressure within the carotid body, if the different parameters change? We have tried to study this question by measurements *in situ* as well as in the isolated hemoglobin-free perfused carotid body.

The influence of changing flow on tissue P_{O_2}

Earlier experiments have shown that tissue P_{O_2} can remain constant, if blood pressure changes considerably (Acker & Lübbers, 1973). This could

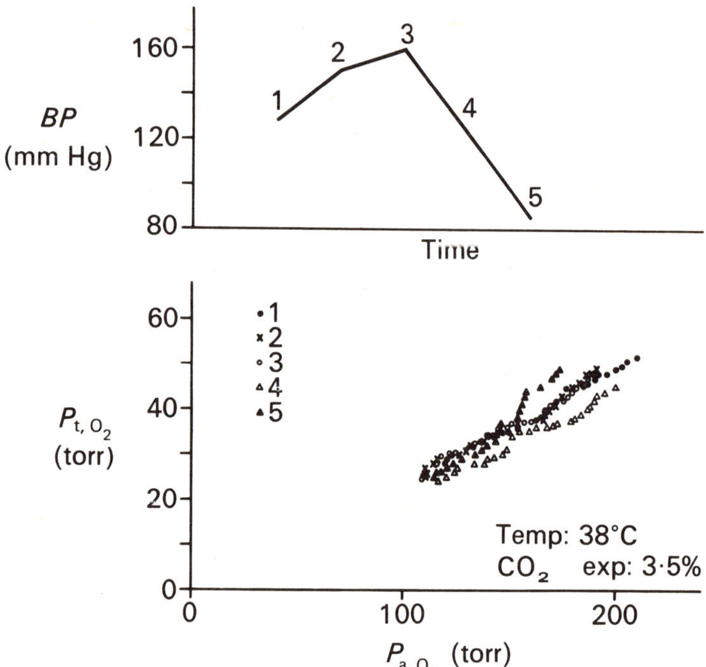

Fig. 9. Dependence of tissue P_{O_2} (P_{t,O_2}) of the carotid body on changes in blood pressure (*BP*). In the upper part of the figure the time course of blood pressure changes (1–5) is shown in an arbitrary scale. In the lower part these blood pressure changes are indicated with different symbols. The relationship between arterial P_{O_2} (P_{a,O_2}) and tissue P_{O_2} (P_{t,O_2}) remained nearly uninfluenced by the blood pressure changes. During this experiment, the point at which P_{t,O_2} is measured remained unchanged.

be achieved by flow autoregulation, i.e. a blood pressure-independent flow. Fig. 8 shows that it is possible to observe such an autoregulatory step in the isolated perfused carotid body. Local flow is measured with hydrogen clearance and total venous flow by weighing the outflowing droplets with a micro-balance, the signal of which is electrically measured. The upper part of the figure shows that with both local and venous flow, an autoregulatory step develops between 100 and 150 mm Hg perfusion pressure for venous flow at 40 μl min^{-1} and for local flow at 600 ml^{-1} 100 g^{-1} min^{-1}. By adding pentobarbital or papaverine it is possible to destroy this autoregulatory step as shown in the lower part of Fig. 8. Neither for local flow nor for the venous outflow can the autoregulatory step be seen. This

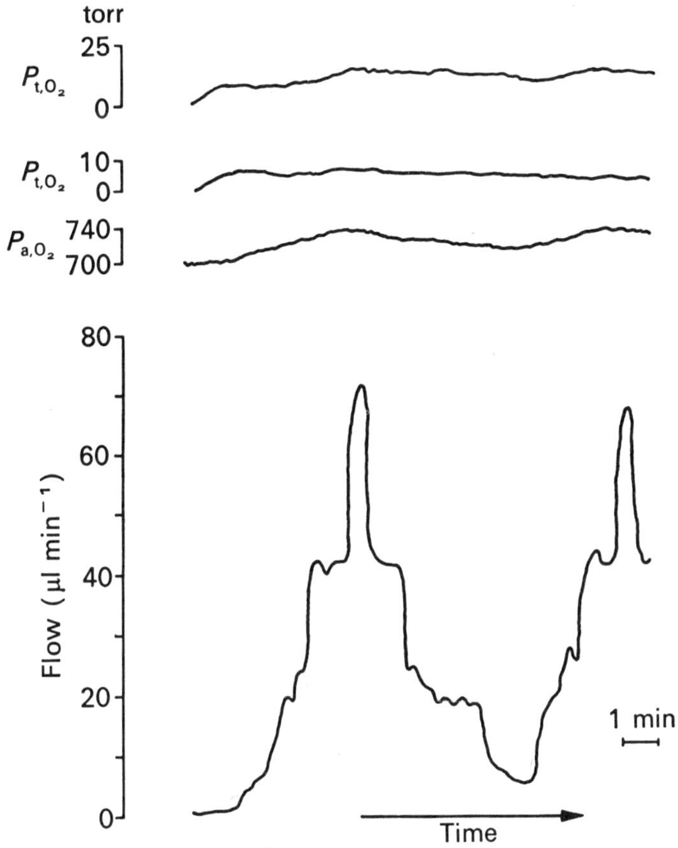

Fig. 10. Influence of flow variations on tissue P_{O_2} measured ih the hemoglobin-free perfused carotid body. From top to bottom: two traces of tissue P_{O_2} measured with needle electrodes, P_{a,O_2} in the arterial inflow and venous outflow measured with a micro balance; t = time in the direction of the arrow.

substantiates the investigations of McCloskey & Torrance (1965, 1971) who have found that pentobarbital has a decisive influence on the flow autoregulation of the carotid body. Fig. 9 shows measurements done *in vivo* with pentobarbital narcosis. We have changed blood pressure stepwise – and herewith also local flow. During the individual steps arterial P_{O_2} and the corresponding tissue P_{O_2} vary only within a narrow range. The steepness of these response curves is similar for blood pressure ranges from 80 to 150 mm Hg. This means that, with flow variations of about 200 %, tissue P_{O_2} and its dependence on arterial P_{O_2} remain fairly constant.

Fig. 10 shows the same phenomenon with the isolated hemoglobin-free perfused carotid body. We have inserted two oxygen microelectrodes into the tissue, recorded as the upper two traces in Fig. 10. The third trace corresponds to the arterial P_{O_2} of the inflowing medium. On the lower part venous flow (μl min^{-1}) is plotted. The record shows that large variations in flow have little influence upon tissue P_{O_2}. Thus simultaneous measurements of tissue P_{O_2} and P_{a,O_2} reveal that *in vivo* under pentobarbital narcosis, i.e. with the linear relationship between pressure and flow as well as *in vitro* without narcosis, i.e. with the flow autoregulatory step, tissue P_{O_2} is independent of local flow over a large range of blood pressure. Thus, flow autoregulation cannot be the reason for the constant tissue P_{O_2}. We have named this phenomenon P_{O_2} autoregulation. Since flow variation must produce changes in tissue P_{O_2}, the carotid body must in some way compensate and hold tissue P_{O_2} steady. The degree of compensation is attenuated by removing calcium ions.

Influence of changing the O_2 *transport capacity of the inflowing medium on tissue* P_{O_2}

Local tissue P_{O_2} remains in the same range when perfusion of blood is changed to Krebs–Henseleit solution provided the same arterial P_{O_2} is maintained. This means that the carotid body possesses a mechanism which is able to compensate for the changed oxygen transport capacity of the arterial medium.

Influence of changing the oxygen consumption on tissue P_{O_2}

Our experiments with changing calcium concentrations show that this change produces a pronounced change in tissue P_{O_2}. The change is brought about by a changed local oxygen consumption.

In conclusion, we may say that local tissue P_{O_2} of the carotid body can be influenced by changing arterial P_{O_2} and local oxygen consumption, but that it is rather independent of changes in local flow and oxygen transport capacity of the arterial perfusion medium. At the moment we are not able to explain these compensatory mechanisms. Perhaps in P_{O_2} autoregulation, the following reactions may proceed. Increasing flow rates 'secrete' more calcium ions into the carotid body tissue thus stimulating oxygen consumption, and vice versa. As a working hypothesis we may also assume that the normal Ca^{++} level within the cell could be related to vesicles and catecholamines. But further experimental work is necessary to understand the normal oxygen tension field of the carotid body and its regulatory processes.

REFERENCES

ACKER, H. (1973). Das Verhalten der Sauerstoff- und Flussautoregulation im isoliert perfundierten Glomus caroticum der Katze. *Pflügers Arch. ges. Physiol.*, **339**, R32.

ACKER, H., KELLER, H.-P., LÜBBERS, D. W., BINGMANN, D., SCHULZE, H. & CASPERS, H. (1973). The relationship between neuronal activity of chemoreceptor fibers and tissue P_{O_2} of the carotid body of the cat during changes in arterial P_{O_2} and blood pressure. *Pflügers Arch. ges. Physiol.*, **343**, 287–96.

ACKER, H., & LÜBBERS, D. W. (1973). Die Veränderung des Gewebesauerstoffdruckes im Glomus caroticum der Katze bei Änderung des arteriellen Sauerstoffdruckes (P_{O_2} (t)-Antwortkurve) und ihre Beeinflussung durch den Blutdruck. *Nova Acta Leopoldina*, **211**, 277–86.

ACKER, H., LÜBBERS, D. W. & PURVES, M. J. (1971). Local oxygen tension field in the Glomus caroticum of the cat and its change at changing arterial P_{O_2}. *Pflügers Arch. ges. Physiol.*, **329**, 136–55.

ACKER, H., LÜBBERS, D. W., WEIGELT, H., BINGMANN, D., SCHÄFER, D. & SEIDL, E. (1973). The oxygen barrier in the carotid body of cat and rabbit. In *Oxygen Transport to Tissue*. Vol. **37A**, 609–15. New York and London: Plenum Press.

BISCOE, T. J. (1971). Carotid body. Structure and function. *Physiol. Rev.* **51**, 437–95.

HEYMANS, C. & NEIL, E. (1958). *Reflexogenic areas of the cardiovascular system*, London: Churchill.

HOWE, A. & NEIL, E. (1972). Arterial Chemoreceptors. In *Enteroceptors, Handbook of Sensory Physiology*, vol. III, part 1, pp. 47–80, Springer-Verlag: Berlin–Heidelberg–New York.

HUXLEY, H. E. (1963). Electron microscope studies on the structure of natural and synthetic protein filaments from striated muscle. *J. molec. Biol.*, **7**, 281–308.

JOELS, N. & NEIL, E. (1963). The excitation mechanism of the carotid body. *Brit. Med. Bull.*, **19**, 21–4.

KELLER, H.-P. & LÜBBERS, D. W. (1972). Flow measurement in the carotid body of the cat by the hydrogen clearance method. *Pflügers Arch. ges. Physiol.*, **336**, 217–24.

LENIGER-FOLLERT, E. & LÜBBERS, D. W. (1973). Determination of local myoglobin concentration in the Guinea pig heart. *Pflügers Arch. ges. Physiol.*, **341**, 271–80.

LÜBBERS, D. W., ACKER, H., KELLER, H.-P. & SEIDL, E. (1973). The influence of blood pressure and blood flow on the local tissue P_{O_2} of the carotid body. In *Neurohumoral and Metabolic Aspects of Injury*, eds KOVACH, A. G. B., STONER, H. B. and SPITZER, J. J., pp. 463–73, New York: Plenum Publishing Corporation.

LÜBBERS, D. W. & BAUMGÄRTL, H. (1967). Herstellungstechnik von palladinierten Pt-Stichelektroden (1–5 μ Aussendurchmesser) zur polarographischen Messung des Wasserstoffdruckes für die Bestimmung der Mikrozirkulation. *Pflügers Arch. ges. Physiol.*, **294**, R39.

MCCLOSKEY, D. I. & TORRANCE, R. W. (1965). Autoregulation of blood flow in the carotid body. *J. Physiol.*, **179**, 37–39P.

MCCLOSKEY, D. I. & TORRANCE, R. W. (1971). Autoregulation of blood flow in the carotid body. *Respir. Physiol.*, **13**, 23–35.

PURVES, M. J. (1970). The effect of hypoxia, hypercapnia and hypotension upon carotid body blood flow and oxygen consumption in the cat. *J. Physiol.*, **209**, 396–416.

THEWS, G. (1968). The theory of oxygen transport and its application to gaseous exchange in the lung. In *Oxygen Transport in Blood and Tissue*, eds LÜBBERS, D. W., LUFT, U. C., THEWS, G. & WITZLEB, E., pp. 1–20, Stuttgart: Thieme Verlag.

TORRANCE, R. W. (1968). Prolegomena. In *Arterial Chemoreceptors*, pp. 1–40, ed. TORRANCE, R. W., Oxford: Blackwell.

DISCUSSION

EYZAGUIRRE I did not fully understand what you thought calcium ions are doing in the carotid body.

ACKER The calcium ions influence oxygen consumption.

EYZAGUIRRE Do they increase it or decrease it?

ACKER If Ca^{++} activity rises, oxygen consumption of the carotid body also rises and vice versa. The evidence for this is as follows. If Ca^{++} is given, tissue P_{O_2} falls without change of arterial P_{O_2} and blood flow and the opposite occurs when Ca^{++} ions are removed.

MITCHELL I understand that you place the oxygen microelectrode in the carotid body usually in the same area. I have two questions. First, have you observed a decrease in P_{O_2} as you pass through the central zone towards the other side and secondly, how does the tissue P_{O_2} profile look as you withdraw the electrode and how does it compare with the entry profile? I am obviously concerned about a pressure artifact from the electrode tip on capillary flow.

ACKER To puncture the carotid body of cat or rabbit is a difficult matter. We found that it was most satisfactory to puncture towards the sinus nerve pole so that in general, the deep part of the structure was explored. The distribution of P_{O_2} is quite characteristic and differs markedly from brain, for example. We were very much aware of the possibility of pressure artifact – giving an abnormal reading from the electrode on the one hand and compressing capillaries on the other. We therefore compared, as you suggested, the profile of P_{O_2} on entry with that coming out and although as would be expected, there were individual differences, the pattern was similar: that is, it was high on the surface, very low in the 50–100 μm below the surface and then progressively rose towards the centre. We also tested the possibility of compression or damage to tissue locally by recording tissue P_{O_2} in parallel with a fast response oxygen electrode placed in the carotid sinus while the animal was given high oxygen to inhale. We found that despite known small differences in response time of the electrodes, the dynamic response of the two electrodes *in situ* were virtually identical. This suggested to us that the diffusion pathways for oxygen between blood vessel and electrode tip cannot have been significantly affected. We also observed that while recording from the low P_{O_2} zone below the surface and when we gave a large dose of antimycin, the P_{O_2} increased to values near air; this suggested that the low P_{O_2} was not due simply to compression or similar artifact. We have not systematically pushed the electrode right through the carotid body to see whether P_{O_2} falls on the other side, as it were.

MILLS What is the oxygen consumption of the Krebs–Henseleit perfused carotid body compared to the blood perfused carotid body?

ACKER I cannot give the value for the whole carotid body because in our hands, the method using flow and the arterio-venous oxygen difference has not been successful and I may add that we have found that mixed venous P_{O_2} can be affected by the P_{O_2} of the surrounding bathing solution. Oxygen consumption in the haemoglobin-free perfusion was low. The comparison to the oxygen-consumption values measured by flow and A–V oxygen difference is difficult because of uncertainties about weight.

PURVES Can I add something to Dr Mitchell's question about the oxygen profile coming out and going in. I do not know about more recent experiments, but when we originally did this work, we drew largely on values obtained while withdrawing the electrode so as to be more certain about excluding pressure artifacts. It is important to emphasize this because when I took the opportunity to study Dr Whalen's technique with him last autumn, it was clear that he was having similar difficulties to those we originally encountered in penetrating the capsule of the carotid body. That is, the electrode at first dimpled the surface and if the electrode didn't break it suddenly entered the body and it was therefore difficult for him in all cases to say anything definite about the P_{O_2} values in the outer 100 μm. In the deep tissue, near the centre, he found, like us, fairly high values though certainly less than arterial and in fact, if you compare our accumulated results and remove the superficial values from ours, there is a fair correspondence. I do not know if this is the full explanation of the differences between our two papers: to resolve them will require further joint experiments. But it is worth pointing out the area of agreement which may be of physiological importance – namely that there is a marked discrepancy between mixed venous and tissue P_{O_2}.

MITCHELL This question of the P_{O_2} profile is of great importance. If the profile is as you describe it that is, low superficially rising to say 40–50 mm Hg in the centre – then the receptors in these areas must be exposed to markedly different oxygen tensions. I have rarely, if ever, encountered recruitment with hypoxia and this suggests either that the receptors must have markedly different thresholds or that there are areas with very few receptors and many receptors in other parts.

MILLS We have discussed in previous papers here (Eyzaguirre, Pallot, this Symposium) that one afferent fibre may innervate many, possibly up to ten, receptor cells. If you only had one afferent fibre going to one cell, you cannot get a chemoreceptor response curve simply by changing the area of the tissue which becomes hypoxic. But if you have one fibre supplying ten receptors, which may be scattered throughout this area, then this might explain the type of response curve which you obtain.

PURVES But I must point out that there is clear evidence from our own and other people's work that the threshold of the chemoreceptor response to hypoxia does vary considerably from fibre to fibre. Strictly, of course, this is not a threshold but a point of inflection in the response curve where the mean firing rate starts to increase rapidly. We have certainly found this in numerous single fibre response curves and it is seen particularly well if you are lucky to record from two identifiable fibres at the same time during the hypoxic period. The curves are quite different (Biscoe, Bradley & Purves, 1970).

BINGMANN I agree with Dr Purves that the shape of the oxygen-response curves can vary to a great extent. The differences between these curves are even more marked when the discharge rate is plotted versus tissue P_{O_2} rather than arterial P_{O_2}. Such differences are particularly detectable in a range between 50 and 150 mm Hg in arterial P_{O_2}.

TORRANCE Nevertheless, the basic mystery of the matter is this. The tissue P_{O_2} in the many records we have seen is low, 20, 30, 40 mm Hg but the arterio–venous oxygen difference is half a volume per cent – and if arterial P_{O_2} is 100 torr, you would expect a venous P_{O_2} of about 80 torr. So unless there is some sort of counter-current or other mechanism, there should be P_{O_2} in vessels of at least 80 torr. If these electrodes are not distorting the tissue and if it is true, as De Castro said that about one-quarter of the volume of the carotid body is blood, then a quarter of the values obtained should have been 80 torr or above. Now Whalen finds a high P_{O_2} near the surface and this falls towards the centre: you are agreed on values at the centre but there is a clear contradiction about the P_{O_2} near to the surface in published reports of experiments.

It occurs to me that the low values you obtained at the periphery may have been because the electrode tip was in fibrous tissue. How are you certain you are in glomus tissue?

ACKER All workers who have measured P_{O_2} in tissue in any organ have found that in some places P_{O_2} can be the same or lower than mixed venous P_{O_2}.

TORRANCE No, I am saying that if you are sampling randomly in an organ, 25 % of which is blood, you should get some values from flowing blood: in this case 80 torr or higher.

ACKER We think that there are several problems here. First, it may be that blood flow is not homogeneously distributed and we have some evidence for the existence of A–V anastomoses. It is quite possible that although the carotid body is very vascular, a sizeable proportion of arterial blood is shunted away from glomus tissue and the volume occupied by these shunts could be quite small. The true venous drainage from glomus tissue which might reasonably account for a much larger proportion of vessels is likely to have a lower P_{O_2} – certainly lower than 80 torr. In consequence, with random sampling, we might be more likely to find tissue P_{O_2} of say 10–20 torr and this is in fact what we did find.

But there is another point for which there is now some evidence. Under normal conditions, we think from the α^*/α quotient that more plasma than red cells perfuse glomus tissue.

TORRANCE Where do the red cells go? We have seen red cells in the vessels in some of Dr Seidl's sections.

ACKER Yes, but there is a suggestion that the distribution of red cells depends on the oxygen pressure. In hypoxia, there are masses of red cells scattered throughout vessels in glomus tissue. In normal conditions we have had sections in which it has been impossible to find a red cell.

But may I come back to the surface oxygen tension again. When we measure P_{O_2}, it is high – very close to arterial: but if we go 20–50 μm below the surface, the P_{O_2} is very low and it is possible to disrupt this barrier by various means.

MILLS This type of barrier is not unique to the carotid body. Duling (1972) has been carrying out work in the mesenteric pouch in the hamster and superfusing the vessels with solutions of varying P_{O_2}. He also finds that if his electrode tip is just below the surface of the vessels, he has great difficulty in showing a rise in P_{O_2} as he increases the P_{O_2} of the perfusate. Furthermore, in another study (Whalen & Nair, 1970) tissue P_{O_2} in skeletal muscle actually fell when the P_{O_2} of the superfusate was raised.

ACKER Yes, we have published a report (Acker & Lübbers, 1971) in which we have measured tissue P_{O_2} in the wall of the internal and external carotid artery in the cat and we have shown that at a certain depth below the surface, it is impossible to affect the tissue P_{O_2} by changing the external P_{O_2}. This is not simply a question of the time constant of the diffusion. Hydrogen diffuses into the artery and into the carotid body for example, without difficulty. This barrier can be disturbed in a number of ways and 20 minutes after the death of the animal, it is possible to raise the P_{O_2} in the carotid body easily by raising external P_{O_2}.

TORRANCE So your essential point is that there is complete plasma skimming in the carotid body and that glomus tissue is perfused by plasma, and that is why you get a low tissue P_{O_2}.

ACKER I should not put it as strongly as that. I am saying that it is a possibility and that there is some evidence for it from Dr Seidl's histological studies and from the α^*/α quotient.

REFERENCES

ACKER, H. & LÜBBERS, D. W. (1971). Eine sauerstoffundurchlässige Zone in der Wand der Arteria carotis der Katze. *Experientia*, **27**, 394.

BISCOE, T. J., BRADLEY, G. W. & PURVES, M. J. (1970). The relation between carotid body chemoreceptor discharge, carotid sinus pressure and carotid body venous flow. *J. Physiol.*, **208**, 99–120.

DULING, B. R. (1972). Microvascular responses to alterations in oxygen tension. *Circulation Res.*, **31**, 481–9.

WHALEN, W. J. & NAIR, P. (1970). Skeletal muscle P_{O_2}: effect of inhaled and topically applied O_2 and CO_2. *Am. J. Physiol.*, **218**, 973–80.

Activity of chemoreceptors in the carotid body of the cat in relation to changes of venous pressure

By D. BINGMANN, H. SCHULZE AND H. CASPERS
Institute of Physiology, University of Münster

We have observed that when a cat is being artificially ventilated, switching off the pump causes a change in carotid body chemoreceptor discharge which differs from the response of these receptors to changes in P_{a,O_2} and P_{a,CO_2}. We have carried out the following experiments to analyse this discrepancy.

The cats were anaesthetized with sodium pentobarbitone, paralysed with gallamine triethiodide and ventilated mechanically. Arterial blood gas tensions were recorded continuously by means of a thermo-stabilized loop system inserted into the femoral artery. Tissue oxygen tension (P_{t,O_2}) was measured in the carotid body from which activity was being recorded with platinum micro-electrodes using the technique of Acker, Lübbers & Purves (1971). Because of a difference in the latency of the tissue and blood electrodes and because of the difference in circulation times from lungs to carotid body and femoral artery, signals from the arterial electrodes occurred some 5–10 seconds after the signals from the carotid body. Activity from the sinus nerve was recorded from single- or few-fibre preparations and was usually displayed as a ratemeter signal.

In Fig. 1 is shown a typical response of chemoreceptors to the asphyxia which occurred after the pump had been switched off. After a latency of 3–4 seconds, chemoreceptor discharge started to rise although there had been no change in P_{t,O_2} or arterial gas tensions. This early rise in discharge rate which was abrupt was followed by a steady rise in discharge and in this latter phase, chemoreceptor activity behaved as would be expected from the recorded changes in arterial blood gas tensions. When the pump was restarted and the animal ventilated with air; there was a similar discrepancy in timing between chemoreceptor discharge and tissue or arterial P_{O_2}.

This was also observed when the respiratory pump was stopped during ventilation with 100 % oxygen instead of air. If the animal had been venti-

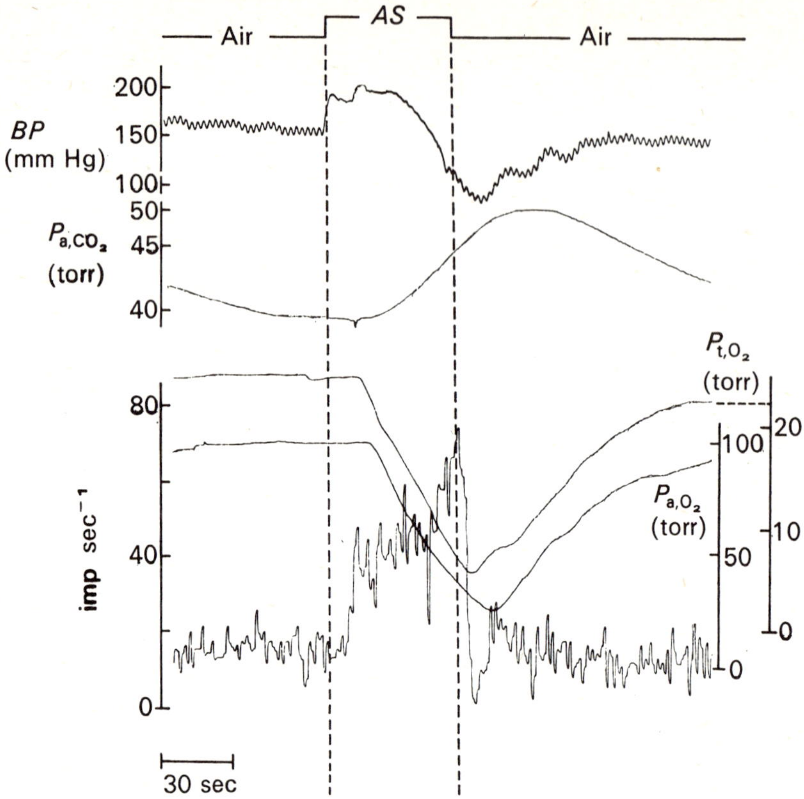

Fig. 1. Reactions of chemoreceptors to asphyxia (AS) when the respiratory pump was turned off. *BP*: arterial blood pressure; P_{a,O_2} and P_{a,CO_2}: partial pressure O_2 and CO_2 in the arterial blood measured in a femoral arterio-venous loop. P_{t,O_2}: tissue P_{O_2}, recorded in the carotid body. Imp sec^{-1}: discharge rate of action potentials recorded from a few-fibre preparation of the carotid sinus nerve.

lated with 100% oxygen for more than 30 minutes and the oxygen container left connected to the trachea (Caspers & Speckmann, 1971), the following changes in chemoreceptor discharge and blood and tissue oxygen tension occurred. Tissue oxygen tension remained at its initial level for at least 5 minutes while P_{a,O_2} slowly decreased and P_{a,CO_2} slowly increased (Bingmann *et al.*, unpublished observations). In contrast to these relatively slow chemical changes, switching off the pump or a sudden reduction in tidal volume is associated with a rapid increase in chemoreceptor discharge. As is shown in Fig. 2*a* (upper line), both the reduction in amplitude of tidal volume and switching off of the pump are associated with a reduction in tracheal pressure and there is a corresponding rise in chemoreceptor activity. Conversely, as is shown in Fig. 2*b*, if the lungs are inflated and

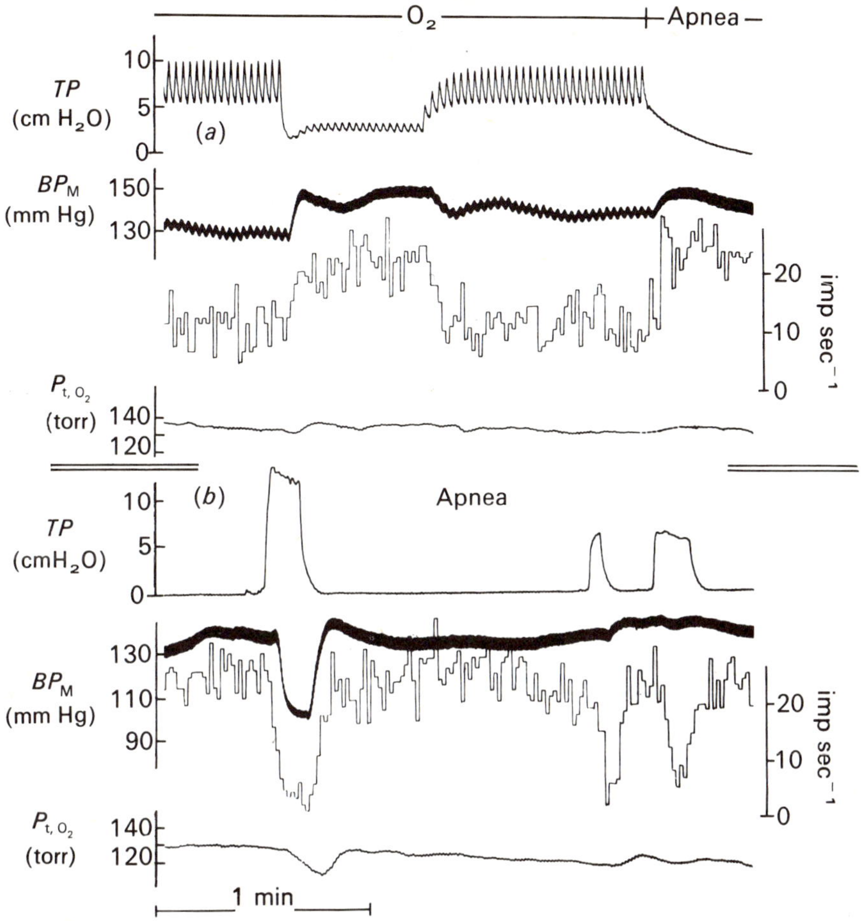

Fig. 2. Response of the rate (imp. sec $^{-1}$) of chemoreceptor afferent discharge to (*a*) a decrease of the pressure in the trachea (*TP*) and to (*b*) a rise of tracheal pressure. BP_M: Mean blood pressure; P_{t,O_2}: tissue P_{O_2}, measured in the carotid body.

pressure increased, receptor activity is rapidly inhibited although there was actually a decline in P_{t,O_2}.

These observations raise the question: How is it that discharge from the carotid body chemoreceptors rises with deflation of the lungs and falls with inflation of the lungs? We have entertained three possible mechanisms: (i) changes in blood gas tensions, (ii) reflex changes and (iii) some form of haemodynamic mechanism.

Changes in blood gas tensions

It is well known that inflation of the lungs constitutes a form of breath-holding and is associated with characteristic changes in blood gas tensions and that these in turn are accompanied by a rise in chemoreceptor discharge. In order to distinguish these changes with the inhibition seen with inflation, the following experiments have been carried out. First, we interposed an artificial delay between lungs and carotid body and observed that this did not affect the latency of the chemoreceptor response to lung inflation. Secondly, P_{a,O_2} and P_{a,CO_2} could be varied over a wide range without affecting this type of receptor inhibition. Thirdly, we observed that if the lungs were inflated with air, nitrogen or carbon dioxide, there was invariably a monophasic steep reduction of P_{O_2} in blood and carotid body and rise in P_{a,CO_2}. Yet the inflation was immediately followed by a rapid reduction in chemoreceptor discharge and this discharge only started to rise when hypoxic or hypercapnic levels had been reached. Finally, an oxygenator was inserted into the common carotid artery which ensured that any fluctuations of blood gas tensions occasioned by inflation or deflation of the lungs were suppressed. Even under these conditions, inhibition of receptor discharge by lung inflation was observed.

In summary, fluctuations of arterial blood gas tensions induced by inflation of the lungs may modify the chemoreceptor response but these fluctuations cannot by themselves explain the inhibitory reaction of the chemoreceptors to such inflation.

Reflex changes

It was clearly possible that the changes in chemoreceptor discharge to lung deflation or inflation could be mediated by a reflex pathway from lungs to carotid body. To test this, we isolated the carotid body anatomically and pharmacologically. Thus even if the carotid sinus region was totally isolated, i.e. connected to the body by plastic tubes only, lung inflation brought about the typical receptor response. Further, it was possible to show that neither atropine nor alpha- or beta-blocking agents had any marked effect upon the inhibitory effect of inflation.

Haemodynamic mechanisms

As is shown in Fig. 2, inflation or deflation of the lungs was accompanied by changes in arterial pressure which had approximately the same time course as the changes in chemoreceptor discharge. However, the inhibition of receptors also occurred when there was no change in arterial pressure (Fig. 2*b*) or when changes in arterial pressure were eliminated by the insertion of a roller pump into a loop in the common carotid artery. These experiments with the roller pump, moreover, showed that inhibition with inflation was proportionately enhanced if pressure in the carotid sinus was raised. Conversely, if carotid sinus pressure was reduced to approximately

80 mm Hg, the inhibitory reaction was abolished. The reappearance of this phenomenon occurred as arterial pressure was raised again but it did not if the hypotension had been caused by a great loss of blood. In this case, reappearance of the inhibitory response only occurred if the volume of blood was replaced. This finding pointed to the venous system and its influence on receptor activity in the carotid body.

Microscopic observation of the veins in the immediate vicinity of the carotid body showed that in parallel with lung inflation, these vessels were dilated. This response was quantified by measuring pressure in the transverse pharyngeal vein and a typical record is shown in Fig. 3. When the

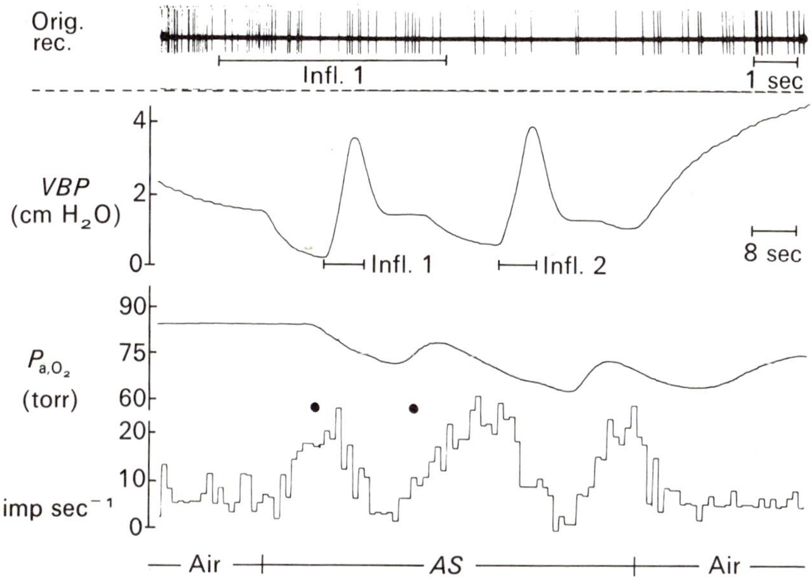

Fig. 3. Responses of the venous blood pressure (*VBP*), of chemoreceptor afferent activity (imp sec^{-1}) and of the arterial P_{O_2} (P_{a,O_2}) to a period of asphyxia (AS) and to two inflations of the lungs. The duration of the inflations (Infl. 1; Infl. 2) is marked by horizontal bars. The neurogram, above, is from a single chemoreceptor and was recorded over the first lung inflation as indicated by the dots.

respiratory pump was switched off, venous pressure decreased in parallel to the rise in discharge from a single chemoreceptor. Similarly, an increase in venous pressure induced by lung inflation is accompanied by a fall in chemoreceptor discharge. The close temporal relation between these changes suggested the possibility of cause and effect and this was tested by increasing venous pressure locally by close injections of blood, dextran (Haemaccel) or Ringer's solution. These injections were followed by a decrease in chemoreceptor activity provided that a dilatation of the veins was achieved which was similar in degree to that observed with lung inflation. A typical experiment is illustrated in Fig. 4. A cat had been venti-

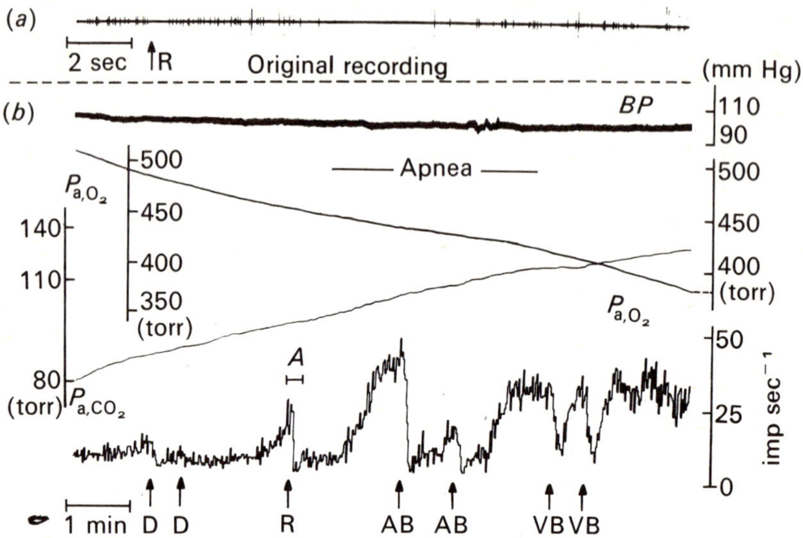

Fig. 4. Changes in chemoreceptor activity induced by injections of Dextran (D), Ringer's solution (R), arterial and venous blood (AB, VB) into the transverse pharyngeal vein (marked by arrows). These changes in chemoreceptor activity occurred in addition to a slow change in discharge in response to the steady fall in P_{a,O_2} and rise in P_{a,CO_2} shown in (*b*).

(*a*) Original recording of the chemoreceptor discharge showing the response to an injection of Ringer's solution. The duration of the record (*a*) is related to the curve of the discharge frequency (bottom graph of *b*) by the horizontal bar.

lated with 100% oxygen and 2 minutes before the tracing start, the pump had been switched off. As expected, P_{a,O_2} is gradually falling and P_{a,CO_2} rising and there is a corresponding slow increase in chemoreceptor discharge. This increase in discharge is interrupted by successive injections into the transverse pharyngeal vein of Haemaccel, Ringer's solution, arterial blood and venous blood and the latency of the chemoreceptor response is similar to that after lung inflation. These observations and the finding that if all the veins of the carotid body are cut, the inhibition of receptor discharge with lung inflation is abolished substantiates our assumption that a rise in venous pressure plays an important part in the inhibitory effect of inflation.

At the moment, it is not clear how changes in venous pressure affect chemoreceptor discharge within the carotid body. Preliminary experiments in which tissue P_{O_2} and local blood flow have been measured have not yet indicated what mechanisms might be involved. But two findings at least suggest that changes in the balance of local blood flow at the receptor site may be responsible: (i) the intensification of the receptor inhibition with lung inflation observed with high arterial pressure and (ii) the diminished inhibition which was observed when arterial pressure was raised at the same time as venous pressure.

REFERENCES

ACKER, H., LÜBBERS, D. W. & PURVES, M. J. (1971). Local oxygen tension field in the Glomus Caroticum of the cat and its change at changing arterial P_{O_2}. *Pflügers Arch. ges. Physiol.*, **329**, 136–55.

CASPERS, H. & SPECKMANN, E.-J. (1971). Gleichspannungsverschiebungen an der Hirnrinde bei Asphyxie. Cortical DC shifts associated with asphyxia. *Ärztl. Forsch.*, **25**, 241–55.

DISCUSSION

TORRANCE In your first slide, it is disturbing to see that tissue P_{O_2} apparently changed before P_{a,O_2}.

BINGMANN I should have made it clear when I described the slide that this is almost certainly due to the fact that (*a*) the tissue microelectrode responds more rapidly than the blood electrode and that (*b*) because we measured P_{a,O_2} from a loop in the femoral artery, there was a considerable delay before blood of new P_{a,O_2} reached the electrode. In fact the tissue microelectrode started to respond not much later than the 2–3 second transit time from lungs to carotid body.

MILLS What was the time constant of the ratemeter which you used to display the chemoreceptor discharge rate?

BINGMANN One second.

MILLS So this is a parallel recording between discharge from a few-fibre strand and tissue P_{O_2}. If you probe around in the carotid body with your O_2 microelectrode, is the response in discharge the same to changes in tissue P_{O_2} at different sites?

ACKER Perhaps I could comment on this now, although I shall be dealing with this problem in more detail in my paper later on. Certainly there is a systematic distribution of P_{O_2} in the carotid body tissue and there also appears to be differences in the discharge/P_{O_2} curves. With regard to the response time of the P_{O_2} tissue electrode I would like to mention that we have related tissue P_{O_2} to arterial P_{O_2} – measured by a platinum electrode of an only slightly longer response time than the tissue electrode. In these experiments the tissue P_{O_2} reacted at the same time as the arterial P_{O_2}. The P_{O_2} maximum was reached slightly earlier by the tissue electrode than by the arterial electrode.

MILLS Right: so Dr Bingmann's figure is from one site only and this measurement may be different from the P_{O_2} in tissue immediately surrounding the nerve ending from which he is recording. This is all I wished to establish.

BINGMANN It should be stressed, however, that the discrepancies between changes in tissue P_{O_2} and chemoreceptor discharge shown in my slides appeared in all experiments independent of the varying position of the P_{O_2} micro-electrode.

WILLSHAW Do you think that the extent of fluctuation of venous pressure which you have seen would be enough to cause the fluctuation of chemoreceptor discharge within each respiratory cycle?

BINGMANN Yes, I think so. In some experiments we observed unusually great oscillations of chemoreceptor activity which were strongly related to the respiratory cycles and to the time course of the local venous blood pressure. These oscillations of receptor activity persisted even if the arterial P_{O_2} was enhanced to hyperoxic values of about 600 torr (Fig. 1).

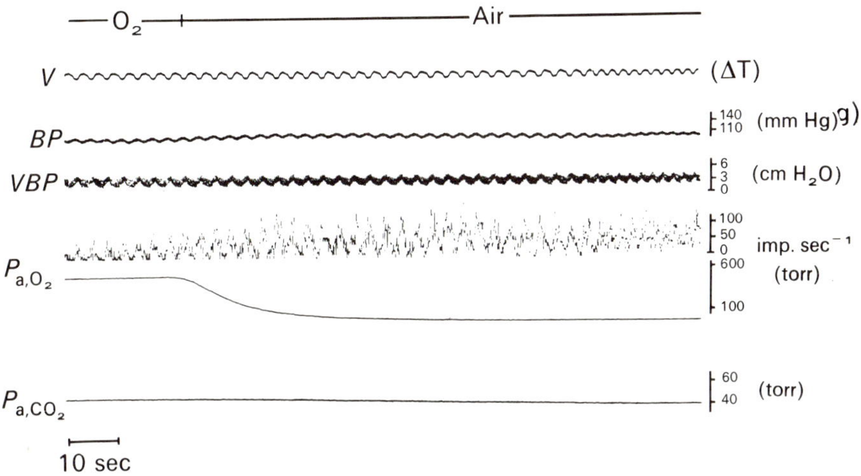

Fig. 1. This demonstrates the relation between the time course of respiratory cycles (V), arterial and venous blood pressure (BP, VBP) on the one hand and of the frequency of chemoreceptor afferent discharge (imp sec^{-1}) on the other hand at hyperoxic and normoxic levels of P_{a,O_2}. Arterial P_{CO_2} (P_{a,CO_2}) constant throughout.

Small respiratory fluctuations of P_{a,O_2} in this range of hyperoxia have no significant effects upon chemoreceptor discharge rate. This observation therefore might indicate that besides effects of changes in gas tensions fluctuations in local venous blood pressure may cause these oscillations of receptor discharge (see Fig. 1).

TORRANCE The interpretation of your results depends very much on the adequacy of the method which you use for measuring tissue P_{O_2}. The example which you give is of a tissue P_{O_2} of 20 torr, which I think you will agree is one of the things about the carotid body which is most difficult to explain. One possibility that comes to mind, of course, is that there was some compression of capillaries particularly if you bear in mind that the one other worker who has used this method obtained considerably higher values for P_{O_2} (Whalen & Nair, 1973).

So if you do have a low P_{O_2}, as a result of compression of capillaries, this may mean that, in the region from which you are recording P_{O_2}, you are at risk of excluding whatever mechanism is involved in the response to a change of venous pressure. You have very small changes in venous pressure and if you are squashing some capillaries, it is unlikely that that change in venous pressure will be transmitted to those capillaries. Whereas in another part of the carotid body, for example, in the part where the nerve ending is, from whose fibre you are recording, this artifact may not be involved.

BINGMANN We were aware of this possible artifact and perhaps I had better leave that part of the discussion to Dr Acker, whose technique was applied in our experiments. He will be considering a number of technical aspects of tissue P_{O_2} recording in his paper. But, coming back to the original question, you can certainly see fluctuations of chemoreceptor discharge in parallel with ventilation measured with a thermocouple, fluctuations of blood pressure and of venous pressure. So I think there is clearly a connection between these parameters.

WILLSHAW But these changes are with large fluctuations of venous pressure. What about a correlation with smaller changes in venous pressure?

BINGMANN This type of correlation has not really been possible to do precisely because of the difficulty of regulating the size of changes in venous pressure. But my impression is that the size of the venous fluctuation and the discharge fluctuation do correlate.

BULLER My impression from one or two of your earlier slides was that there is a difficulty in identifying the response of individual fibres from a few- or multi-fibre preparation. My question is, do you think there are some fibres which respond primarily to changes in pressure and some to a chemical stimulus: and can you distinguish them?

BINGMANN Generally we recorded from two to five fibres and it was certainly possible to show that these fibres responded in a characteristic way to changes in oxygen or carbon dioxide: in other words that they were not pressoreceptors.

BULLER Yes, but do you think there may have been different sensitivities of individual receptors to pressure or chemical change?

BINGMANN It did not appear so. All the fibres which responded to a chemical change, responded to changes in venous pressure in the same way.

HEISTAD If you maintain a change in venous pressure for a minute or so, does the discharge remain depressed for all of that time: or is there some recovery?

BINGMANN I cannot answer that because we found it very difficult to keep the venous pressure elevated for that length of time exactly. But my impression is that when venous pressure is first raised, it is answered by a fall in discharge and over shorter periods, this gradually rises although pressure is maintained.

MITCHELL I think perhaps I can answer that. Dr Lübbers came to our laboratory earlier this year and told us of the work of Dr Bingmann. I wanted to determine if the depression might be due to dopamine release from the Type I cells. We first carried out chemoreceptor response curves and we then did the actual experiments with a P_{a,O_2} of 40 mm Hg so that the receptors were firing near maximally. We raised venous pressure and showed that chemoreceptor discharge fell, but gradually rose towards normal, and that a further rise of venous pressure by 5 cm H_2O caused a second fall in discharge (Fig. 2). The carotid body was totally denervated

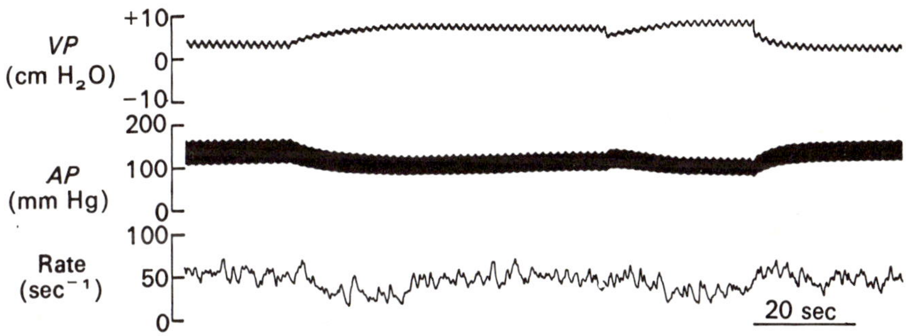

Fig. 2. Effect of increasing venous pressure in the jugular veins (*VP*) on systemic arterial pressure (*AP*) and on the impulse frequency of a chemosensory fibre from the carotid body of a cat. For further details, see text.

and the cat had been atropinized and given dihydroergotamine. We concluded that the response was only transient, and did not depend on the inhibitory modulation we have previously described, nor was it a neural reflex.

PURVES With regard to Dr Willshaw's question, there seem to be two pieces of evidence which argue against fluctuations of venous pressure having a great involvement in the production of fluctuations of chemoreceptor discharge. First, I've been back over the records which Biscoe and I obtained some years ago when we measured discharge before and after opening the chest. I have noted that there was no change in the amplitude of the fluctuations despite a presumed change in the amplitude of intrathoracic pressure swings. Secondly, I understand from the experiments of Band, Semple and others that they can effectively abolish the fluctuating discharge from chemoreceptors by diverting arterial blood through appropriate mixing chambers. They have clear evidence that this abolishes within-breath fluctuations of arterial pH but I cannot see that this manoeuvre could affect carotid body venous pressure in the slightest.

TORRANCE We can certainly confirm your observation – that raising venous pressure sometimes causes the chemoreceptor discharge to fall. Dr Mitchell's observation that this effect passes off with time is interesting because of the possibility that carotid body blood flow autoregulates. It has been shown in the intestine that you have an autoregulating vascular bed and if you raise pressure on the venous side, this is a very effective way of calling into play autoregulation.

BINGMANN We did do some experiments in which the carotid body was perfused at constant flow and oxygen tension from an oxygenator in the common carotid artery. Even with this, a rise in venous pressure still caused a fall in discharge.

CASPERS It is also worth recalling that the fluctuations in discharge could come from other sources, for example, the efferent pathway. So it may be a very complex matter.

REFERENCE

WHALEN, W. J. & NAIR, P. (1973). Microelectrode measurement of tissue P_{O_2} in the carotid body of the cat. In *Oxygen Supply*, pp. 236–7, eds KESSLER, M., BRULEY, D. F., CLARK, L. C., LÜBBERS, D. W., SILVER, I. A. & STRAUSS, J., München: Urban & Schwarzenberg.

On the speed of response of the chemoreceptors

By J. C. M. R. PONTE and M. J. PURVES
Department of Physiology, University of Bristol

At first sight it might be thought a simple matter to measure the speed of response of the carotid body chemoreceptors to their natural stimuli in arterial blood. The lingual, or some other suitable artery, is cannulated and a catheter passed so that its tip lies in the carotid sinus for the injection of a pulse of blood or cell-free medium equilibrated with gas mixtures to produce P_{O_2}, P_{CO_2} or pH significantly different from arterial. Activity from a single afferent fibre in the sinus nerve is continuously measured before, during and after the injection of such a stimulus, and the time course of the response determined.

There are in fact formidable problems to be overcome before accurate measurements can be made. Thus, since the receptor discharge is irregular and of low frequency and since it is difficult to inject the stimulus in a fluid medium without transmission delay and distortion, in order to measure the latent period of the receptor response and the time to peak response precisely, it is necessary to have an independent and accurate method of monitoring the profile of the input signal and a stimulus of sufficient power. This may be achieved either by increasing the size of the stimulus well beyond the physiological range, e.g. by equilibrating blood or Ringer–Locke solution with high carbon dioxide or by injecting the stimulus repetitively and averaging the response. It has been found difficult to achieve the latter method without some form of automatic injection (Fitzgerald, Leitner & Liaubert, 1969). Secondly, the rapid injection of blood or other fluid medium may cause alterations in perfusion pressure which themselves may alter the speed of response of the chemoreceptors; while the equilibration of blood in a tonometer may release vasoactive substances which may seriously affect the sensitivity of the receptor complex (Biscoe, Bradley & Purves, 1970). Thirdly, if large stimuli are used, the response may be distorted because of non-linearities of the stimulus/response curve.

A number of attempts to measure the speed of response of the receptors have been made and, we believe, are susceptible to one or more of the shortcomings outlined above. These will be discussed later. As a result, there is still some uncertainty about such matters as the latent period and time-to-peak response of the receptors to individual chemical stimuli, and relatively little is known in quantitative terms about other dynamic properties.

Finally, it is likely that even if the response of the receptors to step functions is measured precisely, this by itself will not yield information on such important points as: how far are fluctuations of chemoreceptor discharge within each breath at normal respiratory frequencies attenuated?; or, how great is the resonance exhibited by the receptors to changes in P_{CO_2}, P_{O_2} and pH and at what range of frequency is it to be seen? Without this information, it is difficult to predict whether the dynamic properties of the receptors are made use of in respiratory control. Although such information can be obtained from the step response, it is more conveniently obtained by measurement of the frequency response and in this paper, we give the results of experiments designed to do just this.

THE METHOD

The method has been described in some detail elsewhere (Ponte & Purves, 1974*a*,*b*). Briefly, it consists of perfusing the carotid body in a cat via a valve, operated by DC solenoids which allows arterial blood from recipient and donor cats to pass and which returns it to these cats. The arterial blood of the donor differs slightly with respect to P_{CO_2}, P_{O_2} or pH to that of the recipient, and there is also a slight difference in temperature (<0.5 °C) which allows a fine constantin thermocouple, implanted into the carotid sinus of the recipient cat to be used as an accurate monitor of the chemical input function. By switching the valve between donor and recipient arterial blood streams, the input signal can be a step function, one chemical variable only being tested at any given time; single or repetitive square wave pulses of any desired length or period or a random binary sequence. To obtain responses, to step functions, the discharge from single chemoreceptor afferent fibres is sampled at 5 kHz, accumulated over 25 or 50 sweeps in bins and the count demodulated.

Representative responses to step functions of P_{a,CO_2}, P_{a,O_2} and pH are shown in Fig. 1. This confirms that the speed of response is in the order $CO_2 > O_2 > pH$; that the latent period for carbon dioxide and oxygen is 0.2 seconds, that for pH, 0.5 seconds; that the time to peak is approximately 1.5–2.0 seconds; that even with accumulated sweeps, the responses are irregular and that in the carbon dioxide response, there appears to be

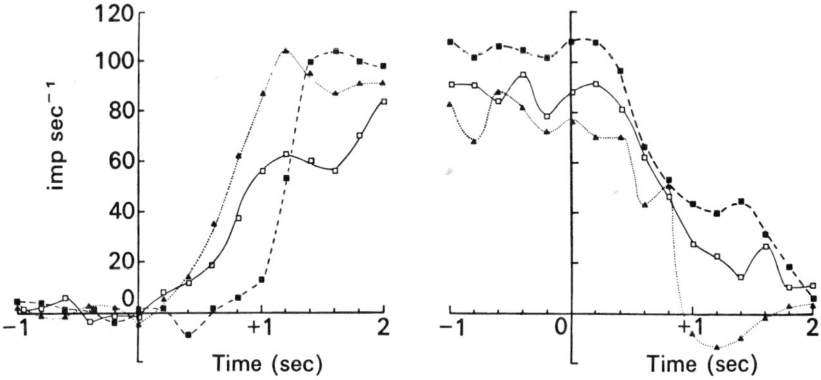

Fig. 1. Representative responses of single chemoreceptors to step changes in P_{a,O_2} (open square), P_{a,CO_2} (closed triangles) and pH_a (closed square). The plots were obtained by demodulating the discharge accumulated in 0.2 second bins from at least 20 switches between binary levels of the chemical variables, the switching period being 12 to 20 seconds. The responses have been standardized by taking $t = 0$ as the initial response of the thermocouple and eliminating differences in the delay between valve and thermocouple and by scaling the observed discharge, zero and 100 impulses sec^{-1} being assigned to the discharge at steady levels of the chemical stimulus. Perfusion pressure was 85 mm Hg throughout.

some evidence of adaptation though this interpretation should be treated with caution since it is uncertain whether the signal *inside the carotid body* remains constant or itself decays with time due to some other, e.g. vascular, component.

In Fig. 2, we give the results of two further series of tests which show that the speed of response varies with (1) perfusion pressure; (2) the magnitude of the chemical signal and (3) the nature of the transmission medium. Since these variables induce changes in the speed of response of the receptor of some size, they cannot be ignored. In general, these values agree with those obtained by other workers. Where differences occur, they can be almost certainly explained in terms of differences in the methods used, e.g. whether blood or cell-free perfusate was used, whether the signal consisted of hundreds of mm Hg P_{CO_2} or whether as in the present experiments, the signal consisted of <8 mm Hg P_{a,CO_2} and P_{a,O_2} and <0.05 pH unit. These technical differences may also explain the very different degree of 'adaptation' seen in response to carbon dioxide by McCloskey (1968) and ourselves.

It is theoretically possible to obtain the frequency response of a receptor from its step response but this is difficult in the case of the chemoreceptor because of the noisy output and of low frequency and because considerations of linearity between stimulus and response impose a severe limit on the size of the input signal. Because determination of the frequency response of

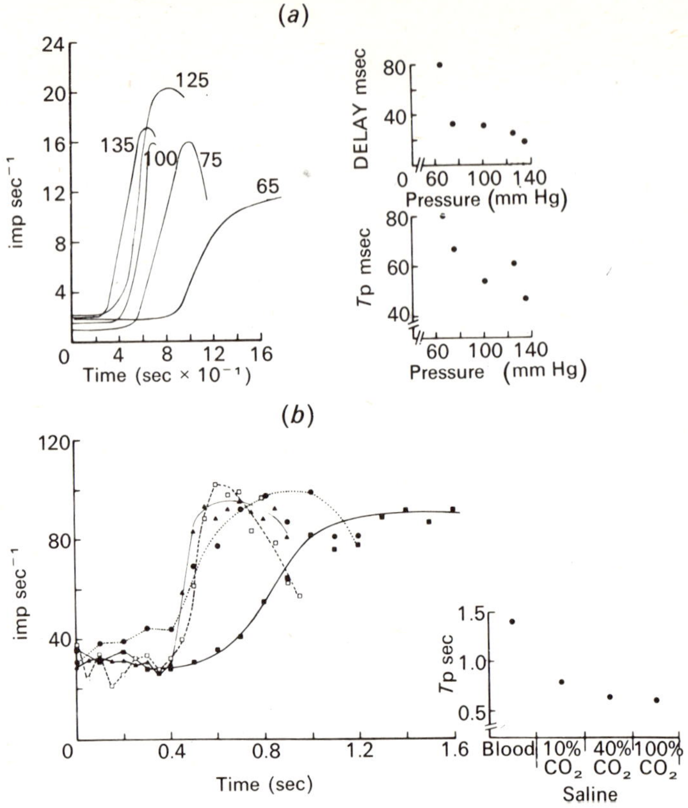

Fig. 2. (*a*) The effect of perfusion pressure upon the speed of response a of single chemoreceptor to a 0.5 second pulse of blood of P_{a,CO_2} 40 mm Hg, control 30 mm Hg. Perfusion pressure was set at five values between 65 and 135 mm Hg. Also shown are the relation between delay, i.e. transit delay between valve and thermocouple in the carotid sinus and perfusion pressure and the relation between time-to-peak response (T_p) and perfusion pressure.

(*b*) The 'on' response of a single chemorecptor, the discharge being demodulated as described in Fig. 1, to 0.5 second pulses of blood, $P_{a,CO_2} > 600$ mm Hg (filled squares); saline equilibrated with 10% CO_2 (filled cirles), 40% CO_2 (open squares) and 100% CO_2 (filled triangles). On the right, the relation between time-to-peak response and the nature of fluid perfusing the carotid body.

a system yields a more complete description of its dynamic characteristics, we have attempted this using two techniques. First, we have used the conventional technique of applying regular square pulses of <8 mm Hg P_{a,CO_2} or P_{a,O_2} over a range of frequencies; counting the action potentials as described above, and comparing the demodulated result at each of 10 to 12 frequencies with the mean discharge rate observed at DC at the two levels of the chemical stimulus. The array of demodulated results from one such experiment together with the Bode plot, is shown in Fig. 3. This shows

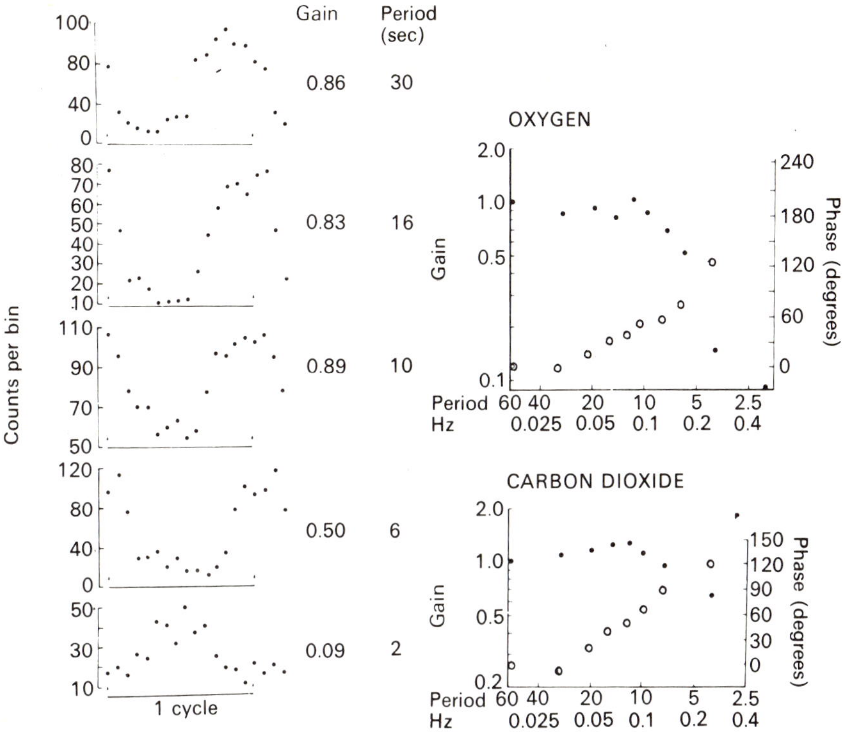

Fig. 3. On the left an array of demodulated results showing the response of a single chemoreceptor to square wave stimulation at binary levels of P_{a,O_2} 104 and 78 mm Hg. Each plot shows the relation between counts per bin and cycle period, the period varying between 2 and 30 seconds and is taken from the accumulated discharge over 18 minutes. On the right (above) the Bode plot constructed from the data on the left showing the relation between log gain (filled circles) and phase angles (open circles) and period (sec) and frequency (Hz). On the right (below) a Bode plot similarly obtained from another experiment showing the chemoreceptor response to binary levels of P_{a,CO_2} 29 and 34 mm Hg.

first that over the range of frequencies tested, there is an approximately 180° phase shift. Secondly, it shows that in response to oxygen, the frequency response of the chemoreceptor behaves like a low pass filter circuit: there is no resonance and at the normal respiratory frequency in the cat (0.5 Hz), the response is considerably attenuated. Thirdly, although the accompanying response to carbon dioxide does show resonance over the range 0.05 Hz, this is not great and the response is similarly but not so markedly as for oxygen, attenuated at 0.5 Hz.

Although this method has been successfully used with other receptor systems, e.g. the retinal ganglion cell (Hughes & Maffei, 1966), arterial baroreceptors (Spickler & Kezdi, 1967), mammalian muscle spindles (Matthews & Stein, 1969), it is of less value with the chemoreceptors because of the time necessary to make even one measurement. Thus in order to cover 10 to 12 frequencies, it may take up to five hours and during this time it is difficult to guarantee the stability of the receptor (Biscoe *et al.*, 1970). Further, this precludes a second frequency response in order to measure the effect of some manoeuvre.

We have accordingly developed an alternative method in which the input is a pseudo-random-binary-sequence (PRBS) which is generated in a shift register with feedback through exclusive 'OR' gates. The logic switching is operated by a clock pulse so that switching of the valve only occurs at multiples of the clock period. The random sequence repeats itself (hence the term 'pseudo-') after $2^N - 1$ clock pulses, N being the length of the register in bits.

With this input, the profile of which is monitored again with the thermocouple, two cross-correlations are made. The first is between receptor output (single fibre discharge) and a signal marking the switch of the valve which gives the impulse response of the whole system between valve and nerve. The second cross-correlation is between output of the thermocouple and PRBS signal to give the impulse response of the valve/transmission system. Frequency responses are obtained by Fourier transform and division of the response of the whole system by that for the valve/transmission system yields the frequency response of the chemoreceptor complex corrected for any distortion of the input signal. The method has been extensively tested *in vitro* using an electronic model of the chemoreceptors (Ponte & Purves, 1974*a*) and *in vivo* where it was shown that the frequency response of arterial baroreceptors using this method was virtually identical with that obtained by conventional methods (Spickler & Kezdi, 1967).

The pooled frequency responses to carbon dioxide, oxygen and pH are shown in Fig. 4. These frequency responses, first, confirm some of the predictions which might be made from the step responses. The response to carbon dioxide shows resonance over the range 0.05 to 0.25 Hz and at a frequency of 0.5 Hz (30 breaths per minute), gain fell to 0.88. Less marked resonance was seen in the oxygen response and at 0.5 Hz, gain was 0.41. No resonance was seen in the pH response and gain at 0.5 Hz was 0.25. Only in the response to carbon dioxide was there any evidence of phase lead and it is to be observed that at 0.5 Hz, the phase lags for carbon dioxide, oxygen and pH were 77, 104 and 119° respectively. The pooled data conceal a fairly wide variation in the degree of resonance observed: in the carbon dioxide experiments, two out of ten showed no resonance while in two others the peak gain value was 1.65 and 1.72.

Secondly, these results confirm that at normal respiratory frequencies in

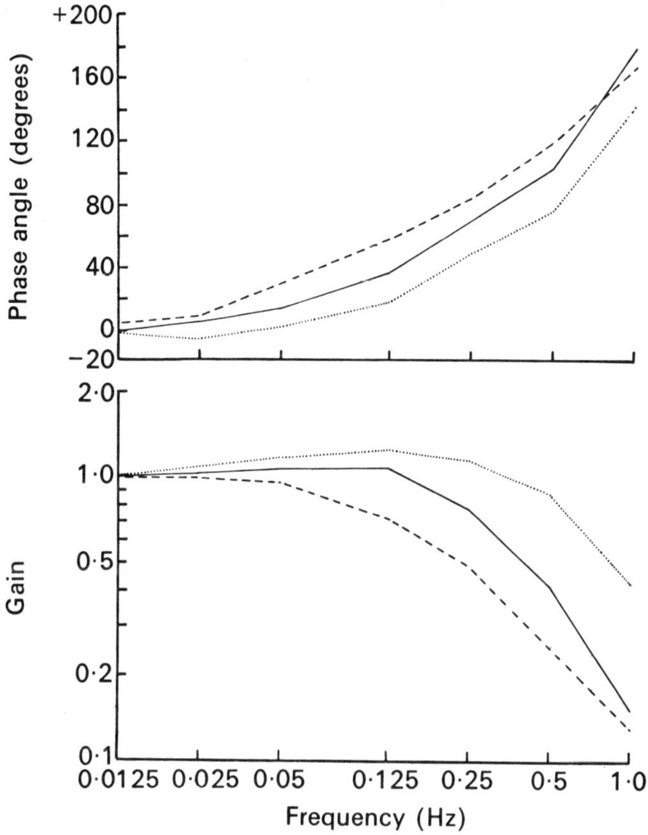

Fig. 4. Bode plots obtained from the averaged results for the response to CO_2 (dotted line), 10 experiments; O_2 (solid line), 14 experiments; pH (dashed line), 3 experiments.

the cat (30–40 min^{-1}), the response to all three chemical variables is attenuated to some degree and that there is a significant phase lag. Thirdly, from these frequency responses, it is possible to make some estimate of the filtering characteristics of the chemoreceptor complex. This can be carried out formally by assuming a simple RC electrical circuit which behaves dynamically in a similar fashion to the receptor. A simple Fortran computer program is used to iterate the variables of the equation describing the transfer function in order to obtain a best fit to the observed data. It turns out that the responses to carbon dioxide and oxygen are best represented by a series of three low pass and one resonant filters, the filters representing the carbon dioxide response having the lower time constants. The response to pH is represented by three low pass filters in series. Although at the moment, none of these filters can be equated with any known process within the chemoreceptor complex, the method indicates that this may be a valuable form of frequency analysis in the future when, for example, the time constants of individual feed-back loops are examined.

Finally, these frequency responses can be used as a basis for estimating the effects of various manoeuvres upon the receptor response. We have studied the effect of (a) cutting the sympathetic nerve supply to the carotid body, (b) stimulating these nerves, (c) injecting papaverine close to the

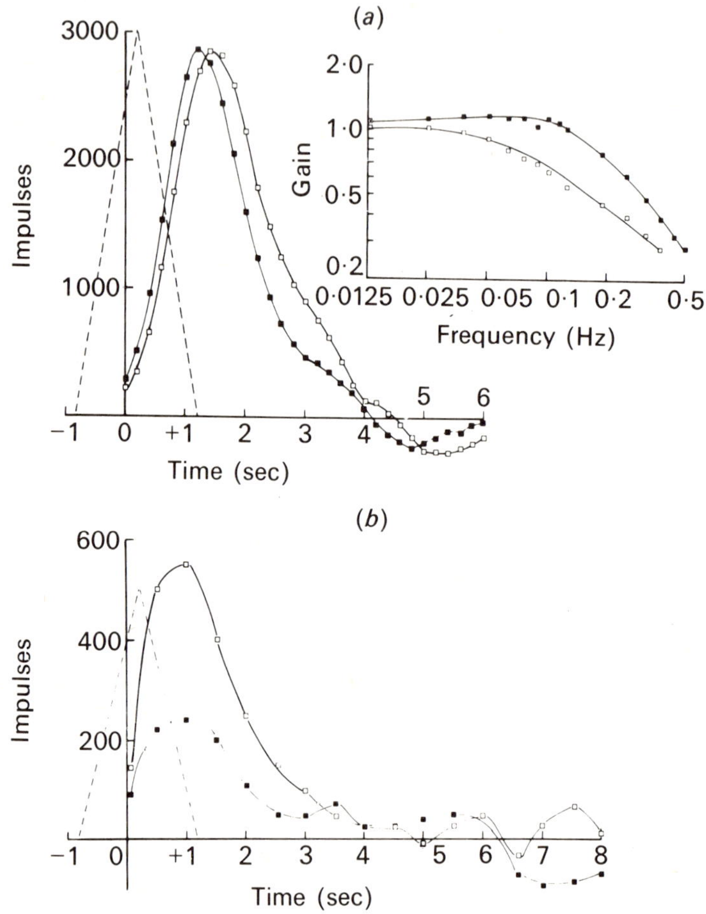

Fig. 5. (*a*) The impulse response of a single chemoreceptor fibre to CO_2 before (closed squares) and after (open squares) section of the post-ganglionic sympathetic fibres to the carotid body and, in the inset, the corresponding frequency responses. The triangle bounded by dashed lines represents the autocorrelation function of the PRBS input function displaced from $t = 0$ by the measured delay between valve and thermocouple response.

(*b*) The effect of papaverine 1 mg injected into the common carotid artery upon the impulse response of the ipsilateral carotid body to CO_2. Control, open squares; immediately following injection of papaverine, closed squares.

carotid body, and (d) injecting acetazolamide. By mechanisms which are very different, these manoeuvres caused (i) a reduction in the amplitude of the discharge fluctuation in response to the binary chemical signal; (ii) increased attenuation of gain at all frequencies tested. An example of the impulse and frequency responses obtained, before and after section of post-ganglionic sympathetic fibres, is shown in Fig. 5*a* and the impulse response before and during the infusion of papaverine in Fig. 5*b*. Among other things, these figures show that in these and in all other tests, it was possible to check that the peak response of the chemoreceptor complex was reached in 1–3 seconds.

The speed and nature of the response of the peripheral arterial chemoreceptors to their natural stimuli, or approximating to them, has been studied by a number of workers. The results have tended to show two rather different types of response. On the one hand, the chemoreceptors have been shown, as in the present paper, to respond to step changes in chemical stimuli with a latent period of 0.2 to 0.5 seconds, with a peak response reached in 1–3 seconds, with evidence of a relatively slight adaptation to carbon dioxide whose speed is uncertain but is certainly not rapid. On the other hand, other workers have obtained similar values for the latent period and time-to-peak response but very different values for the size and rate of adaptation. For example, in the studies of McCloskey (1968), Black, McCloskey & Torrance (1971) the peak response to step changes of carbon dioxide were up to five times the subsequent level obtained during steady stimulation, and adaptation was virtually complete in 10–15 seconds. Such variations as there are in the observed values for latent period and time-to-peak can almost certainly be explained in part by technical considerations, i.e. precision of delivering and monitoring the input stimulus; in part, by the accuracy of estimating the start and rate of change of receptor discharge where the discharge is irregular and of low frequency and in part, by the amplitude of the signal, the nature of the transmission medium and perfusion pressure.

With regard to the size and rate of receptor adaptation, there are several problems. In the first place, we have no direct evidence that this represents true adaptation of the receptor. It is difficult, for example, to exclude the possibility that the decay of discharge after the initial peak represents some alteration in the transmission line between arteriole and receptor, e.g. a vascular readjustment; alterations in the rate of gaseous diffusion, which may alter the nature of the stimulus. In this context, it may be pertinent that the resonant characteristics of the frequency response and the apparent adaptation of the receptor were reduced or abolished by sympathectomy and administration of carbonic anhydrase. Whatever the mechanism of adaptation, it is probably pertinent that adaptation has been most marked where the stimulus has been large, e.g. blood equilibrated with 30% carbon dioxide (McCloskey, 1968) or where a cell-free medium for the stimulus was used. That there should be a relation between the degree of adaptation and the velocity or size of the stimulus is consistent with what is known of slowly-adapting mechanoreceptors (Catton, 1970). It may therefore be unwise to infer dynamic properties of the chemoreceptors quantitatively and under normal conditions from large, single or repeated step changes in the chemical stimulus.

Measurement of the frequency responses allows two further observations to be made with respect to the role of the chemoreceptors in the regulation of respiration. Yamamoto (1962) proposed that the rate of change of chemoreceptor discharge as well as the mean level might be important in the regulation of respiration. The results of the present study provide no support for this idea for it will be observed that in the most rapid response, i.e. that to carbon dioxide, the frequency range at which resonance is observed is approximately an order lower than normal respiratory frequency in the cat while at respiratory frequencies, the response is considerably attenuated.

It has further been suggested that the timing of the peak of the chemoreceptor fluctuation within each breath is important; if the peak or rising phase coincides with inspiration, the subsequent breath is enhanced (Black *et al.*, 1971; Eldridge, 1972). Our results suggest that although there is certainly evidence of a change in membrane potential of central respiratory neurones within each respiratory cycle and which might indicate changes in receptive sensitivity (Salmoiraghi & von Baumgarten, 1961; R. A. Mitchell, personal communication) and although it is possible to demonstrate such altered sensitivity with large input chemical stimuli, electrical stimulation of the sinus nerve, etc., it is unlikely that this relation is of great physiological importance, for two reasons at least. First, we have repeated the experiments of Black *et al.*, using small repetitive pulses of blood some 4–7 mm Hg P_{a,CO_2} higher than control and injected over 10–20 respiratory cycles at each of 0.2 second bins of 4 second respiratory cycles. The averaged results showed no correlation whatever between the phase of the respiratory cycle at which the pulse injections were made and phrenic nerve discharge, used as the output of the respiratory system.

Secondly, consideration of the phase lags which are inherent in chemoreceptor complex together with the larger lags associated with the lung-to-carotid circulation time (>2 seconds in the anaesthetized cat, Biscoe & Purves, 1967) indicates that under resting conditions or with exercise, it is unlikely that there can be any fixed relation between any phase of the fluctuation of chemoreceptor discharge and the respiratory cycle. Thus when these lags are added and at a respiratory frequency of 0.5 Hz, the information reaching the respiratory neurones from the carotid bodies is well over a full cycle out of phase. Further, if we consider that the resting frequency of respiration in the cat varies between 22 and 45 breaths per minute, this represents a change in phase lags of 60° for carbon dioxide, 40° for oxygen and 30° for pH. Should respiratory frequency increase to 1 Hz with hyperthermia or exercise, then the change in phase lags could be well over 180° for these three variables. An even greater change in phase can be due to the lung–carotid circulation time. If this is assumed to be 2 seconds and cardiac output is constant, an increase of respiratory frequency from 22 to 45 min^{-1} represents a change from 270° to 550° in phase lag. Unless cardiac output increases so as to bring the lung-to-carotid delay down to 0.9 seconds, the total phase lag of the receptor discharge in response to respiratory fluctuations of blood gases can vary by 340° (550 − 270 + 60) for carbon dioxide and 320° (550 − 270 + 40) for oxygen. In addition, it should be recalled that in normal life successive breaths are rarely exactly the same and there are additional variations with various types of breatholding, e.g. phonation, eating, drinking. It is therefore probable that the profile of blood gas changes in arterial blood and the modulated discharge from chemoreceptors follow a random pattern.

Finally, we believe that the measurement of the frequency response may be an important method of determining and identifying individual processes within the carotid body. So far, it has been possible to confirm first that the inhibition of carbonic anhydrase seriously affects the ability of the carotid body chemoreceptors to follow rapid changes in P_{a,CO_2}. Secondly we have shown that if the reactivity of vascular smooth muscle in the carotid body is reduced, as with sympathectomy or administration of papaverine, the ability of the chemoreceptors to follow chemical transients is similarly reduced. This suggests that the vascular component, far from imposing an hydraulic delay upon the receptor response, is in some way an essential element in ensuring the high sampling rate which is an outstanding feature of the carotid body complex.

REFERENCES

Biscoe, T. J., Bradley, G. W. & Purves, M. J. (1970). The relation between carotid body chemoreceptor discharge, carotid sinus pressure and carotid body venous flow. *J. Physiol.*, **208**, 99–120.

Biscoe, T. J. & Purves, M. J. (1967). Observations on the carotid body chemoreceptor activity and the cervical sympathetic discharge in the cat. *J. Physiol.*, **190**, 413–24.

Black, A. M. S., McCloskey, D. I. & Torrance, R. W. (1971). The responses of carotid body chemoreceptors in the cat to sudden changes of hypercapnic and hypoxic stimuli. *Respir. Physiol.*, **13**, 36–49.

Catton, W. T. (1970). Mechanoreceptor function. *Physiol. Rev.*, **50**, 297–318.

Eldridge, F. L. (1972). The importance of timing on the respiratory effects of intermittent carotid body chemoreceptor stimulation. *J. Physiol.*, **222**, 319–34.

Fitzgerald, R. S., Leitner, L. M. & Liaubert, M. J. (1969). Carotid chemoreceptor response to intermittent or sustained stimulation in the cat. *Respir. Physiol.*, **6**, 395–402.

Hughes, G. W. & Maffei, L. (1966). Retinal ganglion cell response to sinusoidal light stimulation. *J. Neurophysiol.*, **29**, 333–52.

McCloskey, D. I. (1968). In *Arterial Chemoreceptors*, Wates Foundation Symposium, ed. Torrance, R. W., pp. 279–92, Oxford: Blackwell.

Matthews, P. B. C. & Stein, R. B. (1969). The sensitivity of muscle spindle afferents to small sinusoidal changes of length. *J. Physiol.*, **200**, 723–43.

Ponte, J. C. M. R. & Purves, M. J. (1974*a*). Receptor dynamics using correlation techniques with application to the carotid sinus baroreceptors. *Med. Biol. Engng* (in press).

Ponte, J. C. M. R. & Purves, M. J. (1974*b*). The frequency responses of carotid body chemoreceptors in the cat to P_{a,CO_2}, P_{a,O_2} and pHa. *J. appl. Physiol.* (in press).

Salmoiraghi, G. C. & von Baumgarten, R. (1961). Intracellular potentials from respiratory neurones in brainstem of cat and mechanism of rhythmic respiration. *J. Neurophysiol.*, **24**, 203–18.

Spickler, J. W. & Kezdi, P. (1967). Dynamic response characteristics of carotid sinus baroreceptors. *Am. J. Physiol.*, **212**, 472–6.

Yamamoto, W. S. (1962). Transmission of information by the arterial blood stream with particular reference to carbon dioxide. *Biophysical J.*, **2**, 143–59.

DISCUSSION

TORRANCE You have shown previously (Biscoe & Purves, 1967) that the chemoreceptor discharge fluctuates with respiration and that the amplitude of this oscillation is large – I seem to recall that it oscillated between 100 and 20 impulses per second. Now if you compare this with steady state receptor responses which have been reported by Hornbein, Griffo & Roos (1961), and yourself (Biscoe, Bradley & Purves, 1970), the size of the fluctuation is greater than you would expect. We have interpreted this as evidence of adaptation of the receptor and this would presumably come out in your frequency response relation as resonance and there would be resonance at about respiratory frequencies. But you are suggesting that at respiratory frequencies, the amplitude of the oscillation in output is less than would be predicted by the steady state response curves. Have I got it right?

PURVES Partly. It is certainly true that we have shown fluctuations in chemoreceptor discharge with respiration and have also constructed steady-state response curves. But the two are not comparable since the fluctuations were in response to changes in P_{a,O_2}, P_{a,CO_2} and pH_a and the steady state response to one variable at a time. In my view, it is therefore impossible to say whether the fluctuations in discharge were smaller or larger than one would predict. This is because there may be a true multiplicative effect at receptor level between a rising P_{a,CO_2} and a falling P_{a,O_2}. We have not seen this (Biscoe *et al.*, 1970) though we looked hard for it: but Bingmann reported in his paper earlier today that he had.

TORRANCE But you can use Fitzgerald & Parks' (1971) curve for response to carbon dioxide and oxygen combined and compare their prediction with the amplitude of oscillation which Goodman has got – 50% or more of the mean which is similar to the oscillations which you showed. The prediction from the steady state curve is so much less than the observation that we would say that there is adaptation. I'm therefore astonished that you found resonance at such low respiratory frequencies only.

PURVES Again, I'm not sure that it is valid to use one set of static response curves – especially if I recall correctly from Fitzgerald & Parks' data the receptor discharge was from multi-fibre preparations and expressed as a percentage of some arbitrary 100% maximum discharge – and another set of dynamic responses and compare the two. This is especially so since we have shown that the response curves under steady conditions to both carbon dioxide and oxygen vary very considerably. I think that you can only make these kind of judgements if in the same experiment and under the same conditions you measure the discharge from one or a few receptors to two steady levels of a chemical stimulus and you then cause this stimulus to fluctuate between these levels at different frequencies. This is in fact precisely what we did in the first series of experiments summarized in Fig. 3 and the results obtained by this method were virtually identical to those obtained by the PRBS method.

BULLER Why did you use that particular frequency as fundamental? Why, for example, didn't you choose the maximum value for gain, use that as unity and go on from there?

PURVES We chose this particular frequency – that is with a period of 80 seconds – after considerable trial and error. If we chose a higher frequency as fundamental, we found that the response became very noisy at about the frequencies we were interested in, i.e. 0.5 to 1.0 Hz. So we tended to use lower frequencies which more closely approximated to the original test signal – that is switching between the binary chemical levels once per minute. This meant that we lost information from noise at high frequencies but that we could explore the values for gain and phase shift at respiratory frequencies and some way below.

TORRANCE Nonetheless, I come back to the point that at normal respiratory frequencies, you and other workers have shown that there is an oscillation of discharge.

PURVES Yes and that of course is quite compatible with the frequency response which we have obtained. All we are saying is that if changes in carbon dioxide in arterial blood are mainly the cause of this oscillation of discharge, then at normal respiratory frequencies, the amplitude of this oscillation will be modestly attenuated – by a little over 10 %. If a change in P_{a,O_2} potentiates a change of P_{a,CO_2} in the opposite direction – as in the natural situation – then, for all I know, the attenuation may be even less.

SAMPSON I am confused. It seems to me that in normal respiration, all three chemical variables are changing. So in separating these components, you are in a sense stacking the cards against yourself. The whole point of the observations in the first place – that is, to say whether or not these oscillations are important in respiration – is made obscure.

PURVES I accept that but would just make two points. The first is that from these frequency responses, it is of course only possible to make limited predictions about how such an oscillating discharge would affect respiration. But it is possible to say for example, that at normal respiratory frequencies, the chemoreceptor is not responding to carbon dioxide as a rapidly adapting mechanoreceptor would to its natural stimulus: so, unless there is a large multiplicative effect between oxygen and carbon dioxide, I should not have thought that the oscillating as opposed to the mean chemoreceptor discharge had a great effect upon respiration. The second point is that we did not originally set out to answer this particular question. We were more interested in first determining as accurately as possible the frequency characteristics of the chemoreceptor response to either carbon dioxide or oxygen and then to see how this was affected by various manoeuvres in the hope that this might tell us more about how changes in oxygen or carbon dioxide affected the receptor.

WILLSHAW We find regularly that the chemoreceptor discharge leads the chemical signal. We have used the same technique as Band and his colleagues – that is we inject a pH signal in carotid arterial blood and as far as I can recollect, we usually obtain a phase lead of about 60°.

PURVES That seems to be rather more consistent than the results which Dr Band has published (Band, Saunders & Wolff, 1971). I have discussed this matter with Dr Band and there does seem to be an important difference between the methods which we use. As I understand it, he first of all injects a slug of saline equilibrated with 100 % carbon dioxide at the base of the aorta and uses this to determine the time lag between his pH electrode, used as a marker, in the carotid artery and the carotid body chemoreceptor. He then uses this lag to correct for the phase difference between output of electrode and chemoreceptors. The difficulty about this, as we have shown, is that a saline slug of high carbon dioxide is likely to give a considerably different latency and time to peak response of the chemoreceptors than will the much smaller natural stimulus in blood. So if you use a response to saline as a correction for a response to changes of pH in blood, it is easy to see how an artifactual phase lead can be obtained.

WILLSHAW We use a small bore catheter to inject the signal at the base of the aorta and we find on average that the pH change at the electrode in response to the testing signal is of the order of 0.05 pH unit. So I would guess that neither of these factors would account for the distortion which you mention.

PURVES I cannot be sure either but it seems to me to be a point worth checking – that is whether the time-to-peak response is the same whether you inject equal quantities of blood or saline with the same pH values. I don't think that this is being pernickety since it is worth recalling that at normal respiratory frequencies, a difference of a fraction of a second in the estimate for the time lag could account for substantial phase shifts.

NAIL I was wondering what quantitative differences you observed between the responses of individual receptors.

PURVES Considerable. I showed for the sake of simplicity the pooled frequency responses. If you look at individual responses to carbon dioxide for example, there were increases in gain of up to 1.7. However, even in these, the frequency at which gain was maximal was within the same range as for the group as a whole, i.e. 0.05 to 0.1 Hz. In no case – even in those receptors which responded most rapidly to step changes in carbon dioxide – was there any evidence to suggest rapid adaptation of the receptor which would have shown on the frequency response as an increasing gain with frequency. On the contrary, the responses suggest a mixture of high and low pass filters for the response to carbon dioxide and predominantly low pass filters for the response to oxygen and pH.

TORRANCE I must confess that I am more at home with the response to a sudden and sustained change in the stimulus. Could you tell me from your type of analysis what the receptor response would be to such a step change of oxygen, carbon dioxide or bicarbonate?

PURVES The speed of response would be greatest to carbon dioxide, then to oxygen and then to pH. As was seen in each impulse response from which the frequency responses are derived, the latency to a step change is about 0.2 second, the peak response to carbon dioxide is achieved in 1.0 to 2.2 seconds: that to oxygen in up to 2.6 seconds to pH in up to 3.4 seconds. If the step change in P_{a,CO_2} is small, i.e. <8 mm Hg, there is a measurable overshoot in the response to carbon dioxide, less obvious and less frequent in the response to oxygen and never in the response to pH.

TORRANCE But your response to carbon dioxide does show some evidence of adaptation.

PURVES Insofar as an overshoot of a receptor response with subsequent decline to a new steady level invariably means adaptation, yes. But I avoided the use of the word adaptation because, as I mentioned, I cannot be certain how stable the internal chemical environment of the carotid body is after a step change in arterial chemical stimulus. That is, I cannot be as certain as I would be if I were testing a baroreceptor with a more easily defined change in pressure.

REFERENCES

BAND, D. M., SAUNDERS, K. B. & WOLFF, C. B. (1971). The relation between chemoreceptor discharge and respiratory fluctuation of arterial pH in the anaesthetized cat. *J. Physiol.*, **218**, 73P.

BISCOE, T. J., BRADLEY, G. W. & PURVES, M. J. (1970). The relation between carotid body chemoreceptor discharge, carotid sinus pressure and carotid body venous flow. *J. Physiol.*, **208**, 99–120.

BISCOE, T. J. & PURVES, M. J. (1967). Observations on the rhythmic variation in the cat carotid body chemoreceptor discharge which has the same period as respiration. *J. Physiol.*, **190**, 398–412.

FITZGERALD, R. S. & PARKS, DOROTHY C. (1971). Effect of hypoxia on carotid chemoreceptor response to carbon dioxide in cats. *Respir. Physiol.*, **12**, 218–29.

HORNBEIN, T. F., GRIFFO, Z. J. & ROOS, A. (1961). Quantitation of chemoreceptor activity interrelation of hypoxia and hypercapnia. *J. Neurophysiol.*, **24**, 561–8.

Metabolic aspects of chemoreceptor function

By E. MILLS

Department of Physiology and Pharmacology, Duke University Medical Center, Durham, North Carolina 27710, USA

A characteristic of the arterial chemoreceptors is that the frequency of afferent impulse discharge increases progressively as P_{a,O_2} is lowered below 150–200 torr. The mechanism of stimulation has long been thought to have as the first step a decrease in cellular respiration (Anichkov & Belenkii, 1963; Joels & Neil, 1968). Since the major oxygen utilizing respiratory pathway in mammalian cells is the mitochondrial electron transport chain, it is appropriate to look first at this pathway to study the connection between cellular respiration and receptor activity. By examining the relationship between P_{O_2} and oxygen uptake of mitochondrial preparations from other tissues we can get estimates of how low P_{O_2} must fall within the carotid body before there is a depression of cellular respiration mediated through the mitochondrial electron transport chain. A very high value for P_{O_2} at the onset of decreased oxygen uptake in mitochondrial preparations is about 5 torr and for a 50 % decrease in oxygen uptake, the value is about 0.8 torr (Jöbsis, 1972). If a fall in oxygen uptake by a mitochondrial respiratory chain with similar oxygen affinity in the carotid body is the first step in the stimulation mechanism, then the P_{O_2} at the receptor site in the carotid body should be 5 torr or less at the onset of the progressive increase in chemoreceptor impulse frequency. The arterial P_{O_2} at the onset of the increased discharge is 150–200 torr. At a more physiological P_{a,O_2} of 90–100 torr when chemoreceptor afferents are discharging well above the baseline rate, I would anticipate that there would be many sites within the Type I, II cell/nerve ending complex at which P_{O_2} would be lower than 5 torr. Acker, Lübbers & Purves (1971) found sites in the outer connective tissue layer of the carotid body where P_{O_2} was 5 torr or less when P_{a,O_2} was 90–100 torr. They suggested that an oxygen barrier and possibly a countercurrent flow system in the cat carotid body accounted for the low P_{O_2} in the outer layer and for the observation that P_{O_2} was higher in the more central part of the glomus. There was no evidence, however, that P_{O_2} was 5 torr or less at sites in the Type I, II cell/nerve ending complex when P_{a,O_2} was 90–100 torr. Whalen and Nair (1972), employing a different type of oxygen micro-electrode measured much higher values for P_{O_2} in both outer and central regions of the cat carotid body. At a P_{a,O_2} between 90–100 torr, they found the mean P_{O_2} to be about 80 torr in the outer layer and about 40 torr in the more central region. The weight of the evidence indicates to me that P_{O_2} in the receptor complex is not low enough to depress oxygen uptake in a mitochondrial respiratory chain with an oxygen affinity as high as that found in mitochondria from other tissues.

Mills & Jöbsis (1972) studied the mitochondrial electron transport chain in the intact cat carotid body perfused with haemoglobin free solutions. We found that there was a compound or compounds present in the carotid body with light absorption characteristics identical to those of cytochrome a_3 in mitochondria from other tissues. One component with these absorption characteristics started to become more reduced when the perfusate P_{O_2} was lowered below 150 torr; it was 50 % reduced at a perfusate P_{O_2} of 90 ± 35 torr, and became maximally reduced as P_{O_2} approached 10 torr (Fig. 1). An increased reduction of cytochrome a_3 as P_{O_2} is lowered signals

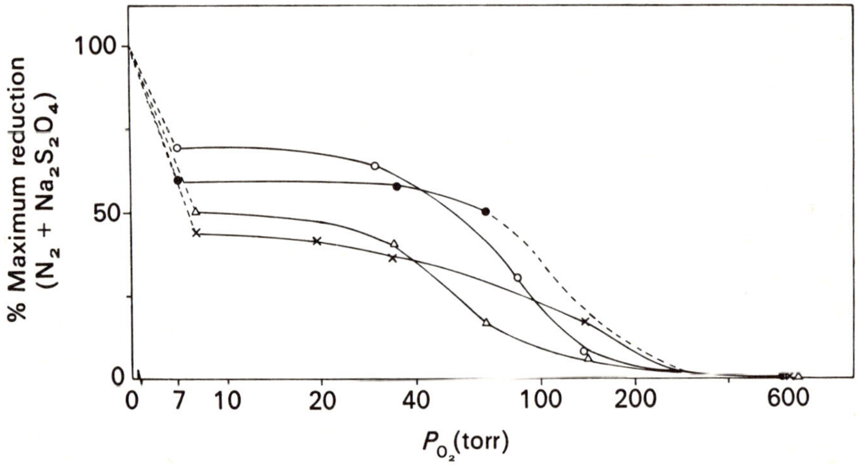

Fig. 1. Per cent of total cytochrome a_3 in the reduced state as a function of perfusate P_{O_2}. Carotid body of the cat. The percentage of cytochrome a_3 in the reduced state is considered to be a minimum of *c.* 600 torr and maximum during perfusion with N_2 equilibrated perfusate containing the reductant $Na_2S_2O_4$. Curves from four different experiments. Note that each curve approaches an initial upper limit as P_{O_2} is lowered towards 7–9 torr. There is a further increase in percent of total cytochrome a_3 in the reduced state when P_{O_2} is lowered to 0 torr by addition of $Na_2S_2O_4$. The cytochrome a_3 component which is already partially reduced at $P_{O_2} = 150$ torr has an unusually low oxygen affinity. The component which is reduced only on adding $Na_2S_2O_4$ is considered to have a high oxygen affinity like that of cytochrome a_3 in other tissues. (From Mills & Jöbsis, 1972, with permission.)

that the oxygen concentration at the mitochondrial level has been lowered sufficiently to limit the rate of electron transfer along the respiratory chain to the acceptor, oxygen. In other words, there is a depression of cellular respiration. The observation that increased reduction of cytochrome a_3 commences at 150 torr rather than 5 torr and that 50% reduction is attained at 90 ± 35 torr rather than 0.8 torr means that the oxygen affinity of this respiratory chain is much lower than that measured in the mitochondrial preparations from other tissues. The criticism may be raised that there is a large P_{O_2} drop from capillary to mitochondria in the preparation so that P_{O_2} at the mitochondrial level might in fact be closer to 5 torr than it is to the perfusate P_{O_2} of 150 torr when reduction commences and that it would be closer to 0.8 torr than to 90 ± 35 torr when cytochrome a_3 is 50% reduced. However, at a perfusate P_{O_2} of 7–9 torr, 33–67% of the total cytochrome a_3 in the carotid body is still in the oxidized state. Since there must be some oxygen available to the mitochondria to maintain this oxidized state at this low perfusate P_{O_2}, the P_{O_2} drop from capillary to mitochondria must be less than 7–9 torr. Unless some special mechanism operates to lower P_{O_2} at the mitochondrial level as perfusate P_{O_2} is raised, it is correct to conclude that the mitochondrial respiratory chain with these properties has an unusually low oxygen affinity.

The changes in NADH/NAD$^+$ ratio which occur in the carotid body when perfusate P_{O_2} is lowered demonstrate clearly that in some part of the receptor complex there is a stimulation rather than a depression of oxidative metabolism. As illustrated in Fig. 2, there is a progressive decrease in NADH/NAD$^+$ ratio as perfusate P_{O_2} is lowered from 600 torr to 7–9 torr. This decrease occurs because there is increased ATP utilization in part of the receptor complex when afferent impulse discharge is increased in response to the fall in perfusate P_{O_2}. The rise in ADP concentration acts to facilitate electron transfer along the mitochondrial respiratory chain to oxygen so that mitochondrial NADH/NAD$^+$ ratio declines. If the P_{O_2} in this part of the receptor complex was low enough to limit electron transfer along the respiratory chain, then the NADH/NAD$^+$ ratio should increase rather than decrease. Although the NADH experiments give no information about the absolute value of P_{O_2} in the receptor complex when perfusate P_{O_2} is lowered to 7–9 torr, they do demonstrate that it is above the critical level for depression of cellular respiration in part of the receptor complex. The cytochrome a_3 experiments show that in another part of the receptor complex there are cells in which respiration is depressed as perfusate P_{O_2} is lowered below 150 torr. If for a moment we put aside the finding that the oxygen affinity of a compound with light absorption characteristics like cytochrome a_3 is unusually high in the carotid body, then what

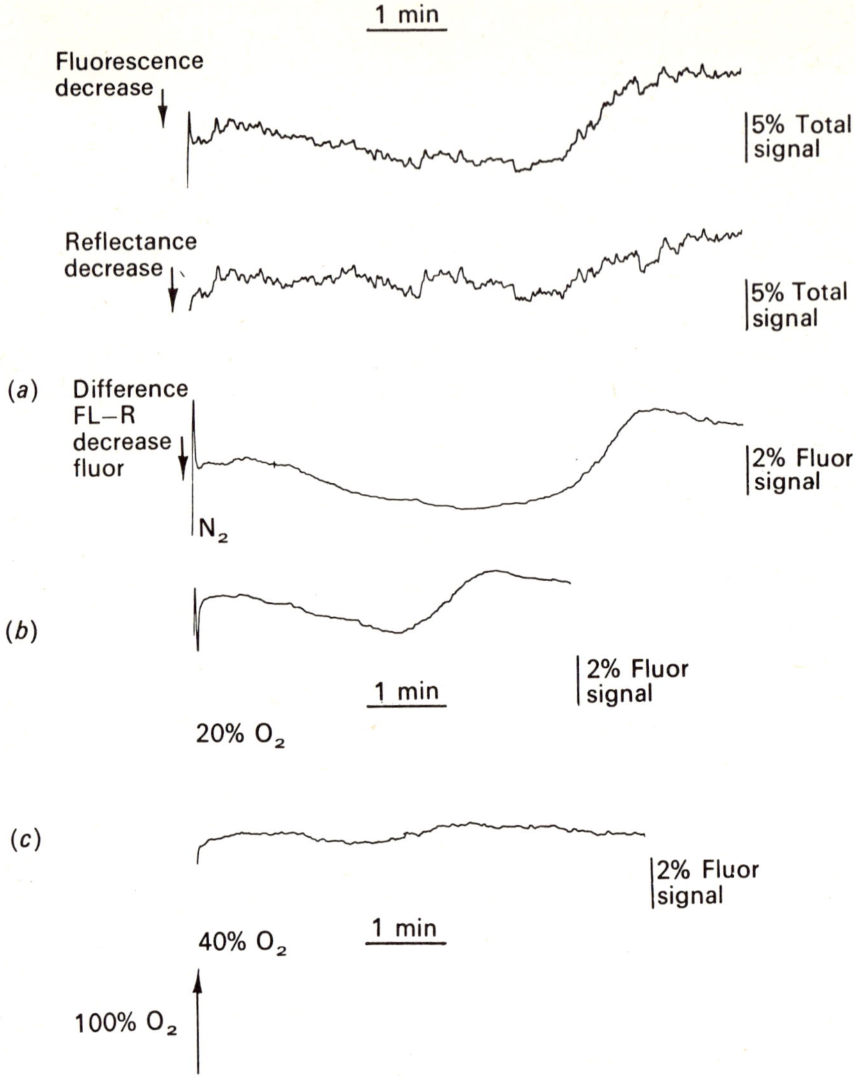

Fig. 2. Effect of perfusate oxygen concentration on fluorescence of the carotid body. All traces from same experiment. Fluorescence trace: excitation wavelength, 366 nm; measuring wavelength, 450–480 nm. Difference traces (*a*)–(*c*): fluorescence minus reflectance. The fluorescence and reflectance records shown at the top were subtracted to give the difference trace shown in (*a*). A downward deflection of the FL – R trace indicates a decrease in fluorescence intensity. This signals a decrease in the $NADH/NAD^+$ ratio, oxidation of NADH. An upward deflection of the trace signals an increase in the $NADH/NAD^+$ ratio, reduction of NAD^+. Perfusion with 100 % O_2-equilibrated solution before deflections (upward arrow) at the beginning of traces. At the deflection in (*a*) switch to N_2-equilibrated perfusate, P_{O_2} = 8 mm Hg; (*b*) switch to air-equilibrated perfusate; (*c*) switch to 40 % O_2-equilibrated perfusate. Calibration: change in fluorescence intensity as percentage of total fluorescence signal. (From Mills & Jöbsis, 1972, with permission.)

other mechanism might explain how a fall in P_{O_2} initiates receptor stimulation? There would have to be some mechanism, perhaps vascular, which acts to lower P_{O_2} below the critical level for cellular respiration in one part of the Type I, II cell/nerve ending complex while maintaining it above this level in an adjacent part of the receptor complex. The painstaking vascular reconstructions of Dr Seidl presented at this workshop indicate that there are several different types of vascular shunts in the carotid body which can affect regional blood flow within the carotid body. However, at present, it is not apparent to me how these vascular shunts could affect the distribution of oxygen within the Type I, II cell/nerve ending complex. The observation reported here by Dr Acker that there is a P_{O_2} dependent oxygen consumption in the cat carotid body supports the idea that there is a respiratory pathway in the carotid body which has an unusually low oxygen affinity. The experiments of Mills & Jöbsis (1972) indicate that this respiratory pathway is the mitochondrial electron transport chain.

Recently, I have used the fluorometric technique to look for P_{O_2} related changes in the reduction state of flavoproteins in the perfused carotid body. In other tissues there are flavoproteins involved in electron transfer from NADH to cytochrome *b* in the respiratory chain, and several other mitochondrial components which are not a direct part of the respiratory chain (Jöbsis, 1964; Chance & Williams, 1955). These include, for example, flavoproteins involved in oxidation of succinate and alphaglycerophosphate. There is, in addition, an important microsomal component involved in the oxidation of NADH by cytochrome b_5. Finally, there are a number of soluble auto-oxidizable flavoproteins, that is flavoproteins reacting directly with oxygen, whose contribution to total oxygen consumption in other tissues is not known. These flavoproteins could be of particular interest if they were present in the carotid body since they have, as a group, low oxygen affinity. The idea of the experiments was to see if there was a flavoprotein component in the carotid body which became reduced at high perfusate P_{O_2}. As illustrated in Fig. 3, there is an oxidation rather than reduction of flavoprotein in the carotid body when perfusate P_{O_2} is lowered from 600 torr to 7–9 torr. This response is the type expected from the flavoprotein component in the mitochondrial electron transport chain when oxidative metabolism is stimulated. The observation is consistent with the previous finding that the mitochondrial $NADH/NAD^+$ ratio decreases when perfusate P_{O_2} is lowered (Mills & Jöbsis, 1972).

In the experiments in which $NADH/NAD^+$ ratio was measured, it was not possible to test the effect of lowering perfusate P_{O_2} from 7–9 torr to zero torr because the reductant employed absorbs light at the excitation wavelength used to measure NADH. As illustrated in Fig. 4, addition of the

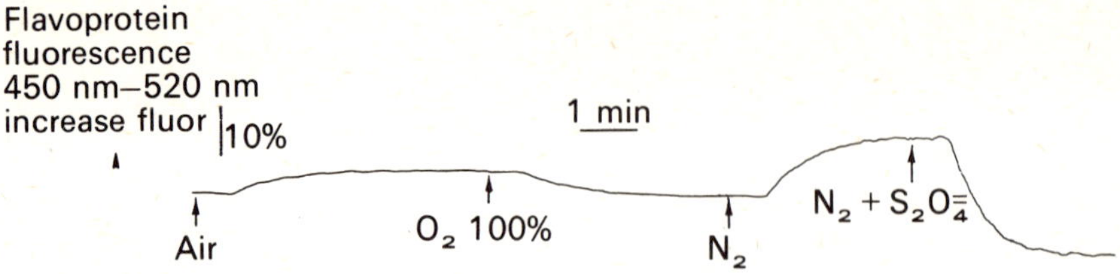

Fig. 3. Cat carotid body perfused *in vitro* with Lockes solution containing albumin and dextran. Flavoprotein fluorescence measured at 520 nm. Excitation wavelength at 450 nm. Total fluorescence signal was that measured during perfusion with Lockes solution (glucose 5.5 mM) equilibrated with 100 % O_2. Recording made at increased gain so that 25 % change in fluorescence gives full-scale deflection. The light reflected from the preparation was simultaneously measured at 450 nm and subtracted electrically from the fluorescence signal measured at 520 nm to compensate for instrumental artifact common to both channels of information. An upward deflection of the trace signals a decrease in the per cent of total flavoprotein in the reduced state, i.e. oxidation of flavoprotein. At the first arrow, air-equilibrated perfusate, P_{O_2} = 142 torr, replaced 100 % O_2-equilibrated perfusate, P_{O_2} = 630 torr. The P_{O_2} of N_2-equilibrated perfusate was 7 torr. Note that transition from P_{O_2} = 630 to P_{O_2} = 142 and from P_{O_2} = 630 to P_{O_2} = 7 torr causes oxidation of flavoprotein, and that the oxidation is greater with the transition from P_{O_2} = 630 torr to P_{O_2} = 7 torr. Note also the reduction of flavoprotein which occurs when P_{O_2} is lowered to 0 torr by addition of the reductant $Na_2S_2O_4$.

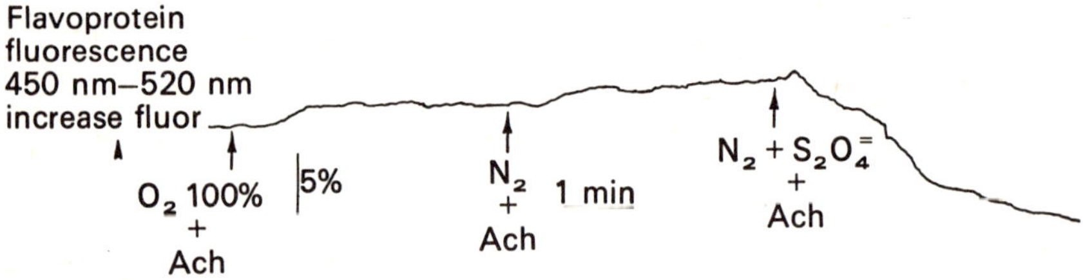

Fig. 4. Cat carotid body, flavoprotein fluorescence. Methods same as for experiment shown in Fig. 2, except full-scale deflection = 10 % of total fluorescence. At the first arrow, addition of acetylcholine to 100 % O_2-equilibrated perfusate, P_{O_2} = 627 torr, to give concentration of 10^{-4} M. At second arrow transition to N_2-equilibrated perfusate, P_{O_2} = 8 torr, containing acetylcholine, 10^{-4} M. At the third arrow transition to N_2-equilibrated perfusate containing the reductant, $Na_2S_2O_4$ and acetylcholine 10^{-4} M. Note that first addition of acetylcholine causes oxidation of flavoprotein and that the presence of acetylcholine does not prevent the further oxidation of flavoprotein with transition to N_2 perfusate and reduction with transition to $N_2+S_2O_4$ perfusate.

reductant S_2O_4 to perfusate with a P_{O_2} of 8 torr causes a large reduction of flavoprotein. This observation is consistent with the earlier finding that 33–67 % of the total cytochrome a_3 in the carotid body is in the oxidized state when perfusate P_{O_2} is 7–9 torr and that it becomes reduced with addition of S_2O_4. This flavoprotein, I believe, is a component of the respiratory chain in the receptor complex which has a normal, high oxygen affinity (Mills & Jöbsis, 1972). A persistent question is to determine where in the Type I, II cell/nerve ending complex the stimulation of oxidative metabolism signalled by NADH and now flavoprotein oxidation occurs. NADH oxidation, you may recall, was produced by stimulating the cut, peripheral end of the sinus nerve as well as by a decrease in perfusate P_{O_2} (Mills & Jöbsis, 1972). The interpretation at the time was that this effect could arise in afferent and efferent axon terminations and in the Type I cell. However, as Dr Goodman has reported at this meeting, there does not appear to be synaptic excitation of any cells in the carotid body when the carotid sinus nerve is stimulated nor is there any change in membrane potential following injection of acetylcholine (Eyzaguirre & Koyano, 1965; Eyzaguirre, Leitner, Nishi & Fidone, 1970; Goodman & McCloskey, 1972). During the recent measurements of carotid body flavoproteins, I tested the effect of adding acetylcholine to the perfusing solutions. As illustrated in Fig. 5, addition of acetylcholine 10^{-4} M to perfusate equilibrated with 100 % oxygen causes oxidation of flavoprotein. Also, the presence of acetylcholine in the perfusing solutions does not prevent the further oxidation of flavoprotein which occurs when perfusate P_{O_2} is lowered to 7–9 torr. It appears then that two interventions, sinus nerve stimulation and addition of acetylcholine, which do not depolarize the Type I or II cells cause fluorescence changes in the carotid body which signal stimulation of oxidative metabolism. Unless metabolic stimulation occurs in the Type I cell without change in membrane potential, then the fluorescence changes must originate solely in the axon terminations in the carotid body. If this is so, then the interpretation that the respiratory chain with low oxygen affinity is located in the Type II cell requires revision (Mills & Jöbsis, 1972). The interpretation was based on anatomical evidence that Type I but not Type II cells could be activated synaptically through the sinus nerve (Biscoe, 1971) and on the observation that the changes in $NADH/NAD^+$ ratio caused by nerve stimulation were those expected from a respiratory chain with normal, high oxygen affinity. More specifically there was a decrease in $NADH/NAD^+$ ratio during nerve stimulation even when perfusate P_{O_2} was lowered to 7–9 torr. Since there is now some question of whether the Type I cell can be activated synaptically through the sinus nerve, we can no longer exclude the possibility that the respiratory chain with low oxygen affinity is in the Type I cell. However, the interpretation that the cell containing the respiratory chain with low oxygen affinity functions as the oxygen sensor part of the receptor complex does not require modification. It appears probable to me that the stimulation of oxidative metabolism observed when perfusate P_{O_2} is lowered occurs at least in part in the afferent axons as a consequence of increased impulse discharge. Certainly this is a site at which a variety of metabolic inhibitors used to excite or to 'exhaust' receptor output might function. The stimulation and cessation of chemoreceptor discharge observed with prolonged exposure to metabolic inhibitors or to anoxia could well reflect exhaustion of the Na/K pump mechanism in the axons rather than exhaustion of the supply of an excitatory transmitter substance.

One metabolic inhibitor that we use often to stimulate chemoreceptors is NaCN, and the assumption has been that it acts by combining with cytochrome a_3 to inhibit electron transfer to oxygen (Anichkov & Belinkii, 1963; Joels & Neil, 1968). However, in the monograph by Heymans and Neil (1958, p. 192) there is the statement that chemoreceptor stimulation by NaCN is decreased during inhalation of 100% oxygen. Since cyanide inhibition of the cytochrome a_3 reaction with oxygen is non-competitive, the observation, if correct, suggests that cyanide acts by some other mechanism to increase receptor discharge.

I would like to conclude this discussion of metabolic aspects of chemoreceptor function by pointing out some interesting connections between monoamine metabolism and oxygen. There are several enzymes involved in synthesis or metabolism of norepinephrine (NE), and 5-hydroxytyramine (5-HT) which use molecular oxygen as an electron acceptor. One group of enzymes which may be of particular interest in the carotid body are the monoamine oxidases. These enzymes act on monoamines such as NE and 5-HT to convert them to the corresponding aldehyde. This pathway is of relatively minor importance for NE metabolism in other tissues because of the presence of a second enzyme, catechol-*O*-methyl transferase. It is, however, of major importance in metabolism of 5-HT. This group of enzymes has a very low oxygen affinity. For example, the rate of oxidation of tyramine in 100% oxygen is reported to be three times the rate in room air and the rate of oxidation in 5% oxygen is one-third the rate in air (Kohn, 1937).* It is possible then that the rate of metabolism of 5-HT in the carotid body could be a function of local P_{O_2}. In the synthesis of 5-HT, the enzyme, tryptophan hydroxylase which converts tryptophan to 5-hydroxytryptophan, also requires molecular oxygen. For this enzyme, there are measurements of the oxygen affinity of the purified preparation (Ikeda, Fahien & Udenfriend, 1966). The enzyme is reported to be saturated with respect to oxygen at 20% oxygen and the K_{m,O_2} is reported to be at 6% oxygen. In the pathway for NE synthesis, tyrosine hydroxylase requires molecular oxygen as does dopamine-β-oxidase which converts dopamine to NE.

Norepinephrine, dopamine and 5-hydroxytyramine are present in the carotid body and these monoamines affect chemoreceptor impulse discharge. It may be of more than passing interest that changes in P_{a,O_2} in the

* I have not found more recent measurement of K_{m,O_2} for purified preparations.

range which influence chemoreceptor discharge also may influence the activity in biosynthetic and metabolic pathways which regulate the concentration of these amines in the carotid body.

REFERENCES

Acker, H., Lübbers, D. W. & Purves, M. J. (1971). Local oxygen tension field in the glomus caroticum of the cat and its change at changing arterial P_{O_2}. *Pflügers. Arch. ges. Physiol.*, **329**, 136–55.

Anichkov, S. V. & Belenkii, M. L. (1963). *Pharmacology of the Carotid Body Chemoreceptors*, Oxford: Pergamon.

Biscoe, T. J. (1971). Carotid body: Structure and function. *Physiol. Rev.*, **51**, 427–85.

Chance, B. & Williams, G. R. (1955). Respiratory enzymes in oxidative phosphorylation. II. Difference spectra. *J. biol. Chem.*, **217**, 395–407.

Eyzaguirre, C. & Koyano, H. (1965). Effects of electrical stimulation on the frequency of chemoreceptor discharges. *J. Physiol.*, **178**, 438–63.

Eyzaguirre, C., Leitner, L. M., Nishi, K. & Fidone, S. (1970). Depolarization of chemsensory nerve endings in carotid body of the cat. *J. Neurophysiol.*, **33**, 685–96.

Goodman, N. W. & McCloskey, D. I. (1972). Intracellular potentials in the carotid body. *Brain Research*, **39**, 501–4.

Heymans, C. & Neil, E. (1958). *Reflexogenic Areas of the Cardiovascular System*, Boston: Little, Brown & Company

Ikeda, M., Fahien, L. A. & Udenfriend, S. (1966). A kinetic study of bovine adrenal tyrosine hydroxylase. *J. biol. Chem.*, **241**, 4452–6.

Jöbsis, F. F. (1964). Basic processes in cellular respiration. In *Handbook of Physiology*, vol. I, sec. 3, Respiration, pp. 63–124, Washington, D.C.: American Physiological Society.

Jöbsis, F. F. (1972). Oxidative metabolism at low P_{O_2}. *Fedn Proc.*, **31**, 1404–13.

Joels, N. & Neil, E. (1968). The idea of a sensory transmitter. In *Arterial Chemoreceptors*, ed. Torrance, R. W., pp. 153–78, Oxford: Blackwell.

Kohn, H. I. (1937). Tyramine oxidase. *Biochem. J.*, **31**, 1693–1704.

Mills, E. & Jöbsis, F. F. (1972). Mitochondrial respiratory chain of carotid body and chemoreceptor response to changes in oxygen tension. *J. Neurophysiol.*, **5**, 405–28.

Whalen, W. J. & Nair, P. (1972). P_{O_2} in the carotid body. *Fedn Proc.*, **31**, 371.

DISCUSSION

MITCHELL Have you studied the denervated carotid body?

MILLS I have some experiments going on right now to find out where the fluorescence change occurs and to see if that cytochrome a_3 is still present. I presume you would like to see the a_3 in the axon termination. All I can say is that I have tried stimulating back down the nerve doing the same as the fluorescent experiments and tried the spectrophotometric measurement for a_3 and have not been able to demonstrate any reduction of a_3.

ACKER I should like to tell you of the recent experiments of Starlinger & Lübbers (1973) where they have measured mitochondrial P_{O_2} and cytochrome a_3 and have found that the critical mitochondrial P_{O_2} is about 0.05 torr. So if you show in your curves the relation between P_{O_2} and the reduction state of cytochrome a_3 after the addition of $Na_2S_2O_4$ you should be at the critical mitochondrial P_{O_2} level and therefore you should get a complete reduction of cytochrome a_3. The first reduction curves of cytochrome a_3 which you showed may be caused by absorption of light by FADH because you measure the cytochrome in the region of 450 Å or so. Another possibility is that you are measuring indirectly the spatial change of the low P_{O_2} zone which we have measured by the micro-P_{O_2}-electrode.

MILLS Yes, but that is taken care of by the measurement because the reference wavelength is kept constant. Thus, when FAD becomes reduced, there is a decrease in absorption in the 440 nm region and since we are measuring cytochrome a_3 at 445 nm relative to 465 nm, increased reduction of FAD would cause a decrease in optical density at 445 nm relative to 465 nm. Consequently, this would not interfere with the cytochrome a_3 measurement.

ACKER An alternative way you could check this would be to measure cytochrome a_3 at 600 nm or so because this is a second peak where FADH absorbs light. Have you done this?

MILLS No, we haven't measured it there at all because we have a problem with haemoglobin contamination: this seems to be a difficulty no matter how much we try to get rid of it.

TORRANCE If the haemoglobin is in a blood vessel, then presumably that blood vessel is not being perfused. So the cytochrome a_3 that is normally supplied by that blood vessel must be being supplied by other blood vessels over a long diffusion distance. And this may possibly be an explanation of why you have two categories of cytochrome a_3. The other point is that you have the same peak for two cytochromes which have quite different affinities for oxygen. I would have thought that if you had quite different affinities, you would have two different peaks. Put together, these two points worry me.

MILLS As to the first question, if we progressively increase the P_{O_2} of the perfusate, we do in fact find that we get a maximally oxidized state before we get to 100% oxygen. Certainly this doesn't mean that there is no impairment of diffusion: on the other hand, we do find complete oxidation under our experimental conditions and this has suggested to us that if there is impaired diffusion, it cannot be of much importance. Another point is that the response curve of the receptors to P_{O_2} is not obviously different to that seen in the intact, that is blood perfused, carotid body, and if there were significant areas where there was impaired diffusion, I would expect to see (1) a further decrease in the reduction state of cytochrome a_3 as I increased perfusate O_2 from say 40% to 100% and (2) the chemoreceptor response curve should be shifted over towards the right.

TORRANCE Where do you think the haemoglobin is?

MILLS In the blood vessels. We have confirmed this with histological sections.

TORRANCE Could you comment on the apparent identity of the two cytochrome peaks?

MILLS There is no real expectation that they are structurally different. It really has to do with the relative oxygen turnover in the chain. That is, you could get the chain becoming reduced at very high oxygen tensions if there is some other metabolic pathway that is consuming oxygen locally and dropping P_{O_2}. All this is to say that a difference in oxygen affinity does not necessarily mean a structural difference. For example, Hersey & Jöbsis (1969) showed that during H^+ secretion in gastric mucosa, cytochrome a_3 gets reduced – in fact the whole chain gets reduced. They measured P_{O_2} with a micro-electrode and it was very high. But this chain still by definition is generally considered to have an unusually low oxygen affinity.

EYZAGUIRRE I understand that you haven't studied the denervated carotid body?

MILLS There should be a batch with the sinus nerves degenerating now – I hope!

EYZAGUIRRE The carotid body has so many other things beside glomus cells, blood vessels, etc.: have you thought of the possibility of making tissue slices and using micro-spectrophotometry to aim your beam at different structures?

MILLS Well, first, it doesn't really matter what kinds of junk there are in the carotid body: all you measure is from mitochondria. I don't think this is technically possible yet: one difficulty is that the signal may be so poor that you have to use averaging techniques. It has been done on single cells but we are not set up to do this.

ACKER There is an interesting point with the carotid body. If you measure NADH in other haemoglobin-free tissues and give nitrogen, you always get reduction: but in the carotid body, we have also found that you first get oxidation. Now this may mean that the substrate hydrogen has an unusual pathway. In particular, what do you think that collagen may hydrolyse and so keep up the hydrogen? This might be a pathway.

MILLS I think I differ with you in your interpretation. Certainly in other tissues, if you cut off oxygen, you get NADH. But this is a tissue where oxygen is stimulating activity some place. So you have this weird situation. It is not like taking mitochondria and taking the oxygen away and you enter stage 5. It is a situation where you are taking oxygen away and that is the stimulus for activity. And all our observations show that down to about 7–9 torr P_{O_2} in the perfusate, there is enough oxygen to allow the mitochondria to go oxidized. As to your last point, about connective tissue, I really have no way of saying whether this is possible or not.

MCDONALD In what other tissues have you looked for this cytochrome?

MILLS I haven't looked for it elsewhere but Hersey & Jöbsis, as I said, found that, in gastric mucosa, cytochrome a_3 goes reduced when it secretes hydrogen ion. I agree with Dr Acker to some extent about this question of critical oxygen tension. I think there may be some significant differences between the characteristic states of the electron transport chain measured in a cuvette and that *in vivo*. I think there is evidence accumulating about that.

PURVES Two questions. First a technical one. What kind of penetration can you get with fluorimetry: how deep from the surface are you recording?

MILLS It is difficult to be certain but I think about 80 to 90 % of the signal comes from the outer 800 μm or so. Now in the carotid body it is not very much thicker than that. So we should be seeing the bulk of the tissue.

PURVES Second, with all these oxygen-dependent mechanisms in the carotid body, do you think there is room for the oxygen-dependent system such as Biscoe has suggested in the small nerve fibre to explain activation with hypoxia?

MILLS No, the difficulty I come up with concerning that idea is that if you assume that the nerve ending is the transducer, you are turning the thing on, as it were, by lowering P_{O_2}. And the oxygen tension, unless there is something very special, has also to be low at the point where the impulse is discharged and at that point you have to have some energy to run the pump. It would therefore seem to me that the small nerve ending would run down when stimulated by low P_{O_2}. Possibly the pump has a greater capacity than is apparent: but that is one of my main difficulties with that hypothesis.

O'REGAN The preparation which you use is perfused with saline which is notoriously unreliable.

MILLS What do you mean, notoriously unreliable?

O'REGAN Well, if you start perfusing the carotid body with saline the oxygen uptake is similar to that shown by Daly *et al.* (1954) and within 15–20 minutes it has fallen to 20 % of that value. If you readmit blood, the oxygen consumption rises to its original level. I know you have no alternative but this fact requires emphasizing and in some way should be incorporated in your interpretation.

MILLS I have acknowledged this point when I have written about it and it remains of interest to know why the oxygen consumption of the saline perfused organ should be so much lower than the blood perfused carotid body. However, if you get the colloid osmotic pressure up to normal, you can obtain apparently normal chemoreceptor responses for a long time. So, whatever it is that you have done by perfusing with saline, it doesn't seem to be too fundamental to the chemoreceptor response to oxygen.

REFERENCES

DALY, M. DE BURGH, LAMBERTSON, C. J., SCHWEITZER, A. (1954). Observations on the volume of blood flow and oxygen utilization of the carotid body in the cat. *J. Physiol.*, **125**, 67–89.

HERSEY, S. J. & JÖBSIS, F. F. (1969). Redox changes in the respiratory chain related to acid secretion by the intact gastric mucosa. *Biochem. biophys. Res. Commun.*, **36**, 243–50.

STARLINGER, H. & LÜBBERS, D. W. (1973). Polarographic measurements of the oxygen pressure performed simultaneously with optical measurements of the redox state of the respiratory chain in suspensions of mitochondria under steady-state conditions at low oxygen tensions. *Pflügers Arch. ges. Physiol.*, **341**, 15–22.

CHEMOREFLEXES AND THE CIRCULATION

Role of the arterial chemoreceptors in the control of the cardiovascular responses to breath-hold diving

By M. de BURGH DALY and JENNIFER E. ANGELL JAMES

Department of Physiology, Medical College of St. Bartholomew's Hospital, Charterhouse Square, London, EC1M 6BQ

The arterial chemcreceptors situated in the carotid and aortic bodies play an important role in the reflex control of the cardiovascular system (see Daly, 1972) as well as in the regulation of breathing (see Heymans & Neil, 1958). Studies of their cardiovascular effects, however, have given rise to difficulties of interpretation in the intact organism on account of the fact that their direct or primary effects on the heart and blood vessels can be modified by secondary mechanical, chemical and reflex effects. These secondary mechanisms include changes in pulmonary ventilation through alterations in arterial blood P_{O_2} and P_{CO_2} and in activity of certain pulmonary vagal receptors, changes in afferent input from receptors innervated by the trigeminal nerves, and changes in respiratory 'centre' activity. The central nervous system integrates the input from several cardiovascular and respiratory receptor groups. As a result of imposing a disturbance, the modification of the autonomic output is a reflection of variable changes in a number of individual inputs rather than a change in a single input. It is the purpose of this communication to describe the integrative control of the cardiovascular system by arterial chemoreceptors taking the responses observed in breath-hold diving as an example.

Submergence in water causes apnoea leading to asphyxia and changes in arterial haemoglobin saturation, P_{O_2} and P_{CO_2} (see Andersen, 1966; Scholander, 1940; also Elsner, 1969; Clausen & Ersland, 1970; Ferrante,

1970; Jones & Purves, 1970*a*). The changes in blood gas tensions are such that they would be expected to cause an increasing discharge from arterial chemoreceptors. But before dealing in detail with the role of these receptors, it is necessary to give a brief account of the cardiovascular responses associated with breath-hold diving.

CARDIOVASCULAR RESPONSES TO BREATH-HOLD DIVING

These responses have been reviewed by Andersen (1966) and more recently by Angell James & Daly (1972*b*). In every diving vertebrate studied, a profound bradycardia occurs immediately on submersion which often becomes more evident with time. It is most pronounced in aquatic animals but occurs, albeit less strikingly, in terrestrial species as well, such as the dog (Elsner, Franklin, Van Citters & Kenney, 1966) and man (Harding, Roman & Whelan, 1965; Strømme, Kerem & Elsner, 1970; Moore, Lin, Lally & Hong, 1972).

Equally dramatic are the changes in cardiac output. Immediately on diving or immersion of the head in water a reduction in output occurs (Murdaugh, *et al.*, 1966; Folkow, Nilsson & Yonce, 1967; Elsner, Franklin & Van Citters, 1964; Kawakami, Natelson & Dubois, 1967), which is approximately proportional to the fall in heart rate since there is either no change or a fall in stroke volume (Elsner *et al.*, 1964; Murdaugh *et al.*, 1966; Folkow *et al.*, 1967). In spite of these changes in flow the mean arterial pressure is relatively well maintained at the pre-dive or pre-immersion level (Van Citters *et al.*, 1965; Elsner *et al.*, 1964, 1966) and even in those species, such as the duck (Butler & Jones, 1971), in which a fall in pressure may occur, it is in no way comparable to the reduction in cardiac output. This means that the systemic vascular resistance must increase.

It has been established that vasoconstriction occurs in skin, muscle, gastro-intestinal tract and kidney (see Andersen, 1966) and that it is due to activation of sympathetic adrenergic fibres (Kobinger & Oda, 1969; Butler & Jones, 1971). Constriction of large arteries as well as resistance vessels (Heistad, Abboud & Eckstein, 1968; Folkow, Fuxe & Sonnenschein, 1966) and capacitance vessels (Djojosugito, Folkow & Yonce, 1969) also occurs. But the resistance in the coronary and cerebral circulations is relatively unaffected (Butler & Jones, 1971; Van Citters *et al.*, 1965) so that during diving the reduced cardiac output is directed towards those tissues, the the brain and heart muscle, which are least able to tolerate periods of hypoxia. In man too, breath-hold face immersion in water leads to vasoconstriction in the limbs (Elsner *et al.*, 1966; Craig, 1963; Campbell, Gooden, Horowitz, 1969).

The mechanisms involved in the production of these cardiovascular responses are numerous. As already indicated the responses are more marked in the marine mammals and diving birds but unfortunately it is in these two groups of animals that our knowledge of the physiology of their cardiovascular control mechanisms is still very meagre. Consequently it is necessary to fall back on the more detailed information, particularly about the integration of cardiovascular reflexes, which has been obtained from experiments on terrestrial mammals in attempting to explain some of the mechanisms underlying the responses occurring in diving. The results obtained on one animal species can only be directly transferred to another with considerable caution but it is with this proviso in mind that the following outline of some of the potential mechanisms will be presented.

ARTERIAL CHEMORECEPTORS

Hollenberg & Uvnäs (1963) and Holm & Sørensen (1972) found in the unanaesthetized duck that asphyxia produced by immersing the beak or head in water caused bradycardia and a rise of blood pressure, both effects being abolished by combined denervation of the carotid baroreceptors and chemoreceptors. Jones & Purves (1970*a*) were to first demonstrate unequivocally the role of the arterial chemoreceptors in the production of the bradycardia of head immersion in the duck. The carotid bodies appear to be the only functional group of peripheral arterial chemoreceptors in this species and regulate respiration reflexly in a way similar to that observed in non-diving animals (Jones & Purves, 1970*b*). In ducks whose carotid bodies were denervated, Jones & Purves (1970*a*) showed that although the latency of onset of the cardiac response was unaffected, the rate of the initial reduction in heart rate and the size of the response were considerably diminished (Fig. 1). The heart rate during head immersion in the carotid body denervated animals fell only 10% compared to 76% in the control group. The carotid bodies therefore make a considerable contribution to the bradycardia of head immersion and help to maintain the arterial blood P_{O_2} at a higher level during asphyxia than would otherwise be the case (Jones & Purves, 1970*b*).

No work has yet been carried out on the contribution of the chemoreceptors to the cardiovascular responses to diving in marine mammals, although Cetacea and Pinnipedia have been shown histologically to possess carotid bodies similar in structure of those of terrestrial mammals (de Kock, 1959, 1960). It seems pertinent therefore to inquire further into the

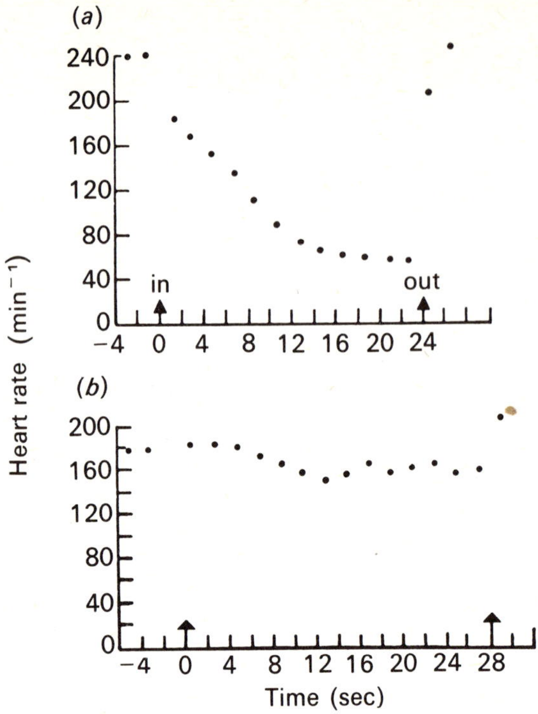

Fig. 1. The effect of submergence of the head of a duck in water on heart rate. (*a*) Control; (*b*) 6 days after denervation of the carotid bodies (Jones & Purves, 1970*a*).

potential chemoreceptor mechanisms that might operate in bringing about the cardiovascular responses during diving from evidence obtained from studies on terrestrial mammals.

Apnoeic asphyxia causes bradycardia and intense vasoconstriction in the limbs and muscle in the dog, sheep and seal (Irving, 1938; Nahas, 1956; Elsner, Hammond & Parker, 1970) but an increase in blood flow through the brain (Irving, 1938). These responses are similar to the primary cardio-vascular responses elicited from the carotid bodies (for references, see Daly, 1972; also Calvelo, Abboud, Ballard & Abdel-Sayed, 1970; Ponte & Purves, 1974). Although there is still some doubt about the control of heart rate by the aortic bodies (Daly, 1972), combined stimulation of the carotid and aortic bodies induces bradycardia (Angell James & Daly, 1969*a*). The predominant vascular response to aortic body stimulation is constriction (Daly, Hazzledine & Howe, 1965; Daly, Hazzledine & Ungar, 1967; Dawes *et al.*, 1968). That the cardio-vascular response and increased systemic vascular resistance of apnoeic asphyxia are due largely to stimulation of the carotid and aortic bodies has been demonstrated by Angell

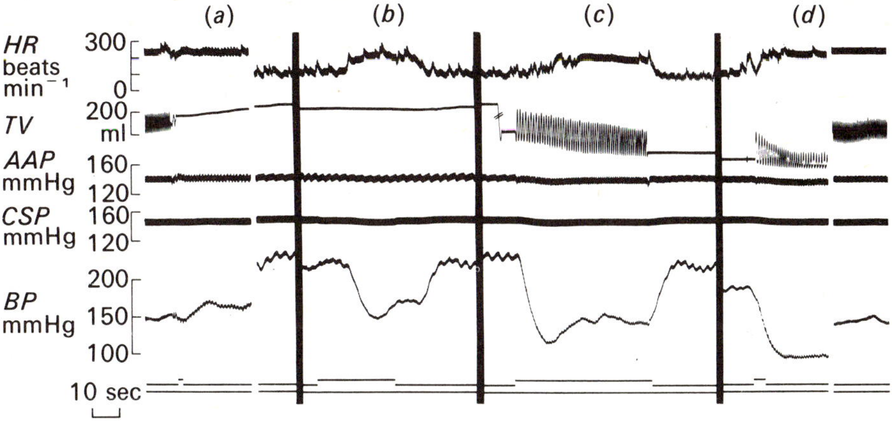

Fig. 2. The effects of apnoeic asphyxia, of withdrawing the arterial chemoreceptor 'drive' and of re-establishing lung movements during apnoeic asphyxia. Dog, male, 13.4 kg: morphine–chloralose–urethane anaesthesia. Perfusion of the systemic circulation at constant blood flow with separate perfusions of the isolated carotid sinuses and isolated aortic arch. No pulmonary circulation; systemic venous blood oxygenated in isolated perfused lungs of a donor animal with closed chest; spontaneous respiration. Carotid sinus, aortic arch and arterial blood P_{O_2} 133 mm Hg, P_{CO_2} 45 mm Hg, pH 7.34.

(*a*) At signal, apnoeic asphyxia carried out by reducing the tidal volume of the isolated perfused donor lungs from 400 to 75 ml., and at the same time creating a bilateral open pneumothorax in the recipient animal, the lungs being held in a semi-inflated position by occluding the trachea. Break in record lasted 3.5 minutes.

(*b*) Apnoeic asphyxia continued (carotid sinus, aortic arch and arterial blood P_{O_2} 64 mm Hg, P_{CO_2} 70 mm Hg, 7.19). During signal, carotid and aortic body chemoreceptor 'drive' withdrawn by substituting oxygenated blood (P_{O_2} 140 mm Hg, P_{CO_2} 44 mm Hg, pH 7.39).

(*c*) Apnoeic asphyxia continued (carotid sinus, aortic arch and arterial blood P_{O_2} 74 mm Hg, P_{CO_2} 70 mm Hg, pH 7.19). During signal pneumothorax reduced and spontaneous breathing temporarily re-established.

(*d*) At signal, end of period of apnoeic asphyxia. Spontaneous breathing re-established and tidal volume of isolated perfused donor lungs restored to 400 ml. Break in record of 4 min.

HR, heart (atrial) rate; *TV*, tidal volume (inspiration upwards); *CSP*, carotid sinus pressure; *AAP*, aortic arch pressure; *BP*, systemic arterial perfusion blood pressure. (Angell James & Daly 1969*a*.)

James & Daly (1969*a*). In Fig. 2*a* apnoeic asphyxia caused a reduction in heart rate and a rise in systemic blood pressure which, since the blood flow was maintained constant, indicates an increase in systemic vascular resistance. Withdrawal of chemoreceptor 'drive' during the period of of apnoeic asphyxia produces tachycardia and a reduction in vascular resistance, whereas reintroduction of hypoxic hypercapnic blood to the chemoreceptors reverses this effect (Fig. 2*b*).

These findings, therefore, lend support to the view that in diving, apnoeic asphyxia is an important mechanism contributing to the cardiac response (Andersen, 1963*a*,*b*; Feigl & Folkow, 1963), and probably also to the vasoconstriction that occurs in a large proportion of the systemic vascular bed, and that these responses are largely reflexly engendered through stimulation of the arterial chemoreceptors.

If we are correct in our analysis so far, then it remains to be explained how it is that when during breath-hold diving the arterial chemoreceptors are being stimulated to such an extent that they contribute reflexly to the production of the cardiovascular response, they do not produce a respiratory 'break-through' and force the animal to 'surface' earlier than in fact occurs. Before attempting to answer this question it will be necessary to discuss another important mechanism contributing to the respiratory and cardiovascular responses of breath-hold diving, that is, reflexes arising from receptors in the nose and on the face.

Trigeminal nerve reflex

Powerful respiratory and cardiovascular responses can be elicited by stimulation of the nasal mucous membrane as first described in the rabbit by Kratschmer (1870). These include apnoea in the expiratory position, bradycardia and vasoconstriction in skin, muscle and intestine, and also in the splenic and renal circulations (seal: Irving, Scholander & Grinnell, 1942; dog: Angell James & Daly, 1969*b*, 1972*a*; rabbit: Tomori, Javorka & Stransky, 1972; White & McRitchie, 1973). Other effects include contraction of the capsule of the spleen (dog: Angell James & Daly, 1972*a*), a reduction in cardiac output (dog: Angell James & Daly, 1969*b*) and an increased secretion of suprarenal catecholamines (rabbit: Allison & Powis, 1971). The afferent pathway for all these effects is the trigeminal nerve, and the efferent pathway for the cardiac response is the vagus and for the vascular responses the sympathetic adrenergic fibres.

Receptors around the nose and on the face can also elicit bradycardia and vasoconstriction during immersion of the head in water in man (Elsner *et al.*, 1966; Whayne & Killip, 1967; Heistad & Wheeler, 1970; Moore *et al.*, 1972), in the seal (Scholander, 1940) and in the duck (Andersen, 1963*a*; and others). These responses are accompanied by apnoea in the expiratory position which is reflexly engendered (Huxley, 1913), but they can occur independently of changes in breathing.

Insofar as the cardio-inhibitory response is concerned, the bradycardia due to face immersion is the result of a number of factors: breath-holding *per se*, the temperature of the water, the depth to which the rest of the body is immersed in water, and the level of the alveolar P_{O_2} (Kawakami *et al.*, 1967; Moore *et al.*, 1972).

Chemoreceptor-trigeminal nerve reflex interactions

We can now return to the problem of the apparent depression of the chemoreceptor–respiratory reflex during apnoeic asphyxia in diving. In experi-

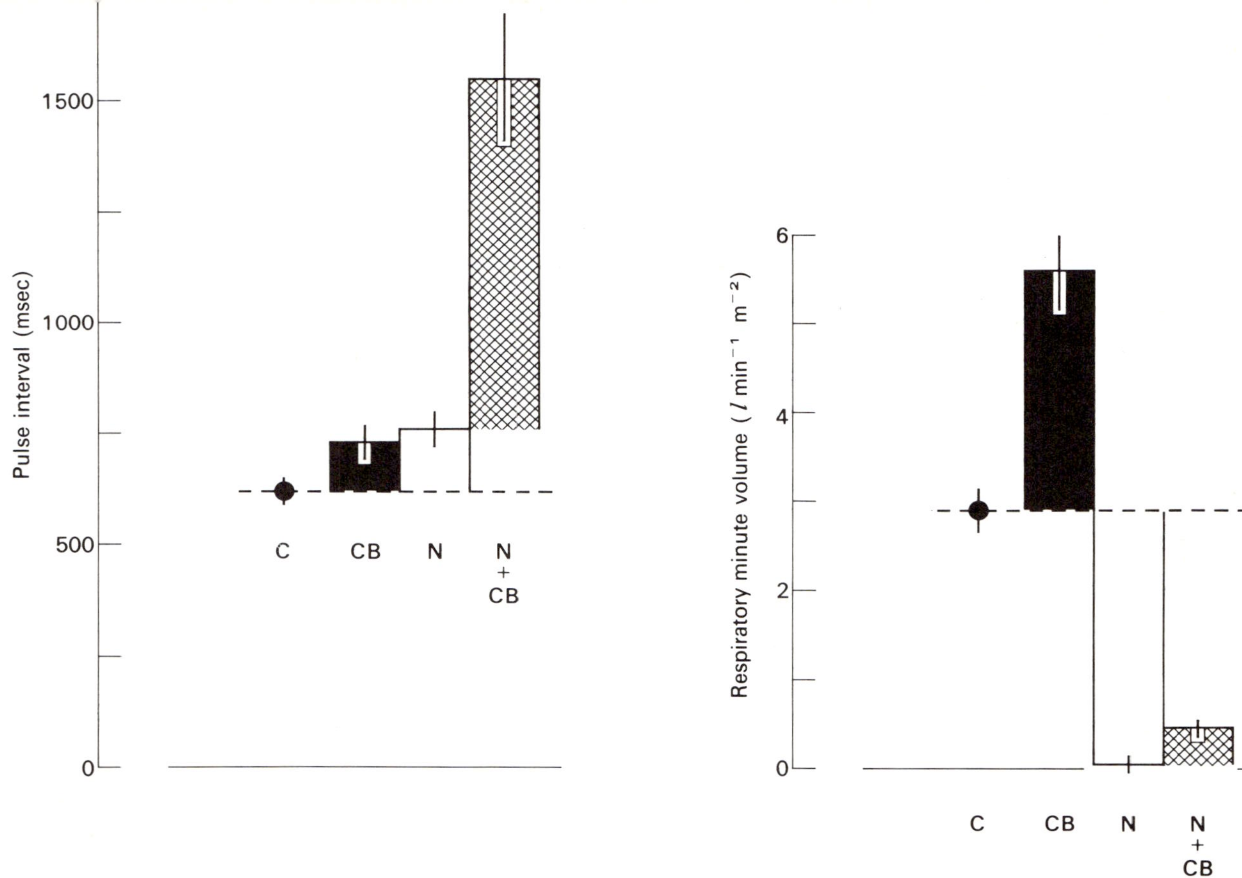

Fig. 3. The effects of stimulation of the carotid body chemoreceptors alone (CB, filled blocks) and during excitation of receptors in the nose (N+CB, cross-hatched blocks) on pulse interval and respiratory minute volume. Open blocks (N), stimulation of the nose alone. The filled circles (●), C, are the control values. The values represent the means ± S.E. of mean from 22 tests in 7 animals. (Angell James & Daly, 1973.)

ments designed specifically to throw light on this problem Angell James & Daly (1973) showed that there is an interaction between the reflexes elicited from the nose on the one hand, and the carotid body–respiratory and carotid body–cardio-inhibitory reflexes on the other. Whereas excitation of the nasal reflex inhibited the carotid body–respiratory reflex, it facilitated the carotid body–cardio-inhibitory reflex (Fig. 3). Thus stimulation of the chemoreceptors in the control state caused an increase in respiratory minute volume which was abolished or reduced in size when the test was repeated during excitation of the nasal mucous membrane by water. On the other hand, the chemoreceptor–cardio-inhibitory response was enhanced when combined with nasal stimulation. A similar conclusion concerning the chemoreceptor–cardiac reflex was reached by Holm & Sørensen (1972) on the basis of their experiments on the unanaesthetized duck. These findings are also compatible with those of Jones & Purves (1970*a*) who observed in unanaesthetized ducks that a level of arterial hypoxaemia sufficient to cause a three-fold increase in pulmonary ventilation due to stimulation of the carotid bodies (Jones & Purves, 1970*b*) failed to produce a respiratory 'break-through' during an apnoea elicited by head immersion.

The powerful inhibitory effect on respiration of the trigeminal nerve afferent input is therefore an important mechanism preventing early respiratory 'break-through' during apnoeic asphyxia.

OTHER ARTERIAL CHEMORECEPTOR INTERACTIONS

There are some other mechanisms which may not only contribute to the cardiovascular responses to breath-hold diving but may also modify the chemoreceptor component in such a way as to unmask the primary chemoreceptor cardiovascular effects.

Pulmonary vagal reflex

Cardiovascular responses are initiated by changes in the degree of stretch or inflation of the lungs and probably also by the rate of change of lung volume. Inflation of the lungs causes tachycardia due largely to a reduction in vagal tone (Anrep, Pascual & Rössler, 1936*a*,*b*; Daly & Scott, 1958) and a decrease in systemic vascular resistance (Salisbury, Galletti, Lewin & Rieben, 1959; Daly *et al.*, 1967; Daly & Robinson, 1968; Ott & Shepherd, 1971; Ott, Lorenz & Shepherd, 1972) due to vasodilatation in skin, muscle and the splanchnic vascular bed through a reduction in activity of sympathetic adrenergic fibres (Daly *et al.*, 1967; Daly & Robinson, 1968). This is a low threshold, slowly adapting reflex with its afferent pathway lying in the vagus nerves (Anrep *et al.*, 1936*a*; Daly & Scott, 1958; Daly *et al.*, 1967; Daly & Robinson, 1968; Angell James & Daly, 1969*a*).

A reduction in lung volume or collapse of the lungs produces the opposite cardiovascular effects to lung inflation and evidence has been presented that these responses result from a reduction in activity of the pulmonary receptors (Anrep *et al.*, 1936*a*; Daly *et al.*, 1967). This reflex could therefore be a mechanism contributing to the bradycardia and vasoconstriction of breath-hold diving.

Chemoreceptor-pulmonary vagal reflex interaction

A pulmonary vagal inflation reflex has been shown to inhibit the primary cardiac response from the carotid bodies (Daly & Hazzledine, 1963) and the primary vascular responses from the carotid and aortic bodies (Daly & Ungar, 1966; Angell James & Daly, 1969*a*). Such interactions have been demonstrated under conditions in which chemoreceptor stimulation is accompanied by hyperventilation. It follows that during apnoea when the activity of the pulmonary receptors is reduced, their opposing influence on the chemoreceptor-cardiovascular reflexes would be removed or at least considerably reduced. The experiment from which Fig. 2 is taken demonstrates this point. In (*a*) apnoeic asphyxia caused bradycardia and an increased systemic vascular resistance due to stimulation of the arterial chemoreceptors (*b*). In (*c*), the respiratory movements of the lungs were temporarily restored during the period of asphyxia and without altering the arterial blood P_{O_2} and P_{CO_2}, and this resulted in tachycardia and a reduction in vascular resistance to approximately their control values. Cessation of lung movements reversed these effects. This over-riding influence of the lung movements is abolished by denervation of the lungs (Angell James & Daly, 1969*a*).

Trigeminal nerve-pulmonary vagal reflex interaction

If the lungs are inflated during the apnoeic period of stimulation of the nasal mucous membrane, the cardio-inhibitory and vasoconstrictor responses due to the trigeminal nerve reflex are temporarily reduced in size. This opposing influence of lung inflation is abolished by denervation of the lungs (J. E. Angell James and M. de B. Daly, unpublished observations).

Respiratory 'centre' activity

Impulses arising from the respiratory 'centres' during the phase of inspiration were thought by Anrep *et al.* (1936*b*) to 'irradiate' to the cardio-inhibitory 'centre' causing its inhibition and hence tachycardia. It can be demonstrated in paralysed animals during diffusion respiration (Joels & Samueloff, 1956). In the experiment from which Fig. 4 is taken, the lungs

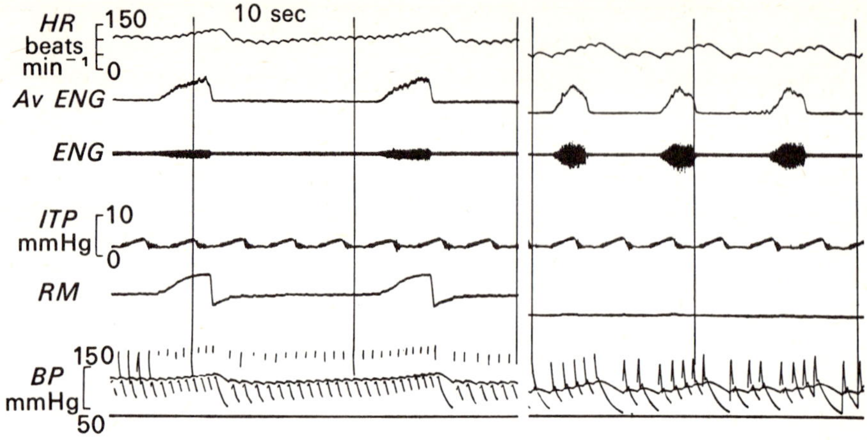

Fig. 4. Dog, female, 14.5 kg: morphine–chloralose–urethane anaesthesia: open chest: positive pressure ventilation: both lungs denervated (Daly & Scott, 1958). Records of: *HR*, heart rate; *Av ENG*, averaged phrenic electroneurogram; *ENG*, phrenic electroneurogram (efferent activity); *ITP*, intratracheal pressure; *RM*, rib movements (inspiration upwards); and BP, arterial blood pressure (mean and phasic). Between the two records, decamethonium, 0.25 mg kg^{-1}; time marker, 10 seconds.

were denervated to prevent their movements by artificial respiration from influencing the respiratory 'centres'. It may be noted that inspiratory movements of the ribs and efferent activity in fibres of the phrenic nerve are accompanied by acceleration of the heart. The same relationship between phrenic nerve activity and tachycardia persists after injecting a neuromuscular blocking agent showing that it is not the result of a mechanical effect of respiratory movements on the circulation.

It has also been postulated that during expiration, activity of the experitory 'centre' causes stimulation of the cardio-inhibitory 'centre' and contributes to the slowing of the heart that occurs during this phase of the respiratory cycle (Cordier & Heymans, 1935). The interaction between respiratory 'centre' activity and the pulmonary vagal inflation reflex in the control of heart rate have been discussed by Anrep *et al.* (1936*b*). The cardio-inhibitory 'centre' is under two influences from the lungs during respiration – a direct effect inhibiting the 'centre' and causing tachycardia, and an indirect one through the respiratory 'centres' whereby at the height of inspiration lung inflation inhibits the respiratory 'centres' which in turn reduce the inhibition on the cardio-inhibitory 'centre' causing slowing of the heart.

It follows that cessation of respiratory 'centre' activity during breath-hold diving may contribute to the slowing of the heart, though probably only to a small extent.

Chemoreceptor-respiratory 'centre' interaction

The question now arises as to whether the primary cardiac response from the carotid bodies can be influenced by respiratory 'centre' activity as it is by the pulmonary vagal inflation reflex.

It has been observed previously (Daly & Scott, 1958) that although the average heart rate slowed in response to carotid body stimulation, sinus arrhythmia sometimes became more evident. In the experiment from which Fig. 5 is taken, the anaesthetized dog was artificially ventilated and both

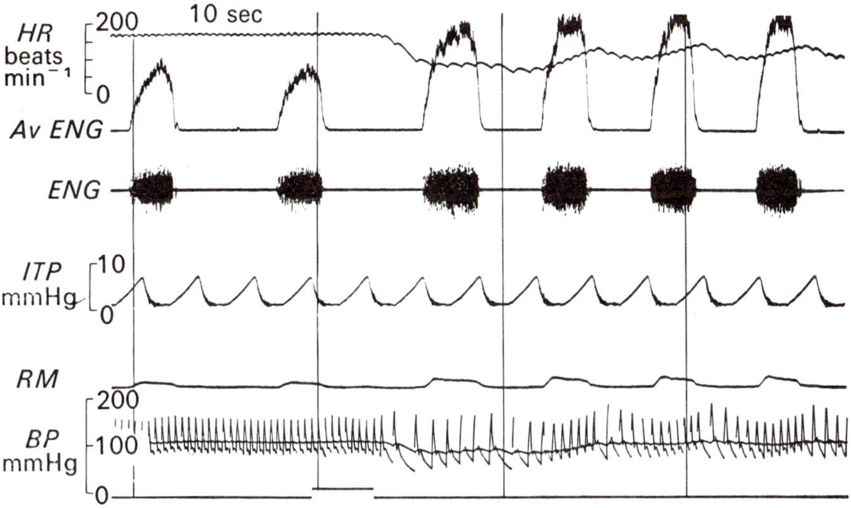

Fig. 5. Dog, male, 10.2 kg: morphine–chloralose–urethane anaesthesia: open chest: positive pressure ventilation: both lungs denervated (Daly & Scott, 1958). Records of: *HR*, heart rate; *Av ENG*, averaged phrenic electroneurogram; *ENG*, phrenic electroneurogram (efferent activity); *ITP*, intratracheal pressure; *RM*, rib movements (inspiration upwards); *BP*, arterial blood pressure (mean and phasic). At signal, intracarotid injection of 5 $\mu g\ kg^{-1}$ sodium cyanide; time marker, 10 seconds.

lungs were denervated to exclude the participation of pulmonary reflexes. Stimulation of the carotid body by intracarotid cyanide caused increased phrenic nerve activity and slowing of the heart, but it may be noted that periods of tachycardia were now superimposed on the slowing and were associated with the burst of activity in the phrenic nerve. A plot of the relationship between the averaged phrenic nerve impulse activity and pulse interval for a number of spontaneous 'respiratory' cycles before and during excitation of the carotid bodies is shown in Fig. 6. Sinus arrhythmia was absent during the control state, but became evident, as indicated by the presence of a 'loop' when the pulse interval had increased from *A* to *B*. As the effect of the cyanide wore off, the position of the 'loops' shifted towards the original control length of the pulse interval.

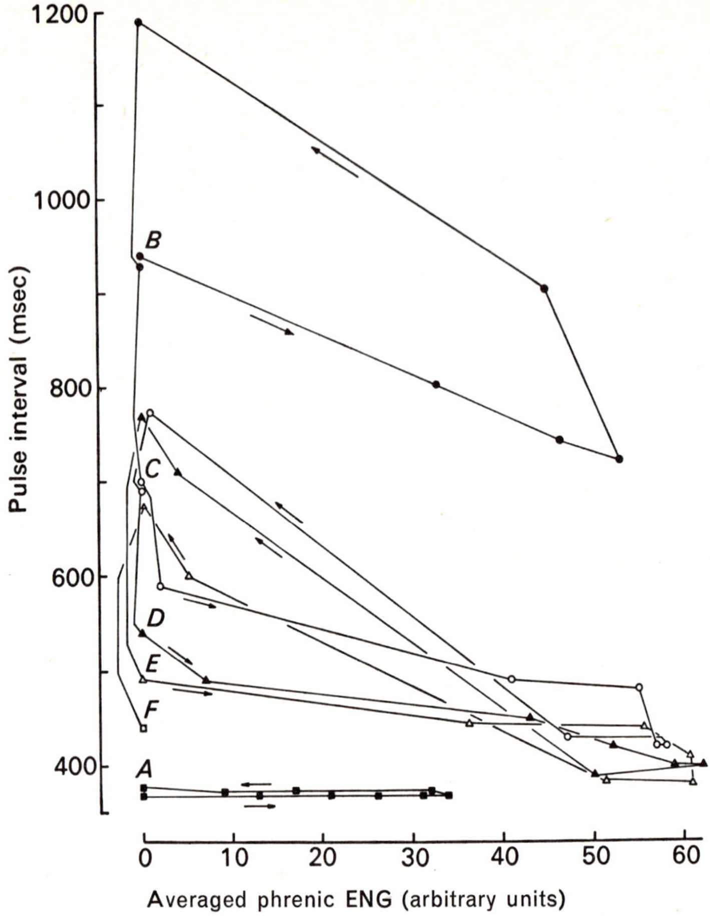

Fig. 6. Data plotted from experiment illustrated by Fig. 5, showing relationship between pulse interval and averaged phrenic electroneurogram. *A* (■), control; *B* (●), *C* (○), *D* (▲) and *E* (△) represent the beginning of each loop respectively after the injection of cyanide.

Although the evidence is not yet available it might be anticipated from these results that during the stimulation of the carotid bodies, as in breath-hold diving, the size of the cardio-inhibitory response would be greater in the absence of respiratory 'centre' activity produced, for example, by activation of the trigeminal nerve reflex (Angell James & Daly, 1973).

These findings suggest that the absence of rhythmic breathing *per se* is an important integrative mechanism whereby the bradycardia of breath-hold diving becomes evident through other mechanisms now being unopposed. We have attempted to obtain evidence that a similar situation occurs in man.

The following experiment was made on four subjects whose respiration was measured through a respiratory valve and mouth-piece by a spirometer connected to a bag-in-box apparatus. The typical experiment is illustrated by Fig. 7. This subject had a sinus arrhythmia lying in the

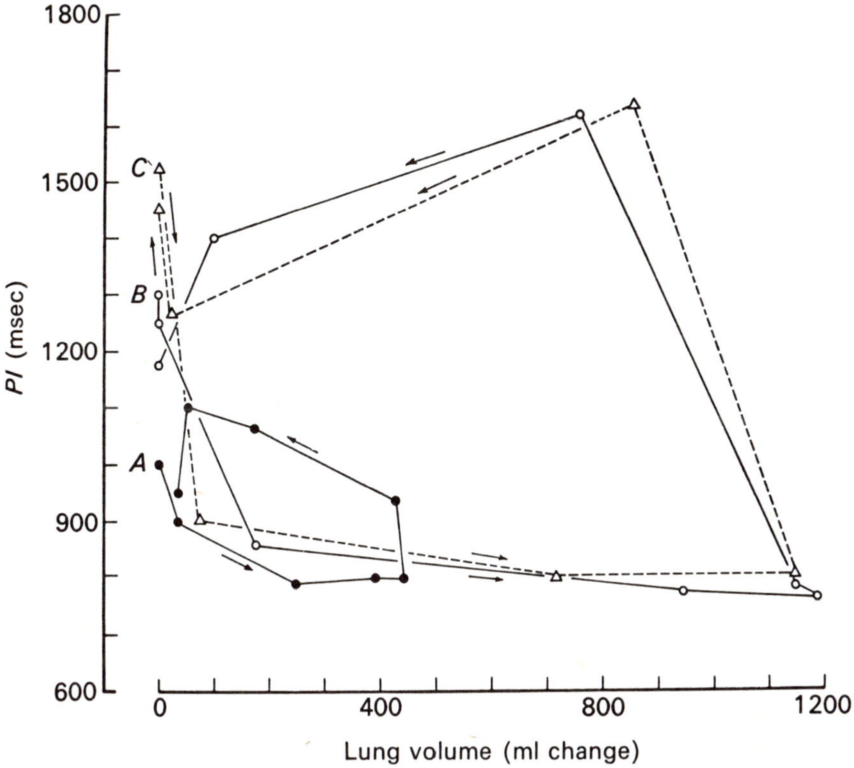

Fig. 7. Subject JB in prone position. Respiration measured by spirometer and bag-in-box method. The relationship between pulse interval (*PI*) and change in lung volume, during single breaths, zero being taken as the end-expiratory level. *A* (●), normal air breathing, end-tidal P_{O_2} 100.5 mm Hg, P_{CO_2} 38.8 mm Hg. Between *A* and *B*, apnoeic face immersion in water at 26 °C, 'breaking point' end-tidal P_{O_2} 76.3 mm Hg, P_{CO_2} 45.6 mm Hg. *B* (○), at 'breaking point', single breath of air while still submerged. End-tidal P_{O_2} 95 mm Hg, P_{CO_2} 41 mm Hg. Between *B* and *C*, second period of apnoeic face immersion. *C* (△), at 'breaking point', single breath of a gas mixture containing 10.5 % O_2, 6.5 % CO_2 and 83 % N_2. End-tidal P_{O_2} 81 mm Hg, P_{CO_2} 45.1 mm Hg. Points *A*, *B* and *C* represent the beginning of each loop respectively.

prone position and the relationship between pulse interval and change in lung volume during a single respiratory cycle is illustrated by the 'loop' commencing at point *A*. It may be noted that at a given lung volume the pulse interval is shorter during the inspiratory than during the expiratory phase of the cycle. Face immersion in water combined with breath-holding at the end-expiratory level caused an increase in pulse interval (B Fig. 7). At the 'breaking-point' one breath of air was taken which caused a transient tachycardia related to the change in lung volume. From the time relationship of this response, it is unlikely to be due to withdrawal of the chemoreceptor 'drive' consequent upon restoration of the end-tidal gas tensions towards their pre-immersion level. Support for this view is provided by the results of another test illustrated by Fig. 7. The subject, at or near his 'breaking-point', inhaled a gas mixture of 10.5 % oxygen and 6.5 % carbon dioxide in nitrogen which maintained nearly constant his end-tidal P_{O_2} and P_{CO_2} at the 'breaking-point' level. A cardio-accelerator response ('loop' C) similar to that produced by breathing air ('loop' *B*) was obtained. It is concluded that under these conditions the over-riding effect on heart rate associated with changes in lung volume is of nervous origin. But it is not possible to establish from these experiments whether this response is due to an effect of respiration, such as respiratory 'centre' activity, a pulmonary vagal inflation reflex or afferent nerve activity from receptors in the chest wall, or is secondary to a change of arterial blood pressure and therefore mediated by a baroreceptor reflex. Further experiments are required to elucidate this point. Whatever the mechanism, it might be anticipated from Fig. 7, 'loop' *C*, that at levels of end-tidal P_{O_2} and P_{CO_2} at the 'breaking-point' combined with face immersion in water, the heart rate will be slower when associated with apnoea than in the presence of breathing at a level of alveolar ventilation commensurate with the degree of asphyxia.

ARTERIAL BARORECEPTORS

Arterial chemoreceptor–baroreceptor interaction

The possibility that the arterial baroreceptors contribute to the diving bradycardia has been discussed by Andersen (1966) and by Kobinger & Oda (1969). The last-mentioned authors concluded that they were of secondary importance and this view would be in keeping with the fact that the mean arterial pressure often undergoes little change. The fact, however, that bradycardia and vasoconstriction occur at relatively constant blood pressure means that the normal relationships between arterial blood pressure and heart rate, and between arterial blood pressure and vascular resistance, are altered. That is to say, the baroreceptor–cardiac *reflex* is reset towards slowing of the heart and the baroreceptor–vasomotor *reflex* is reset towards an increased vascular resistance. This resetting could be accounted for by a central interaction resulting from stimulation of the trigeminal nerve reflex and of the arterial chemoreceptors. In this connection it has been shown that stimulation of the nasal mucous membrane at constant arterial blood pressure causes bradycardia and vasoconstriction in a number of vascular territories (Angell James & Daly, 1972*a*). At constant pressure in the carotid sinuses and aortic arch, combined stimulation of the carotid and aortic bodies results in bradycardia and an increase in systemic vascular resistance (Angell James & Daly, 1969*a*). Whether resetting of the baroreceptor *reflexes* is accompanied by changes in their sensitivity is at present not known.

A summary of some of the mechanisms discussed above are shown diagrammatically in Fig. 8.

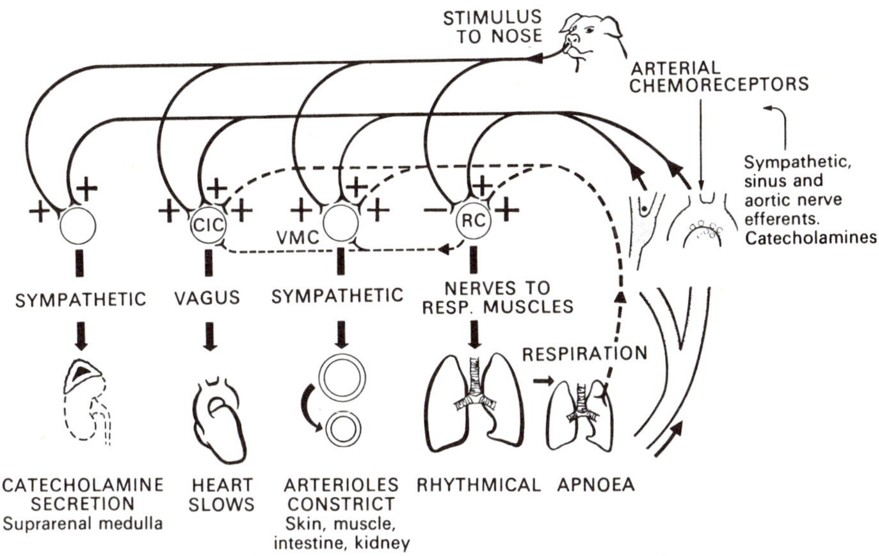

Fig. 8. Diagrammatic representation of proposed sequence of events occurring in mammals during submersion. Stimulation by water of trigeminal nerve receptors in and around the nose and on the face leads to reflex apnoea in the expiratory position, bradycardia, vasoconstriction and secretion of suprarenal catecholamines (*open lines*). The apnoea in turn causes a reduction in activity of a pulmonary vagal reflex resulting in bradycardia and vasoconstriction (*solid lines*). Cessation of activity of the respiratory 'centres' slows the heart, but the effect, if any, on the vasomotor 'centre' is not known (*thin interrupted lines*). Then arterial hypoxia and hypercapnia stimulate the arterial chemoreceptors which reflexly reinforce the cardio-inhibitory, vasoconstrictor and humoral responses (*thick interrupted lines*). CIC, cardio-inhibitory 'centre'; VMC, vasomotor 'centre'; RC, respiratory 'centres'. +, stimulation; −, inhibition. For further details, see text.

EMERGENCE FROM WATER

The end of a dive and emergence from water is accompanied by hyperventilation, tachycardia and vasodilatation. From the foregoing discussion the following mechanisms probably contribute to these cardiovascular effects:

(1) withdrawal of the chemoreceptor 'drive' (Angell James & Daly, 1969*a*);
(2) abolition of the trigeminal nerve reflex;
(3) initiation of the pulmonary vagal inflation reflex through hyperventilation causing cardio-acceleration and vasodilatation (Anrep *et al.*, 1936*a*; Daly & Scott, 1958; Daly *et al.*, 1967), and at the same time inhibition of the primary cardiovascular responses from the arterial chemoreceptors (Daly & Scott, 1958; Daly, 1972);
(4) respiratory 'centre' activity causing tachycardia (Anrep *et al.*, 1936*b*; Joels & Samueloff, 1956; this paper) and partly over-riding the chemoreceptor primary cardiac response.

In this connection Andersen (1963*a*) observed in the duck that the conspicuous post-dive tachycardia was brought about during the first respiratory effort.

SUMMARY

Evidence is presented that chemoreceptor reflexes contribute to the cardiovascular responses to breath-hold diving.

Breath-hold diving in vertebrates causes bradycardia, a reduction in cardiac output, and a diminution in blood flow accompanied by vasoconstriction in skin, muscle, intestine and kidney. The cerebral blood flow is relatively well maintained, as is the mean arterial blood pressure. The following mechanisms are probably involved in the initiation and maintenance of these responses:

(1) the trigeminal nerve reflex initiated by face immersion in water;
(2) cessation of respiratory 'centre' activity;
(3) a reduction in the activity of a pulmonary vagal stretch reflex, particularly if the lungs are held in the expiratory position, and
(4) stimulation of peripheral arterial chemoreceptors by the progressively increasing hypoxia and hypercapnia.

The chemoreceptor contribution, however, depends on a number of integrative mechanisms. Bradycardia and vasoconstriction represents the primary cardiovascular responses to stimulation of the carotid bodies. Vasoconstriction also occurs as a result of stimulation of the aortic bodies. The appearance of these chemoreceptor effects is dependent on the absence of secondary antagonistic mechanisms, such as respiratory 'centre' activity and an increase in the vagal input from the lungs, which themselves contribute to the diving response. Thus apnoea associated with asphyxia during diving may play an important role in the integrative mechanisms controlling the circulation whereby the primary cardiovascular responses from chemoreceptors are prepotent and unopposed.

During breath-hold diving the normal chemoreceptor-respiratory reflex is inhibited by concomitant excitation of a trigeminal nerve reflex by water on the face or in and around the nostrils. The chemoreceptor–cardio-inhibitory reflex, however, is facilitated.

Evidence is presented that the normal relationships between arterial blood pressure and heart rate, and between arterial blood pressure and vascular resistance, is altered implying that resetting of the arterial baroreceptor–cardiac and baroreceptor–vasomotor *reflexes* occurs towards bradycardia and vasoconstriction respectively by a central integration engendered by stimulation of a trigeminal nerve reflex and of the arterial chemoreceptors.

We wish to express our thanks to Mr D. R. Bacon for expert technical assistance. Our work cited was generously supported by grants from the Medical Research Council, Royal Society and St Bartholomew's Hospital Endowment Fund. Figures 1, 2 and 3 are reproduced by kind permission of the editors of the *Journal of Physiology*.

REFERENCES

ALLISON, D. J. & POWIS, D. A. (1971). Adrenal catecholamine secretion during stimulation of the nasal mucous membrane in the rabbit. *J. Physiol.*, **217**, 327–40.

ANDERSEN, H. T. (1963*a*). Factors determining the circulatory adjustments to diving. I. Water immersion. *Acta physiol. Scand.*, **58**, 173–85.

ANDERSEN, H. T. (1963*b*). Factors determining the circulatory adjustments to diving. II. Asphyxia. *Acta physiol. Scand.*, **58**, 186–200.

ANDERSEN, H. T. (1966). Physiological adaptations in diving vertebrates. *Physiol. Rev.*, **46**, 212–43.

ANREP, G. V., PASCUAL, W. & RÖSSLER, R. (1936*a*). Respiratory variations of the heart rate. I. The reflex mechanism of the respiratory arrhythmia. *Proc. R. Soc. B.*, **119**, 191–217.

ANREP, G. V., PASCUAL, W. & RÖSSLER, R. (1936*b*). Respiratory variations of the heart rate. II. The central mechanism of the respiratory arrhythmia and the inter-relations between the central and the reflex mechanisms. *Proc. R. Soc. B.*, **119**, 218–30.

ANGELL JAMES, J. E. & DALY, M. DE B. (1969*a*). Cardiovascular responses in apnoeic asphyxia: role of arterial chemoreceptors and the modification of their effects by a pulmonary vagal inflation reflex. *J. Physiol.*, **201**, 87–104.

ANGELL JAMES, J. E. & DALY, M. DE B. (1969*b*). Nasal reflexes. *Proc. R. Soc. Med.*, **62**, 1287–93.

ANGELL JAMES, J. E. & DALY, M. DE B. (1972*a*). Reflex respiratory and cardiovascular effects of stimulation of receptors in the nose of the dog. *J. Physiol.*, **220**, 673–96.

ANGELL JAMES, J. E. & DALY, M. DE B. (1972*b*). Some mechanisms involved in the cardiovascular adaptations to diving. *Soc. exp. Biol. symp.*, **26**, 313–41.

ANGELL JAMES, J. E. & DALY, M. DE B. (1973). The interaction of reflexes elicited by stimulation of carotid body chemoreceptors and receptors in the nasal mucosa affecting respiration and pulse interval in the dog. *J. Physiol.*, **229**, 133–49.

BUTLER, P. J. & JONES, D. R. (1971). The effect of variations in heart rate and regional distribution of blood flow on the normal pressor response to diving in ducks. *J. Physiol.*, **214**, 457–80.

CALVELO, M. G., ABBOUD, F. M., BALLARD, D. R. & ABDEL-SAYED, W. (1970). Reflex vascular responses to stimulation of chemoreceptors with nicotine and cyanide.

Activation of adrenergic constriction in muscle and noncholinergic dilatation in dog's paw. *Circulation Res.*, **27**, 259–76.

Campbell, L. B., Gooden, B. A. & Horowitz, J. D. (1969). Cardiovascular responses to partial and total immersion in man. *J. Physiol.*, **202**, 239–50.

Clausen, G. & Ersland, A. (1970). Blood O_2 and acid-base changes in the beaver during submersion. *Respiration Physiol.*, **11**, 104–12.

Cordier, D. & Heymans, C. (1935). *Le Centre Respiratoire*, Paris.

Craig, A. B. Jr (1963). Heart rate responses to apnoeic underwater diving and to breath holding in man. *J. appl. Physiol.*, **18**, 854–62.

Daly, M. de B. (1972). Interaction of cardiovascular reflexes. *Lect. scient. Basis Med.* 307–32.

Daly, M. de B. & Hazzledine, J. L. (1963). The effects of artificially induced hyperventilation on the primary cardiac reflex response to stimulation of the carotid bodies in the dog. *J. Physiol.*, **168**, 872–89.

Daly, M. de B., Hazzledine, J. L. & Howe, A. (1965). Reflex respiratory and peripheral vascular responses to stimulation of the isolated perfused aortic arch chemoreceptors of the dog. *J. Physiol.*, **177**, 300–22.

Daly, M. de B., Hazzledine, J. L. & Ungar, A. (1967). The reflex effects of alterations in lung volume on systemic vascular resistance in the dog. *J. Physiol.*, **188**, 331–51.

Daly, M. de B. & Robinson, B. H. (1968). An analysis of the reflex systemic vasodilator response elicited by lung inflation in the dog. *J. Physiol.*, **195**, 387–406.

Daly, M. de B. & Scott, M. J. (1958). The effects of stimulation of the carotid body chemoreceptors on heart rate in the dog. *J. Physiol.*, **144**, 148–66.

Daly, M. de B. & Ungar, A. (1966). Comparison of the reflex responses elicited by stimulation of the separately perfused carotid and aortic body chemoreceptors in the dog. *J. Physiol.*, **182**, 379–403.

Dawes, G. S., Lewis, B. V., Milligan, J. E., Roach, M. R. & Talner, N. S. (1968). Vasomotor responses in the hind limbs of foetal and new-born lambs to asphyxia and aortic chemoreceptor stimulation. *J. Physiol.*, **195**, 55–81.

Djojosugito, A. M., Folkow, B. & Yonce, L. R. (1969). Neurogenic adjustments of muscle blood flow, cutaneous A-V shunt flow and of venous tone during 'diving' in ducks. *Acta. Physiol. Scand.*, **75**, 377–86.

Elsner, R. (1969). Cardiovascular adjustments to diving. In *The Biology of Marine Mammals*, ed. Andersen, H. T., pp. 117–45, New York and London: Academic Press.

Elsner, R. W., Franklin, D. L. & van Citters, R. L. (1964). Cardiac output during diving in an unrestrained sea lion. *Nature, Lond.*, **202**, 809–10.

Elsner, R., Franklin, D. L., van Citters, R. L. & Kenney, D. V. (1966). Cardiovascular defence against asphyxia. *Science*, **153**, 941–9.

Elsner, R., Hammond, D. D. & Parker, H. R. (1970). Circulatory responses to asphyxia in pregnant and fetal animals: a comparative study of Weddell seals and sheep. *Yale J. biol. Med.*, **42**, 202–17.

Feigl, E. & Folkow, B. (1963). Cardiovascular responses in 'diving' and during brain stimulation in ducks. *Acta physiol. Scand.*, **57**, 99–110.

Ferrante, F. L. (1970). Oxygen conservation during submergence apnea in a diving mammal, the nutria. *Am. J. Physiol.*, **218**, 363–71.

Folkow, B., Fuxe, K. & Sonnenschein, R. R. (1966). Responses of skeletal musculature and its vasculature during 'diving' in the duck: peculiarities of the adrenergic vasoconstrictor innervation. *Acta physiol. Scand.*, **67**, 327–42.

Folkow, B., Nilsson, N. J. & Yonce, L. R. (1967). Effects of 'diving' on cardiac output in ducks. *Acta. physiol. Scand.*, **70**, 347–61.

Harding, P. E., Roman, D. & Whelan, R. F. (1965). Diving bradycardia in man. *J. Physiol.*, **181**, 401–9.

Heistad, D. D., Abboud, F. M. & Eckstein, J. W. (1968). Vasoconstrictor response to simulated diving in man. *J. appl. Physiol.*, **25**, 542–9.

Heistad, D. D. & Wheeler, R. C. (1970). Simulated diving during hypoxia in man. *J. appl. Physiol.*, **28**, 652–6.

Heymans, C. & Neil, E. (1958). *Reflexogenic Areas of the Cardiovascular System*, Churchill: London.

HOLLENBERG, N. K. & UVNÄS, B. (1963). The role of the cardiovascular response in the resistance to asphyxia of avian divers. *Acta physiol. Scand.*, **58**, 150–61.

HOLM, B. & SØRENSEN, S. C. (1972). The role of the carotid body in the diving reflex in the duck. *Acta physiol. Scand.*, **15**, 302–9.

HUXLEY, F. M. (1913). On the reflex nature of apnoea in the duck in diving. I. The reflex nature of submersion apnoea. *Q. J. exp. Physiol.*, **6**, 147–57.

IRVING, L. (1938). Changes in the blood flow through the brain and muscles during the arrest of breathing. *Am. J. Physiol.*, **122**, 207–14.

IRVING, L., SCHOLANDER, P. F. & GRINNELL, S. W. (1942). The regulation of arterial blood pressure in the seal during diving. *Am. J. Physiol.*, **135**, 557–66.

JOELS, N. & SAMUELOFF, M. (1956). The activity of the medullary centres in diffusion respiration. *J. Physiol.*, **133**, 360–72.

JONES, D. R. & PURVES, M. J. (1970*a*). The carotid body in the duck and the consequences of its denervation upon the cardiac responses to immersion. *J. Physiol.*, **211**, 279–94.

JONES, D. R. & PURVES, M. J. (1970*b*). The effect of carotid body denervation upon the respiratory response to hypoxia and hypercapnia in the duck. *J. Physiol.*, **211**, 295–310.

KAWAKAMI, Y., NATELSON, B. H. & DUBOIS, A. B. (1967). Cardiovascular effects of face immersion and factors affecting diving reflex in man. *J. appl. Physiol.*, **23**, 964–70.

KOBINGER, W. & ODA, M. (1969). Effects of sympathetic blocking substances on the diving reflex of ducks. *Europ. J. Pharmacol.*, **7**, 289–95.

KOCK, L. L. DE (1959). The arterial vessels of the neck in the pilot-whale (*Globicephala Meleana Traill*) and the porpoise (*Phocaena phocaena* L.) in relation to the carotid body. *Acta anat.*, **36**, 274–92.

KOCK, L. L. DE (1960). The carotid body of a half term finwhale (*Balaenoptera physalus* (L)). *Acta. anat.*, **43**, 321–36.

KRATSCHMER, F. (1870). Über Reflexe von der Nasenschleimhaut auf Athmung und Kreislauf. *Sber Akad. Wiss. Wien.*, **62**, 147–70.

MOORE, T. O., LIN, Y. C., LALLY, D. A. & HONG, S. K. (1972). Effects of temperature, immersion, and ambient pressure on human apneic bradycardia. *J. appl. Physiol.*, **33**, 36–41.

MURDAUGH, H. V., ROBIN, E. D., MILLEN, J. E., DREWRY, W. F. & WEISS, E. (1966). Adaptations to diving in the harbor seal: cardiac outputd uring diving. *Am. J. Physiol.*, **210**, 176–80.

NAHAS, G. S. (1956). Heart rate during short periods of apnea in curarized dogs. *Am. J. Physiol.*, **187**, 302–6.

OTT, N. T., LORENZ, R. R. & SHEPHERD, J. T. (1972). Modification of lung-inflation reflex in rabbits by hypercapnia. *Am. J. Physiol.*, **223**, 812–19.

OTT, N. T. & SHEPHERD, J. T. (1971). Vasodepressor reflex from lung inflation in the rabbit. *Am. J. Physiol.*, **221**, 889–95.

PONTE, J. C. M. R. & PURVES, M. J. (1974). The role of carotid chemoreceptors in the baboon in the regulation of cerebral blood vessels. In *Peripheral arterial chemoreceptors*, ed. PURVES, M. J., Cambridge University Press.

SALISBURY, P. F., GALLETTI, P. M., LEWIN, R. J. & RIEBEN, P. A. (1959). Stretch reflexes from the dog's lung to the systemic circulation. *Circulation Res.*, **7**, 62–7.

SCHOLANDER, P. F. (1940). Experimental investigations on the respiratory function in diving mammals and birds. *Hvalrådets Skrifter; Norske Videnskaps-Akad., Oslo*, **22**, 1–131.

STRØMME, S. B., KEREM, D. & ELSNER, R. (1970). Diving bradycardia during rest and exercise and its relation to physical fitness. *J. appl. Physiol.*, **28**, 614–21.

TOMORI, Z., JAVORKA, K. & STRANSKY, A. (1972). Cardiovascular, respiratory and glottal changes during the apnoeic reflex of olfactory-trigeminal and vagal origin. *Physiol. Bohemoslov.*, **21**, 45–52.

VAN CITTERS, R. L., FRANKLIN, D. L., SMITH, O. A., JR, WATSON, N. W. & ELSNER, R. W. (1965). Cardiovascular adaptations to diving in northern elephant seal *Mirounga Angustirostris*. *Comp. Biochem. Physiol.*, **16**, 267–76.

WHAYNE, T. F., Jr & KILLIP, T. III. (1967). Simulated diving in man: comparison of facial stimuli and response in arrhythmia. *J. appl. Physiol.*, **22**, 800–7.

WHITE, S. W. & MCRITCHIE, R. J. (1973). Nasopharyngeal reflexes: integrative analysis of evoked respiratory and cardiovascular effects. *Aust. J. expt. biol. Med. Sci.*, **51**, 17–31.

DISCUSSION

BULLER I'm not quite sure about the point you were making in your human experiments. What was apparent was that after immersion of the face the pulse started on a higher level interval but that the volume was very much greater, and if I remember correctly, you can bring out sinus arrhythmia in a normal individual by breathing more deeply.

DALY You are quite right in saying that during face immersion there is slowing of the heart. If while the subject is still immersed in water, he takes a breath of air, we find that he does get a marked sinus arrhythmia. What we were interested in determining is whether a chemical factor is involved; so instead of breathing air, if the subject breathed a gas mixture which kept his alveolar gas tensions at the 'breaking point' level, then we still obtained the same response. So it seems unlikely that the sinus arrhythmia obtained when taking a breath during immersion was due to any chemical change; we suggest it is due to a reflex mechanism.

BULLER Yes, I see that but how would that breath, the first after the breaking point with immersion, compare with a similar breath of the same volume before immersion? I would have guessed that this would bring out a greater change of pulse frequency during the cycle.

DALY No, it didn't usually and in fact I chose this slide because the subject did show sinus arrhythmia normally: most of the other subjects did not and even when they did take a deep breath, they did not obviously develop a more marked sinus arrhythmia.

ABBOUD About this question of resetting the baroreceptors. I am proposing that this could be adjustment of the chemoreceptors and that you don't need to invoke the baroreceptors at all.

DALY No, you certainly don't need to. But many people are looking at this type of adjustment of the circulation and are relating arterial pressure to pulse interval and they suggest that if this curve is shifted, it indicates resetting of the baroreflex. It may just be a way of describing the situation without necessarily indicating what pathways are involved. We could of course stimulate the chemoreceptors in future experiments to see whether the baroreceptor reflexes are modified. This might substantiate your point.

WILLSHAW Could these subjects breathe while they were submerged?

DALY Yes.

WILLSHAW So you just asked them to hold their breath as long as possible? And did you stimulate the face in any other way?

DALY Yes to the first question and no to the second. There are other ways of stimulating facial receptors but we just used immersion in water.

MILLS In the dog experiments, I understand that the extent of interaction between the baroreceptor and chemoreceptor reflexes differs depending on what anaesthetic you use.

DALY We have used a mixture of chloralose and urethane; but you still obtain this type of interaction with pentobarbitone.

MILLS In your analysis of the interaction between trigeminal and carotid body responses, what stimulus did you use for the carotid body?

DALY A mechanical infusion of sodium cyanide for 3 seconds.

MILLS So the gas tensions of blood going to the head was unaffected?

DALY Yes.

MILLS What happens in a situation in a diving animal where with asphyxia, the P_{O_2} of cerebral blood falls and P_{CO_2} rises?

DALY I do not know.

PURVES In the diving animal, there seem to be two possibilities. They are able to hold their breath for a comparatively long period of time with appropriate change in blood gas tensions and there may be some change in the sensitivity of the chemoreceptors to these changes or what you are suggesting is that there is some form of central inhibition which in some way allows the respiratory neurones to ignore the large progressive rise in chemoreceptor discharge.

DALY I don't think we can exclude a change in chemoreceptor discharge. Certainly you obtain the same type of response after sympathetic denervation of the chemoreceptors but we have not, of course, been able selectively to interrupt the efferent pathway to them. But if you argue that the reduction in the respiratory effect of stimulation of the chemoreceptors when you excite the trigeminal reflex is partly due to a change in sensitivity of the chemoreceptors, then it means that the facilitation of the chemoreceptor–cardiac reflex is that much greater.

PURVES So the diving animal is in a dilemma. It wants to remain apnoeic so that it can search for its food: on the other hand, it wants to get the advantages of the redistribution of cardiac output.

DALY Exactly.

WILLSHAW I don't think the efferent pathway could be involved in this response. In your diagram, you indicate that the sinus nerve is passing information which affects three modalities and if the efferent pathway was involved, it would surely affect all three. Clearly, it does not. So it is more likely to be a central problem rather than affecting the sensitivity of the peripheral chemoreceptors.

MITCHELL I might mention that in looking for a function for the efferent pathway, asphyxia during diving was one of the situations we considered. So we measured efferent activity in the sinus nerve of cats while immersing their face. However, we did not observe any change in discharge. It was rather disappointing.

The role of the carotid body chemoreceptors in the control of the cerebral blood vessels

By M. J. PURVES and J. C. M. R. PONTE
Department of Physiology, University of Bristol

During the latter part of the nineteenth century and for the whole of the present century, there has been controversy over the question of whether the cerebral blood vessels are innervated and if so, whether these nerves have any physiological function. In recent years, the first question has been clearly answered by electron microscopic studies (Dahl & Nelson, 1964; Iwayama, 1970; Iwayama, Furness & Burnstock, 1970; Lavrentieva, Mchedlishvili & Plechkova, 1968) and histochemical studies (Falck, Nielsen & Owman, 1968; Ohgushi, 1968; Kajikawa, 1968) which together with earlier studies (Chorobski & Penfield, 1932) make it clear that cerebral vessels certainly down to a calibre of 10 μm are supplied by adrenergic fibres originating in the superior cervical ganglion and cholinergic dilator pathway carried in part by the seventh cranial nerve and in part by the greater superficial petrosal nerve. The second question has been less satisfactorily resolved. Some workers have shown unequivocally that stimulation of sympathetic fibres causes cerebral vasoconstriction (James, Millar & Purves, 1969; Harper, Deshmukh, Rowan & Jennet, 1971; D'Alecy & Feigl, 1972) but other workers have found either little or no response. In any case, as some workers have observed (Folkow & Neil, 1972; Raper, Kontos, Wei & Patterson, 1972) it is difficult to see what physiological function such vasoconstriction would serve. The dilator pathway has been much less fully studied, possibly because of its relative inaccessibility but it has been shown to be important in maintaining a relatively high blood flow during progressive hypotension, in hypoxia and in hypercapnia (James *et al.*, 1969). We have therefore looked at a possible connection between the carotid chemoreceptors and baroreceptors in order to see whether these could be responsible for initiating an important series of reflexes responsible for protecting the cerebral nervous system against deficient blood or oxygen supply.

The methods used to isolate the left carotid sinus and to perfuse the head artificially in the baboon have been described in detail elsewhere (Ponte & Purves, 1974). Briefly, the left carotid sinus is vascularly isolated so that it and the carotid body can be perfused naturally with arterial blood at systemic blood pressure, or artificially with venous or arterial blood at any desired pressure. The left sinus nerve is preserved but the right carotid and both aortic bodies are denervated. Preliminary tests after this surgery, cannulation, etc. showed that cerebral blood flow, measured by the ^{133}Xenon desaturation technique (Høedt-Rasmussen, Sveinsdottir & Lassen, 1966) increases normally with carbon dioxide and that the characteristic independence of flow and pressure is preserved. Stimulation of the left carotid body with venous blood causes an average increase in cortical blood flow of 70–75% of control which is approximately the same as raising P_{a,CO_2} from control (40 mm Hg) to 55 mm Hg (Fig. 1). If the

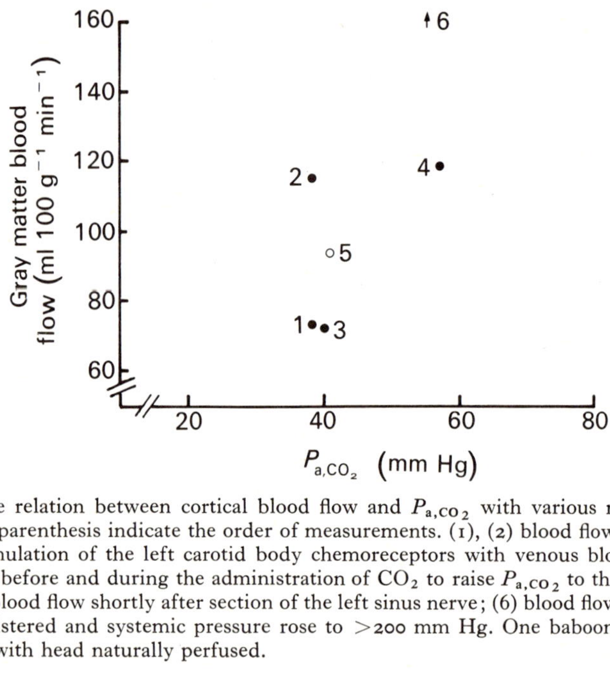

Fig. 1. The relation between cortical blood flow and P_{a,CO_2} with various manoeuvres. Figures in parenthesis indicate the order of measurements. (1), (2) blood flow before and during stimulation of the left carotid body chemoreceptors with venous blood; (3), (4) blood flow before and during the administration of CO_2 to raise P_{a,CO_2} to the level indicated; (5) blood flow shortly after section of the left sinus nerve; (6) blood flow when CO_2 was administered and systemic pressure rose to >200 mm Hg. One baboon, artificially ventilated with head naturally perfused.

remaining (left) sinus nerve is cut, there is an abrupt rise of systemic pressure to >180 mm Hg and accompanying this is a rise in cerebral blood flow. If the baboon is again given carbon dioxide to inhale and P_{a,CO_2} raised to *c.* 55 mm Hg, there is a further rise in arterial pressure to >200 mm Hg and blood flow increases so greatly to >170 ml 100 g^{-1} min^{-1}

that it can no longer be counted with any accuracy. Further, if the animals, which are paralysed and artificially ventilated, are made hypoxic, P_{a,O_2} *c.* 45 mm Hg, there is a marked and rapid fall in cerebral blood flow requiring equally rapid resuscitation.

For these reasons – because of the danger of premature death of the animal with hypoxia and because it was not possible to distinguish whether the marked increase in blood flow after total vasosensory denervation was due to the rise in carbon dioxide or pressure, a second series of experiments was carried out in which the preparation in the neck was as before but, in addition, the head was perfused artificially with a pump with blood from the femoral arteries via the innominate artery, right vertebral and both common carotid arteries. It was obviously possible that with the additional cannulation and action of the pump the reactivity of the cerebral blood vessels could have been impaired. In fact it was possible to show (i) by subtraction angiography that there was no obvious spasm of the internal carotid artery or its larger branches, (ii) that the vascular response to carbon dioxide was not different to control and (iii) that over the physiological range, flow was independent of perfusion pressure. A point of some interest was that if pressure was raised artificially to >180, i.e. 30–40 mm Hg above the control range, blood flow was significantly above control in seven animals so tested.

Under these conditions, the following tests were made and the results are summarized in Table 1.

Stimulation of the chemoreceptors

Stimulation of the chemoreceptors with venous blood invariably caused a rise in grey and white matter blood flow. If the ipsilateral sympathetic nerve had been cut, the increase in flow was not significantly different from the control response. If both seventh cranial nerves were cut intracranially, the vascular response was completely abolished. If the remaining sinus nerve was cut, blood flow increased but stimulation of the chemoreceptors was without effect upon cerebral vascular flow.

Effect of hypoxia

Systemic hypoxia caused the expected rise in cerebral blood flow. If the remaining sinus nerve was cut, and blood flow maintained artificially, hypoxia caused changes in blood flow which were within the experimental method.

Effect of hypercapnia

The average increase in cortical blood flow when P_{a,CO_2} was raised was

TABLE 1. *Summary of the effects of stimulating the chemoreceptors and other stimuli upon grey matter blood flow under various conditions*

	Tests (n)		Grey matter blood flow (ml 100 g^{-1} min^{-1}) Control	Chemoreceptor stimulation*	Hypercapnia P_{a,CO_2} 55 ± 3 mm Hg	Hypoxia P_{a,O_2} 38–45 mm Hg	Per cent change
With sinus nerve intact							
1. Natural perfusion of head	21	Mean	74.0	127.0	—	—	+72
(MAP, 125–140 mm Hg)		± SE	0.67	2.8			
	28		76.5	—	121.0	—	+58
			0.78	—	4.8	—	—
	17		73.4	—	—	100.8	+37
			0.65	—	—	3.9	—
After section of ipsilateral	7		76.8	136.4	—	—	—
cervical sympathetic nerve			0.7	2.9	—	—	—
After bilateral intracranial	6		71.4	75.4	—	—	—
section of VIIth nerves			0.8	2.4	—	—	—
2. Artificial perfusion of the head	14		71.5	131.0	—	—	+73
			0.78	2.7	—	—	—
	14		72.8	—	118.4	—	+63
			0.7	—	3.2	—	—
	8		72.5	—	—	97.0	+34
			0.73	—	—	2.1	—

With left sinus nerve cut						
1. Natural perfusion of the head	5	94.7	95.6	—	—	+1
		2.3	1.9	—	—	—
	2	93.0	—	> 170†	—	—
	2	91.5	—	—	< 35‡	—
2. Artificial perfusion of the head	7	78.3	—	91.9	—	+17
(*MAP* 120–140 mm Hg)		2.1	—	1.7	—	—
	5	82.3	—	—	80.7	< 1
		1.7	—	—	2.8	—
(*MAP* 155–165 mm Hg)	5	88.4	—	105.4	—	+19
		1.8	—	2.2	—	—

* Left carotid sinus perfused with venous blood.

† Mean arterial pressure increased to 200–215 mm Hg.

‡ Falling systemic arterial pressure and measurements abandoned after 3–4 minutes.

58% of control and the average slope of the relation between flow and P_{a,CO_2} was 3.2 ml 100 g^{-1} min^{-1} per mm Hg P_{a,CO_2}. Following section of the left sinus nerve, the vascular response to hypercapnia was substantially reduced, being on average +17% of control. These tests were carried out at a cephalic inflow pressure of 125 to 150 mm Hg and they were repeated at one higher pressure in the range 155 to 165 mm Hg. The average increase in grey matter flow was 19% and this suggests that the large increase in flow with hypercapnia following section of the sinus nerve in the experiments where the head was naturally perfused was in fact due to the accompanying rise in systemic pressure. The results also suggest that in the intact animal, approximately two-thirds of the vascular response to carbon dioxide is reflexly determined. The mechanism of the residual intrinsic response remains unknown and was not investigated in the present study.

Effects of altering carotid sinus pressure

We confirmed that when the head was artificially perfused and when inflow pressure was altered over the range 80 to 160 mm Hg, flow remained independent of pressure as shown in Fig. 2. When cerebral inflow pressure was held constant within the range 125–140 mm Hg, if carotid sinus pressure was independently raised to 160 to 190 mm Hg, there was invariably a fall in grey matter blood flow, average −28% of control (n = 8). Under the

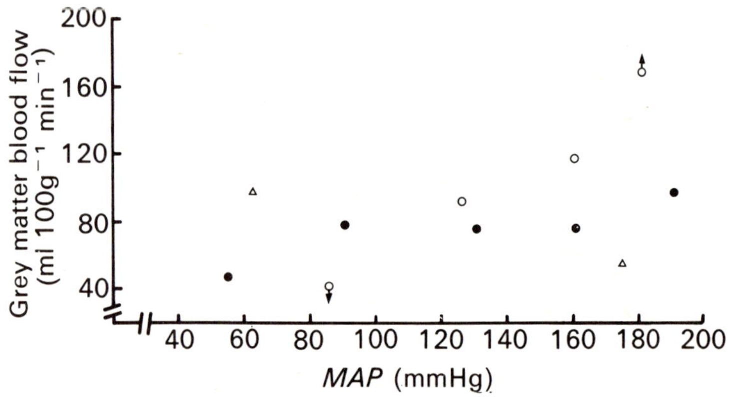

Fig. 2. Relation between cortical blood flow and mean arterial pressure (*MAP*) in one baboon. Values represented by solid circles are those obtained when the head was naturally perfused at the pressures indicated after bleeding and replacing blood, the left sinus nerve being intact. Values represented by triangles are those obtained when the head was perfused at the pressures indicated and when the carotid sinus was perfused at 175 mm Hg and at 65, left sinus nerve being intact. Open circles represent values for blood flow when the head was perfused at the pressures indicated shortly after the left sinus nerve had been cut. Values at pressures of 85 and 185 mm Hg are approximate only.

same conditions and if carotid sinus pressure was reduced to the range 45 to 65 mm Hg, grey matter blood flow invariably increased by +39% of control ($n = 8$).

When the left and remaining sinus nerve was cut and cephalic inflow pressure varied over the range 85 to 180 mm Hg, the response seen was as shown in Fig. 2. At 85 mm Hg inflow pressure, cortical flow was so low that it could not accurately be measured, Fig. 3*b*. At 180 mm Hg inflow

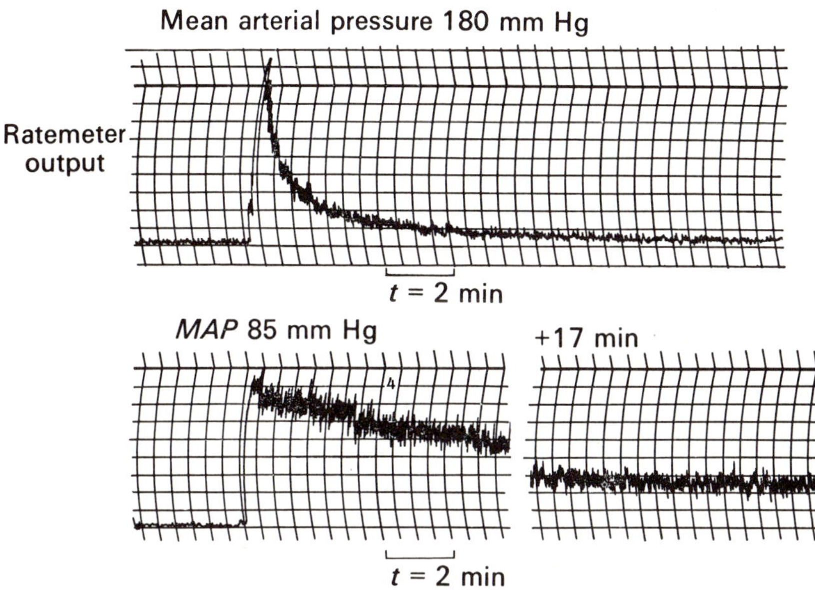

Fig. 3. Examples of the ^{133}Xenon desaturation curves measured over the left parietal cortex following denervation of all baroreceptors and chemoreceptors. In the upper trace, the mean inflow pressure was 185 mm Hg and grey matter flow computed to be >175 ml 100 g^{-1} min^{-1}. In the lower trace, mean inflow pressure was 85 mm Hg and blood flow was so slow that after 17 minutes, activity was still some way from having returned to background. Grey matter blood flow was computed to be less than 40 ml 100 g^{-1} min^{-1}.

pressure, flow was so great that it was virtually impossible to assign a value for the fast component – and flow has been designated as being >175 ml 100 g^{-1} min^{-1}, Fig. 3*a*. Measurements of flow were made at intermediate levels of pressure but because of the uncertainty about levels of flow at the extremes of pressure tested, the relation between flow and pressure after vasosensory denervation is difficult to characterize. An example of this relation is shown in Fig. 2 which was typical of the responses observed.

These results raise two questions – how do they compare with those of earlier workers and what insights do they give about the regulation of cerebral blood flow. It is of interest that similar results were obtained by earlier workers. Thus Anrep & Starling (1925) found that if aortic pressure in the dog was reduced, there was a substantial rise in cerebral blood flow and this response was abolished by bilateral cervical vagotomy. Bouckaert & Heymans (1935) showed that in cross-circulation experiments in dogs cerebral blood flow varied inversely with carotid sinus pressure. One difficulty about these experiments is that contamination by blood from extracerebral sources through large and variable anastomoses was not excluded and for this reason it is realized that cerebral venous or carotid arterial flow may be inexact indices of cerebral blood flow. Similarly, it was shown by a number of workers (Forbes & Wolff, 1928; Cobb & Finesinger, 1932; Ask-Upmark, 1935) that if the carotid sinus nerve was stimulated, the commonest type of variable responses observed was pial dilatation. This again is difficult to interpret because of the uncertainty as to whether chemoreceptor or baroreceptor afferent fibres are predominantly being stimulated.

Fog (1937) by contrast showed that pial vessels still dilated in response to graded reductions of systemic pressure even after vagus, cervical sympathetic and sinus nerves had been cut in the neck.

For a variety of reasons, it is difficult to infer quantitative changes in flow from changes in the calibre of pial, or indeed any other blood vessel. In consequence, it is difficult to be certain from Fog's results whether or not there had been a significant alteration in the pressure/flow relation in cerebral vessels after vasosensory denervation. Equally it should be stated that in the present experiments, some form of intrinsic response of cerebral vessels to changes in pressure is not excluded: the results merely indicate that a reflex initiated by baroreceptors can affect cerebral blood flow and that this seems to be of particular importance in the regulation of flow at the extremes of pressure tested.

A study by Rapela, Green & Denison (1967) also appeared to deny a significant role of baroreceptor reflex action in the response of cerebral vessels to changes in systemic pressure for they showed that although the pressure/flow relation in the cerebral vascular bed was altered after carotid sinus nerve section in the dog, the independence of blood flow and pressure over the physiological range persisted. It should be remarked that in this study values of cerebral blood flow were very low, i.e. *c.* 10 ml min^{-1} which suggests that either a substantial volume of venous blood was bypassing the flowmeters or that the vessels had been damaged by the earlier surgery. Further, these workers produced graded changes in cerebral perfusion pressure by occlusion of the carotid artery – a procedure which would affect carotid body chemoreceptor afferent activity as well as that of the baroreceptors, as would, of course, section or stimulation of the sinus nerve while in all their experiments, the aortic nerves were intact so that it is quite possible that cerebral vessels were affected by any changes in systemic pressure which occurred.

Some recent experiments have provided some confirmation of the present results. Thus Mchedlishvili & Nikolaishvili (1970) showed in rabbits that pial vessels responded to graded reductions in systemic pressure by dilating but that this response was abolished if atropine was given, while James, MacDonell, Hall & Brant (1973) have shown in cross-circulation experiments in dogs that hypercapnic or hypoxic blood perfusing the carotid sinus of the recipient dog causes a large increase in cerebral blood flow (measured with ^{85}Kr) and that cerebral blood flow varies inversely with carotid sinus pressure.

These experiments have been reviewed in some detail because it has seemed to us that the question as to whether or not peripheral arterial receptors can affect cerebral blood vessels reflexly has been delayed in the answering largely for reasons of the technique used. We believe that the question is important enough for the present experiments to be repeated, preferably using different methods for the measurement of cerebral blood flow.

If the present results are confirmed, then it will no longer be possible to regard the regulation of cerebral vessels as previously when it was considered that the cerebral vascular bed was regulated entirely by intrinsic mechanisms. It is clear, for example, that the reflex control of cerebral vessels is of particular importance in hypoxia and this may be thought of as a mechanism which ensures that tissue and venous P_{O_2} are maintained at a high level and which may also cause the opening of latent capillaries, reduce inter-capillary distance and so ensure adequate oxygenation of central nervous tissue. The response to carbon dioxide is complex since it would appear that it is partly reflex and partly intrinsic; and this would be consistent with the fact that pial vessels dilate in response to carbon dioxide being blown across the surface of the brain (Shalit, Shimojyo & Reinmuth, 1967) and the local application of H^+ (Wahl *et al.*, 1970). In general, the effect of such a cerebral vasodilatation will be to stabilize extracellular pH in central nervous tissue but it should be recalled that in life, outside the laboratory, hypercapnia occurs most commonly as part of asphyxia: in this context, hypercapnia could be considered as a further protective mechanism for the brain.

The evidence is by no means complete but it appears that stimulation of the chemoreceptors in both hypoxia and hypercapnia leads to the activation of both constrictor and dilator pathways to cerebral blood vessels, the net effect being one of vasodilatation. Precisely how these pathways interact is unknown but it is reasonable to regard the action of the sympathetic vasoconstrictor pathway as a fine control on the dilator action of the reflex and intrinsic mechanism. A further unsolved problem is whether the aortic receptors also exert a similar action upon cerebral vessels. If they do and if the responses of the chemoreceptors to changes in systemic pressure are the same in species other than the cat, then it would be predicted that they would be more sensitive to changes in pressure (Lee, Mayou & Torrance, 1964) than the carotid body chemoreceptors (Biscoe, Bradley & Purves, 1970) and in consequence, they would augment the dilator action of hypotension upon the baroreceptors.

The present experiments do not clarify any more than the earlier ones (James *et al.*, 1969) how it is that when systemic pressure is raised, there is cerebral vasoconstriction. The present experiments clearly implicate the carotid sinus baroreceptors since if they are denervated, cerebral blood flow alters with perfusion pressure. But whether baroreceptor stimulation leads to a fall in tonic dilator activity or an increase in sympathetic discharge remains to be determined. With hypotension, the position is clearer. In the experiments of James *et al.* (1969), it was shown that if the dilator pathway was interrupted, blood flow to the brain fell markedly when systemic pressure was reduced by even modest amounts. The present experiments indicate that inhibition of the baroreceptors by hypotension activates the dilator pathway in the intact animal since there is the same marked fall in cerebral blood flow with relatively mild hypotension after the receptors have been denervated.

All these points emphasize the dominant role of the dilator pathway in the response of cerebral blood vessels to such stimulation as hypoxia, hypercapnia and hypotension and it is probably the possession of such a pathway that causes the responses of the cerebral blood vessels to be distinguished from other vascular beds. The fact that the sympathetic vasoconstrictor pathway has what appears to be a secondary role in these responses may explain why previous workers have had difficulty in demonstrating its physiological function and why, in consequence, there has been prolonged debate about the more general question of the neural control of the cerebral circulation.

The present results suggest that the cerebral vessels take part in the co-ordinated response of the whole cardiovascular system to disturbances of the internal environment, in particular those in which the blood or oxygen supply to the brain is threatened. In this reaction, the peripheral chemoreceptors have a central role. Indeed, their function of ensuring adequate oxygenation of the brain may be the most important one they possess.

REFERENCES

ANREP, G. V. & STARLING, E. H. (1925). Central and reflex regulation of the circulation. *Proc. R. Soc. B*, **97**, 463–87.

ASK-UPMARK, E. (1935). The carotid sinus and cerebral circulation. An anatomical, experimental and clinical investigation. Including some observations on rete mirabile caroticum. *Acta psychiat. Kbh. Suppl.*, **6**, 1–374.

BISCOE, T. J., BRADLEY, G. W. & PURVES, M. J. (1970). The relation between carotid body chemoreceptor discharge, carotid sinus pressure and carotid body venous flow. *J. Physiol.*, **208**, 99–120.

BOUCKAERT, J. J. & HEYMANS, C. (1935). On the reflex regulation of the cerebral blood flow and the cerebral vasomotor tone. *J. Physiol.*, **84**, 367–80.

CHOROBSKI, J. & PENFIELD, W. (1932). Cerebral vasodilator nerves and their pathways from the medulla oblongata with observations on pial and intracerebral plexus. *Archs Neurol. Psychiat.*, **47**, 77–96.

COBB, S. & FINESINGER, J. E. (1932). Cerebral Circulation: XIX. The vagal pathway of the vasodilator fibres. *Archs Neurol. Psychiat.*, **28**, 1243–56.

DAHL, E. & NELSON, E. (1964). Electron microscopic observations on human intracranial arteries. *Archs Neurol, Chicago*, **10**, 158–64.

D'ALECY, L. G. & FEIGL, E. O. (1972). Sympathetic control of cerebral blood flow in dogs. *Circulation Res.*, **31**, 267–83.

FALCK, B., NIELSEN, K. C. & OWMAN, C. (1968). Adrenergic innervation of the pial circulation. *Scand. J. clin. Lab. Invest., Suppl.*, **102**, 96–9.

FOG, M. (1937). Cerebral Circulation. The reaction of the pial arteries to a fall in blood pressure. *Archs Neurol. Psychiat.*, **37**, 351–64.

FOLKOW, B. & NEIL, E. (1972). *The Circulation*, Oxford University Press.

FORBES, H. S. & WOLFF, H. G. (1928). Cerebral Circulation. III. The vasomotor control of cerebral vessels. *Archs Neurol. Psychiat., Chicago*, **19**, 1057–86.

HARPER, A. M., DESHMUKH, V. D., ROWAN, J. O. & JENNET, W. B. (1971). Studies on neurogenic control of cerebral circulation. *Panminerva Medica*, **13**, 172–3.

HØEDT-RASMUSSEN, K., SVEINSDOTTIR, E. & LASSEN, N. A. (1966). Regional cerebral blood flow in man determined by intra-arterial injection of radioactive inert gas. *Circulation Res.*, **18**, 237–47.

IWAYAMA, T. (1970). Ultrastructural changes in the nerves innervating the cerebral artery after sympathectomy. *Z. Zellforsch. mikrosk. Anat.*, **109**, 465–80.

IWAYAMA, T., FURNESS, J. B. & BURNSTOCK, G. (1970). Dual adrenergic and cholinergic innervation of the cerebral arteries of the rat: an ultrastructural study. *Circulation Res.*, **26**, 635–46.

JAMES, I. M., MILLAR, R. A. & PURVES, M. J. (1969). Observations on the extrinsic neural control of cerebral blood flow in the baboon. *Circulation Res.*, **25**, 77–93.

JAMES, I. M., MACDONELL, L., HALL, S. & BRANT, P. C. (1973). The role played by the peripheral baroreceptors and chemoreceptors in the control of the cerebral circulation. *Clin. Sci.*, **45**, 7P.

KAJIKAWA, H. (1968). Fluorescence histochemical studies on the distribution of adrenergic nerve fibres to intracranial blood vessels. *Arch. Jap. Chir.*, **37**, 473–84.

LAVRENTIEVA, N. B., MCHEDLISHVILI, G. I. & PLECHKOVA, E. K. (1968). Distribution and activity of cholinesterase in the nervous structures of the pial arteries (A histochemical study). *Bull. Exp. Biol. Med.*, **64**, 110–13.

LEE, K. D., MAYOU, R. A. & TORRANCE, R. W. (1964). The effect of blood pressure upon chemoreceptor discharge by hypoxia and the modification of this effect by the sympathetic-adrenal system. *Q. Jl. exp. Physiol.*, **49**, 171–83.

MCHEDLISHVILI, G. I. & NIKOLAISHVILI, L. S. (1970). Evidence of a cholinergic nervous mechanism mediating the autoregulatory dilatation of the cerebral blood vessels. *Pflügers Arch. ges. Physiol.*, **315**, 27–37.

OHGUSHI, N. (1968). Adrenergic fibres to the brain and spinal cord vessels in the dog. *Arch. Jap. Chir.*, **37**, 294–303.

PONTE, J. C. M. R. & PURVES, M. J. (1974). The role of the carotid body chemoreceptors and carotid sinus baroreceptors in the control of cerebral blood vessels. *J. Physiol.*, **237**, 315–40.

Rapela, C. E., Green, H. D. & Denison, A. B. (1967). Baroreceptor reflexes and autoregulation of cerebral blood flow in the dog. *Circulation Res.*, **21**, 559–68.

Raper, A. J., Kontos, H. A., Wei, E. P. & Patterson, J. L. (1972). Unresponsiveness of pial precapillary vessels to catechol amines and sympathetic stimulation. *Circulation Res.*, **31**, 257–66.

Shalit, M. N., Shimojyo, S., & Reinmuth, O. M. (1967). Carbon dioxide and cerebral circulatory control. I. The extravascular effect. *Arch. Neurol. Chicago*, **17**, 298–303.

Wahl, M., Deetjen, P., Thurau, K., Ingvar, D. H. & Lassen, N. A. (1970). Micropuncture evaluation of the importance of perivascular pH for the arteriolar diameter on the brain surface. *Pflügers Arch. ges. Physiol.*, **316**, 152–63.

DISCUSSION

ACKER How fast are the changes which you observe in the cerebral circulation?

PURVES This is not known with great certainty and in fact Dr Heistad and I have been carrying out experiments on this topic recently. You will appreciate that no information on this point is given by the xenon desaturation technique. But we have found that in response to chemoreceptor stimulation, a change in vessel diameter is detectable after a delay of 2–3 seconds and peak response is achieved in 10–15 seconds. This is similar to results obtained by Raper, Kontos & Patterson (1971) and by Severinghaus & Lassen (1967).

McDONALD How does the carotid body input to the brain influence cerebral blood vessels? What are the neural links?

PURVES The outputs are fairly clear. There is a sympathetic output and fibres which supply cerebral vessels are exclusively derived from the superior cervical ganglion on each side. There is also a dilator pathway which may well be parasympathetic which is carried in the seventh cranial nerve as far as the geniculate ganglion and from this the dilator fibres run in the greater superficial petrosal nerve across the floor of the skull to mingle with the sympathetic fibres around the carotid vessels and be distributed to the peripheral vessels. The position in the CNS is less well known. It is known that stimulation of selected sites near the hypothalamus will cause cerebral vasoconstriction or dilatation and it is also known that selected lesions in the medulla and pons will greatly reduce the cerebral vascular response to carbon dioxide. But the precise connections of input and output in the CNS require much more careful study.

McDONALD Don't the autonomic nerves which you describe affect the cerebral vessels before they enter the brain – say in the sub-arachnoid space?

PURVES Well, the difficulty is that we do not know precisely where they act. As you know, certainly better than I do, the nerve terminals have been demonstrated on vessels down to 10 μm in diameter in both pial vessels and penetrating vessels: and of course they supply the larger vessels at the base of the brain. It has been shown by direct observation that with stimulation of the sympathetic nerve, pial vessels of all sizes constrict but a difficulty is that there does not appear to be a pre-capillary sphincter mechanism as in other vascular beds. So the precise extent of nerve action remains uncertain.

DALY Could you enlarge on the relative roles of the sympathetic vasoconstrictor and parasympathetic dilator fibres? Does the dilatation occur because of withdrawal of sympathetic tone?

PURVES I'm not certain how these fibres work together. The evidence is that in hypoxia or hypercapnia, activity in both sympathetic and dilator fibres increases and of course the net effect is one of dilatation. So it clearly is not a reciprocal type of action and it is possible in these circumstances, the sympathetic acts as some form of fine control.

DALY Then is the action of the dilator fibres affected by atropine?

PURVES Yes, there is some evidence for this. Mchedlishvili & Nikolaishvili (1970) showed that the dilatation of pial vessels which you get in response to graded haemorrhage and fall in blood pressure is abolished by atropine.

SHEPHERD You showed that stimulation of parasympathetic fibres caused an increase in cerebral blood flow, and that this increase depended on the level of carbon dioxide. At 40 mm Hg, it was very small; at 55 mm Hg, it was large. What are your thoughts about this?

PURVES The lesson I draw from these results is that cerebral vessels are rather different from these in other vascular beds, except possibly the coronary vessels. In the cerebral vascular bed, it would appear that the importance of neural influences becomes progressively greater as you deviate from the physiological range and with normal blood gas tensions and blood pressure, there seems to be relatively little neural influence. I think it may be this one fact which has led to the long controversy about the action of nerves on cerebral vessels because most workers have simply cut or stimulated the nerves within this range and the results have been either small or negligible.

ABBOUD You talk about increases or decreases in flow. I presume you have simultaneous measurements of arterial pressure so that we can say whether the vessels dilate or not?

PURVES I have tried to avoid using the terms constriction and dilatation when describing data because of the difficulty of estimating changes in vascular resistance in a non-linear system. But I can say, for example, in the response to carbon dioxide, that the increase in arterial pressure is trivial compared to the increase in flow: so this must clearly be dilatation.

ABBOUD Are you suggesting that hypoxia and hypercapnia exert their effects entirely through a reflex initiated by the carotid body and not by any direct action on blood vessels?

PURVES Almost but not quite. The results suggest that the effect of hypoxia is virtually all reflexly determined though I should point out two things: first, I haven't tested the action of high oxygen in this way and second, I can only say that after carotid body denervation any changes are within the error of the method which is about $\pm 10\%$. On the other hand, there clearly is a residual, presumably intrinsic, vascular response to carbon dioxide after vasosensory denervation. There are a number of hypotheses as to what the mechanisms are but these need not concern us here.

ABBOUD You also observed, I think, a fall in blood flow when you made the animal hypoxic after cutting the sinus and aortic nerves. Could that be because you don't get the normal arterial pressor response?

PURVES Well, not entirely. In fact, of course, after carotid body denervation, hypoxia usually leads to hypotension and we presume that when the head is naturally perfused, the drastic fall in flow is related to this, because if the head is artificially perfused, there is very little change in cerebral flow with hypoxia. But I think the evidence suggests clearly that under normal circumstances, the rise in blood flow with hypoxia is not simply due to the pressor effect: there is also active dilatation and this fails after denervation as is shown by the marked fall in blood flow in the two animals: one animal died and one had to be resuscitated at speed.

WILLSHAW You were measuring only changes in cortical blood flow?

PURVES We measured sub-cortical or white matter flow as well but for simplicity, I only presented changes in cortical flow.

WILLSHAW Because there is a paper by Pannier *et al.* (1972) in which they perfused the ventricles with different types of CSF and I think I'm right in saying that when they perfused with alkaline CSF, they did not observe much difference in cortical flow but flow in the caudate nucleus actually fell.

PURVES I should make it clear that the xenon or other similar desaturation methods gives an average flow for each of two compartments – a fast flow related to the cortex and a slower flow related to sub-cortical structures. On the one hand, I never observed the white matter flow to change in an opposite direction to cortical although the size of the change was always much smaller. But this does not exclude the possibility that flow in individual structures does change in the opposite direction.

MILLS Is there a possibility that in the presence of anaesthetics, etc. the autoregulatory responses of cerebral vessels are attenuated? Could that explain any of your findings?

PURVES I don't think so. I have shown that the independence of flow and pressure over the physiological range and some way above and below it is preserved under control conditions – that is when the animal is simply anaesthetized with pentobarbitone and this type of relation is the same as that seen in other types of experiments with different anaesthetics. This type of independence of flow and pressure was also preserved with artificial perfusion of the head. Incidentally, I'm not sure that when the nerves have been shown to affect the regulation of flow that autoregulation is an appropriate term.

SAMPSON Could you clarify this question of the sympathetics. Are you suggesting a direct effect of sympathetic fibres in the vessels or are you suggesting an indirect effect of the sympathetic upon blood vessels in the carotid body?

PURVES I assume that the sympathetic is affecting the blood vessels of the carotid body although I certainly haven't looked at this point and I don't think anyone has in this context. No: the sympathetic action I have been discussing is a direct one in fibres originating in the superior cervical ganglion.

MCDONALD Because you are measuring the net effect on blood flow, could there not be subtle changes in blood flow in other parts of the brain?

PURVES I'm sure there are and this I acknowledge is a drawback of this method. This is true both with regard to individual structures and to other parts of the brain. To my mind, it is quite conceivable, indeed probable, that with hypoxia or hypocapnia, there is a redistribution of blood flow. I can only speak for net changes in one part of the brain.

BINGMANN I recall some previous studies (Fleisch, 1931) in which changes in carotid sinus pressure affected venous/vascular resistance. Do you think that the changes which you observed could have been due to irritation of the carotid sinus and stimulation of the baroreceptors?

PURVES Well, this possibility was tested separately. If you perfuse the carotid sinus pressure with arterial blood at raised pressure, you get a fall in cerebral flow and if you lower pressure in the carotid sinus, you get a rise in flow. Again, I cannot explain this – that is at high pressure, because the cerebral vessels behave exactly in an opposite direction to those in other vascular beds and there must therefore be something odd about the baroreceptor/sympathetic pathway. It is just possible that the dilator pathway is involved at high pressure for it certainly seems to be involved at low carotid sinus pressure. If this pathway is divided and the animal made hypotensive, there is a marked fall in blood flow.

WILLSHAW What about changes in chemoreceptor discharge being involved in the flow changes with changes in pressure as well as the baroreceptors?

PURVES I don't know for certain because I haven't done static chemoreceptor discharge versus oxygen tension measurements in the baboon. But if they are the same as in the cat, then I would expect that with hypotension, the chemoreceptor discharge would be augmented especially from the aortic chemoreceptors and so hypoxia would augment the dilator action of hypotension and vice versa.

REFERENCES

FLEISCH, A. (1931). Venomotorenzentrum und Venenreflexe II Mitteilung, Blutdruckzügler und Venenreflexe. *Pflügers Arch. ges. Physiol.*, **226**, 393–410.

MCHEDLISHVILI, G. I. & NIKOLAISHVILI, L. S. (1970). Evidence of cholinergic nervous mechanism mediating the autoregulatory dilatation of the cerebral blood vessels. *Pflügers Arch. ges. Physiol.*, **315**, 27–37.

PANNIER, J. L., WEYNE, J., DEMEESTER, G. & LEUSEN, I. (1972). Influence of changes in the acid–base composition of the ventricular system on cerebral blood flow in cats. *Pflügers Arch. ges. Physiol.*, **333**, 337–51.

RAPER, A. J., KONTOS, H. A. & PATTERSON, J. L. Jr (1971). Response of pial precapillary vessels to changes in arterial carbon dioxide tension. *Circulation Res.*, **28**, 518–23.

SEVERINGHAUS, J. W. & LASSEN, N. (1967). Step hypocapnia to separate arterial from tissue P_{CO_2} in the regulation of cerebral blood flow. *Circulation Res.*, **20**, 272–8.

Differential responses of the coronary circulation and other vascular beds to chemoreceptor stimulation

BY FRANÇOIS M. ABBOUD, DONALD D. HEISTAD, ALLYN L. MARK AND PHILLIP G. SCHMID

Cardiovascular Division, Department of Internal Medicine, University of Iowa College of Medicine, Iowa City, Iowa 52242

GENERAL INTRODUCTION

For several years we have been interested in contrasting the reflex autonomic control of various vascular beds and have been impressed by the degree of selectivity with which different efferent pathways may be activated reflexly. Both quantitative and qualitative differences have been noted not only between vascular beds but also between consecutive vascular segments of the same bed. Specifically, we have recently examined responses of vessels in skeletal muscle, in paw (which is predominantly a cutaneous bed) and in the coronary circulation of dogs in response to stimulation of chemoreceptors. The results to be presented indicated to us that there is (1) a vasoconstrictor response in skeletal muscle, (2) a vasodilator response in the paw; both these responses are actively mediated through the sympathetic system, and (3) a coronary dilator response independent of metabolic changes and mediated through a vagal cholinergic pathway. Data bearing on the possible role of coronary and myocardial receptors as true chemoreceptors will be presented also.

The studies will be described under three major headings:

(i) Reflex activation of adrenergic constriction in muscle and non-cholinergic dilatation in paw by chemoreceptors.

(ii) Reflex activation of vagal cholinergic innervation to coronary vessels by chemoreceptors.

(iii) Are there coronary chemoreceptors?

REFLEX ACTIVATION OF ADRENERGIC CONSTRICTION IN MUSCLE AND NON-CHOLINERGIC DILATATION IN PAW BY CHEMORECEPTORS

Introduction

Although the changes in total peripheral vascular resistance during stimulation of aortic or carotid chemoreceptors indicate that the predominant and net peripheral response is a vasoconstrictor one (Daly & Ungar, 1966; Stern & Rapaport, 1967), studies on individual beds suggest that a vasodilator response may occur in certain vascular beds simultaneously with the increase in total resistance (Bernthal & Schwind, 1945; Korner & Uther, 1969). In these experiments we have studied vascular effects of stimulation of aortic and carotid chemoreceptors with nicotine and cyanide and with hypoxic and hypercapnic blood.

Methods

Male mongrel dogs were anaesthetized with chloralose (50 mg kg^{-1}) and urethane (500 mg kg^{-1}), paralysed with decamethonium bromide and ventilated with room air and supplemental oxygen at 1–2 l min^{-1} to maintain normal systemic arterial blood gases throughout the study.

The cranial tibial artery to a hindpaw was exposed and collateral arteries at that level were ligated. The ipsilateral gracilis artery was also exposed and the gracilis muscle was dissected free from surrounding tissues except for the artery, vein and nerve. After intravenous heparin (5 mg kg^{-1}), the cranial tibial and gracilis arteries were perfused separately at constant flow with arterial blood withdrawn from the femoral artery and perfusion pressure to each bed was measured. In some experiments venous pressures were measured by cannulating retrogradely a small dorsal metatarsal vein and a small vein draining the caudal end of the gracilis. Changes in perfusion pressure reflect changes in vascular resistance and changes in venous pressure reflect changes in venous resistance downstream from the points of pressure measurement in these vascular beds. Femoral vein pressure did not change significantly during the interventions.

The carotid bifurcations were isolated and perfused at constant flow (100 to 120 ml min^{-1}), with blood that had passed through a bubble 'oxygenator' which permitted adjustment of P_{O_2} and P_{CO_2}. The occipital artery distal to the carotid body and the internal carotid, lingual and ascending pharyngeal arteries were ligated. The external carotid arteries were cannulated and the outflow from them was passed through a Starling resistor to regulate arterial pressure in the carotid bifurcations. The oxygenator was primed with 500 ml of dextran (av. mol. wt 70 000) and a gas mixture of 5 % carbon dioxide and 95 % nitrogen was bubbled through it to produce hypoxic and hypercapnic blood. The temperature of the blood perfusing the carotid arteries was maintained at 37 °C with a heat exchanger.

Chemoreceptors were also stimulated with injections of nicotine bitartrate (0.5 to 10 μg kg^{-1}) and sodium cyanide (0.05 to 0.4 mg kg^{-1}) in the ascending aorta downstream from the coronary ostia or in the carotid arteries. In some experiments an arrangement of delay coils similar to that used by Comroe & Mortimer (1964) to stimulate the aortic and carotid bodies separately or simultaneously was employed as described in detail in a previous publication (Calvelo, Abboud, Ballard & Abdel-Sayed, 1970).

Results

Nicotine and cyanide

Stimulation of chemoreceptors with nicotine or cyanide caused pronounced vasoconstriction in the muscle but not in the paw (Fig. 1). In contrast, in the paw some interventions (nicotine–carotid or cyanide–aortic and carotid) caused reflex vasodilatation (Table 1). The dilatation

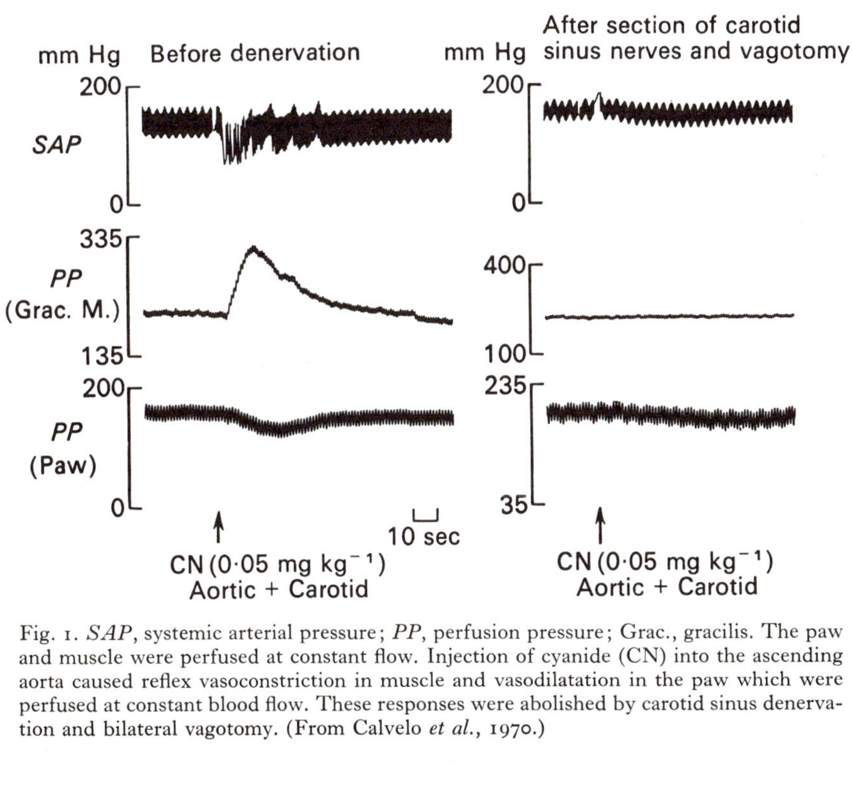

Fig. 1. *SAP*, systemic arterial pressure; *PP*, perfusion pressure; Grac., gracilis. The paw and muscle were perfused at constant flow. Injection of cyanide (CN) into the ascending aorta caused reflex vasoconstriction in muscle and vasodilatation in the paw which were perfused at constant blood flow. These responses were abolished by carotid sinus denervation and bilateral vagotomy. (From Calvelo *et al.*, 1970.)

Table 1. *Reflex responses in muscle and paw during stimulation of chemoreceptors*

Chemoreceptors	Stimulus	Δ Perfusion Pressure Muscle	Paw
		mm Hg	
Carotid	Nicotine (2.5 μg kg^{-1})	52±6*	−30±4*
	Cyanide (0.1 mg kg^{-1})	46±9*	2±13
	Hypoxia & Hypercapnia (P_{O_2} 32±2, P_{CO_2} 73±5 mm Hg)	55±12*	−10±3*
Aortic	Nicotine (5 μg kg^{-1})	20±5*	14±12
	Cyanide (0.2 mg kg^{-1})	40±8*	−14±15
Aortic & Carotid	Nicotine (5 μg kg^{-1})	65±15*	−4±12
	Cyanide (0.2 mg kg^{-1})	85±17*	−33±7*

* Statistically significant responses. Responses to nicotine and cyanide were published earlier (Calvelo *et al.*, 1970) and are shown here to contrast with the response to carotid hypoxemia and hypercapnia. Reflex vasoconstriction in muscle was consistently observed with all stimuli but the dilator response in paw was significant only with carotid–nicotine, carotid–hypoxia and aortic+carotid–cyanide.

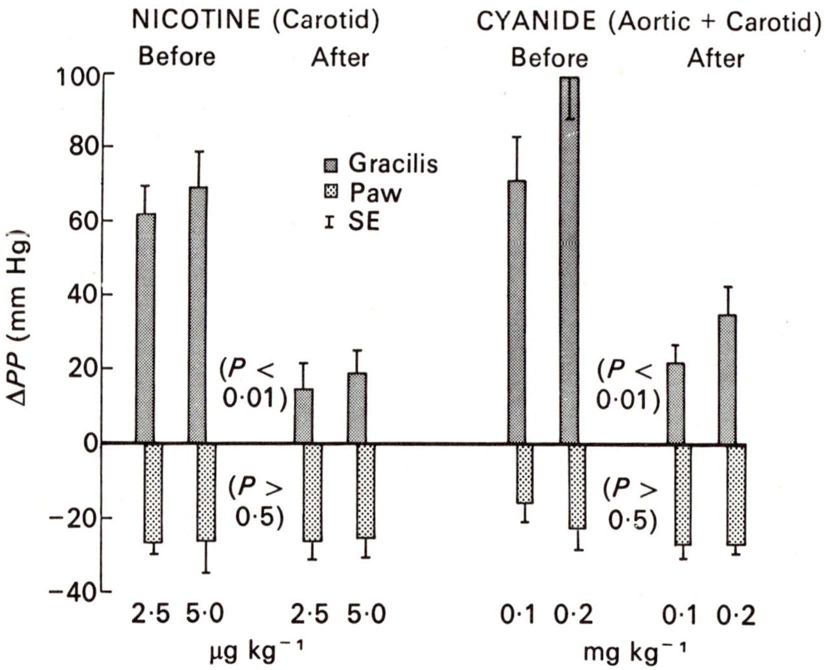

Fig. 2. The bar graphs show the average increases in perfusion pressure in muscle (vasoconstriction) and decreases in perfusion pressure in the paw (vasodilatation) in eight dogs in response to two doses of nicotine and two of cyanide before and after alpha-adrenergic blockade with phentolamine (0.5 to 1 mg intra-arterially). Phentolamine reduced the constrictor but not the dilator responses; the latter tended to increase indicating that the reflex dilatation was active and not caused by withdrawal of sympathetic tone.

was not reduced and was even unmasked or augmented in some experiments by alpha-receptor blocking agents (Fig. 2). When seen the dilator response was not blocked by atropine, tripelennamine, propranolol or carboxypeptidase (Calvelo *et al.*, 1970). Venous responses were small and inconsistent in the gracilis but in the paw there were small but statistically significant increases in venous pressure seen only during stimulation of aortic and carotid chemoreceptors with large doses of nicotine (10 mg kg^{-1}) or cyanide (0.4 mg kg^{-1}). The increases in small vein pressure averaged 7.4 ± 2 and 4.0 ± 0.6 mm Hg respectively. The absence of large venous

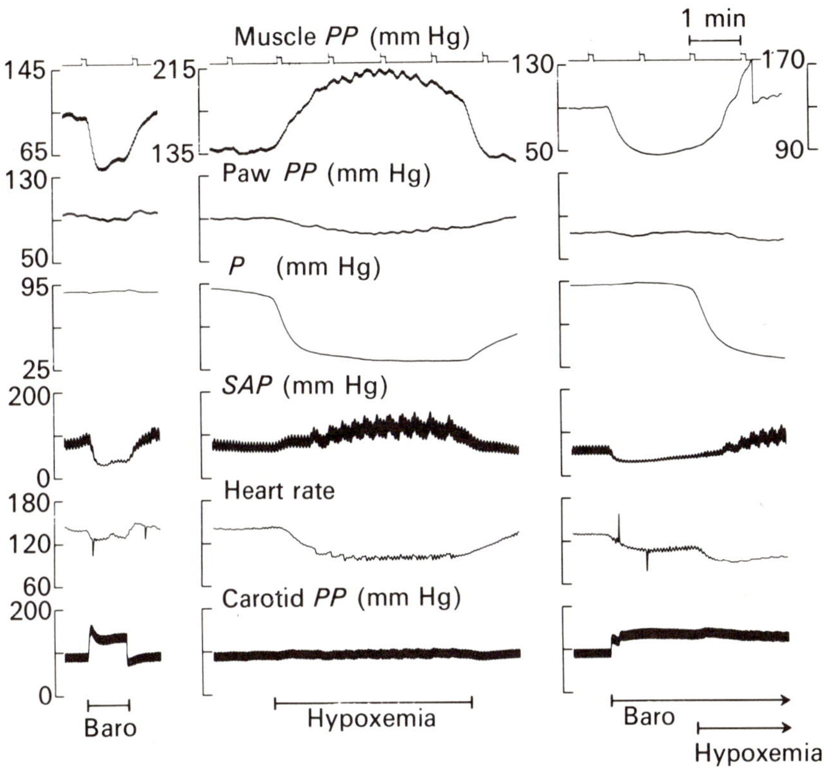

Fig. 3. The perfusion of hypoxaemic blood into both carotids ($P_{O_2} = 25$ mm Hg) at constant pressure caused vasoconstriction in muscle, vasodilatation in paw, hypertension and marked bradycardia (middle panels). Elevation of carotid perfusion pressure by increasing the pressure in the Starling resistor caused significant dilatation in muscle and minimal dilatation in the paw (left and right panels). When carotid hypoxaemia was superimposed on the elevated carotid pressure (right panels), the dilatation in muscle was reversed to intense constriction and additional dilatation was noted in the paw. If one assumes that at a carotid perfusion pressure of 150 mm Hg in this experiment sympathetic tone to peripheral vessels was withdrawn then the additional dilator response in paw must be caused by activation of dilator fibres.

responses, despite the rich innervation of venous segments especially in the hindpaw, emphasizes the selectivity of activation of efferent pathways.

Hypoxia and hypercapnia

Perfusion of the carotids with hypoxic, hypercapnic blood caused reflex vasoconstriction in muscle and vasodilatation in the paw (Fig. 3). The vasodilator response was not antagonized by pharmacological blockers (Fig. 4).

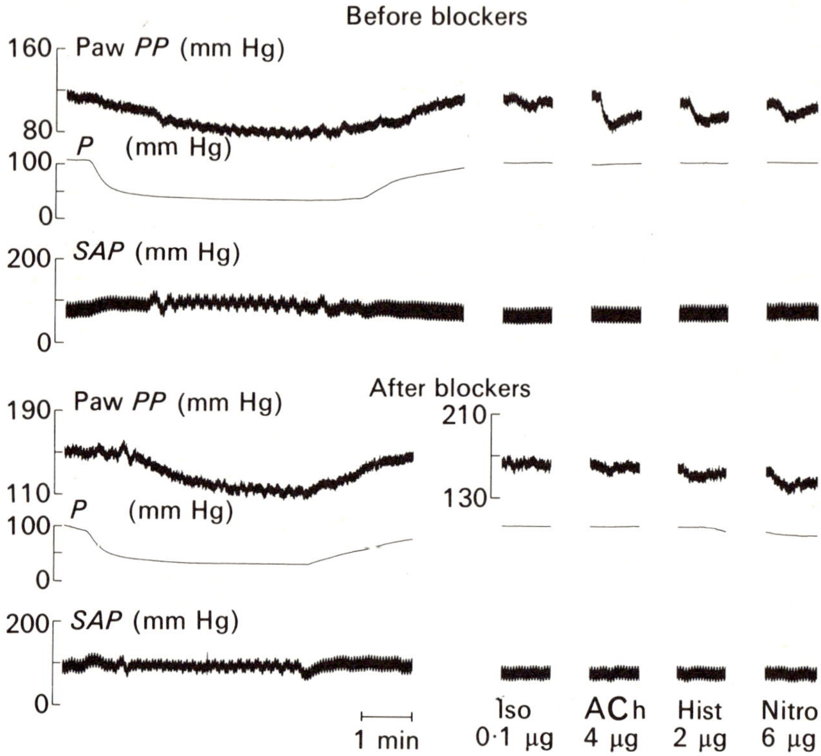

Fig. 4. The vasodilator responses seen in the paw during carotid hypoxaemia (P_{O_2} 40 mm Hg) and injections of isoproterenol (Iso), acetylcholine (ACh) and histamine (Hist) are shown in the top tracings before the administration of blockers. After intra-arterial administration of propranolol (1 mg), atropine (0.1 mg) and tripelennamine (5 mg), the agonists were blocked but the reflex dilator effect of hypoxaemia persisted.

Contribution of baroreceptor activity to the chemoreceptor reflex

In order to exclude the possibility that the vascular responses to stimulation of chemoreceptors were caused in part by activation of baroreceptors with nicotine or by the changes in arterial pressure concomitant with the chemoreceptor reflex, we did the following experiment. The carotid sinus nerve fibres were carefully dissected as close as possible to the carotid sinus bulb (to avoid fibres originating in the carotid body) and stimulated electrically at supramaximal frequencies (60 Hz). This caused hypotension, bradycardia and vasodilatation in both muscle and paw probably through withdrawal of sympathetic tone to these two beds (Fig. 5). When in the

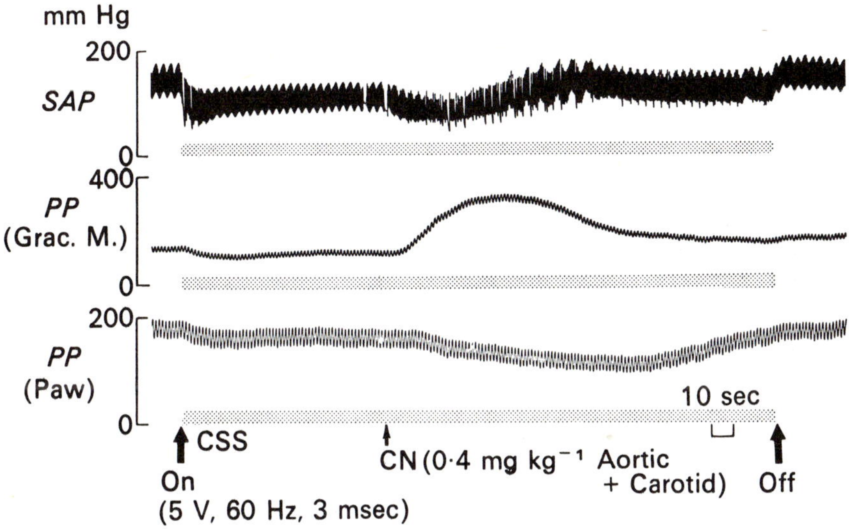

Fig. 5. During sustained electrical stimulation of the carotid sinus nerve (CSS) close to the bulb to activate exclusively baroreceptor nerves there was reflex hypotension and dilatation in muscle and paw caused by withdrawal of sympathetic tone. The electrical stimulus was sustained for a few minutes (as indicated by the horizontal bar under each tracing) and during the stimulus cyanide (CN) was injected into the ascending aorta to activate chemoreceptors. Activation of chemoreceptors triggered vasoconstriction in muscle and additional dilatation in paw. (From Calvelo *et al.*, 1970.)

same animals the effects of nicotine or cyanide on the ipsilateral carotid body, or on carotid + aortic chemoreceptors, were tested during electrical stimulation of the carotid sinus nerve, additional hypotension was noted, the vasodilator response in muscle was reversed and the vasodilator response in the paw was augmented (Fig. 5). The response to stimulation of chemoreceptors was similar whether or not the baroreceptor nerves were being stimulated. Thus, the chemicals activate a chemoreceptor reflex which differs from the baroreceptor in its vascular effects and which can over-ride or reverse the effects of the baroreceptor reflex.

Similar findings were noted when the carotid baroreceptors were activated by elevating carotid artery pressure and subsequently perfusing the carotids with hypoxic and hypercapnic blood (Fig. 3). Activation of the chemoreceptor reflex reversed the vasodilator effect of activation of the baroreceptor reflex in the gracilis and augmented the dilator effect in the paw.

Discussion

Stimulation of chemoreceptors causes selective activation of sympathetic vasoconstrictor adrenergic pathways to vessels in skeletal muscle and of sympathetic vasodilator pathways to vessels in the paw. Since the changes in small vein pressure were minimal the efferent pathways that were activated reflexly must supply pre-venous resistance vessels and most probably the arterioles. The vasodilatation in the paw was not caused by withdrawal of sympathetic adrenergic tone nor by activation of sympathetic cholinergic fibres. The presence of non-cholinergic vasodilator pathways supplying the fore or hindpaw of the dog had been demonstrated earlier (Beck, Pollard, Kayallp & Weiner, 1966; Abboud & Eckstein, 1966; Zimmerman, 1966; Ballard, Abboud & Mayer, 1970) but the results reported here are the first indication of the participation of these efferent neurogenic vasodilator pathways in a physiologic reflex. The neuromediator of this dilatation is unknown.

The opposite effects on the vasculature of muscle and skin may subserve an important function in oxygen conservation during the chemoreceptor reflex by favouring the redistribution of blood flow away from the muscle which has a higher oxygen consumption to the skin which has a low metabolic rate.

The venous responses in the paw were relatively small despite the rich adrenergic innervation of veins, particularly those draining the paws in the dog (Mayer, Abboud, Ballard & Eckstein, 1968). This venous constriction reflects predominantly responses of small veins or venous resistance vessels and contrasts with the venodilatation reported at this workshop by Dr Shepherd which reflects predominantly the large capacitance veins. The difference may be another indication of the selective activation of efferent pathways to consecutive venous segments.

Finally the results indicate that the vascular responses to the chemoreceptor reflex may over-ride or reverse the effects of stimulation of arterial baroreceptors.

REFLEX ACTIVATION OF VAGAL CHOLINERGIC INNERVATION TO CORONARY VESSELS BY CHEMORECEPTORS

Introduction

The differential responses seen in the peripheral vessels with stimulation of chemoreceptors and the lack of information concerning responses of coronary vessels to this stimulus prompted this study. Coronary vascular responses to stimulation of chemoreceptors with nicotine or cyanide were studied in anaesthetized open-chest dogs and compared to responses observed during electrical stimulation of the carotid sinus nerve. The administration of autonomic blockers permitted the identification of the neural pathway mediating the efferent components of the reflexes.

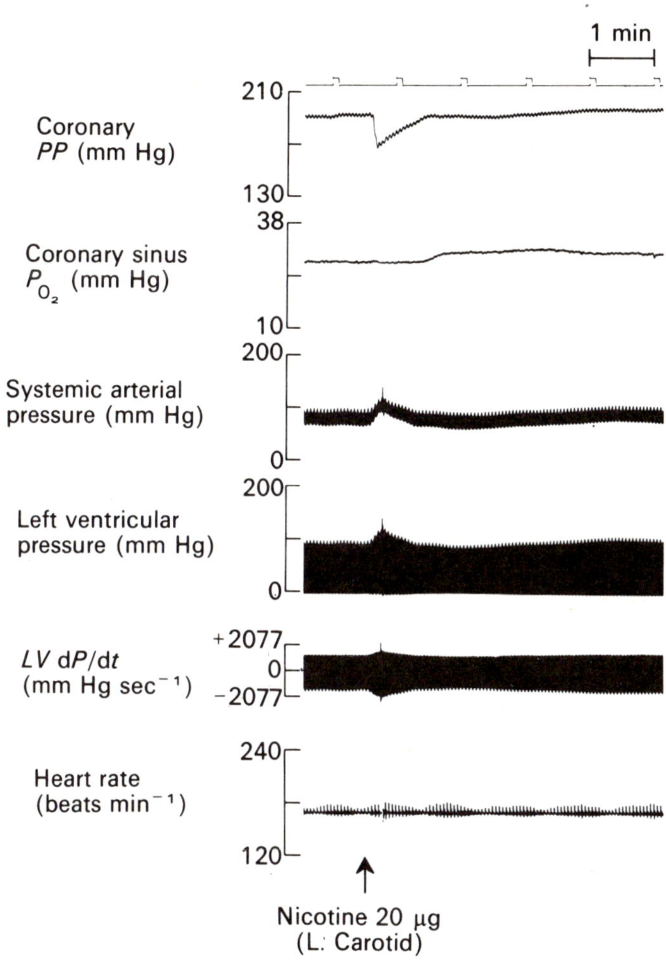

Fig. 6. Injection of nicotine into the left carotid triggered reflex coronary vasodilatation associated with a rise in coronary sinus P_{O_2}. Heart rate was constant (pacing) and practolol (1.0 mg kg^{-1} i.v.) had been administered to minimize increases in cardiac metabolic requirements. The rise in arterial pressure and LV dP/dt probably reflect the systemic vasoconstrictor effect of the reflex.

Methods

We used a preparation similar to that described above and in addition the left circumflex coronary artery was exposed through a thoracotomy, cannulated near its origin and perfused with arterial blood at constant flow (40 to 62 ml min^{-1}). Changes in perfusion pressure reflected changes in vascular resistance. Nicotine (20 μg base) and cyanide (100 μg base) were injected into the carotid arteries or the ascending aorta. To avoid or minimize the indirect effects of changes in heart rate and contractility on coronary resistance the hearts were paced and a cardio-selective beta-receptor antagonist (practolol AY-21,011) was used (McRaven, Mark, Abboud & Mayer, 1971; Mark *et al.*, 1972).

Results

Nicotine or cyanide caused a reflex coronary vasodilatation and small increases in arterial pressure and left ventricular d*P*/d*t* (Fig. 6). The dilator response was abolished by intravenous atropine but the changes in arterial pressure and LV d*P*/d*t* persisted (Hackett *et al.*, 1972). The coronary dilator response to intra-carotid nicotine averaged -18 ± 3.2 mm Hg in seven studies and was accompanied by an increase in coronary sinus P_{O_2} ($+1.1 \pm 0.4$ mm Hg) (Fig. 6). In contrast the coronary dilator response to an increase in metabolic demand induced by pacing of the heart at a rapid

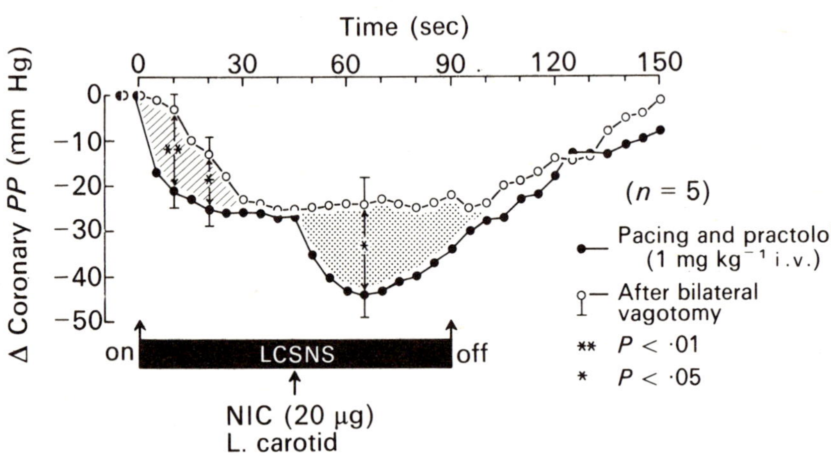

Fig. 7. The dots represent the decreases in perfusion pressure at constant flow (coronary vasodilatation) during electrical stimulation of the left carotid sinus nerve (LCNS) close to the bulb (5 V, 60 Hz, 3 msec.) which is indicated by the horizontal bar. During the stimulus additional dilatation was noted following injection of nicotine (NIC) into the left carotid. Thus, activation of the chemoreflex was superimposed on the baroreflex. The circles represent the responses after bilateral vagotomy. The chemoreflex was abolished and the initial abrupt phase of the baroreflex was delayed.

rate was minimal (-5 ± 1.4 mm Hg) compared to that caused by chemoreceptor stimulation and was accompanied by a decrease in coronary sinus P_{O_2} (-3.8 ± 1.1 mm Hg).

Bilateral vagotomy or denervation of the carotid body by stripping the adventitia off the occipital artery abolished the responses.

Comparison of baroreceptor and chemoreceptor reflexes

The left carotid sinus nerve was stimulated electrically for 90 seconds. This caused an abrupt vasodilatation which rapidly reached a plateau. When chemoreceptors were activated by an injection of nicotine during electrical stimulation an additional vasodilator response was seen; the latter response was abolished by vagotomy but the former response was only attenuated and delayed (Fig. 7).

Discussion

It appears that baroreceptor and chemoreceptor reflexes trigger different efferent pathways to the coronary circulation. The chemoreceptors activate a vagal cholinergic vasodilator response. Although the presence of such a vasodilator innervation to the coronary vessels had been demonstrated by others (Berne, DeGeest & Levy, 1965; Levy, Ng, Lipman & Zieske, 1966; Feigl, 1969) this study represents the first evidence of reflex activation of these pathways. The coronary vasodilatation does not represent a metabolic response for three reasons: (1) heart rate was constant and a beta-blocker minimized the adrenergic stimulus to the heart; (2) coronary sinus P_{O_2} increased during the dilatation caused by the chemoreceptor reflex in contrast to the decrease in P_{O_2} which occurred during the dilatation induced by increased pacing rate; and (3) atropine blocked the dilator response without modifying significantly the increases in pressure and LV dP/dt. The opposite vascular responses seen in skeletal muscle and in the coronary circulation would favour the redistribution of flow to the organ most dependent on aerobic metabolism, namely, the heart during the stimulus triggered by the lack of oxygen.

The baroreceptor reflex also causes coronary vasodilatation but through a different mechanism; by withdrawal of adrenergic vasoconstrictor tone to the coronaries. The early abrupt component of the coronary vasodilator response to stimulation of baroreceptors may be caused by activation of vagal cholinergic pathways.

ARE THERE CORONARY CHEMORECEPTORS?

Introduction

Although the Bezold–Jarisch reflex which causes hypotension and bradycardia represents a chemoreflex originating in the coronary vessels or the myocardium one should question whether this reflex represents activation of chemoreceptors as defined by Dawes & Comroe (1954).

Sleight & Widdicombe (1965) and Thoren (1972) reported that asphyxia is occasionally associated with increased vagal afferent impulses, particularly if severe derangements in cardiac dynamics occur. These results, however, do not exclude the possibility that a chemoreceptor reflex may originate in the coronaries or the myocardium and be transmitted through the cardiac sympathetic nerves (Brown & Malliani, 1971; Peterson & Brown, 1971). Furthermore, asphyxia activates many cardiovascular reflexes which alter hemodynamics drastically and may modify the afferent activity from the heart. The present experiments were done to see if local coronary and myocardial hypoxaemia may activate a reflex vascular response.

Methods

The same preparation described above was employed and nicotine (1.25 to 5.0 μg kg^{-1}) was injected into the coronary artery to produce reflex vasodilatation in the perfused muscle and hindpaw, thus, demonstrating an intact Bezold–Jarisch reflex. The coronary artery was then perfused with hypoxic and hypercapnic blood (P_{O_2} 43 mm Hg and P_{CO_2} 100 mm Hg and pH 6.9 ± 0.13 units) for 90 to 150 seconds and then normoxic, normocapnic perfusion was restored.

Results

The reflex vasodilator response to intracoronary nicotine was seen in muscle and paw and was abolished by bilateral vagotomy. Coronary hypoxia and hypercapnia caused coronary vasodilatation without triggering reflex vascular responses (Fig. 8). The hypoxia was sufficient, however, to cause marked coronary vasodilatation.

Prolonged perfusion of the coronaries with hypoxic blood for 12 to 22 minutes still did not induce reflex responses.

To exclude the possibility that the fall in arterial pressure which occurs during coronary hypoxia might have masked a reflex response the carotid sinuses were perfused at constant pressure in six experiments. This manoeuvre did not unmask a coronary chemoreceptor reflex. In the same preparation, however, perfusion of the carotids with the same hypoxic hypercapnic blood that did not trigger a coronary reflex response caused reflex vasoconstriction in muscle and vasodilatation in the paw as described earlier.

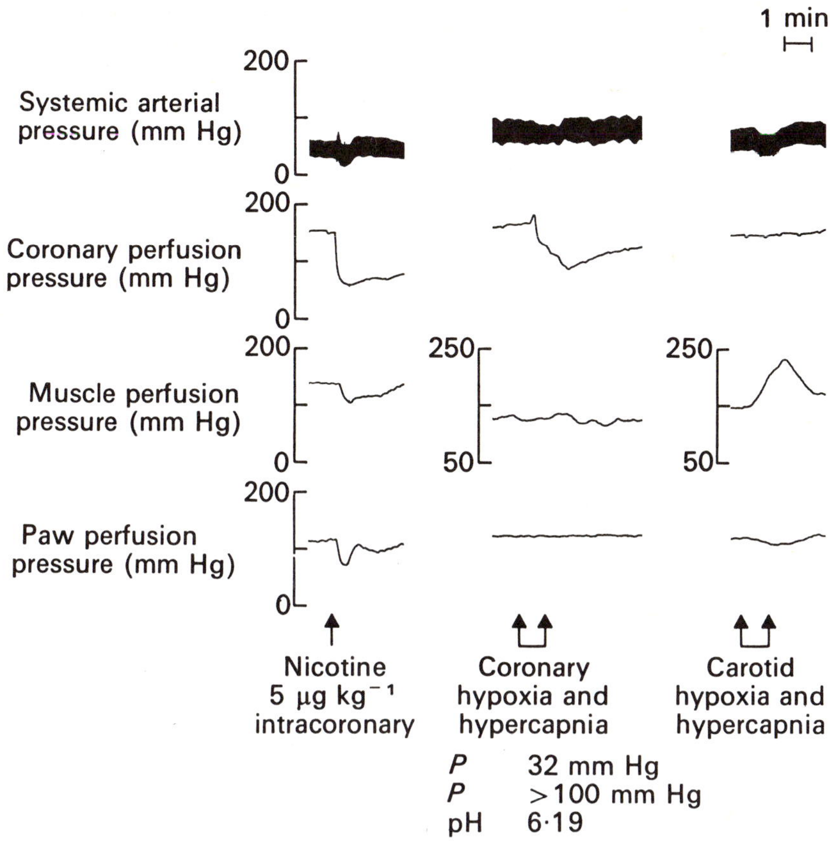

Fig. 8. Intracoronary nicotine activated the 'Bezold-Jarisch' reflex but intracoronary 'hypoxaemia' did not. In ten dogs intracoronary nicotine decreased perfusion pressure in muscle and paw by an average of -28 ± 5 and -13 ± 4 mm Hg respectively. Corresponding values with coronary hypoxia and hypercapnia were -1 ± 3 and -2 ± 2 mm Hg. The same degree of hypoxaemia activated the carotid chemoreceptor reflex response. (From Mark, Abboud, Heistad, Schmid & Johannsen. *Am. J. Physiol.*, (in press), 1974.)

Discussion

The results indicate that coronary hypoxia and hypercapnia *per se* are probably not the stimuli for abnormal reflex responses originating in distal coronary arteries and myocardium with interventions such as coronary artery occlusion (Constantin, 1963; Hanley, Costin & Skinner, 1971). Furthermore, the evidence demonstrates that in the coronary vessels receptors responding to chemical activation by drugs do not necessarily respond to activation by hypoxia and hypercapnia reflecting a selectivity in the chemical activation of true chemoreceptors. There is no evidence of true chemoreceptors in the circumflex coronary vessel or the left ventricular myocardium in the dog despite the fact that chemoreflexes may be activated.

SUMMARY

The changes in total peripheral vascular resistance during stimulation of the chemoreceptors indicate that the predominant peripheral response is vasoconstrictor; but this provides no information about responses of individual vascular beds. In order to examine this point, we have (*a*) measured responses of coronary vessels together with those in skeletal muscle and in skin of dogs to stimulation of aortic and carotid chemoreceptors with nicotine, cyanide, hypoxia and hypercapnia, and (*b*) assessed the role of coronary or myocardial receptors in mediating peripheral vascular responses.

The circumflex coronary artery was perfused at constant flow so that a change in perfusion pressure reflected a change in coronary vascular resistance. Beta-receptors in the myocardium were selectively blocked and the heart was paced in order to minimize the indirect effects of myocardial responses on coronary resistance. Carotid and aortic injections of nicotine caused a fall in coronary perfusion pressure, a response which was blocked by bilateral vagotomy or atropine indicating that stimulation of the carotid body activates vagal cholinergic vasodilator fibres to coronary vessels. Stimulation of carotid chemoreceptors caused vasoconstriction in the isolated perfused gracilis muscle and vasodilatation in the isolated perfused paw, both responses being abolished by carotid sinus denervation or peripheral denervation of these vascular beds. There was no indication that venous segments participated in these responses. This pattern of responses seen in myocardium, skeletal muscle and paw may represent an attempt to distribute and conserve oxygen by diverting blood away from metabolically active skeletal muscle (which can function anaerobically) to a metabolically hypoactive cutaneous paw and to the myocardium which cannot sustain a significant degree of oxygen deprivation.

In other studies we confirmed that direct injection of nicotine into the coronary arteries elicited reflex vasodilatation (the von Bezold–Jarisch effect) but this could not be reproduced by perfusion of hypoxic hypercapnic blood; which indicates that these receptors, though capable of initiating a chemoreflex, are not true chemoreceptors.

This work was supported by research grants from the American Heart Association and from the United States Public Health Service (NHLI) HE-14388 and HE-02644. Some of the results presented at the workshop and reported here have been published from our laboratory in *Circulation Research* (Calvelo, Abboud, Ballard & Abdel-Sayed, 1970; Hackett, Abboud, Mark, Schmid & Heistad, 1972).

REFERENCES

ABBOUD, F. M. & ECKSTEIN, J. W. (1966). Comparative changes in segmental vascular resistance in response to nerve stimulation and to norepinephrine. *Circulation Res.*, **18**, 263–77.

BALLARD, D. R., ABBOUD, F. M. & MAYER, H. E. (1970). Release of a humoral vasodilator substance during neurogenic vasodilatation. *Am. J. Physiol.*, **219**, 1451–7.

BECK, L., POLLARD, A. A., KAYALLP, S. O. & WEINER, L. M. (1966). Sustained dilatation elicited by sympathetic nerve stimulation. *Fedn Proc.*, **25**, 1596–1606.

BERNE, R. M., DEGEEST, H. & LEVY, M. N. (1965). Influence of the cardiac nerves on coronary resistance. *Am. J. Physiol.*, **208**, 763–9.

BERNTHAL, T. G. & SCHWIND, F. S. (1945). Comparison in intestine and leg of the reflex vascular response to carotid–aortic chemoreceptor stimulation. *Am. J. Physiol.*, **143**, 361–72.

BROWN, A. & MALLIANI, A. (1971). Spinal sympathetic reflexes initiated by coronary receptors. *J. Physiol.*, **216**, 685–705.

CALVELO, M. G., ABBOUD, F. M., BALLARD, D. R. & ABDEL-SAYED, W. (1970). Reflex vascular responses to stimulation of chemoreceptors with nicotine and cyanide. *Circulation Res.*, **27**, 259–76.

COMROE, J. H., JR & MORTIMER, L. (1964). Respiratory and cardiovascular responses of temporally separated aortic and carotid bodies to cyanide, nicotine, phenyldiguanide and serotonin. *J. Pharmac. exp. Ther.*, **146**, 33–41.

CONSTANTIN, L. (1963). Extracardiac factors contributing to hypotension during coronary occlusion. *Am. J. Cardiol.*, **11**, 205–17.

DALY, M. DE B. & UNGAR, A. (1966). Comparison of ther eflex responses elicited by stimulation of the separately perfused carotid and aortic body chemoreceptors in the dog. *J. Physiol.*, **182**, 379–403.

DAWES, G. S. & COMROE, J. H. (1954). Chemoreflexes from the heart and lungs. *Physiol. Rev.*, **34**, 167–201.

FEIGL, E. O. (1969). Parasympathetic control of coronary blood flow in dogs. *Circulation Res.*, **25**, 509–19.

HACKETT, J. G., ABBOUD, F. M., MARK, A. L., SCHMID, P. G. & HEISTAD, D. D. (1972). Coronary vascular responses to stimulation of chemoreceptors and baroreceptors. *Circulation Res.*, **31**, 8–17.

HANLEY, H. E., COSTIN, J. C. & SKINNER, N. S., JR (1971). Differential reflex adjustments in cutaneous and muscle vascular beds during experimental coronary artery occlusion. *Am. J. Cardiol.*, **27**, 513–21.

KORNER, P. I. & UTHER, J. B. (1969). Dynamic characteristics of the cardiovascular autonomic effects during severe arterial hypoxia in the unanesthetized rabbit. *Circulation Res.*, **24**, 671–87.

LEVY, M. N., NG, M., LIPMAN, R. I. & ZIESKE, H. (1966). Vagus nerves and baroreceptor control of ventricular performance. *Circulation Res.*, **18**, 101–6.

MCRAVEN, D. R., MARK, A. L., ABBOUD, F. M. & MAYER, H. E. (1971). Responses of coronary vessels to adrenergic stimuli, *J. clin. Invest.*, **50**, 773–8.

MARK, A. L., ABBOUD, F. M., SCHMID, P. G., HEISTAD, D. D. & MAYER, H. E. (1972). Differences in direct effects of adrenergic stimuli on coronary, cutaneous and muscular vessels. *J. Clin. Invest.*, **51**, 279–87.

MAYER, H. E., ABBOUD, F. M., BALLARD, D. R. & ECKSTEIN, J. W. (1968). Catecholamines in arteries and veins of the foreleg of the dog. *Circulation Res.*, **23**, 653–61.

PETERSON, D. F. & BROWN, A. M. (1971). Pressor reflexes produced by stimulation of afferent cardiac nerve fibres in the cardiac sympathetic nerves of the cat. *Circulation Res.*, **28**, 605–10.

SLEIGHT, P. & WIDDICOMBE, J. G. (1965). Action potentials in fibres from receptors in the epicardium and myocardium of the dog's left ventricle. *J. Physiol.*, **181**, 235–58.

STERN, S. & RAPAPORT, E. (1967). Comparison of the reflexes elicited from combined or separate stimulation of the aortic and carotid chemoreceptors on myocardial contractility, cardiac output and systemic resistance. *Circulation Res.*, **20**, 214–27.

THOREN, P. (1972). Left ventricular receptors activated by severe asphyxia and by coronary artery occlusion. *Acta physiol. scand.*, **85**, 455–63.

ZIMMERMAN, B. G. (1966). Sympathetic vasodilatation in the dog's paw. *J. Pharmacol. exp. Ther.*, **152**, 81–7.

DISCUSSION

ACKER It has been found that when the carotid body is stimulated, sodium excretion in the kidney is reduced (Fleming *et al.*, 1973). Do you think that the changes in serum electrolytes could affect the vascular changes which you have observed.

ABBOUD Before I answer that question, I would need to know how great and how fast were the changes in electrolytes. I think, however, that it is unlikely that changes in serum electrolyte would affect the vascular responses to any significant degree. The changes would have to be significant and would have to occur fast enough to account for the acute changes which we report. The vascular responses we are reporting were immediate. Perhaps the electrolyte changes could be involved in long-term vascular adjustments to stimulation of chemoreceptors.

McDONALD You mentioned the changes in flow through the paw and that there is normally dilatation despite the large innervation.

ABBOUD There is an adrenergic vasoconstricting innervation to the vessels of the paw and a non-adrenergic and non-cholinergic vasodilator innervation also. If one stimulates the lumbar sympathetic chain electrically, one gets vasoconstriction in the paw because of the predominant response to activation of the vasoconstrictor adrenergic fibres. When one triggers the chemoreceptor reflex, however, the efferent adrenergic vasoconstricting fibres are not activated but instead the vasodilator sympathetic fibres are. Thus, the reflex activation of efferent sympathetic pathways is selective to different vascular beds and the selectivity probably occurs at a central level.

McDONALD But can I get this clear. These nerves are sympathetic and they are adrenergic and yet they cause dilatation?

ABBOUD They are certainly sympathetic but they probably are not adrenergic, nor are they cholinergic.

McDONALD But why sympathetic?

ABBOUD What other nerves would you suggest supply the blood vessels in the paw?

McDONALD Parasympathetic nerves, for example.

ABBOUD I don't believe that parasympathetic nerves supply vessels in the limb. In the coronaries, they do.

SAMPSON A question on the same line. If you denervate the paw, the vasodilatation is limited, is that so?

ABBOUD Yes.

SAMPSON You seem to imply that activation of the baroreceptors causes the bradycardia when you inject nicotine or cyanide into the carotid sinus.

ABBOUD We had to consider the possibility that activation of baroreceptors might have contributed to the bradycardia and to the vasodilatation seen in the paw, as well as the early hypotension.

SAMPSON Why do you assume activation of the baroreceptors: from the pressure of injection?

ABBOUD No, from nicotine itself. It has been suggested that these chemicals might activate the baroreceptors in addition to their effect on the chemoreceptors.

SAMPSON I would have thought you would want to study inactivation of the baroreceptors because of the concomitant bradycardia and hypotension. That might be equally important. Another point: where were you stimulating the left sinus nerve because you would also be stimulating chemoreceptors electrically.

ABBOUD There were two reasons why we did not think that the electrical stimulation was activating chemoreceptor fibres. The first was the type of reflex response seen which was different from that which one gets from stimulation of the chemoreceptors. With electrical stimulation of the carotid sinus nerve, we obtain vasodilatation in both muscle and paw, whereas with stimulation of the chemoreceptors, we obtain vasoconstriction in muscle and vasodilatation in the paw. The second reason is that during the continuous electrical stimulation of the baroreceptor nerves at high frequency, when both the muscle and paw were dilating, the injection of nicotine into the carotid sinus reversed the response of the muscle from vasodilatation to vasoconstriction and the response in the paw was augmented by an additional dilatation. It was obvious that the chemicals activated a different reflex than was being activated by the electrical stimulation.

SAMPSON You are stimulating at 60 Hz, I think. If you give nicotine, you will get an initial discharge far in excess of that frequency.

ABBOUD You mean by giving nicotine we would activate baroreceptor fibres over and above those that were activated at that level of electrical stimulation?

SAMPSON Yes.

ABBOUD So you believe that with our electrical stimulation, we were stimulating both chemoreceptor and baroreceptor fibres.

SAMPSON I think it is likely. You are getting a mixed effect.

ABBOUD I can only say that the peripheral vascular effect of these two stimuli was entirely different and the response we obtained with the stimulation of the baroreceptor nerves electrically was different from that seen with the chemoreceptor stimulus.

SAMPSON It could be of course that with that frequency of stimulation you have a predominantly baroreceptor effect but you can overcome it with nicotine stimulation of the chemoreceptors.

MILLS With respect to the experiments on the coronary circulation, you give evidence of cutting the vagus or giving atropine. How exactly did you do the comparable experiments with the sympathetic?

ABBOUD I did not carry out a sympathetic denervation; that is technically beyond me. But I gave phentolamine to block the adrenergic vasoconstrictor innervation and was unable to demonstrate a reduction of the vasodilator response to the chemoreceptor reflex. Furthermore, the work of Feigl (1969) would suggest that there are no sympathetic vasodilator fibres to the coronary vessels.

MITCHELL I would like to make some comments about the aortic bodies – we haven't heard much about them. Perhaps they are being kept for another meeting! There are some important differences between the aortic and the carotid bodies. For example, as Dr Sampson (Sampson & Hainsworth, 1972) has shown and we have confirmed, they do not respond to changes in hydrogen ion, but they do respond of course to hypoxia and hypercapnia. Another point of difference which we (Drs J. D. Hatcher and D. Gillis) have found is that part of the cardiovascular response to anaemia

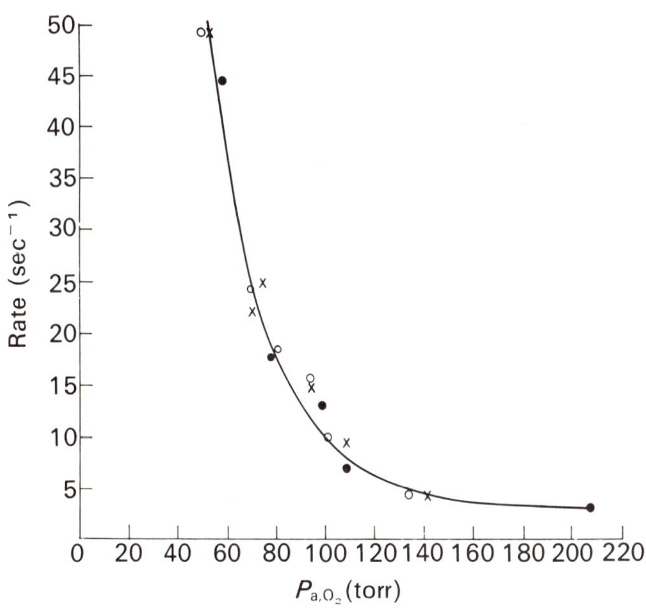

Fig. 1. Plot of carotid chemosensory activity as a function of P_{a,O_2} in a normal cat with initial haematocrit (Hct) of 32.5 % (●); after exchange of blood for dextran, Hct 18.0 % (×) and after reinfusion of packed red cells, Hct 24.5 % (○).

may be mediated through stimulation of the aortic bodies. The first figure shows the effect of graded anaemia, produced by a blood for dextran exchange, on the carotid body chemoreceptor oxygen response curves. There was no consistent change in the response curves which were done 30–90 minutes after the production of anaemia (Fig. 1). However, immediately after the exchange there was an increase in chemoreceptor activity which returned to near control levels within 30–60 minutes. In contrast, the aortic chemoreceptors were consistently shifted upward and to the right by

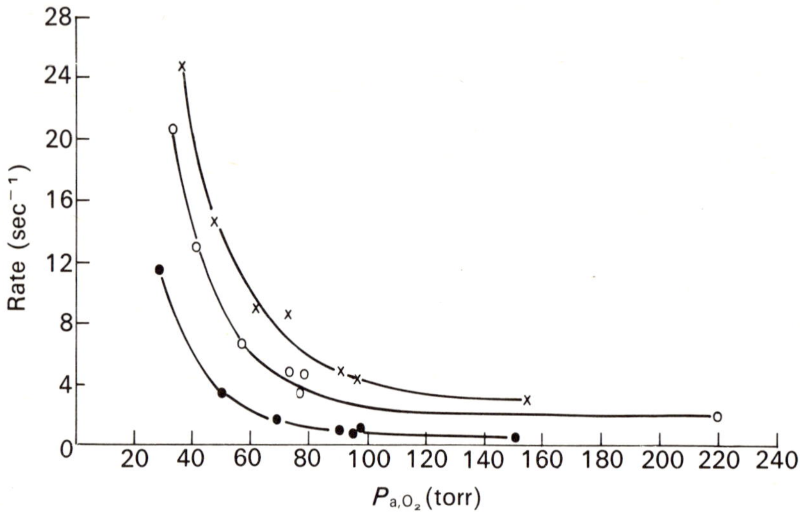

Fig. 2. Plot of aortic chemosensory activity as a function of P_{a,O_2}; in a normal cat with initial Hct 44.5 % (●); after exchange of dextran for blood, Hct 14.5 % (×) and after reinfusion of packed cells, Hct 28 % (○).

anaemia (Fig. 2). Also, the magnitude in the shift was proportional to the degree of anaemia. The difference between the carotid and aortic receptors' response to anaemia was significant, which agrees with Dr Abboud's statement. We then measured the effect of hypoxia and anaemia upon cardiac output and distribution of flow in various vascular beds using radioactive spheres and have found that after section of the aortic nerves there are very marked differences, not only quantitatively, but also in direction.

BULLER Do I take it that this means that the aortic bodies respond to both tension and content of oxygen?

MITCHELL Yes, it would seem so.

MILLS This is not consistent with what Edwards and I (Mills & Edwards, 1968) found in carboxyhaemoglobinaemia. We found that when you lower oxygen content with carbon monoxide, you get stimulation of both aortic and carotid bodies.

MITCHELL I think I mentioned that the carotid bodies did respond but only initially – say over the first half hour. But by the time the response curves were carried out, after an hour or so, this was not apparent. So they do respond transiently.

MILLS You might say therefore that both aortic and carotid bodies are tension receptors but the aortic body responds more obviously to content.

REFERENCES

FEIGL, E. O. (1969). Parasympathetic control of coronary blood flow in dogs. *Circulation Res.*, **25**, 509–19.

FLEMING, H., HONIG, A., PÖTZSCHKE, B., RAUHUT, U., ROLOFF, D. & SCHULTZE, CH. (1973). Renale Natriumausscheidung bei Karotisverschluss vor und nach Ausschalten der Karotischemorezeptoren. *Acta biol. med. germ.*, **30**, 85–93.

MILLS, E. & EDWARDS, M. W. JR. (1968). Stimulation of aortic and carotid chemoreceptors during carbon monoxide inhalation. *J. appl. Physiol.*, **25**, 494–502.

SAMPSON, S. R. & HAINSWORTH, R. (1972). Responses of aortic body chemoreceptors of the cat to physiological stimuli. *Am. J. Physiol.*, **222**, 953–8.

Modulation of the chemoreceptor reflex by changes in baroreceptor activity

DONALD D. HEISTAD, FRANÇOIS M. ABBOUD, ALLYN L. MARK AND PHILLIP G. SCHMID

Cardiovascular Division, Department of Internal Medicine, University of Iowa College of Medicine and the Veterans Administration Hospital, Iowa City, Iowa 52242

INTRODUCTION

Reflexes originating in chemoreceptors and baroreceptors have a major influence on the autonomic control of the circulation. The chemoreceptor reflex tends to conserve oxygen and preserve the myocardium and brain, whereas the baroreceptor reflex tends to maintain arterial pressure. Most experimental studies have dealt with the effects of activation of these reflexes separately. The purpose of this study was to evaluate the effects of simultaneous activation of these two reflexes. We considered the possibility that an interaction between the two reflexes might exist for two reasons. (1) Neurophysiological studies by Miura & Reis (1972) indicate that neurons mediating chemoreceptor and baroreceptor function are distributed in close proximity in the solitary and paramedian reticular nuclei in the medulla so that interneuronal connections might mediate interaction between the two reflexes. (2) In the clinical setting, both reflexes are frequently activated simultaneously. For example, the hypotension following myocardial infarction is often associated with hypoxia and the circulatory adjustment may represent the interaction of reflexes originating in both chemoreceptors and baroreceptors.

This study was done to examine the possibility that responses to stimulation of chemoreceptors might be affected by the level of arterial pressure, and to consider systematically mechanisms which might be involved in such an interaction.

METHODS

Dogs were anaesthetized with chloralose (50 mg kg^{-1}) and urethane (500 mg kg^{-1}), treated with decamethonium bromide (0.3 mg kg^{-1}), and ventilated with room air and supplemental oxygen to maintain normal arterial blood gases.

One carotid artery bifurcation was exposed and a needle was inserted for injection of nicotine to stimulate the carotid body chemoreceptors. Reflex responses to chemoreceptor stimulation by nicotine were observed in the perfused gracilis muscle. The gracilis muscle was dissected free from surrounding tissue except for the gracilis artery, vein, and nerve. The artery was perfused at constant flow and the perfusion pressure was measured. At constant flow an increase in perfusion pressure in the muscle indicates vasoconstriction. Nicotine stimulates carotid chemoreceptors and causes reflex vasoconstriction in the gracilis muscle. We have used the magnitude of the increase in perfusion pressure in the muscle (which indicates vasoconstriction) as an indication of the magnitude of the chemoreceptor reflex. This response is abolished by denervating either the carotid body or the gracilis muscle, which indicates that it is a reflex response arising from the carotid body.

The chemoreceptor reflex was studied in 13 dogs during hypotension induced by bleeding the animals rapidly into a reservoir until mean arterial pressure was between 65 and 85 mm Hg. The observations were repeated after bleeding the animals further to a mean arterial pressure of 40 to 50 mm Hg.

In 8 dogs, a balloon catheter was inserted into the descending aorta and inflated to raise systemic arterial pressure upstream from the balloon by 20 to 40 mm Hg to stimulate baroreceptors. The responses to stimulation of chemoreceptors by nicotine were observed in these animals during elevation of arterial pressure and compared to responses obtained before inflation of the balloon and immediately after deflation.

Because of the possibility that changes in arterial pressure may directly influence the responsiveness of the carotid body, another series of experiments was carried out. The left and right carotid artery bifurcations were exposed in 17 dogs, the vagus nerves were cut to eliminate afferent impulses from the aortic chemoreceptors and baroreceptors, and the common carotid arteries were perfused separately at constant blood flow. Blood flowed out through cannulae in the external carotid arteries and through Starling resistors which allowed adjustment of pressure in each carotid sinus separately. We could thus elevate pressure in one carotid sinus and maintain constant pressure in the other carotid. We reasoned that if elevation of arterial pressure on one side could alter the response to injections of nicotine into the opposite carotid artery (which was perfused at constant pressure), we would then have to conclude that the interaction between the two reflexes was a neural interaction and not the result of changes in arterial pressure within the carotid chemoreceptors.

In seven animals we examined the effects of changes in baroreceptor activity induced in one carotid on the responses to a more physiological stimulus of chemoreceptors on the opposite carotid; namely, the perfusion of that contralateral carotid artery with hypoxic hypercapnic blood at constant perfusion pressure (Fig. 1).

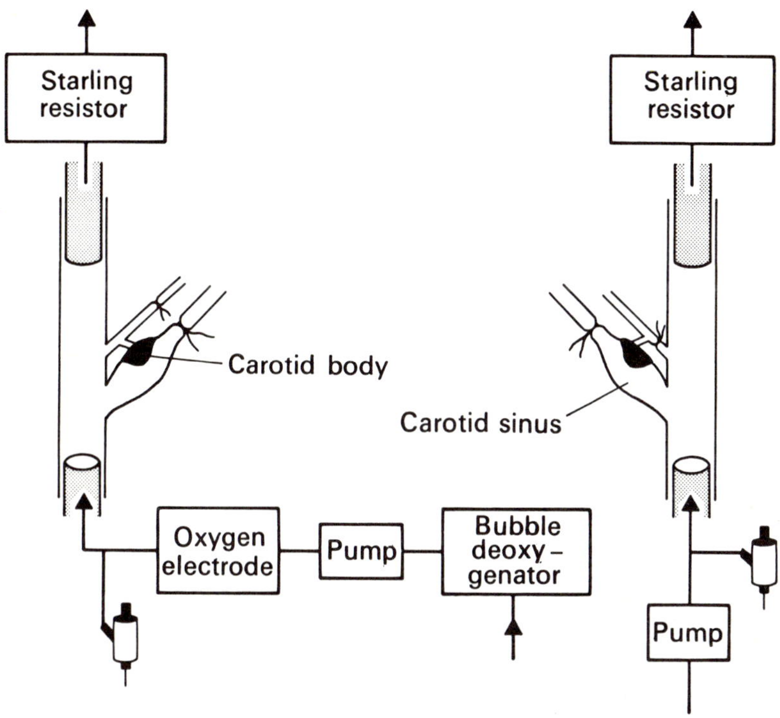

Fig. 1. Diagram of the separate perfusion of both carotid arteries with ligation of the internal carotid and interposition of Starling resistors along the external carotid. Arterial pressure was changed in one carotid artery and maintained constant in the contralateral carotid where activation of chemoreceptors was carried out by perfusing the carotid artery with hypoxic hypercapnic blood; thus, one carotid was perfused with the hypoxic blood mixture at constant pressure whereas the opposite carotid was perfused with normoxic blood at varying pressures. Average values of P_{O_2}, P_{CO_2}, and pH in the hypoxic blood mixture was 33 ± 3.0 mm Hg, 84 ± 5.6 mm Hg, and 7.12 ± 0.04 respectively.

In the last group of studies, ventilatory responses to stimulation of chemoreceptors were measured in spontaneously breathing anaesthetized dogs. Air flow was recorded from a Fleisch pneumotachograph and electrically integrated to measure tidal volume and minute volume. Ventilatory

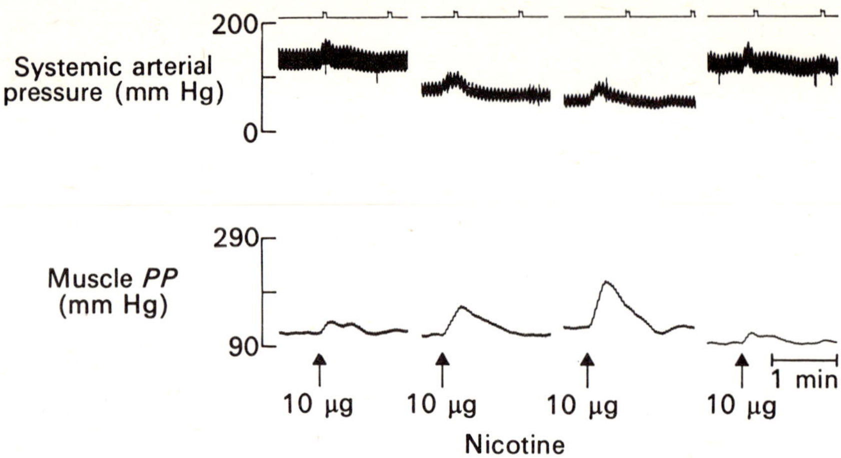

Fig. 2. The effect of stimulating the carotid body chemoreceptors with nicotine at control systemic pressure, two levels of hypotension and after replacement of blood. Gracilis perfusion pressure (Muscle *PP*) measured at constant flow.

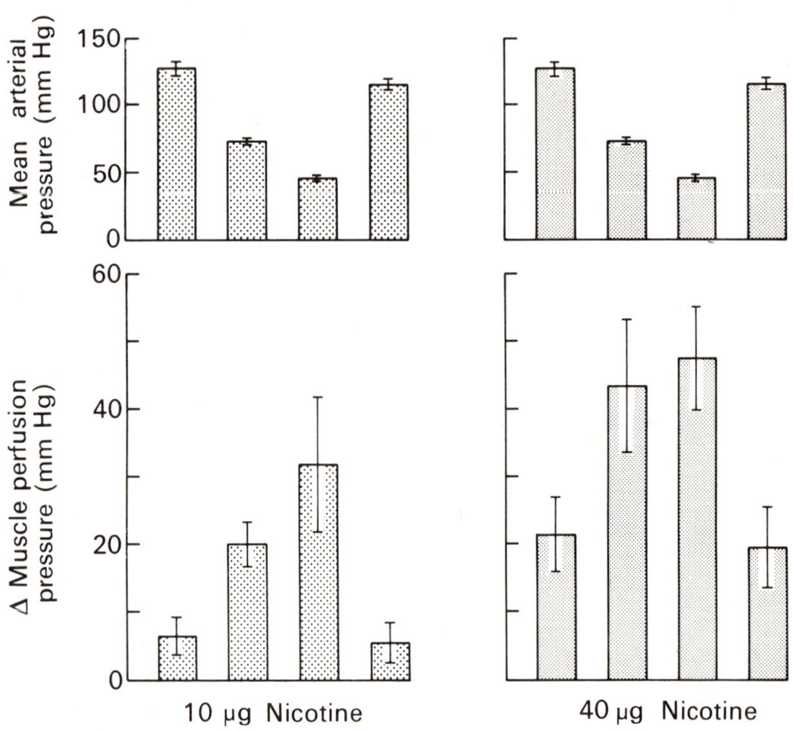

Fig. 3. Summary of effects of haemorrhage upon responses to stimulation of the carotid body chemoreceptors with nicotine, 10 and 40 μg, $n = 13$. Values are mean ±SE.

responses to injection of nicotine into the common carotid artery were observed in 5 dogs during haemorrhagic hypotension and in 4 dogs during elevation of arterial pressure (by aortic occlusion).

The interaction of baroreceptor and chemoreceptor reflexes were evaluated statistically by correlating responses to chemoreceptor stimulation (i.e. the changes in gracilis perfusion pressure) with the level of systemic arterial pressure or carotid perfusion pressure, which reflected the degree of baroreceptor activity. A significant relationship was indicated by a statistically significant regression line which was calculated by fitting a line using the least squares principle (Huntsberger & Leaverton, 1970).

RESULTS

Effects of hypotension

Haemorrhagic hypotension augmented the vasoconstrictor response to stimulation of chemoreceptors by nicotine (Figs. 2 and 3). The potentiation was related to the severity of hypotension.

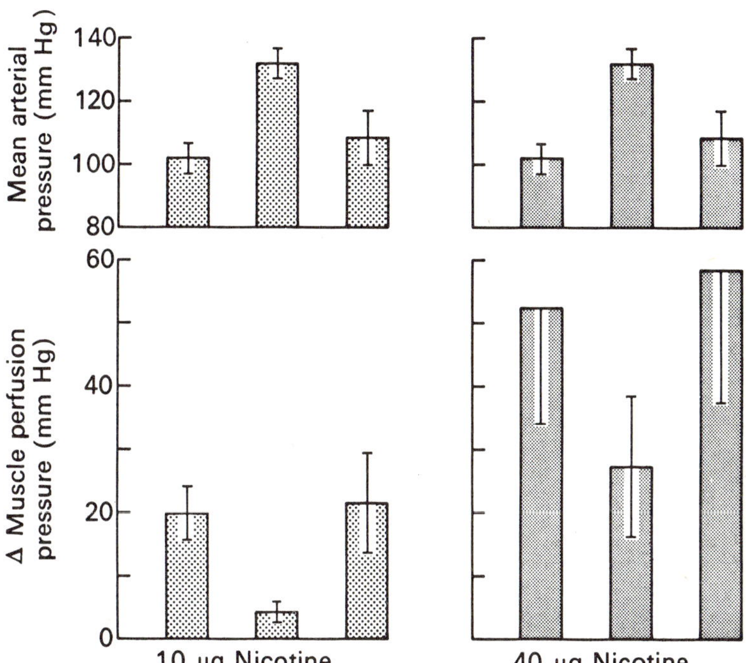

Fig. 4. Summary of effect of occluding the descending aorta upon the response of gracilis muscle perfusion pressure to stimulation of the carotid body chemoreceptors with nicotine, 10 and 40 μg. $n = 8$. Values are mean ±SE.

Effects of elevation of systemic arterial pressure

Transient occlusion of the descending aorta by inflating a balloon raised arterial pressure in vessels upstream from the occlusion by an average of 30 mm Hg. The vasoconstrictor response to stimulation of chemoreceptors by nicotine was inhibited profoundly during elevation of arterial pressure and restored promptly after deflation of the balloon and return of systemic arterial pressure to normal (Fig. 4).

Responses to carotid chemoreceptor stimulation during activation of the contralateral carotid baroreceptors

Vasoconstrictor and systemic pressor responses to stimulation of chemoreceptors by injecting nicotine into one carotid artery perfused at constant pressure were inhibited by elevation of pressure in the contralateral carotid

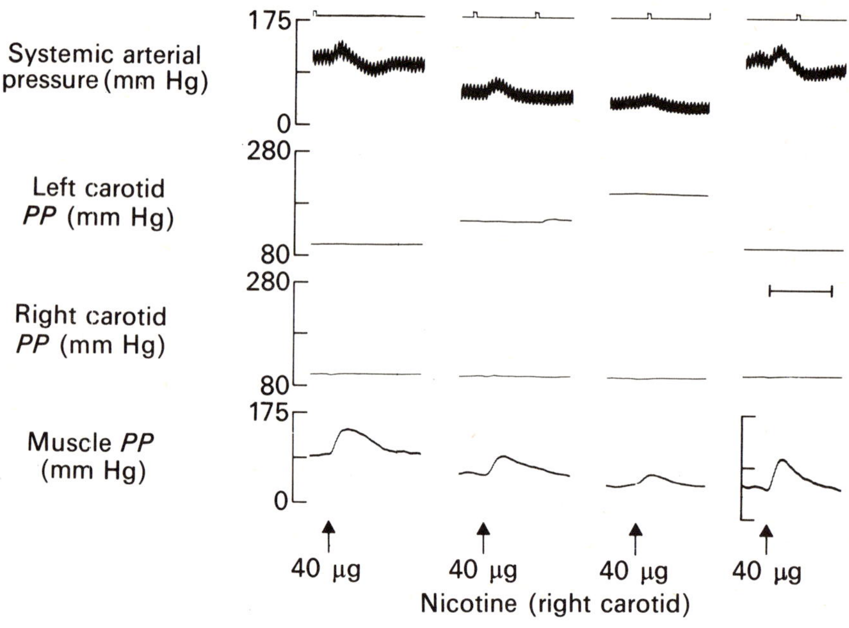

Fig. 5. Experiment as illustrated in Fig. 3 except that as nicotine 40 µg was injected into right carotid sinus perfused at constant pressure, pressure was raised in the left carotid sinus, as indicated: both vagi cut.

baroreceptors (Figs. 5 and 6). Similarly, responses to stimulation of chemoreceptors by hypoxic and hypercapnic blood also were inhibited by elevation of pressure in the contralateral baroreceptors (Figs. 7 and 8).

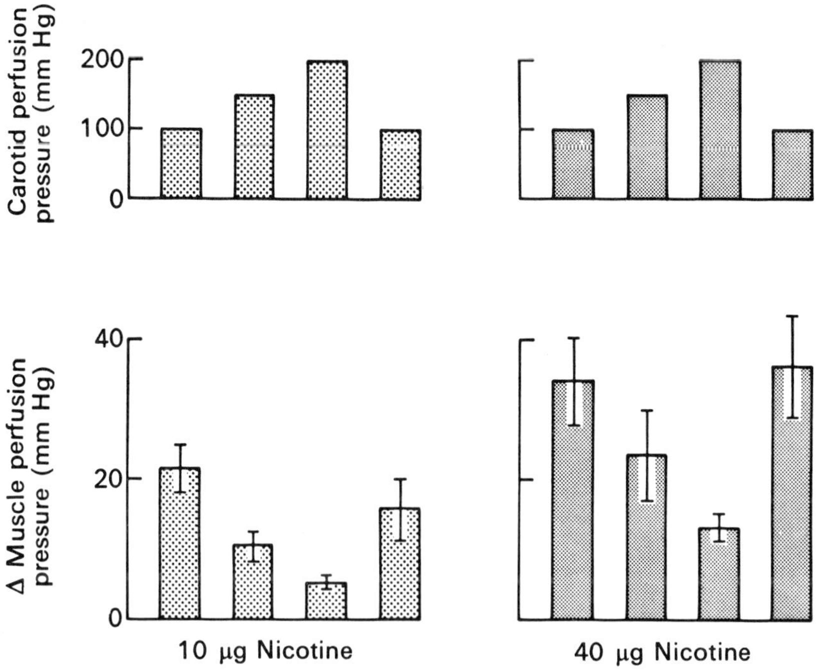

Fig. 6. Summary of results of experiments illustrated in Fig. 5. $n = 17$. Values are mean ±SE.

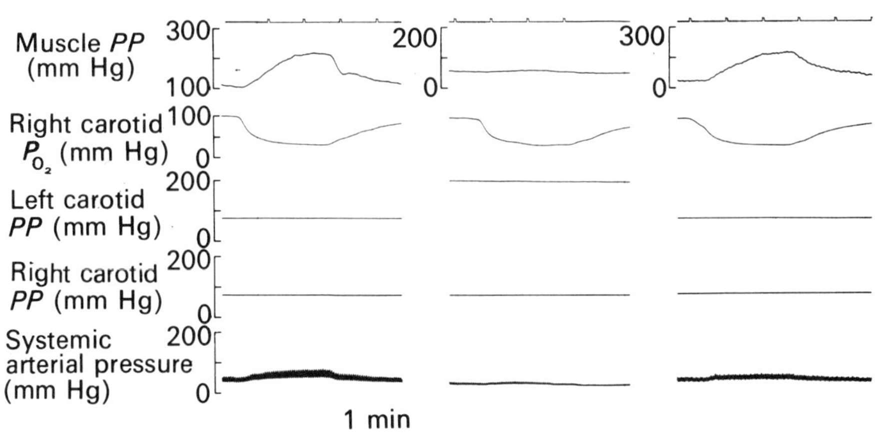

Fig. 7. The effect of perfusing the right carotid body chemoreceptors with hypoxic, hypercapnic blood at constant pressure upon gracilis muscle perfusion pressure and the effect upon this response of raising pressure in the left carotid sinus.

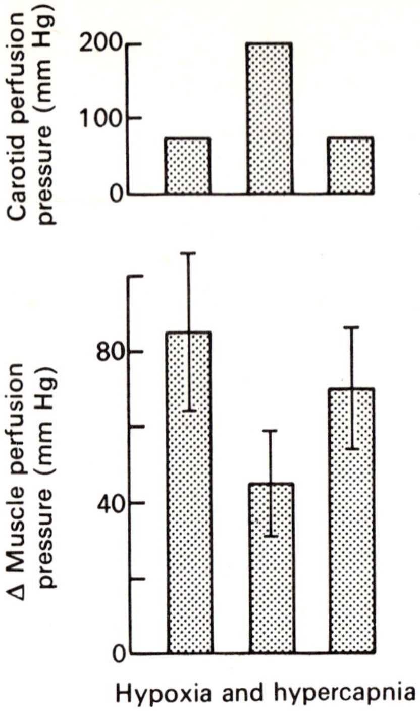

Fig. 8. Summary of results of experiments illustrated in Fig. 7. $n = 7$. Values are mean ± SE.

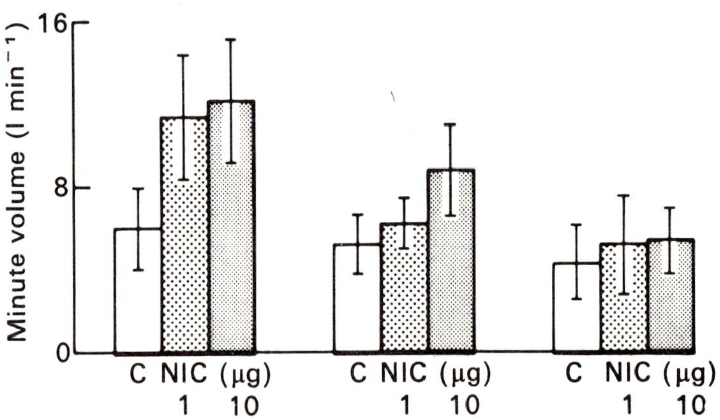

Fig. 9. Summary of results of experiments in which the effect, upon ventilation, of injecting nicotine (C, control; NIC, nicotine 1 and 10 μg) into a common carotid artery was measured when systemic arterial pressure was lowered by haemorrhage (5 dogs) or raised by inflating a balloon in the aorta (4 dogs). Values given as mean ± SE.

Ventilatory responses

Bleeding caused a decrease in systemic arterial pressure (from 130 ± 6 to 68 ± 4 mm Hg) and augmented the ventilatory response to stimulation of chemoreceptors by nicotine (Fig. 9). Ventilatory responses were reduced during elevation of arterial pressure to 149 ± 9 mm Hg by aortic occlusion.

DISCUSSION

This study indicates that the baroreceptor reflex modulates the chemoreceptor reflex. Hypotension potentiates vasoconstrictor responses to chemoreceptor stimulation and transient hypertension inhibits these responses.

We considered the possibility that altered baseline vascular resistance in the muscle might have contributed to altered responses. Vasoconstrictor responses to norepinephrine injected intra-arterially in the perfused gracilis muscle were tested during alteration in arterial pressure and were found to be either unchanged or slightly reduced and not increased during haemorrhagic hypotension. This indicates that the increased responses to chemoreceptor stimulation during hypotension occur despite, and not because of, the increase in resting vascular resistance of the muscle.

There are several mechanisms by which changes in arterial pressure might affect the magnitude of the chemoreceptor reflex. Changes in arterial pressure might affect carotid chemoreceptors directly. Biscoe, Bradley & Purves (1970), in a study which modifies an earlier conclusion (Landgren & Neil, 1951), observed that arterial pressure has little effect on carotid chemoreceptors over the range 60–160 mm Hg. In our study of hypotension, the first level of haemorrhage did not lower arterial pressure below 60 mm Hg in any dogs, so the direct effect on carotid chemoreceptors of changes in arterial pressure were minimal. The second level of haemorrhage lowered arterial pressure below 60 mm Hg, so there may have been a direct effect of pressure on carotid chemoreceptors. An effect of pressure on carotid chemoreceptors is excluded in the study in which changes in pressure were induced in one carotid and nicotine was injected in the other carotid being perfused at constant pressure.

Another possible mechanism for altered responses to chemoreceptor stimulation is an effect of changes in arterial pressure on cerebral perfusion. Ischaemia of the central nervous system has been reported to increase sympathetic discharge (Downing, Mitchell & Wallace, 1963), and it seemed possible that hypotension might cause cerebral ischaemia and potentiate the chemoreceptor reflex. In this study haemorrhagic hypotension augmented the chemoreceptor reflex, but systemic hypotension during elevation of carotid sinus perfusion pressure was associated with depression of the chemoreceptor reflex. Since systemic hypotension and a fall in cerebral perfusion pressure were associated with both increases (during haemorrhage) and decreases (during perfusion of the carotids at high pressure) in the chemoreceptor reflex, the effect of arterial pressure on the chemoreceptor reflex cannot be the result primarily of changes in cerebral perfusion pressure. Of interest is the recent conclusion by Hainsworth & Karim (1973) that cerebral hypotension is of minor importance in vascular control unless it is associated with systemic and cerebral hypoxaemia.

This study indicates that physiologic activation of baroreceptors, by elevation of pressure in one carotid sinus, inhibits the vasoconstrictor response to activation of the contralateral carotid chemoreceptors. We therefore conclude that the primary mechanism for the interaction is a neural one. The neural interaction may be central, as a study by Miura & Reis (1972) suggests; it is possible that the central interaction may occur at the spinal level as well (Gebber, Taylor & Weaver, 1973). A second possibility is that the interaction may be the result of changes in sympathetic tone within the carotid body. Changes in sympathetic tone have been shown to affect chemoreceptor activity (Floyd & Neil, 1952; Purves, 1970). In our studies, changes in arterial pressure and carotid perfusion pressure still altered the chemoreceptor reflex after vagotomy; because low cervical vagotomy probably interrupts sympathetic input to the superior cervical ganglion (Miller, Christensen & Evans, 1964) and carotid body, it appears that changes in sympathetic tone within the carotid body are not the primary mechanism for the interaction which we have observed. A third possibility is that an efferent pathway (other than sympathetic pathways) activated by baroreceptors affects the carotid body. The only one of these three possibilities which currently has direct support from other studies is the interaction within the central nervous system.

The last part of the study indicates that the level of arterial pressure may affect ventilatory as well as circulatory responses to stimulation of chemoreceptors.

In conclusion, we have demonstrated that the baroreceptor reflex modulates the chemoreceptor reflex. Hypotension potentiates responses to chemoreceptor stimulation whereas transient hypertension inhibits these responses. There are important physiological and pathophysiological implications to these findings. A potential benefit of such an interaction would be in providing an optimal circulatory adjustment during hypotension and hypoxia. In addition, the studies point out the need for evaluating physiological responses to interaction of reflexes. Although the classical approach in studies of the chemoreceptor reflex is to attempt to control all other reflexes, it would seem useful also to evaluate effects of simultaneous activation of other reflex mechanisms. Responses to simultaneous activation of two reflex mechanisms might not be predictable from knowledge of individual responses as the possibility of interaction of reflexes at the level of the central nervous system exists.

We wish to thank Leon F. Burmeister, Ph.D., for his assistance with the statistical analyses, and Mr Leonard A. Brooks for his technical assistance.

This study was supported by Clinical Investigatorships from the Veterans Administration, by research grants HL 14388, HL 09835, and HL 02644, and Research Career Development Award HL K4 28749 from the National Heart and Lung Institute, and by a grant from the American Heart Association. Travel funds were provided by the Royal Society of Medicine Foundation.

REFERENCES

BISCOE, T. J., BRADLEY, G. W. & PURVES, M. J. (1970). The relation between carotid body chemoreceptor discharge, carotid sinus pressure, and carotid body venous flow. *J. Physiol.*, **208**, 99–120.

DOWNING, S. E., MITCHELL, J. H., & WALLACE, A. G. (1963). Cardiovascular responses to ischemia, hypoxia, and hypercapnia of the central nervous system. *Am. J. Physiol.*, **204**, 881–7.

FLOYD, W. F. & NEIL, E. (1952). The influence of the sympathetic innervation of the carotid bifurcation on chemoreceptor and baroreceptor activity in the cat. *Archs int. pharmacodyn. Thér.*, **91**, 230–9.

GEBBER, G. L., TAYLOR, D. G. & WEAVER, L. C. (1973). Electrophysiological studies on organization of central vasopressor pathways. *Am. J. Physiol.*, **224**, 470–81.

HAINSWORTH, R. & KARIM, F. (1973). Left ventricular inotropic and peripheral vasomotor responses from independent changes in pressure in the carotid sinuses and cerebral arteries in anaesthetized dogs. *J. Physiol.*, **228**, 139–55.

HUNTSBERGER, D. V. & LEAVERTON, P. E. (1970). *Statistical Inference in the Biomedical Sciences*, p. 190, Allyn and Bacon, Inc.: Boston.

LANDGREN, S. & NEIL, E. (1951). Chemoreceptor impulse activity following hemorrhage. *Acta physiol. scand.*, **23**, 158–67.

MILLER, M. E., CHRISTENSEN, G. C. & EVANS, H. E. (1964). *Anatomy of the Dog.*, p. 634. Saunders Co.: Philadelphia.

MIURA, M. & REIS, D. J. (1972). The role of the solitary and paramedian reticular nuclei in mediating cardiovascular reflex responses from carotid baro- and chemoreceptors. *J. Physiol.*, **223**, 525–48.

PURVES, M. J. (1970). The role of the cervical sympathetic nerve in the regulation of oxygen consumption of the carotid body of the cat. *J. Physiol.*, **209**, 417–31.

DISCUSSION

PURVES It seems to me that with the preparation which you have, you are in a strong position to say whether or not the second possibility which you mention is true. Thus if you stimulate the baroreceptors on one side, you should be able to determine whether in fact what you call augmentation of chemoreceptor response isn't simply an alteration of the sensitivity of the chemoreceptors. By this I mean an alteration of blood flow within the carotid body by an action of the sympathetic. This would be an important extension to your experiments and would provide a plausible alternative to a central interaction between baroreceptors and chemoreceptors.
HEISTAD Yes: I agree it could be an alternative explanation. The persistence of the interaction after vagotomy, which probably interrupts sympathetic input to the carotid body, suggests that the interaction is not the result of changes in sympathetic tone within the carotid body.
MCDONALD Could you amplify your description of the anatomical structures which you think are involved in the central interaction of chemoreceptor and baroreceptor afferent fibres?
HEISTAD Miura & Reis (1972) demonstrated that they could obtain recordings from units in the medulla which could be affected by pressure changes in the carotid sinus and others close by which were affected by lobeline. So the first point is that the projections are close together. Possibly more convincing evidence is that they could make a lesion in the paramedian reticular area which abolished the depressor effect of baroreceptor stimulation and augmented the pressor response from lobeline injections. Since that area is primarily a baroreceptor area they concluded that there had been tonic baroreceptor modulation which was eliminated by this lesion.
MCDONALD Are there other forms of sensory input to the brain which alter these responses?
HEISTAD We have had an example from the diving reflex where the afferent discharge from pulmonary stretch receptors and the trigeminal nerve apparently affected the chemoreceptor response.
MCDONALD Let me be more specific. Does stimulation of somatic nerves alter chemoreceptor responses?
SHEPHERD I can't think of any hard evidence on this point but there must be interaction between a wide variety of afferent inputs to the central nervous system and the chemoreceptor pathway.
HESS That's probably what the brain is for.
BULLER Well, that's what Sherrington said, wasn't it?

MILLS Presumably all these reflexes are affecting the gracilis muscle through the sympathetic, and what you do is to rest one reflex and bring in another one. What is known about the relative changes in sympathetic activity to a vascular bed compared to the peripheral response. Is there any possibility of some form of peripheral interaction?

HEISTAD That could be answered in two ways. One would be able to measure the traffic, in say haemorrhage, and then to impose a second type of stimulus: but it might be difficult to show that that measured traffic in a peripheral sympathetic nerve was responsible for the recorded vascular changes. In any case, one would expect increased nerve traffic during haemorrhage, and one would not expect increased nerve traffic to potentiate another reflex – so this mechanism is unlikely. However, we have considered the possibility that there may be local changes which would augment or reduce the neural effect: for example, during haemorrhage there might be non-specific changes which would affect the response. So we carried out some experiments in which we injected noradrenaline into the artery supplying the gracilis muscle (Fig. 1). If you bleed a dog (top panel)

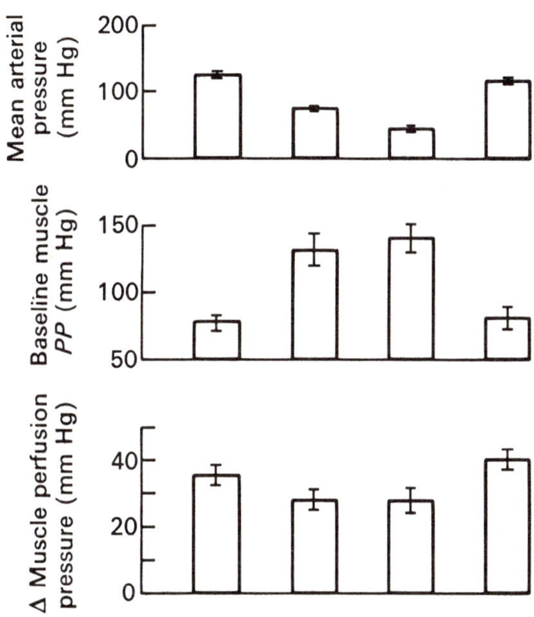

Fig. 1. Summary of results of experiments in which hypotension was induced by haemorrhage in dogs with first, a control value and last, a value during recovery after blood was replaced. The middle panel shows the accompanying changes in gracilis muscle perfusion pressure at constant flow and the lower panel shows the effect of injecting noradrenaline into the muscle arterial supply upon muscle perfusion pressure. $n = 6$. Values given as mean $\pm$ SE.

the perfusion pressure in muscle increases (middle panel). The hypotension, and increase in perfusion pressure, are associated with decreased vasoconstrictor responses to norepinephrine (bottom panel). So the increased responses to chemoreceptor stimulation during hypotension are not the result of a peripheral interaction, but occur despite it.

O'REGAN Landgren & Neil (1951) did this kind of experiment, that is, they measured sympathetic activity during haemorrhage and compared it with the chemoreceptor change which also occurred.

HEISTAD Yes: there are problems with that experiment. They found that blood pressure affected chemoreceptor discharge over a wide range whereas with perhaps a better analysis recently, this was shown not to be so (Biscoe, Bradley & Purves, 1970).

O'REGAN Yes, I agree. But it still remains that if blood pressure is reduced beyond a certain level, chemoreceptor discharge increases.

HEISTAD That is why we always carried out our experiments at two levels of systemic pressure – one above and one below 60 mm Hg because it would seem that this approximately is the point at which chemoreceptor discharge starts to increase.

REFERENCES

BISCOE, T. J., BRADLEY, G. W. & PURVES, M. J. (1970). The relation between carotid body chemoreceptor discharge, carotid sinus pressure and carotid body venous flow. *J. Physiol.*, **208**, 99–120.

LANDGREN, S. & NEIL, E. (1951). Chemoreceptor impulse activity following haemorrhage. *Acta physiol. scand.*, **23**, 158–67.

MIURA, M. & REIS, D. J. (1972). The role of the solitary and paramedian reticular nuclei in mediating cardiovascular reflex responses from carotid baro- and chemoreceptors. *J. Physiol.*, **223**, 525–48.

Carotid chemoreflex and circulatory control

By JOHN T. SHEPHERD and CONRAD L. PELLETIER
Mayo Clinic and Mayo Foundation, Rochester, Minnesota 55901

In the experiments to be described, the objectives were to stimulate the carotid chemoreceptors selectively by changes in P_{O_2}, P_{CO_2}, and pH and to determine (1) the threshold stimulus necessary to activate the resistance vessels in the hindlimb, (2) the stimulus–response curve of these vessels, (3) the reaction of the cutaneous and splanchnic venous beds compared with that of the resistance vessels, and (4) the effect of the interaction between the carotid chemoreflex and baroreflex on the response of the resistance vessels of the hindlimb and the kidney.

METHODS

Dogs, 20 to 30 kg in weight, were anaesthetized with thiopental and chloralose (15 and 60 mg kg^{-1}, respectively) and artificially ventilated at 12 to 15 cycles min^{-1}. Additional doses of chloralose (10 mg kg^{-1}) were administered hourly to maintain an even plane of light anaesthesia. Before cannulation of the blood vessels, gallamine and heparin (3 mg kg^{-1} each) were given intravenously and repeated hourly (1 mg kg^{-1} each). A bilateral cervical vagotomy was performed at the level of the cricoid cartilage to eliminate the effect of the intrathoracic receptors; the cervical sympathetic nerves were cut to prevent their having any vasoconstrictor effects on carotid body blood vessels (Purves, 1970).

Perfusion of carotid bodies

The two carotid bifurcations were isolated by ligating the internal carotid arteries distal to the sinus, the occipital arteries and their branches distal to the carotid bodies, and the ascending pharyngeal, facial, and external carotid arteries (Fig. 1). The excluded carotid bifurcations were perfused with a roller pump at a flow rate of 125 ± 15 ml min^{-1}. Non-pulsatile perfusion pressure was obtained by interposing a depulsator in the inflow circuit to the carotid bifurcations. The pressure in the bifurcations was

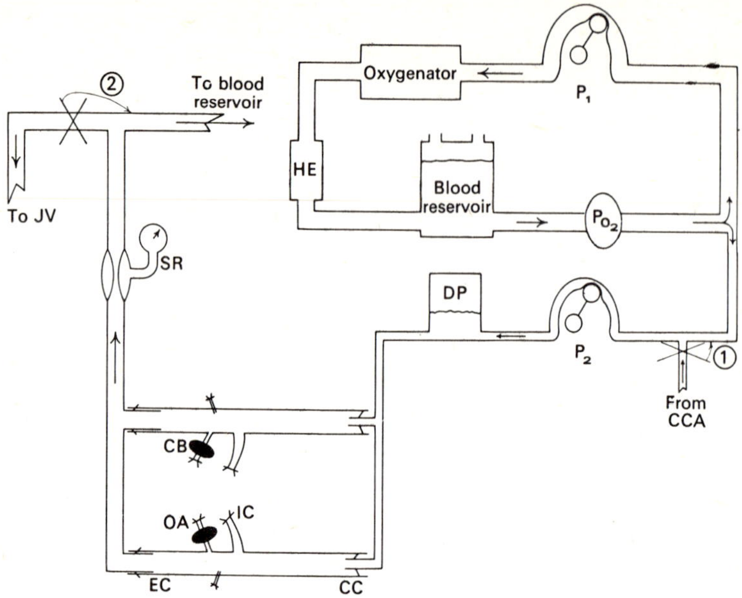

Fig. 1. Technique for perfusion of isolated carotid chemoreceptors. Blood collected in reservoir is continuously pumped (P_1) through membrane oxygenator for equilibration at desired level of gases. Temperature in reservoir is kept at 37 °C by heat exchanger (HE), and P_{O_2} is continously monitored by flow-through polarographic analyser (P_{O_2}). This blood is pumped (P_2) to carotid bodies (CB) through cannulas in common carotid arteries (CC) and returned to reservoir by cannulas in external carotid arteries (EC). Pressure of perfusion is maintained constant by Starling resistance (SR). Between stimulations, blood from common carotid artery (CCA) perfuses bifurcations and is returned to dog via jugular vein (JV). IC, internal carotid artery; OA, occipital artery; DP, depulsator. (From C. L. Pelletier, 1972, *Circulation Res.*, **31**, 431–43, by permission of the American Heart Association Inc.)

adjusted to give a mean pressure similar to the aortic blood pressure before cannulation. The sinus pressure never varied by more than 3 mm Hg during the course of an experiment. When the roller pump was stopped and the outflow circuit was clamped, the systemic pressure increased above 200 mm Hg due to the effect of the baroreceptors; a decrease in pressure in the bifurcations to 40 mm Hg or less was taken as evidence of exclusion of the carotid bodies from the systemic circulation and of the absence of any significant arterial inflow from collateral vessels.

Extracorporeal circuit

A total of 400 ml of autologous blood was collected in a reservoir and replaced by transfusing the dogs with a similar volume of dextran 40. This blood was circulated with a roller pump through a silicone rubber membrane oxygenator.

Blood gas tension and pH

The dogs were artificially ventilated with room air. The tidal volume was adjusted and oxygen was added to maintain the systemic arterial P_{O_2} between 90 and 140 mm Hg (mean, 111 mm Hg) and P_{CO_2} between 30 and 44 mm Hg (mean, 37 mm Hg). Bicarbonate (7.5 %) was administered as needed to keep the arterial blood pH between 7.30 and 7.40 (mean, 7.34).

Gases obtained from tanks were mixed to achieve the desired gas tensions in the blood, using a constant gas flow of 2 l min^{-1} through the oxygenator for a total blood flow of approximately 1 l min^{-1}. The pH was altered by adding either bicarbonate or 0.1 N HCl to the blood. When the desired levels of P_{O_2}, P_{CO_2}, and pH were reached, the blood from the reservoir was used to stimulate the carotid chemoreceptors for 2 to 4 minutes by shifting the inflow to the pump perfusing the bifurcations. This blood was returned to the reservoir by an alternate outflow circuit, and there was no significant transfusion of blood to the dog. This approach made precise variation of the oxygen and carbon dioxide tensions and pH possible at the carotid bodies, while the systemic arterial gases and pH were kept constant and normal. In some experiments the carotid bodies were perfused with blood from a donor dog.

Measurements

Pressures were measured with strain-gauge transducers. One hindlimb or kidney (or both) and one lateral saphenous (cutaneous) vein were perfused (roller pumps) at constant flow with autologous arterial blood, and these perfusion pressures were measured. The splanchnic capacitance bed was studied by using the iso-volumetric spleen technique (Webb-Peploe, 1969).

The efferent activity from multifibre preparations of the renal sympathetic nerve was recorded with a bipolar platinum electrode. The noise level, as judged by the background activity during the quiet periods between bursts of impulses, was eliminated by the use of a discriminator; the frequency of impulses was integrated over a fixed duration to obtain mean activity.

RESULTS

Stimulus–response curves for resistance vessels

Responses to changes in P_{O_2}

The P_{O_2} of the blood perfusing the carotid chemoreceptors was decreased in steps from 104 to 34 mm Hg (means); the P_{CO_2} and pH were kept constant and averaged 36 mm Hg and 7.35, respectively. A record from one experiment is shown in Fig. 2, and the stimulus–response curves for the

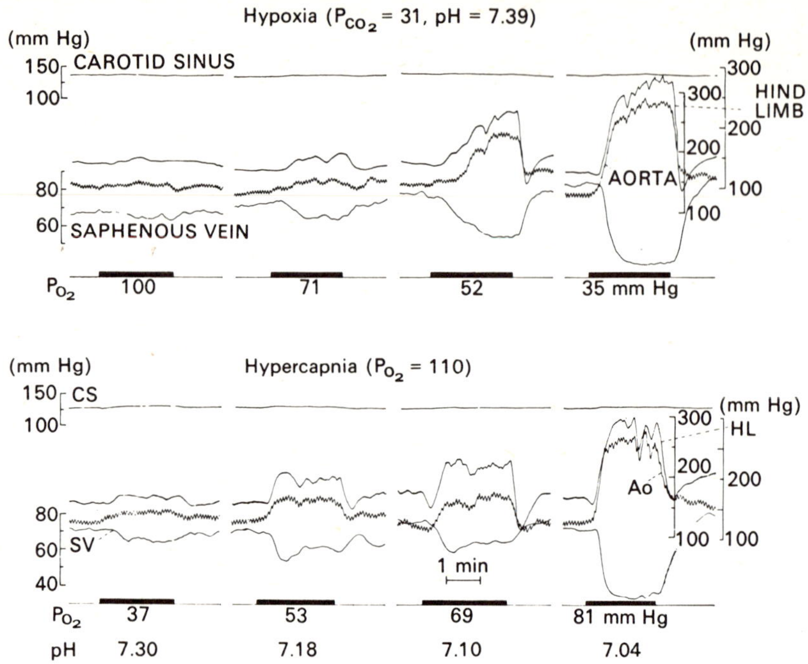

Fig. 2. Changes in mean aortic pressure and in hindlimb and saphenous vein perfusion pressures in response to graded stimulation of carotid chemoreceptors in dog with vagi cut. Systemic arterial P_{O_2}, P_{CO_2}, and pH were normal; carotid sinus pressure was constant. Time of stimulation is indicated by signal at bottom of each record, along with stimulus used. Scales at left refer to carotid sinus pressure and saphenous vein perfusion pressure. Scales at right refer to mean aortic pressure and hindlimb perfusion pressure. (From C. L. Pelletier, 1972. *Circulation Res.*, **31**, 431–43, by permission of the American Heart Association Inc.)

changes in aortic pressure and hindlimb perfusion pressure are shown in Fig. 3. In both aortic and hindlimb curves the inflection occurred at P_{O_2} = 80 mm Hg and the response increased linearly thereafter, with no evidence of flattening at the tops of the curves. The mean ($\pm$SE) maximal increases in aortic and hindlimb pressures were 88 ± 15 and 76 ± 15 mm Hg, respectively, at a mean P_{O_2} of 34 mm Hg (Pelletier, 1972).

The number of impulses in the carotid sinus nerve increases as the P_{O_2} is decreased below 200 mm Hg, but it is not until P_{O_2} has reached 80 to 100 mm Hg that a sharp increase in neural discharges is obtained (Biscoe, Purves & Sampson, 1970). The vascular responses are thus related to the P_{O_2} at which there is a rapid increase in chemoreceptor activity. Even though the nerve impulses increase when the P_{O_2} is decreased from 200 to 100 mm Hg, the rate of increase is small and probably does not reach the threshold necessary to cause reflex vascular responses.

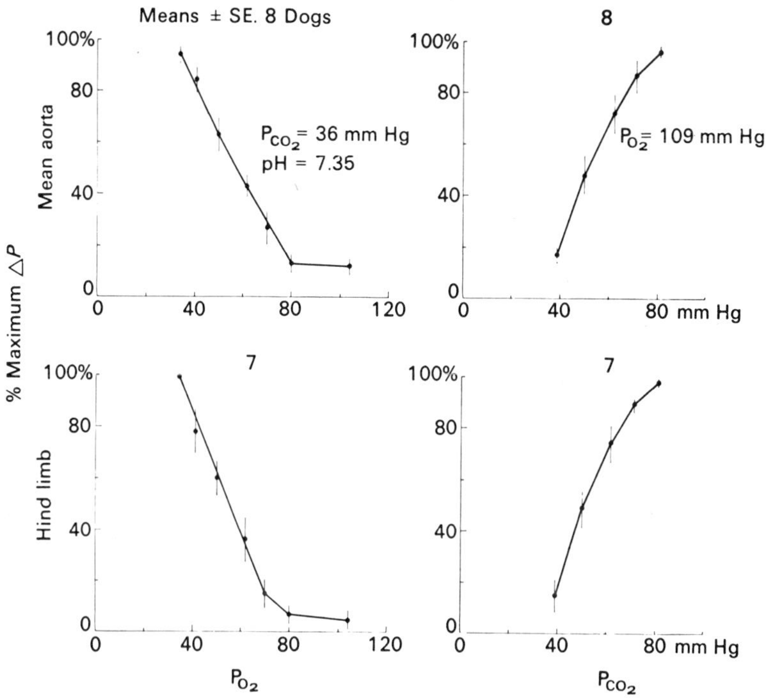

Fig. 3. Increases in aortic pressure and hindlimb perfusion pressure, as percentages of maximal response, in dogs with vagi cut. *Left*, responses to chemoreceptor stimulation by graded hypoxia; mean systemic arterial P_{O_2} was 114 mm Hg; P_{CO_2}, 36 mm Hg; and pH, 7.35. *Right*, responses to chemoreceptor stimulation by graded hypercapnia; mean systemic arterial P_{O_2} was 115 mm Hg; P_{CO_2}, 38 mm Hg; and pH, 7.33. SE for P_{O_2} and P_{CO_2} = 1 mm Hg or less. (From C. L. Pelletier, 1972. *Circulation Res.*, **31**, 431–43, by permission of the American Heart Association Inc.)

Responses to combined changes in P_{CO_2} and pH

The blood P_{CO_2} was increased in steps from 39 to 81 mm Hg while the P_{O_2} was kept constant, averaging 109 mm Hg. The blood pH was allowed to fluctuate spontaneously with the increasing carbon dioxide tension (Fig. 2). A small but significant increase in aortic pressure and hindlimb perfusion pressure occurred at a mean P_{CO_2} of 39 mm Hg, even though this was not different from the average systemic P_{CO_2} of the dogs (Fig. 3). With increasing carbon dioxide tension, there was almost linear increase in the circulatory responses until a P_{CO_2} of 71 ± 1 mm Hg was reached, when the curves started to flatten. The maximal aortic and hindlimb pressure increases (at P_{CO_2} = 81 mm Hg) were not significantly different from the maximal responses to changes in P_{O_2} (Pelletier, 1972). This agrees with the responses of single afferent fibres of carotid body chemoreceptors in the cat; with hypercapnic acidosis, a sigmoid curve was obtained, and afferent discharge was more or less linear between carbon dioxide tensions of 25 and 65 mm Hg (Biscoe *et al.*, 1970).

Responses to separate changes in P_{CO_2} and pH

Hypercapnia with normal pH (mean, 7.38) was obtained by adding bicarbonate to the blood of the reservoir for each step increase in carbon dioxide tension. After P_{CO_2} and pH were returned to normal, the blood pH was decreased in steps by the addition of 0.1 N HCl while the oxygen and carbon dioxide tensions were kept normal (means, 106 and 40 mm Hg, respectively). The effects of the changes in P_{CO_2} only and of pH only are shown in Fig. 4, where they are contrasted with the effects of hypercapnia

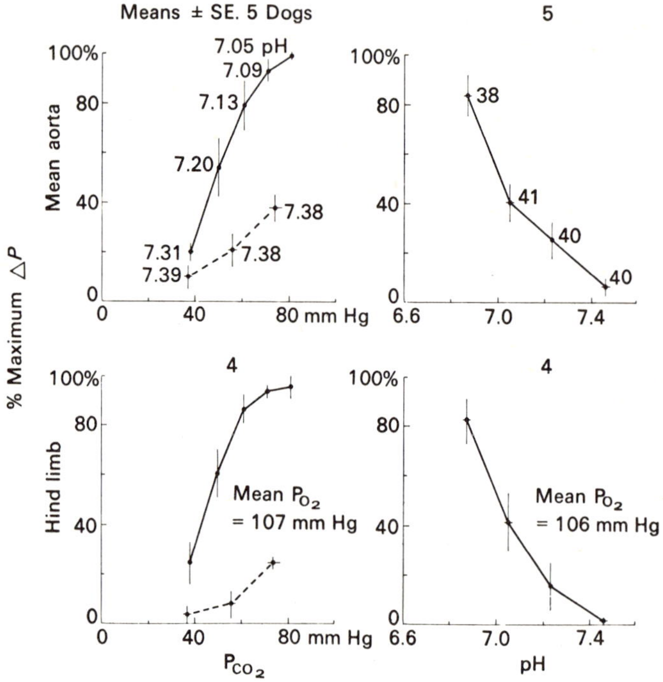

Fig. 4. Increases in mean aortic pressure and hindlimb perfusion pressure as percentages of maximal response to hypercapnic acidosis in dogs with vagi cut. Mean systemic arterial P_{O_2} was 114 mm Hg; P_{CO_2}, 38 mm Hg; and pH, 7.31. *Left*, responses to chemoreceptor stimulation by graded hypercapnia with spontaneously changing pH (solid line) and with normal pH (broken line). *Right*, responses to chemoreceptor stimulation by decreasing pH with normal P_{CO_2}. SE for P_{CO_2} is given only when >1 mm Hg. (From C. L. Pelletier, 1972. *Circulation Res.*, **31**, 431–43, by permission of American Heart Association Inc.)

with spontaneous decrease in pH. When the pH was kept normal, hypercapnia still caused changes in both aortic and hindlimb perfusion pressures, but the curve was depressed to the right of that with hypercapnic acidosis.

With normal pH, the response to the same degree of hypercapnia was decreased by more than 60 %. Acidosis with normocapnia was a potent stimulus, and the maximal responses obtained at a pH of 6.87 were 83 % of the maximal responses to hypercapnic acidosis. However, for equivalent changes in pH, the responses to acidosis alone were approximately 50 % of those obtained with combined hypercapnic acidosis (Pelletier, 1972). Hornbein & Roos (1963) and Gray (1968) reported that changes in carbon dioxide tension are ineffective in activating the chemoreceptors if not accompanied by decreases in pH. However, using the superfused carotid body *in vitro*, Eyzaguirre & Koyano (1965) found that a high carbon dioxide tension with normal pH causes an increase in chemoreceptor discharge, and this was confirmed by Biscoe *et al.* (1970) who recorded from single fibre preparations in the intact cat.

Venous responses

Activation of the chemoreflex by decreasing the P_{O_2} of the blood perfusing the carotid bifurcations, or by increasing the P_{CO_2} and allowing the pH to vary, causes dilatation of the cutaneous veins. This is demonstrated by the decrease in perfusion pressure of the saphenous vein when perfused at constant flow. The dilatation is proportional to the degree of stimulation (Fig. 2), but the variations in tone in this vein during each experiment prevented a quantitative analysis of the response. Obviously, relaxation can only be demonstrated when the vein is constricted prior to stimulation, and studies that have demonstrated no response of the cutaneous veins to chemoreceptor stimulation (Iizuka *et al.*, 1970) can be explained on this basis. For example, anaesthetic agents such as chloralose cause the vein to relax and little further dilatation can be elicited by the chemoreflex (Fig. 5). With the occluded-limb technique to study the reflex responses of the saphenous vein, small increases or large decreases in venous pressure were noted during cyanide stimulation of the chemoreceptors (Browse & Shepherd, 1966). In retrospect, these small increases were probably of mechanical origin from the reflex constriction of the resistance vessels (Pelletier & Shepherd, 1972).

Concerning the mechanism of the venodilatation, previous studies of the cutaneous veins of the dog have shown that the effects of both local and central changes in temperature on venomotor tone are mediated by the adrenergic nervous system (Webb-Peploe & Shepherd, 1968*b*). There is no evidence that these veins possess any cholinergic vasodilator nerve supply (Webb-Peploe & Shepherd, 1968*a*) but they have a β-receptor mechanism, stimulation of which causes the veins to dilate (Webb-Peploe & Shepherd, 1969; Vanhoutte & Shepherd, 1970). The reflex

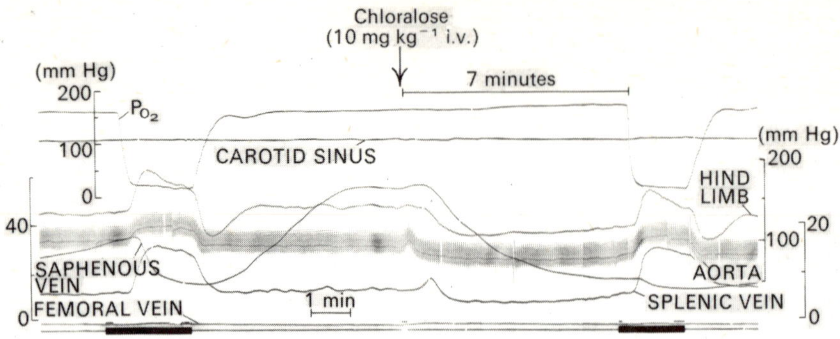

Fig. 5. Effect of change in resting tone of saphenous vein on its reflex response to chemoreceptor stimulation with hypoxic blood in dog with vagi cut. Systemic arterial blood gases were normal. Right common iliac artery (hindlimb) and left saphenous vein were perfused at constant flow with autologous blood, so that changes in perfusion pressure are a measure of changes in vasomotor activity. Blood flow through the spleen was temporarily arrested, trapping a fixed volume of blood in the organ. Under these conditions, changes in splenic vein pressure are a measure of changes in smooth-muscle tension in the splenic capsule and veins. Initially there was marked decrease in saphenous vein perfusion pressure with stimulation; with additional dose of chloralose, perfusion pressure decreased markedly. When same stimulus was repeated, there was similar increase in aortic, hindlimb, and splenic vein pressures but only small decrease in saphenous perfusion pressure. (From C. L. Pelletier & J. T. Shepherd, 1972. *Am. J. Physiol.*, **223**, 97–103.)

venodilatation to chemoreceptor stimulation is unaffected by atropine, antihistaminic agents, or propranolol; thus, the most likely explanation of the reflex venodilatation is the withdrawal of adrenergic activity. Blockade of the α-receptors by phenoxybenzamine causes the vein to relax and no further dilatation occurs with chemoreceptor stimulation.

By contrast, the venous pressure in the vascularly isolated spleen increases, indicating that its smooth muscle is contracting (Fig. 5). The opposite responses of the spleen and the saphenous vein were obtained whatever the stimulus used to activate the chemoreceptor. Thus, the spleen behaves similarly to the resistance vessels of the hindlimb (Pelletier & Shepherd, 1972). Both responses are mediated by increased activity of the adrenergic nerves, because the responses are prevented or much attenuated by blockade of the α-receptors. Previous studies have shown that the changes in venous pressure in the iso-volumetric spleen reflect not only the reaction of the splenic capacitance vessels but also the reflex response of the capacitance vessels throughout the splanchnic bed (Webb-Peploe, 1969). Other differences between the responses of the cutaneous veins and the splanchnic capacitance vessels have been observed. The former do not participate in the high-pressure (Brender & Webb-Peploe, 1969) or low-pressure (Pelletier, Edis & Shepherd, 1971) baroreflexes but have a specific function in temperature regulation (Webb-Peploe, 1969). The latter participate in the baroreflexes but are not influenced by changes in body temperature (Fig. 6).

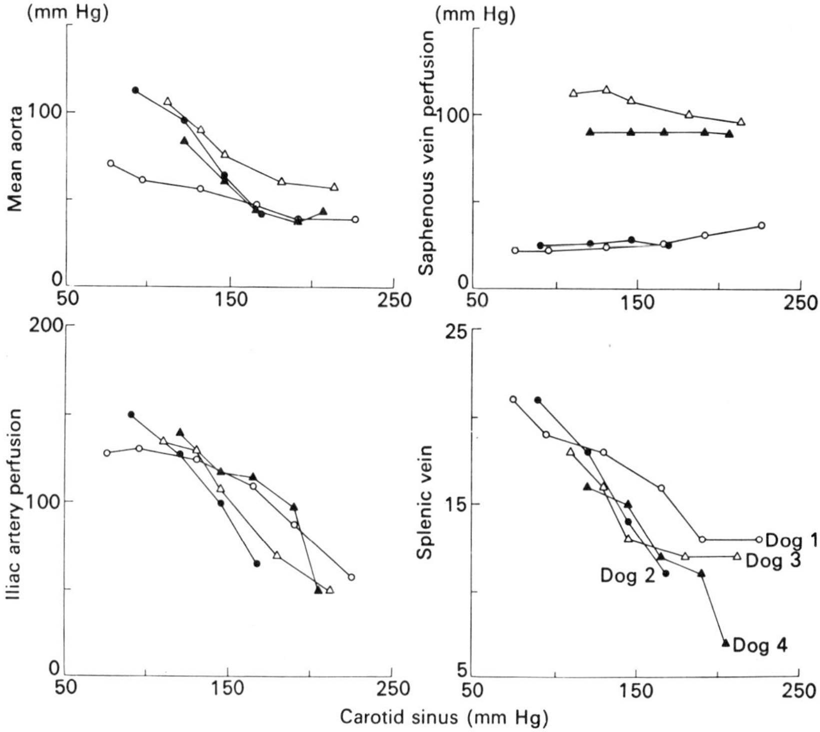

Fig. 6. Vascular effects of stepwise pressure changes in vascularly isolated carotid sinus. Right common iliac artery and left saphenous vein were perfused at constant flow with autologous blood. Blood flow through spleen was temporarily arrested, trapping a fixed volume of blood in organ. With decreases in carotid sinus pressure, aortic, iliac artery, and splenic venous pressures increased, but saphenous vein perfusion pressure (*upper right*) showed little change whether the initial venomotor tone was high or low. (From D. Brender & M. M. Webb-Peploe, 1969. *J. Physiol.*, **205**, 257–74.)

Interaction of carotid baroreflexes and chemoreflexes

In 10 vagotomized dogs with intact carotid sinuses, the increases in aortic blood pressure and in hindlimb and kidney perfusion pressures with bilateral carotid occlusion were studied during normoxia and hypoxia. When the dogs were breathing room air (mean arterial P_{O_2} = 107 mm Hg), the mean pressure increases were 63, 66, and 45 mm Hg, respectively; with 10% oxygen (mean P_{O_2} = 41 mm Hg) the increases were 99, 66, and 132 mm Hg. These increases occurred with no changes in the control pressures before occlusion. After exclusion of the carotid bodies by ligation of the occipital arteries at their origin, the responses to bilateral carotid occlusion were similar during both normoxia and hypoxia (Fig. 7). In

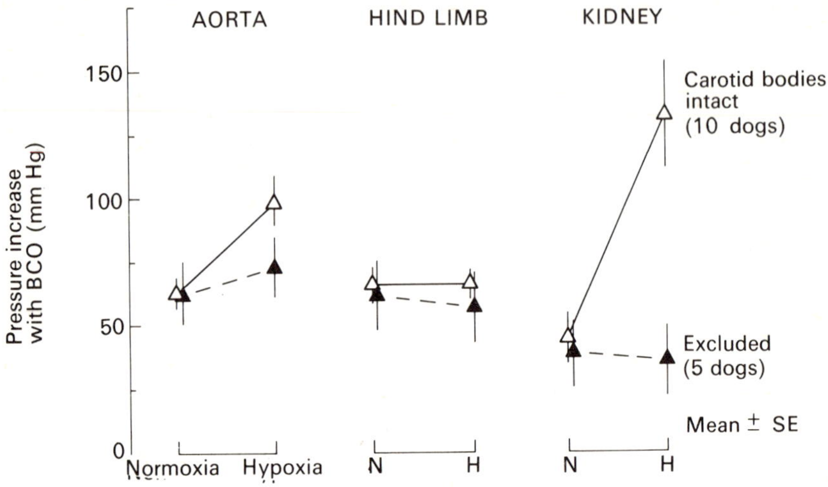

Fig. 7. Effect of systemic hypoxia on vascular responses to carotid baroreflex in vagotomized dogs. Hindlimb and kidney were perfused at constant flow with autologous blood. BCO, bilateral carotid occlusion. Normoxia: systemic arterial P_{O_2} = 100 to 107 mm Hg. Hypoxia: systemic arterial P_{O_2} = 40 to 41 mm Hg.

another six dogs with bilateral vagotomy, neurograms were obtained from multifibre preparations of sympathetic nerves to the left kidney. During ventilation with air (mean arterial P_{O_2} = 100 mm Hg) there was little or no activity in the renal nerve, and only a slight increase in mean activity (12%) occurred during ventilation with 10% oxygen (mean arterial P_{O_2} = 40 mm Hg). Bilateral carotid occlusion caused the mean aortic pressure to increase by 56 mm Hg during normoxia and by 90 mm Hg during hypoxia. The increase in mean renal nerve activity with carotid occlusion averaged 56 ± 10% greater during hypoxia than during normoxia. Ganglionic blockade abolished the changes in aortic pressure and the sympathetic activity in the renal nerves. Thus, the increase in the sympathetic outflow to the kidney caused by withdrawal of carotid baroreceptor activity can be markedly potentiated by the simultaneous stimulation of the carotid chemoreceptors while that to the hindlimb resistance vessels remains unaffected.

DISCUSSION

These experiments were concerned only with the primary vascular response to the chemoreflex mechanism; respiratory reflex changes were prevented by vagotomy and artificial ventilation. The circulatory changes were due solely to activation of the carotid chemoreceptors because they were abolished by blockade of the carotid sinus nerves (Pelletier & Shepherd, 1972) and by surgical exclusion of the carotid bodies. The responses were well sustained throughout the period of stimulation (2 to 4 minutes).

In the cat, afferent fibres from the carotid chemoreceptors continue to discharge at an arterial P_{O_2} of more than 600 mm Hg; as the P_{O_2} is decreased, the response curve is similar to a hyperbola, the impulse frequency increasing first slowly and then rapidly. The oxygen tension at which the curve of rate of discharge against oxygen tension departed from nearly horizontal averaged ($\pm$SD) 190 ± 40 mm Hg (Biscoe *et al.* 1970), but the rate of increase in nerve activity remained low until a P_{O_2} of 80 to 100 mm Hg was reached, at which point the curve turned upward sharply. In the experiments just described in the dog, a constriction of the hindlimb resistance vessels was not detected until the P_{O_2} of the blood perfusing the carotid bifurcations was below 80 mm Hg. Presumably, if the P_{O_2} is above this value, the afferent activity in the sinus nerve does not reach the threshold necessary to cause vascular responses. Below this P_{O_2} value, the curve for the vascular responses is similar to that obtained from afferent chemoreceptor activity – both are linear with no flattening at the top. By contrast, with hypercapnic acidosis, the vascular responses and the increase in neural discharge occur as soon as the P_{CO_2} is varied from normal. The curves are very steep initially, with a decrease in rate of change in pressure or impulse frequency at $P_{CO_2} > 70$ mm Hg.

Some investigators have reported that changes in carbon dioxide tension are ineffective in activating the chemoreceptors if not accompanied by decreases in pH (Hornbein & Roos, 1963; Gray, 1968). More recently, Biscoe *et al.* (1970), using the single fibre recording technique, showed that hypercapnia can cause an increase in nerve impulses when pH is kept normal, even though the response is much attenuated. In agreement, our studies demonstrate that perfusion of the carotid chemoreceptors with hypercapnic blood at normal pH causes reflex vascular responses, but that these are much greater when the pH changes with the increasing carbon dioxide tension.

Conversely, decreases in pH with normal P_{O_2} and P_{CO_2} cause marked vascular responses and increases in nerve firing rate, indicating an important role of pH in the excitation process of the chemoreceptors.

The increase in arterial pressure caused by the chemoreflex is mainly a consequence of the increase in peripheral resistance mediated by the sympathetic nervous system (Daly & Scott, 1962; 1963). However, the increase in sympathetic outflow is not uniformly distributed. Chemoreceptor stimulation has been shown to cause small decreases in heart rate and ventricular contractility, and no increase in nerve discharge was obtained from recordings of cardiac sympathetic fibres (Downing, Remensnyder & Mitchell, 1962; Daly & Scott, 1963; Downing & Siegel, 1963). The resistance vessels in muscle and splanchnic bed (Bernthal & Schwind, 1945) constrict, together with the splanchnic venous bed. The latter can account for the decrease in total venous capacitance during chemoreceptor stimulation (Braunwald *et al.*, 1963); by increasing the venous return to the heart, this should help to maintain or increase the cardiac output. By contrast, the arterial (Calvelo, Abboud, Ballard & Abdel-Sayed, 1970) and venous segments (Pelletier & Shepherd, 1972) of the cutaneous circulation dilate.

In comparing the response of the renal and hindlimb resistance vessels to a decrease in activity of the carotid baroreceptors, it was found that the constriction of the former but not the latter was greatly augmented when the carotid chemoreflex was activated simultaneously. That the hindlimb vessels were capable of further constriction was demonstrated by electrical stimulation of the lumbar sympathetic nerves to the limb. This selective potentiation of the renal response to the carotid baroreflex is of central origin because it also was demonstrated by neurograms from renal sympathetic fibres. This potentiation may be the basis for the rapid deterioration of kidney function when hypoxaemia occurs in patients with systemic arterial hypotension.

The investigations in this laboratory were supported in part by Research Grant HL-5883 from the National Institutes of Health, Public Health Service, and by a grant from the Medical Research Council of Canada.

REFERENCES

Bernthal, T. & Schwind, F. J. (1945). A comparison in intestine and leg of the reflex vascular response to carotid-aortic chemoreceptor stimulation. *Am. J. Physiol.*, **143**, 361–72.

Biscoe, T. J., Purves, M. J. & Sampson, S. R. (1970). The frequency of nerve impulses in single carotid body chemoreceptor afferent fibres recorded *in vivo* with intact circulation. *J. Physiol.*, **208**, 121–31.

Braunwald, E., Ross, J., Jr, Kahler, R. L., Gaffney, T. E., Goldblatt, A. & Mason, D. T. (1963). Reflex control of the systemic venous bed: effects on venous tone of vasoactive drugs, and of baroreceptor and chemoreceptor stimulation. *Circulation Res.*, **12**, 539–50.

Brender, D. & Webb-Peploe, M. M. (1969). Influence of carotid baroreceptors on different components of the vascular system. *J. Physiol.*, **205**, 257–74.

Browse, N. L. & Shepherd, J. T. (1966). Response of veins of canine limb to aortic and carotid chemoreceptor stimulation. *Am. J. Physiol.*, **210**, 1435–41.

Calvelo, M. G., Abboud, F. M., Ballard, D. R. & Abdel-Sayed, W. (1970). Reflex vascular responses to stimulation of chemoreceptors with nicotine and cyanide: activation of adrenergic constriction in muscle and noncholinergic dilatation in dog's paw. *Circulation Res.*, **27**, 259–76.

DALY, M. DE B. & SCOTT, M. J. (1962). An analysis of the primary cardiovascular reflex effects of stimulation of the carotid body chemoreceptors in the dog. *J. Physiol.*, **162**, 555–73.

DALY, M. DE B. & SCOTT, M. J. (1963). The cardiovascular responses to stimulation of the carotid body chemoreceptors in the dog. *J. Physiol.*, **165**, 179–97.

DOWNING, S. E., REMENSNYDER, J. P. & MITCHELL, J. H. (1962). Cardiovascular responses to hypoxic stimulation of the carotid bodies. *Circulation Res.*, **10**, 676–85.

DOWNING, S. E. & SIEGEL, J. H. (1963). Baroreceptor and chemoreceptor influences on sympathetic discharge to the heart. *Am. J. Physiol.*, **204**, 471–9.

EYZAGUIRRE, C. &. KOYANO, H. (1965). Effects of hypoxia, hypercapnia, and pH on the chemoreceptor activity of the carotid body *in vitro*. *J. Physiol.*, **178**, 385–409.

GRAY, B. A. (1968). Response of the perfused carotid body to changes in pH and P_{CO_2}. *Respir. Physiol.*, **4**, 229–45.

HORNBEIN, T. F. & ROOS, A. (1963). Specificity of H ion concentration as a carotid chemoreceptor stimulus. *J. appl. Physiol.*, **18**, 580–4.

IIZUKA, T., MARK, A. L., WENDLING, M. G., SCHMID, P. G. & ECKSTEIN, J. W. (1970). Differences in responses of saphenous and mesenteric veins to reflex stimuli. *Am. J. Physiol.*, **219**, 1066–70.

PELLETIER, C. L. (1972). Circulatory responses to graded stimulation of the carotid chemoreceptors in the dog. *Circulation Res.*, **31**, 431–43.

PELLETIER, C. L., EDIS, A. J. & SHEPHERD, J. T. (1971). Circulatory reflex from vagal afferents in response to hemorrhage in the dog. *Circulation Res.*, **29**, 626–34.

PELLETIER, C. L. & SHEPHERD, J. T. (1972). Venous responses to stimulation of carotid chemoreceptors by hypoxia and hypercapnia. *Am. J. Physiol.*, **223**, 97–103.

PURVES, M. J. (1970). The role of the cervical sympathetic nerve in the regulation of oxygen consumption of the carotid body of the cat. *J. Physiol.*, **209**, 417–31.

VANHOUTTE, P. M. & SHEPHERD, J. T. (1970). Effect of cooling on beta-receptor mechanisms in isolated cutaneous veins of the dog. *Microvasc. Res.*, **2**, 454–61.

WEBB-PEPLOE, M. M. (1969). The isovolumetric spleen: index of reflex changes in splanchnic vascular capacity. *Am. J. Physiol.*, **216**, 407–13.

WEBB-PEPLOE, M. M. & SHEPHERD, J. T. (1968*a*). Response of large hindlimb veins of the dog to sympathetic nerve stimulation. *Am. J. Physiol.*, **215**, 299–307.

WEBB-PEPLOE, M. M. & SHEPHERD, J. T. (1968*b*). Responses of the superficial limb veins of the dog to changes in temperature. *Circulation Res.*, **22**, 737–46.

WEBB-PEPLOE, M. M. & SHEPHERD, J. T. (1969). Beta-receptor mechanisms in the superficial limb veins of the dog. *J. clin. Invest.*, **48**, 1328–35.

DISCUSSION

BULLER Could I make general comment as someone who is more involved in neural mechanisms that there is some sort of analogy in what we have heard today with the Sherringtonian idea of interaction. And he made the point that when you are assessing the effect of two afferent inputs, it is often valuable to make the comparison with respect to time rather than simply to compare the effect after section of one or other input. This approach seems to me, possibly naively, to have paid off in Dr Heistad's experiments where he shifted the input with respect to time. Certainly, this proved to be a valuable way of investigating comparable problems in the spinal cord.

MILLS Exactly how did you exclude the carotid body in your experiments?

SHEPHERD The occipital artery was exposed at its origin and ligated.

ABBOUD I was intrigued by your observation that the response to carbon dioxide seemed to be smallest whereas there appeared to be a greater sensitivity to changes in oxygen or pH. Have any of our colleagues who record afferent activity from the chemoreceptors seen a comparable response of the receptors to oxygen, carbon dioxide and pH?

PURVES I'm not sure that it is possible to answer that question exactly. You can certainly consult the published chemoreceptor response curves and say that over such and such a range of the chemical variable, chemoreceptor discharge will change by so much. But these papers will also show that individual response curves vary very considerably and in consequence, it would be difficult to predict what the mass discharge in the sinus nerve would do. The direct answer to your question would be that with modest deviations of P_{O_2} and P_{CO_2} from normal levels, the discharge would increase most obviously to a rise in carbon dioxide and least to P_{O_2} and it is not until you have a fairly severe degree of hypoxia that chemoreceptor discharge increases in a similar fashion.

SHEPHERD Professor Daly, you have measured vascular and respiratory responses to stimulation of the chemoreceptors. Have you any data which might answer this point?

DALY Not really, because I have not carried out experiments to determine such response curves to separate increases in carbon dioxide or reductions in carbon dioxide. But Professor Neil some years ago suggested that the increase in discharge in response to carbon dioxide appeared to be considerably less than to oxygen and that the response to carbon dioxide rather depended upon the prevailing P_{O_2}; that is, it was augmented with hypoxia.

PURVES Yes but there is not unanimity on this point. For example,

Sampson, Biscoe and I showed that there is a sizeable increase in discharge with carbon dioxide even at high P_{O_2} and that this increase was not obviously greater during hypoxia. Also, there may be a species problem since Dr Cunningham has pointed out that in this respect, there is a difference between Oxford medical students and cats!

GOODMAN Certainly you can see oscillations of chemoreceptor discharge in hyperoxia.

ABBOUD But am I right in saying that the size of the chemoreceptor discharge may not necessarily determine the size of the response? There may be some form of central nervous selectivity or modulation, for example.

SAMPSON But that would assume that some fibres were selectively affected say by carbon dioxide whereas we have found in all fibres so studied that if they respond to carbon dioxide, they all respond to oxygen as well.

BULLER Might you not be both right if there is some kind of divergence centrally. You would then get alternative pathways one of which facilitated and one of which inhibited. So if you suggest that the peripheral sensitivity of the chemoreceptors is the same in all fibres, it doesn't mean that you cannot affect the specific activity centrally if you have alternative routings: that is, if you do not have a one-to-one relation. One-to-one relations are not at all common in the central nervous system.

PURVES It might also be worth pointing out that if you inhale carbon dioxide it could be acting at a number of points in addition to the chemoreceptors – upon vasomotor neurones, for example, or upon the peripheral effector organs. And this could mean that the effect of any change in chemoreceptor discharge could be markedly affected en route. So this may be an additional reason why it is a lost cause trying to establish an exact quantitative relationship between receptor discharge and a vascular – or any other – response.

SHEPHERD Nonetheless, there does seem to be some correspondence. For example, vascular responses to hypoxic stimulation of the carotid bodies are closely related to the number of impulses in the carotid sinus nerve.

MILLS Could I get back to the cutaneous vascular bed. Did I understand that it was not very responsive to baroreceptor stimulation?

SHEPHERD The cutaneous veins are not responsive to stimulation of the carotid or the cardiopulmonary mechanoreceptors.

MILLS And stimulation of the chemoreceptors causes the skin vessels to dilate?

SHEPHERD Yes.

MILLS So if a subject has a myocardial infarct and comes into hospital hypoxic and hypotensive, will his skin be rosy and warm or what?

SHEPHERD The situation here is much more complex, since many reflexes and circulating vasoactive substances may be involved.

ACKER Chemoreceptor stimulation causes the hind-limb vessels to constrict. What is the purpose of this constriction? To centralize blood volume perhaps?

SHEPHERD Well, I think Professor Daly's paper suggested one purpose – namely to conserve oxygen during a dive. In this situation the brain and the heart get most of the oxygen that is available in the blood. The reduced blood flow to the skeletal muscles means that relatively little oxygen will be extracted by these muscles; in the true diving animal, the vasoconstriction can be intense.

DALY I agree. The situation in the diving mammals and birds is clear but the responses are less evident in terrestrial animals. There are some other rather important factors to be considered. For example, during systemic hypoxia, the chemoreceptors will be stimulated and cause, amongst other things, peripheral vasoconstriction through a primary reflex. But this can be overcome by the lung inflation reflex elicited secondarily through hyperventilation. There may also be a local action of hypoxia upon the vessels themselves. So the net effect in the dog at least is opposite from that which you might expect from an analysis of the primary chemoreceptor reflex; that is, you do not obtain vasoconstriction but instead vasodilatation. All this suggests that it is important to analyse individual components of the reflex and also the whole body response.

ACKER Yes, but if we regard the chemoreceptors as initiating a series of reactions which are concerned with oxygen supply, then I still don't understand the significance of the vasoconstriction.

ABBOUD The overall series of reflexes are, as you suggest, concerned with the oxygen supply and particularly with oxygen conservation for the cerebral and coronary circulations. In systemic hypoxia, there may be a limit to the increase in oxygen supply possible by a rise in ventilation and cardiac output; so the alternative left is to redistribute cardiac output from an area such as skeletal muscle which can work anaerobically for some time to brain which can't. So then you ask me why do skin vessels dilate? I do not know, but I suggest that they do so because although the skin has a large vascular capacity its consumption of oxygen is relatively small and the vasodilatation in the skin may balance in part the vasoconstriction in muscle. This would tend to prevent a significant increase in left ventricular afterload and myocardial oxygen consumption.

PURVES On a different topic, one point I would like clarified is this. Throughout this session, we have been concerned with experiments in which the carotid sinus or gracilis muscle preparation and so on have been perfused at steady pressure rather than with pulsatile pressure. I wonder if Dr Angell James could tell us what effect this is likely to have on the baroreflexes.

ANGELL JAMES I cannot say anything about the effects of pulsatile or non-pulsatile perfusion upon the chemoreceptors themselves or upon their reflex responses. As far as the baroreceptors are concerned, when you substitute a pulsatile for a non-pulsatile pressure you obtain a fall in a vascular resistance which has been shown to be due to recruitment of baroreceptor units. There is a family of curves relating mean pressure in the baroreceptor areas to vascular resistance depending on the pulse pressure (Angell James & Daly, 1971).

BULLER Is there a differential element in this or is there simply a uniform increase in the response in all vascular beds?

DALY I don't think we have enough information at present to say whether the reflex vascular responses to changes in pulse pressure occur more prominently in one vascular bed than in another.

PURVES It would seem from the results this afternoon that if we are not yet able to write 'The Integrative Action of the Peripheral Arterial Chemoreceptors', we can say that some of the individual components have been worked out in detail.

REFERENCE

ANGELL JAMES, JENNIFER E. & DALY, M. DE B. (1971). Effects of graded pulsatile pressure on the reflex vasomotor responses elicited by changes of mean pressure in the perfused carotid sinus-aortic arch regions of the dog. *J. Physiol.*, **214**, 51–64.

CONTRIBUTOR INDEX

Bold type indicates start of original article

AUTHOR INDEX

Figures in bold type indicate pages on which references are listed

SUBJECT INDEX